PRENTICE-HALL PHYSICS SERIES

Consulting Editors

Francis M. Pipkin

George A. Snow

PHYSICS FOR LIFE SCIENCE STUDENTS

PHYSICS FOR LIFE SCIENCE STUDENTS

HANS BREUER

PRENTICE-HALL, INC. / Englewood Cliffs, N.J.

Library of Congress Cataloging in Publication Data

Breuer, Hans, date
Physics for life science students.

Includes bibliographies.
1. Physics. I. Title.
QC23.B817 530 74-18388
ISBN 0-13-674150-9

10 9 8 7 6 5 4 3 2 1

Printed in the United States of America

Prentice-Hall International, Inc., *London*
Prentice-Hall of Australia, Pty. Ltd., *Sydney*
Prentice-Hall of Canada, Ltd., *Toronto*
Prentice-Hall of India Private Limited, *New Delhi*
Prentice-Hall of Japan, Inc., *Tokyo*

To Rosi, Klaus, and Hannes

ACKNOWLEDGMENTS

Among those who contributed directly or indirectly to this text, I am especially indebted to my teachers P. Brix, W. Pohlit, and B. Rajewsky. The suggestions and support of William H. Grimshaw and Nicholas Romanelli of Prentice-Hall were invaluable. The close cooperation with J. Warren Blaker of Vassar College is appreciated as are the comments of Arthur Walters. I also remember with pleasure the long discussions with my friends and colleagues John Rawlins and William Hallidy. And, for their stimulating influence, I thank my students.

Appreciation is also expressed to the following publishers for permission to reproduce illustrations: Fig. 1-19, D'Arcy W. Thompson, *On Growth and Form*, Macmillan Publishing Co., Inc., New York; Fig. 2-11, *Nature*, vol. 229, Feb. 1971, Macmillan Journals Ltd., London; Fig. 4-13, Theodore von Karman, *Aerodynamics*, Cornell University Press, Ithaca, N.Y.; Fig. 9-39, Picker Corporation, North Haven, Conn.; Fig. 14-21, Diane M. Ramsey, ed., *Image Processing in Biological Science*, University of California Press, Berkeley, Ca.; Fig. 14-22, Rheinisches Landesmuseum, Bonn, West Germany.

CONTENTS

TO THE STUDENT

Teaching physics to life science students is a difficult and challenging undertaking. If the physics becomes too abstract or too far removed from topics in biology and medicine, the student may lose interest. On the other hand, if the physics becomes so diluted that its clarity and overall consistency are submerged, then the intellectual rigor has been removed but the book does not fill the student's needs. The organization of this book has been dictated by the objective of avoiding both of these extremes. A careful determination of the relative importance of the various branches of physics for life science students was required. As a result some topics are treated only briefly while others are more detailed, this book being an introduction to physics and not to the life sciences.

You may wonder how much mathematics is used and whether it is necessary to know calculus. From my experience working with students of biology, medicine, and physics in Germany, Canada, and the United States, I know that it is not so much the calculus but the algebra that causes mathematical difficulties for the student. Therefore I have shown, in general, every intermediate step in calculations and derivations. You do not need to know calculus when you first open this book. You are introduced to elementary differential and integral calculus "as we go." This is surely simpler for you than surrounding formulas and phenomena with excessive wordiness and confusing analogies when a single line of calculus suffices. The same approach applies to vector notation when introduced. The use of graphical representation also helps to clarify topics—a method that is indispensable to the modern scientist.

Although the use of mathematics underlines and emphasizes the exactness and internal consistency of physics, do remember that the laws of physics are not necessarily the ultimate truth. They are only approximations, even where applied to pure physics. You must, therefore, take a critical stance: "What is the scope of that law?" "Are the basic assumptions fulfilled for these specific applications?" "What limits the accuracy of this particular formula?" Those are some of the questions you should ask yourself constantly. In this text I have stressed such limitations where they are not obvious.

In the life sciences we must be aware of the pitfalls of straightforward application of physical principles. The simplest biological systems have more parameters than the most complex ones in physics. There are also aspects intrinsic to living systems that elude the purely physical approach. For example, one obvious way to calculate a mutation rate is to estimate the radiation damage done by ever-present cosmic radiation to the carriers of genetic information, DNA. Knowing the structure of those double-helix molecules and the flux of cosmic radiation, we could then predict the rate of evolution. The calculation is simple enough, but it leads to an incorrect result. Why? Because there is a biological factor that has been overlooked: whenever one of the identical strands of the double helix is broken by radiation, a mechanism is triggered to repair the damaged strand by using information available from another strand.

Keeping such limitations in mind, you can then use physics as a means of gaining insight into life science phenomena; e.g., why man can see better than the anatomical structure of the retina seems to allow. You can observe conservation principles at work and may even enjoy understanding odd questions such as what an electron, waterbug, and supersonic aircraft have in common.

Some organizational aspects of this book: I have not pretended that you are completely ignorant of physics when you turn to the first page. We all live in a technical world that is largely governed by physical laws, and you are certainly aware of some of them. Besides, you had some introduction to science in high school, some rudimentary knowledge that remains. Therefore, I have used similarities and analogies with material that is covered in detail later in the book; of course, that has to be very basic material.

Throughout the book, the metric convention is used. If space-age organizations like NASA and conservative countries like England "go metric," we should not pretend that inch, pound, slug, and fortnight are still up-to-date units of measurement. However, conversions are given for metric versus nonmetric units. The general style and notation follows the *Style Manual* of the American Institute of Physics and the *Council of Biology Editors Style Manual*.

If you study the table of contents you will find the topics usually covered in an introductory physics text. However, the emphasis is different. Some subjects such as point mechanics, rigid bodies, thermodynamics, magnetostatics, and electrostatics are treated rather briefly. More space and time is devoted to topics that are pertinent for the student of life sciences: deformable media, phases of matter, transport phenomena, and radiation. You will also encounter chapters on unconventional topics such as graphical representation and information handling and processing. Most chapters include a flow chart to serve as a guidepost so that you are always aware of the general structure of the chapter and the interrelationship of topics, even when working at a seemingly small detail. The additional reading material cited at the end of each chapter will serve to give you further insight or more detailed explanations of special features of the chapters. Do not overlook the appendices: part A should be especially useful to you.

It is my hope that you will discover, once having finished this book, that physics is not by nature dry and indigestible to a student of the life sciences.

Hans Breuer

Dakar, Africa

PHYSICS FOR LIFE SCIENCE STUDENTS

chapter 1

GRAPHICAL REPRESENTATION

Before we delve into physics, let me first explain a technique which we shall use throughout the text: graphical representation of formulas, measured quantities, chains of thought, and directional quantities. Since the time of Cartesius* coordinate graphs have been used to convey the functional relation between quantities man encounters in his search and desire to distill the multitude of natural phenomena into a finite set of laws. Within those laws one quantity depends on another. Mathematics describes their interrelation in a precise way. Although mathematical formulation is sufficient to apply the laws, more often than not it is difficult to grasp their meaning, to evaluate rapidly their consequences for given variables. Here the coordinate graph reveals its value as a tool for understanding: with a glance we can recognize how a change in one quantity influences others.

Laws governing the life sciences are very complex; consequently we employ as many tools as possible to find patterns in the seemingly infinite variety of quantities. But that is not our sole aim: by reducing sets of quantities into a graphical form we also learn about the limits imposed on them by nature or our insufficient data. We thus are less likely to extrapolate laws beyond their region of validity. We shall then recognize the fact that interpolations can be invalid if we do not take certain precautions. Once we have plotted a set of experimental data, we can appreciate the problems stemming from the unavoidable uncertainty of such data.

Another kind of graphical representation, in its present form an outgrowth of the art of computer programming, is the *flow chart.* It enables

* René Descartes, 1596–1650, philosopher, physicist, and mathematician. More information about his contributions to science can be found in J. F. Scott, *The Scientific Work of Descartes,* London, 1952.

us to break down a logical chain of thought into its individual steps. The flow chart is especially helpful if we want to understand complex processes without getting confused by a host of details.

Finally, there are quantities having directional characteristics; there the obvious representation is the pointed arrow. Such directional quantities can be handled purely in an algebraical fashion. However, that is not only more complicated and difficult to do, but the inside view of the underlying process gets lost.

You could ask "Why should I spend time working through this chapter"? The answer is simple: The total time you spend on understanding graphical representation is small compared with the saving in time later on, because of the insight gained. Expressed in another way, we can say "the cost-benefit ratio is favorable."

1. COORDINATE GRAPHS

1.1 Terminology

We begin our study of coordinate graphs with the following example.

As shown in Figure 1-1, a general two-dimensional coordinate system (or graph) has two axes. Both axes are perpendicular to each other;

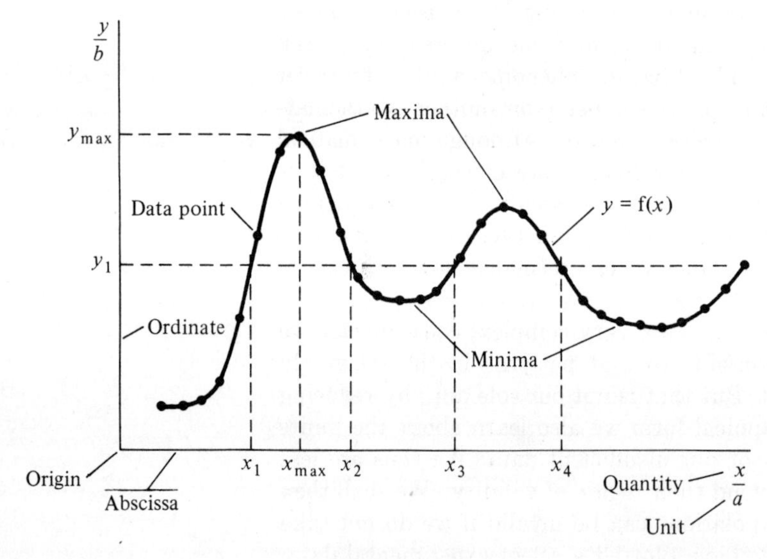

Figure 1-1

hence they are called a *rectangular coordinate system*. Each axis is subdivided into suitable fractions of the numerical values x/a and y/b, which measure the quantity displayed at the respective axis. The quantities could be time, temperature, electric current density, power, distance. Then the appropriate units are seconds, degrees, amperes, watts, meters.

In general, a graph is read from left to right. It may contain data points which are smoothly connected by a curve. This curve is said to demonstrate the functional relationship [$y = f(x)$] between both quantities. The functional relationship can be a succession of points, a curved or straight line, or points connected by a line. Each position on a curve represents a set of two values, one for each displayed quantity. Any point where a curve rises to a peak relative to its immediate neighborhood is called a *maximum;* a relative dip is called a *minimum.*

Section 1.1
Terminology

How to read a graph: If one of the quantities is given, the other one can be read quickly off the graph. If, for example, the value x_{max} is given, what is its corresponding value on the y-axis? Draw a line from x_{max} as indicated in Figure 1-1. Where the line intercepts the curve, draw another line perpendicular toward the ordinate and read the value at interception (here, y_{max}). Or, if the value y_1 is given, then again do as indicated in Figure 1-1. Now the curve is intercepted four times; that means that one value on the ordinate (here, y_1) corresponds to four values (x_1, x_2, x_3, x_4) on the abscissa.

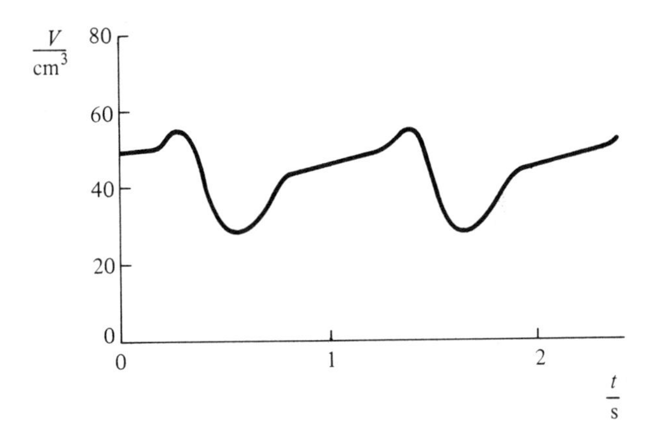

Figure 1-2 Time change of volume of the left ventricle of a dog's heart. The volume V is measured in cm^3, the time t in s. V and t are physical quantities; cm^3 and s are units.

Figure 1-2 shows a graph encountered frequently in medical routine; its notation follows the terminology just described.

1.2 Two-dimensional graphs

Graphs can be used in two ways. If we already have a formula expressing a law in mathematical form, we can convert the formula into graphical form in order to see the interplay of the various quantities and to gather results promptly for a given set of values. But a graph can also be used the other way around. From experiments we can extract some data and display those in a fashion which enables us to find the underlying law.

As an example of the former use of graphs, take the relationship between the volume of an ideal gas and its temperature, if its pressure is kept constant:

$$V = V_0(1 + \tfrac{1}{273}T) \tag{1.1}$$

where

V: gas volume at temperature T,
V_0: gas volume at zero degree Celsius,
T: gas temperature measured in degrees Celsius.

In Equation (1.1), V and T are variables and V_0 is a constant. If we assign a value to one of those values then we can calculate the other.

1.2(a). LINEAR GRAPHS. Such a simple law is, in general, displayed as a two-dimensional graph on rectangular coordinates with linear subdivisions. It is a two-dimensional graph because there are two variables (V and T), one for each axis. The axes are perpendicular to each other (for simplicity), and the divisions on the axes are equally spaced (linear) because Equation (1.1) is, mathematically speaking, linear.

Figure 1-3 shows the relation between the volume and temperature of an ideal gas as displayed on rectangular, linear coordinates, thus giving a simplified qualitative display of Equation (1.1). Figure 1-4 shows the detailed and quantitative representation of that equation on graph paper.

What can we learn from such graphs? First, each axis clearly shows which quantity is displayed and in what units that quantity is measured. Second, the point where both axes intersect is the *origin* and, unless explicitly stated otherwise, is the numerical value zero for both quantities. The gas law is represented by a straight line, called a *linear relationship*.

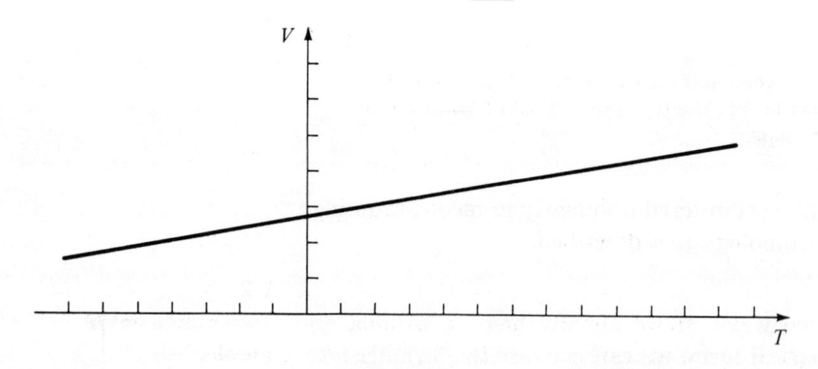

Figure 1-3

Equation (1.1) is a linear relation since a straight line can be expressed by

$$y = mx + a \tag{1.2}$$

and if we observe the correspondence

$$V : y$$

$$T : x$$

$$\frac{V_0}{273} : m$$

$$V_0 : a$$

Then $V_0/273$ is the slope of the straight line and V_0 is the intersection with the volume axis.

For any value of T (since the graph has to be reproduced on a page of finite size, the range of values is limited) we can find the corresponding value without recourse to Equation (1.1). Observe, for example, Figure 1-4. There the dashed lines show how to obtain the volume at a tem-

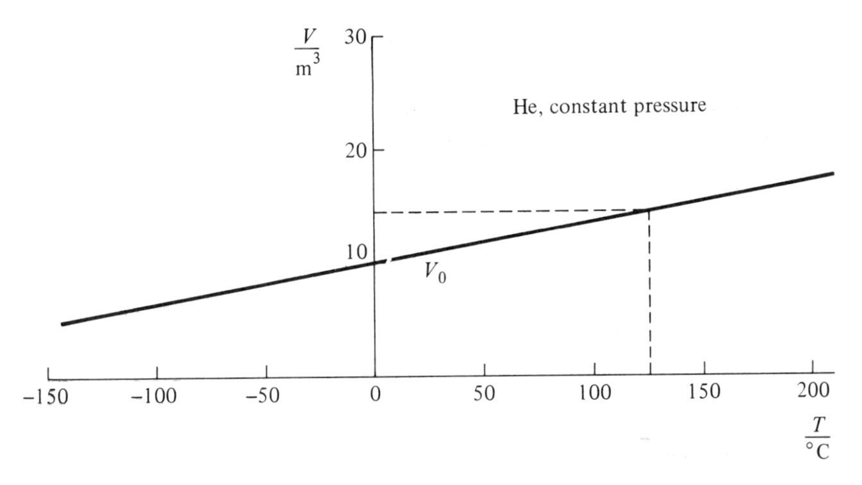

Figure 1-4

perature $T = 125°C$. We draw a line perpendicular to the temperature axis, starting at the position 125. From the point of intercept with the unbroken line we draw another line perpendicular to the volume axis. Where it intercepts this axis, we read the value for V (here 14.5 m) in agreement with the number we get by replacing T in Equation (1.1) by 125°C. Since the drawing is small, the extracted results are necessarily of limited accuracy.

Caution: In many cases it is not safe to extend a graph into regions which are beyond that displayed. The relation may be invalid outside the exhibited region.

Now look into the other problem: Instead of already knowing a relationship between two quantities, we have a set of data; for example, the activity (measured in recorded events per second) of a radioactive sample, determined at intervals of 1 minute. See Table 1.

Table 1

Activity (measured in events per second)	Time (measured in minutes)
20 052	1
14 130	2
10 980	3
7 651	4
4 983	5
4 148	6
2 910	7
2 001	8
1 312	9
1 202	10
861	11
701	12
415	13
296	14

Each pair of values can be plotted on graph paper with linear rectangular coordinates (see Figure 1-5). Point 4, for example, is the one representing the fourth pair of values in Table 1.

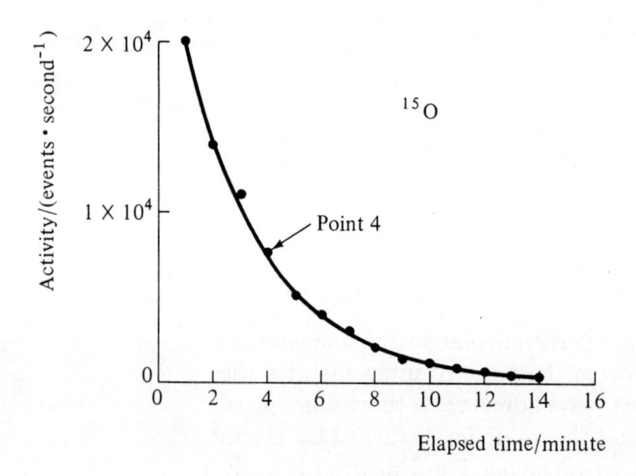

Figure 1-5 Activity of a ^{15}O-source.

1.2(b). Semilogarithmic graphs. If we smoothly fit all the points, we get a line which is not straight but curved. That means no linear relationship between activity and time. To find out the mathematical relation between both quantities, it is helpful to plot the data in such a way that they are represented by a straight line. This can only be done resorting to graphs in which one or both axes are not linearly subdivided. The usual way is to try first a *semilogarithmic* graph, which means that one axis has linear subdivisions and the other axis has intervals spaced logarithmically (like some scales on a slide rule). Assigning the activity to the log-axis, the data from Table 1 appear as shown in Figures 1-6 and 1-7. In Figure 1-6 the data from Table 1 are presented in detail on

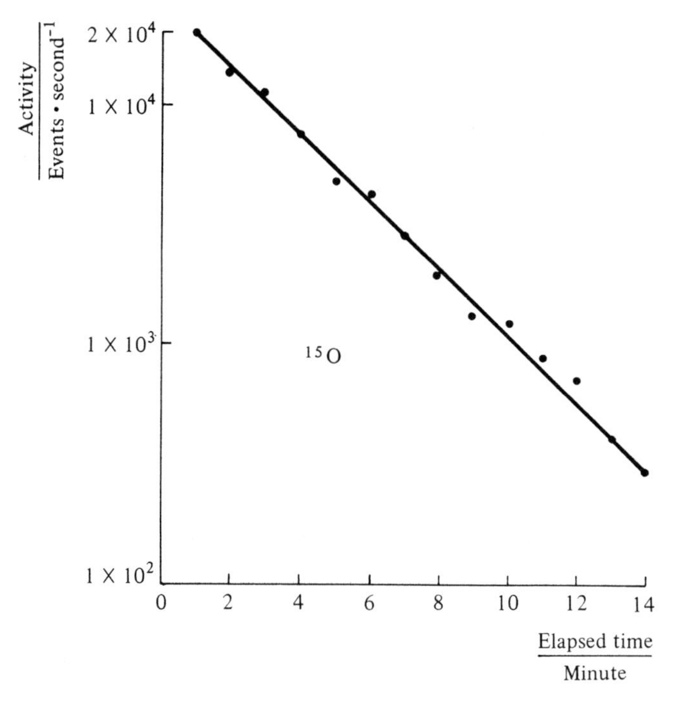

Figure 1-6 Activity of the preceding oxygen-15 source displayed as a semilog plot.

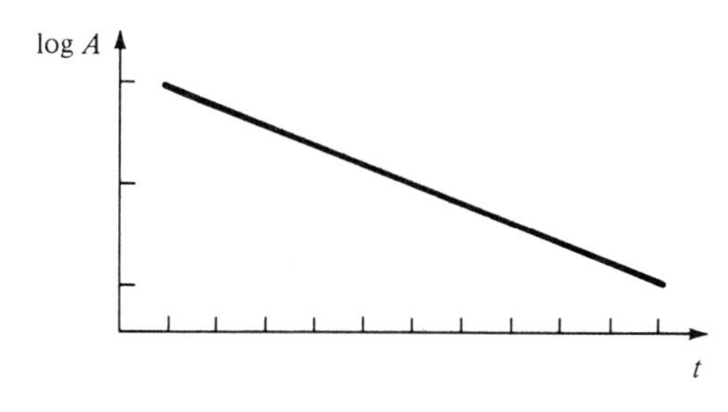

Figure 1-7

semilog graph paper, but Figure 1-7 shows the more general and simplified presentation usually found in a textbook.

Notice that the vertical axis has no zero.

In this type of display the data are fitted best by a straight line; hence the underlying law obeys an exponential form, which is

$$A = A_0 e^{-\lambda t} \tag{1.3}$$

where

t: time,
A: activity at time t,
A_0: activity at time $t = 0$,
λ: a constant, called decay constant.

That Equation (1.3) is represented by a straight line in a semilog plot is not difficult to understand: Taking the logarithm of Equation (1.3) we get

$$\log A = -\lambda t \log e + \log A_0 \tag{1.4}$$

We notice that $\log e = 0.434$; hence

$$\log A = -0.434\lambda t + \log A_0 \tag{1.5}$$

Comparing Equation (1.5) with the standard equation for a straight line, we have

$$y = mx + a$$

And noting the correspondence

$$y{:}\log A$$
$$x{:}t$$

we conclude that Equation (1.5) is the equation of a straight line with the slope $m = -0.434\ \lambda$. Here we meet for the first time a letter from the Greek alphabet, namely λ, representing the decay constant. Those letters are used for traditional reasons and because we run out of symbols if we stick only to the Latin alphabet. Although you will find the complete Greek alphabet in the appendix, a selection of widely used Greek letters follows:

Greek Letter	Greek Name
α	alpha
β	beta
γ	gamma
Γ	gamma (capital)
δ	delta
ϵ	epsilon
η	eta
θ	theta
λ	lambda
μ	mu
ν	nu
π	pi
ρ	rho
Σ	sigma (capital)
σ	sigma
φ	phi
ψ	psi
ω	omega

The semilog representation of an exponential relationship between two quantities not only yields a straight line but has the additional convenience that the logarithmic scale can accommodate a wide range of values.

1.2(c). Double-logarithmic graphs. If we are unable to achieve a straight line by plotting our data on linear or semilogarithmic graph paper, we can try a *double-logarithmic* display, where both axes are subdivided logarithmically. If we now obtain a straight line, then the governing law is a power function. As an example, let us plot the distance l covered by a free-falling body as a function of time t, as shown in Figure 1-8.

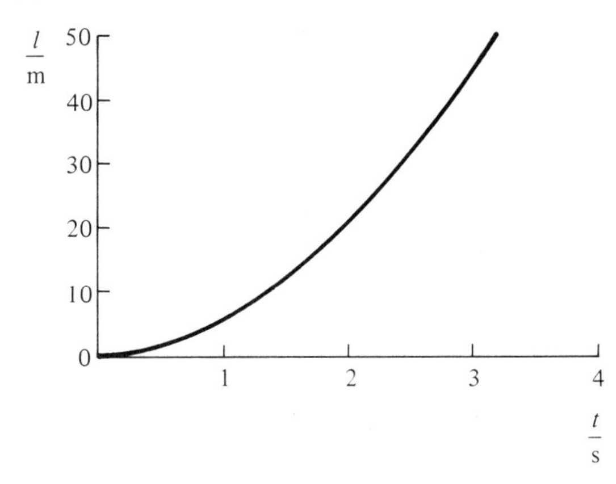

Figure 1-8

The corresponding display on double-log paper is shown in Figure 1-9; it is shown in simplified graphical form in Figure 1-10.

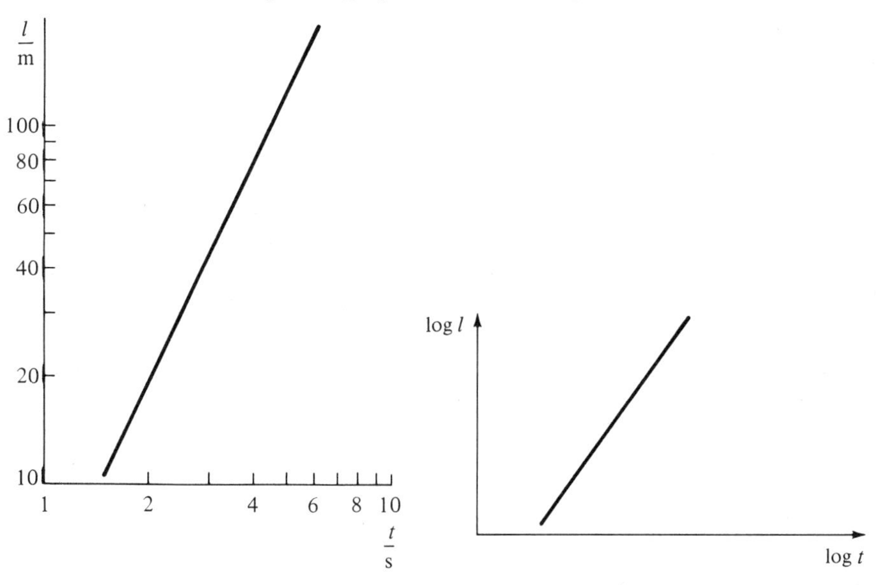

Figure 1-9

Figure 1-10

The free fall (starting from rest) obeys the following law:

$$l = bt^2 \tag{1.6}$$

where b is a constant.

Proof that a power function appears as a straight line in a double-logarithmic display: Take the logarithm of Equation (1.6):

$$\log l = 2 \log t + \log b \tag{1.7}$$

Compare this with the standard equation for a straight line:

$$y = mx + a$$

and identify

$$y\!:\!\log l$$

$$x\!:\!\log t$$

Then Equation (1.7) is the equation of a straight line with the slope $m = 2$.

A double-logarithmic graph not only displays a linear relationship between quantities of a power law; it also encompasses a very wide range of values on both axes. Notice that the point of intersection of the axes does not have the value zero for either axis.

Caution: Due to the compressed form of representation, readings taken from double-logarithmic plots are bound to be inaccurate.

1.2(d). Polar graphs. Another form of graphical display is the *polar* diagram. There the origin of a rectangular coordinate system coincides with the center of a set of concentric rings. The quantities are plotted as angles (called polar angles) with respect to one of the rectangular coordinate axes and as distances measured from the center of the rings. Naturally, laws which depend on an angle are presented in this form.

EXAMPLE: THE DISCRIMINATING VIEWING POWER OF THE HAWK FOR VARIOUS DIRECTIONS

As shown in Figure 1-11 the angle of view increases counterclockwise from 0 degree to 360 degrees. The discriminating power is measured in arbitrary units, the tenth ring representing the value 100.

From this polar graph we learn that the hawk cannot see backward between 150 degrees and 210 degrees; that it sees best at approximately 70 degrees and 290 degrees (about 120 on the arbitrary scale); and that its viewing power reaches 100 looking straight ahead.

Consider another polar diagram, one more closely related to pure physics: It concerns the relative energy emission of a dipole oscillator for different directions (see Figure 1-12).

The maximum amount of energy is emitted at right angles (90 degrees and 270 degrees) to the oscillator axis (0 degree), because the distance of the curve from the origin is directly proportional to the emitted energy. If, for example, the energy emitted at 90 degrees is 7.0 watts, then the energy emitted at 30 degrees is 1.75 watts. The oscillator does not radiate energy in either a forward or backward direction.

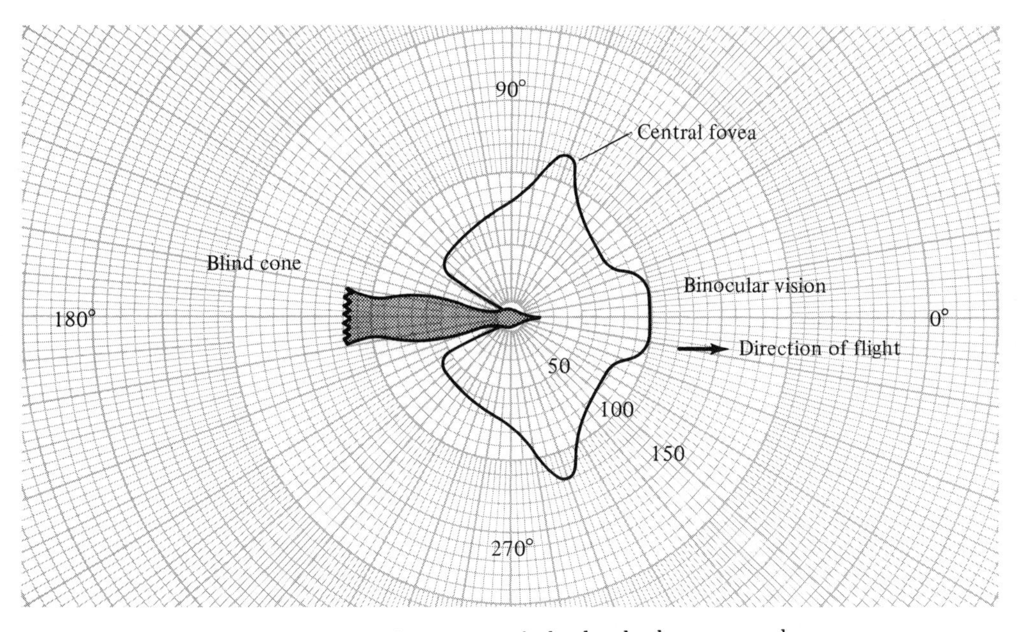

Figure 1-11 The viewing power of the hawk shown on polar coordinates.

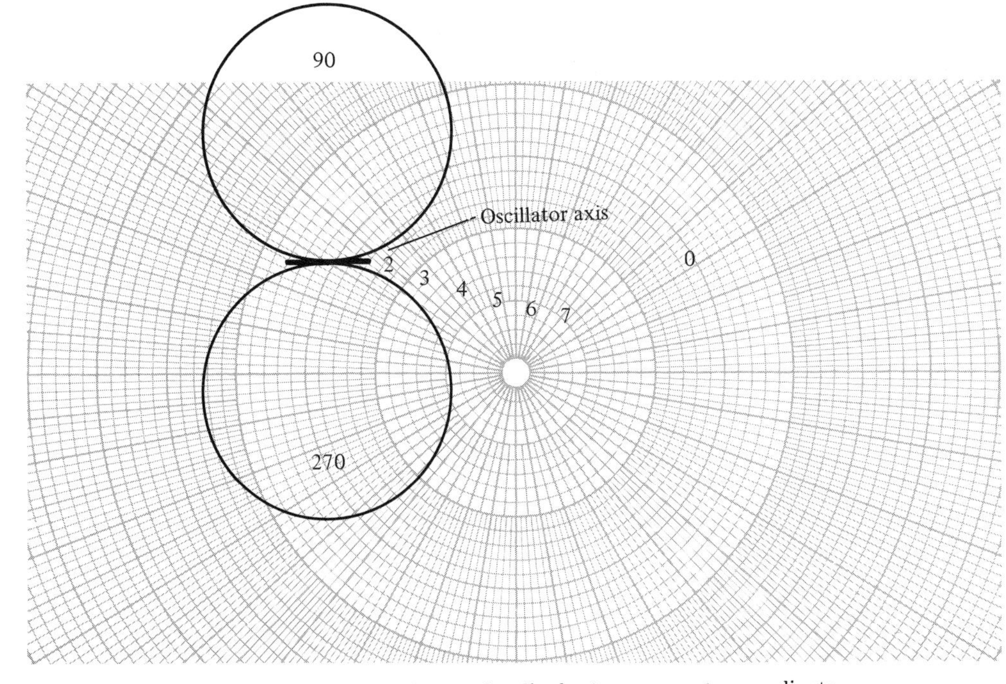

Figure 1-12 Energy emission of a dipole shown on polar coordinates.

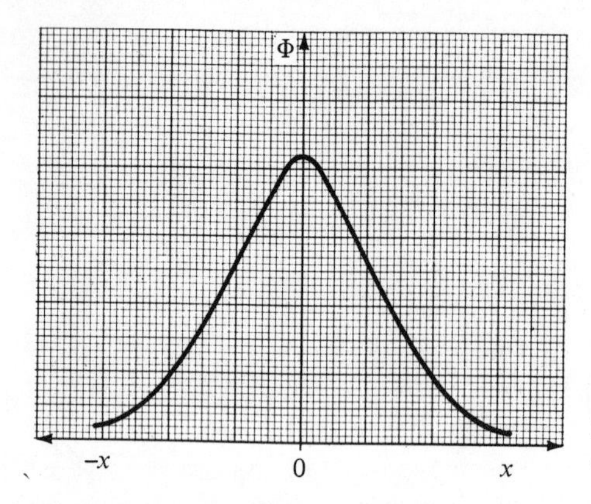

Figure 1-13 Error-function shown on linear graph paper.

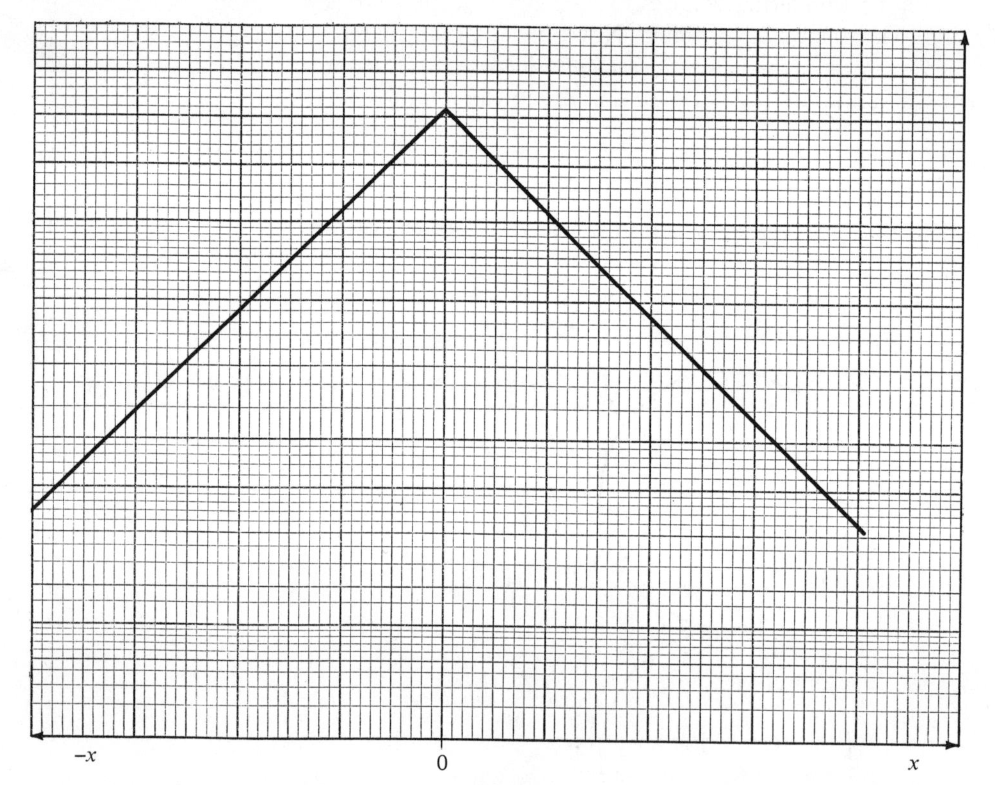

Figure 1-14 Error-function shown on probability graph paper.

There are more specialized kinds of graphs available, but "linear," "semilog," "double-log," and "polar" are sufficient for most practical purposes.

1.2(e). PROBABILITY GRAPHS. One of the more unusual kinds of displays is the *probability graph*. For an example, we shall use the "error or Gauss-function."*

* Karl Friedrich Gauss, 1777–1855, mathematician and astronomer.

Using linear graph paper we obtain the curve shown in Figure 1-13.

But by using probability graph paper we obtain the curve shown in Figure 1-14.

The governing law is

$$\Phi = \frac{1}{\sqrt{2\pi}} e^{-x^2/2} \tag{1.8}$$

Sometimes it is impossible to find by graphical means a mathematical relation between measured quantities. In that case a "best-fit" procedure can be used with the aid of a digital computer. This procedure produces a set of functions (for example, power functions) which describe the data within a well-defined region.

By now you may be wondering what is wrong with a graph that is not a straight line. The answer is: nothing! Nevertheless, a linear representation is convenient because (a) it is simpler to fit data points to a straight line, (b) the mathematical formula describing the relationship between the displayed quantities can be established easily, and (c) interpolating values off a straight line is simple.

1.2(f). Other graphs. There are many other kinds of graphs and specialized two-dimensional graphical displays. From them we choose one which is of importance to the student of the life sciences: the electrocardiogram (ECG). The action potentials (later on we shall understand the full meaning of that expression) of the heart muscles are detected at suitable places and automatically plotted as functions of time. Figure 1-15 simulates an ECG.

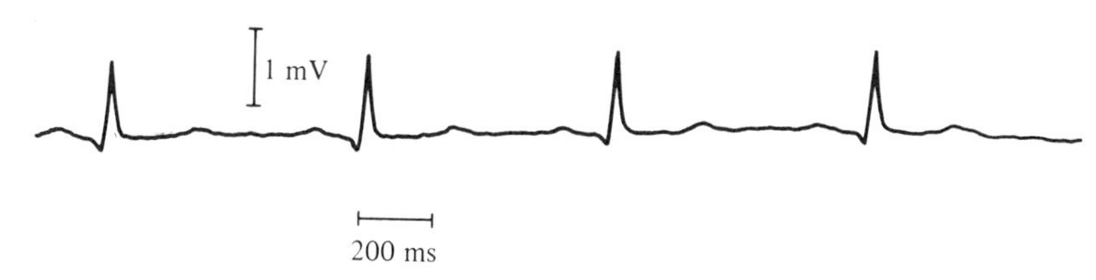

Figure 1-15 Routine electrocardiogram (ECG) taken with an automatic recorder.

It does not make much sense—although it is possible—to express this curve in mathematical terms. Here a comparison with a standard, for example the ECG of a sound average person, is the answer. Figure 1-16 shows such a standard ECG. Comparing the curves in Figures 1-15 and 1-16 is a simple task, but interpreting whether the discrepancies are of significance is another matter. The individual ECGs of healthy persons vary to some extent as does the ECG of the same individual, depending on time and mood. Here the experience of the investigator is of utmost importance.

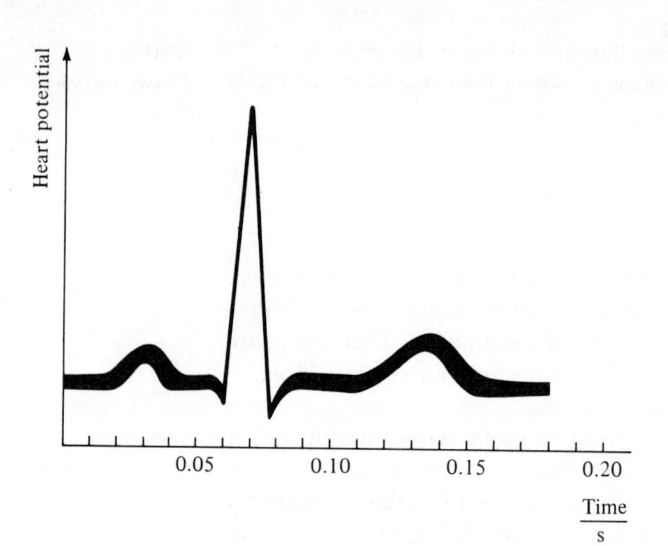

Figure 1-16 Standard electrocardiogram.

1.3 Three-dimensional graphs

Until now we have restricted ourselves to two-dimensional graphs. Those are the most common ones, but it is also possible to display the interrelation of three variables in one graph.

Figure 1-17 shows the distribution of radiation over a brain into which a radioisotope (^{97}Tc) was injected for diagnostic purposes. The

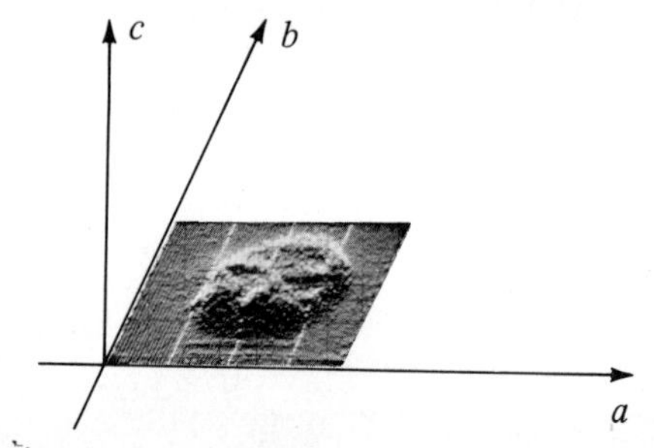

Figure 1-17 Three-dimensional oscilloscope display of the radioactivity of a brain after injection of technetium-97.

outline of the brain is displayed as a two-dimensional object in the horizontal plane. Each position is characterized by a coordinate pair (a, b). The observed radioactivity, as measured in counts per minute, for each (a, b) position is plotted along the axis c. From this three-dimensional graph we can deduce that activity is highest in the front lobes.

1.4 Interpolation and extrapolation

Before we become involved with other types of graphical representations, let us dwell briefly on the concepts of interpolation and extrapolation. *Interpolating* on a graph means that we have a graph with individual points, not connected by a continuous line (see Figure 1-18). We want to know a value which lies between two known ones, A and B, for example. How to do that would seem to be obvious, but do not forget that the interpolation implies that we know the shape of the curve connecting A and B. If both points can be linked by a contour line like the dashed one in Figure 1-18, then we encounter no difficulties inter-

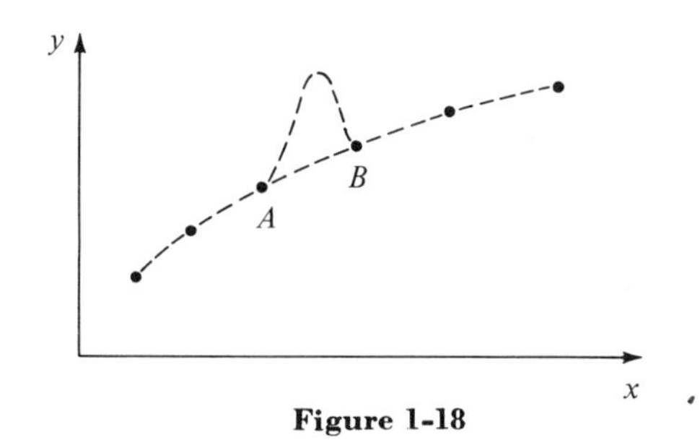

Figure 1-18

polating between points A and B. We just read the value off the graph. But what if, by chance, the curve between A and B must be described by the other (hatched) line? So always be careful when you interpolate.

This problem is enhanced if we extrapolate data or curves. *Extrapolation* means extending the range of data or of a curve into a region beyond the region of checked validity. Observe, for example, Figure 1-4. The volume is displayed in the temperature region from $-150°C$ to $+250°C$. If we want to know the volume of the gas at a temperature of 400°C, we extrapolate the line to the right. For the volume at $-270°C$ we extrapolate to the left. In both cases we leave the region of checked validity. As it turns out, the value for 400°C is correct, but the value obtained by linear extrapolation down to $-270°C$ is grossly wrong: Helium liquifies at $-269°C$. So beware of the dangers of extrapolating, a danger that is more acute the further away we move from the region of established validity.

1.5 D'Arcy Thompson principle

What was said above may sound dry and is hardly exciting, so let us relax for a moment and look at some more stimulating although obscure illustration of the use of coordinate graphs: the D'Arcy Thompson principle. That principle states that any initial biological form, if embedded in a rectangular coordinate system, will yield closely related forms if the original coordinate system is continuously deformed. Figure 1-19 shows a porcupine fish (diodon) in a rectangular coordinate system. If the coordinate system is deformed as in Figure 1-20, a genetically closely related fish (though quite different in appearance) emerges, the sunfish (orthagoriscus). The principle would, for example, enable the zoologist to reconstruct the structures of extinct forms, if a mechanism is specified which underlies the transformation. Although there are many spectacular examples for the D'Arcy Thompson principle, its underlying law still awaits discovery—if it exists at all.

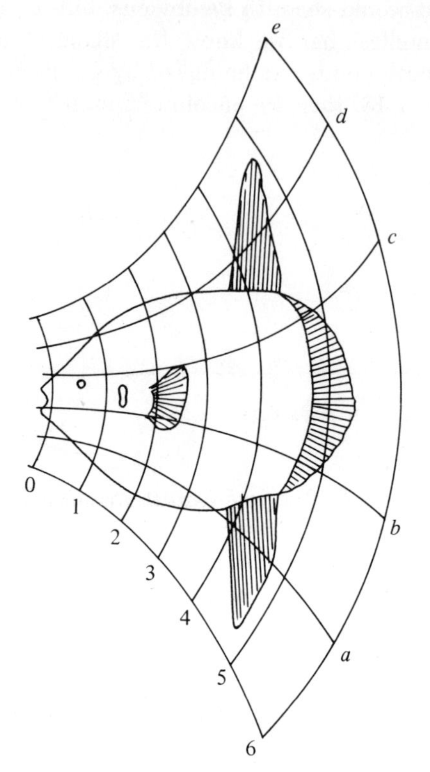

Figure 1-19 **Figure 1-20**

2. VECTORS

2.1 Terminology

In performing experiments, stating results, or describing events, we inevitably produce quantities. Looking close, we find that they consist of two different types: scalar and vector quantities.

A *scalar* is a number, a numerical value. If we use it in connection with a unit, then there are no ambiguities. Examples: 2.5 m^3, 3 hours, and -40°C are quantities, where 2.5, 3, and -40 are scalars and m^3, hour, °C are units. No additional information is necessary to understand these quantities completely. But what about a displacement of 300 m or a wind velocity of 30 km/h? Here we do not have complete information; a vital ingredient is missing: the direction.

A quantity that needs for its complete description a pure number (scalar), a unit, and a direction is called a *vector*.

A vector is presented as an arrow. The length of the arrow is proportional to the numerical value (called magnitude or absolute value of the vector), the tip pointing into its direction.

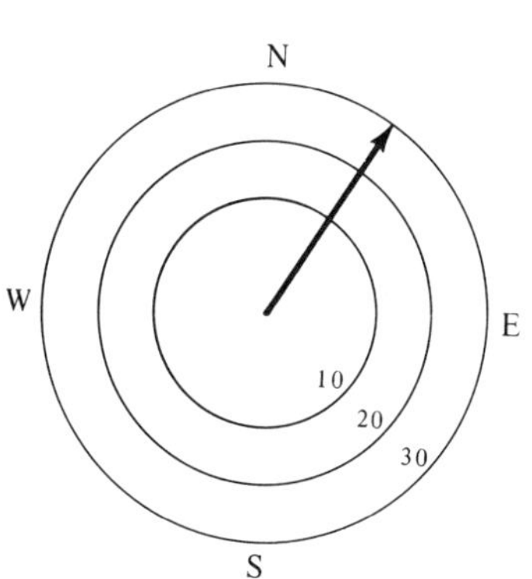
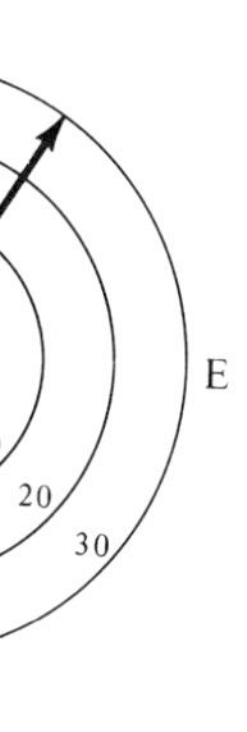

Figure 1-21

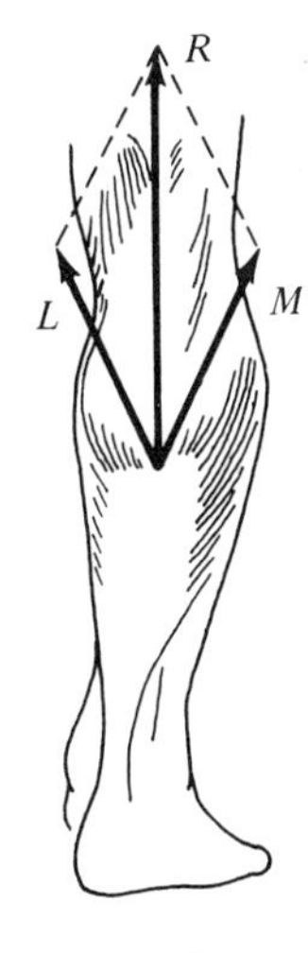

Figure 1-22 Resultant force from two muscles.

EXAMPLES:

The wind blows from NNE with a velocity of 30 km/h as in Figure 1-21.

Or, as in Figure 1-22, two muscles pull at the same tendon.

Not only do we need to know the force exerted by each muscle but also the direction of each force if we want to determine how the limb will move under the combined force of the muscles.

Conventions to note: Vectors are printed in **boldface**. For example,

$$\mathbf{F}$$

Still in use are:

$$\vec{F} \quad \text{or} \quad \mathfrak{F}$$

The absolute value (or magnitude) of a vector, its value regardless of direction, appears as

$$|\mathbf{F}| \quad \text{or} \quad F$$

To work with scalars, we employ normal algebra. Dealing with vectors requires the knowledge of another field, vector algebra. Since we shall encounter vectors solely in connection with simple operations, you need to learn only some vector algebra, consisting of addition, subtraction, and multiplication.

2.2 Addition and subtraction

Addition:

$$\mathbf{a} + \mathbf{b} = \mathbf{c}$$

c is the resultant vector.

We cannot just algebraically add the absolute value of **a** and **b** to get **c**; we must add them vectorially. For example, if Figure 1-23 shows vector **a** and Figure 1-24 shows vector **b**, then we add them by placing the tip of one arrow at the end of the other, shifting the vectors so that their initial directions remain unaltered. The straight line connecting the free end of one of the combined vectors with the free tip of the other yields the resultant vector **c** (dashed arrow) as shown in Figure 1-25.

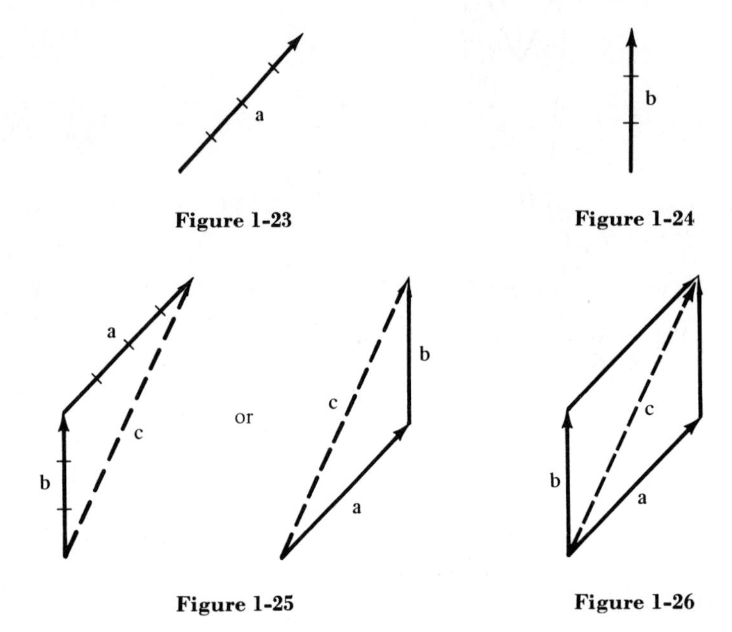

Figure 1-23

Figure 1-24

Figure 1-25

Figure 1-26

Another method employed is to connect vectors **a** and **b** at their beginnings and draw a parallelogram as shown in Figure 1-26. The diagonal represents the resultant vector **c**. Naturally, both methods yield the same resultant.

Subtraction:

$$\mathbf{a} - \mathbf{b} = \mathbf{d}$$

d is the resultant vector.

We can rewrite this operation as

$$\mathbf{a} + (-\mathbf{b}) = \mathbf{d}$$

Vector $-\mathbf{b}$ has the same length (absolute value) as vector **b** but points into the opposite direction. Therefore, we can perform the vector subtraction according to the same rules as vector addition.

If **b** is represented by Figure 1-27 and **a** by Figure 1-28, then $-\mathbf{b}$ is represented by Figure 1-29. Consequently, $\mathbf{a} - \mathbf{b}$ yields the representations shown Figure 1-30.

Following the parallelogram method, we get Figure 1-31.

So far we have only seen how to get resultants from two vectors. We can easily extend vector addition and subtraction to many vectors. If the four vectors **a**, **b**, **c**, and **d** are added, then the resultant is derived

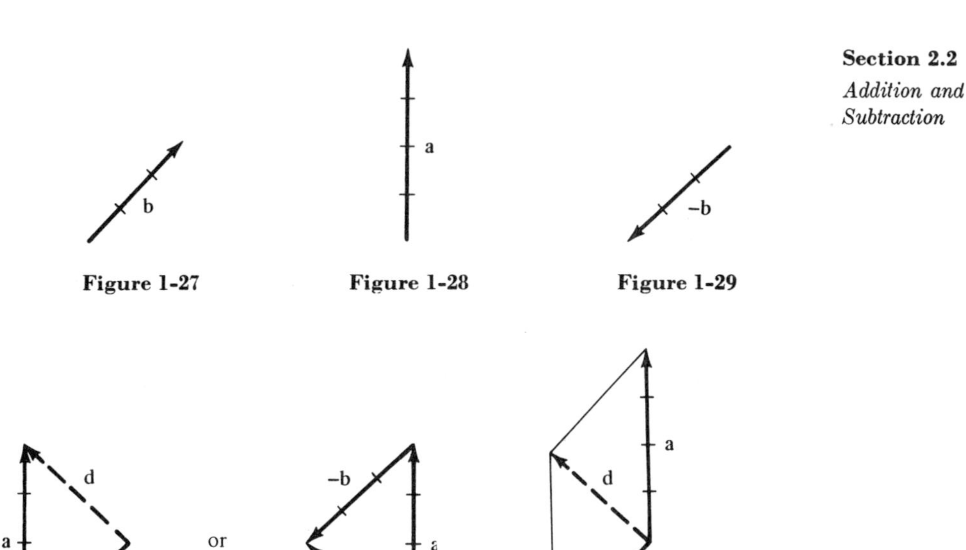

Figure 1-27 Figure 1-28 Figure 1-29

Figure 1-30 Figure 1-31

by connecting all four in the same manner as described under addition. The order of connection does not matter. For

$$\mathbf{s} = \mathbf{a} + \mathbf{b} + \mathbf{c} + \mathbf{d}$$

the resultant **s** is shown in Figure 1-32. Note, that in this instance all vectors are assumed to be in the same plane.

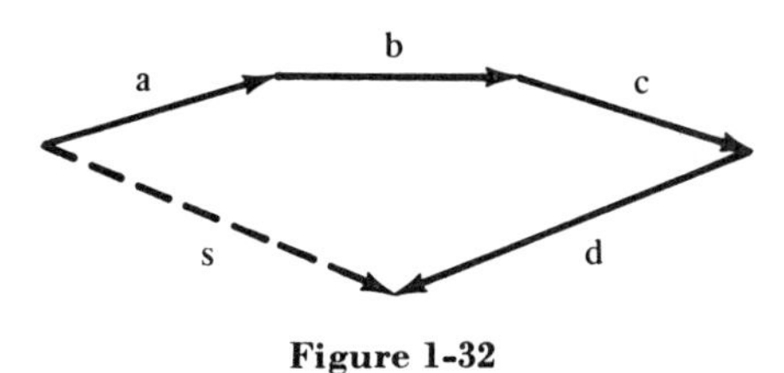

Figure 1-32

Multiple subtraction is carried out accordingly.

Just for an exercise, determine the dominant muscular force at the wrist in Figure 1-33.

Of course, we can abandon the graphical representation altogether and describe vector operations purely algebraically. The absolute value $|\mathbf{c}|$ of the resultant vector **c** from the addition $\mathbf{a} + \mathbf{b}$ is then, according to general trigonometry,

$$|\mathbf{c}| = \sqrt{|\mathbf{a}|^2 + |\mathbf{b}|^2 - 2|\mathbf{a}||\mathbf{b}| \cos \theta} \qquad (1.9)$$

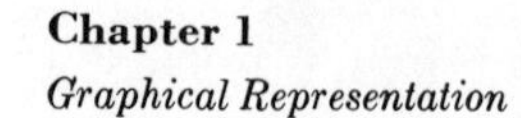

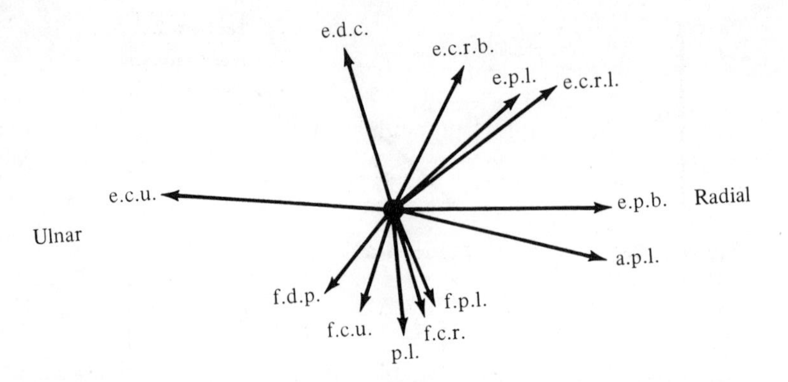

Figure 1-33 Schematic diagram of the muscular forces acting on the wrist.

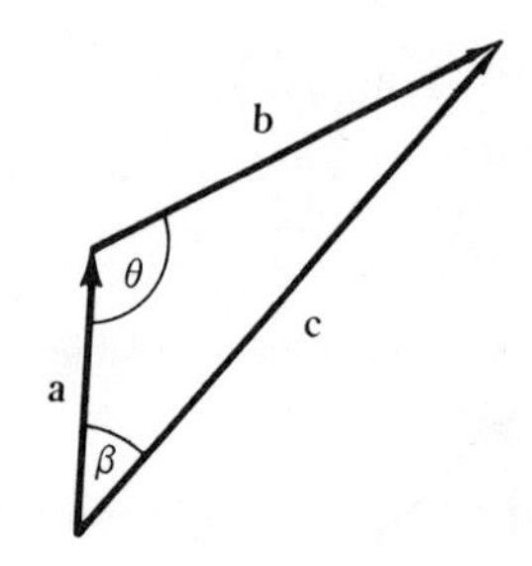

Figure 1-34

where θ is the angle between **a** and **b** (Figure 1-34). The direction of **c** is then

$$\sin \beta = \frac{|\mathbf{b}|}{|\mathbf{c}|} \sin \theta$$

where β is the angle between **a** and **c**.

2.3 Resolution into components

Often it is useful to resolve a vector into two or more components. Obviously, those components (vectors themselves) must add up to the original vector. Normally we resolve a vector into a pair of components which are perpendicular to each other. We place the vector to be resolved into a rectangular coordinate system so that the end of the vector

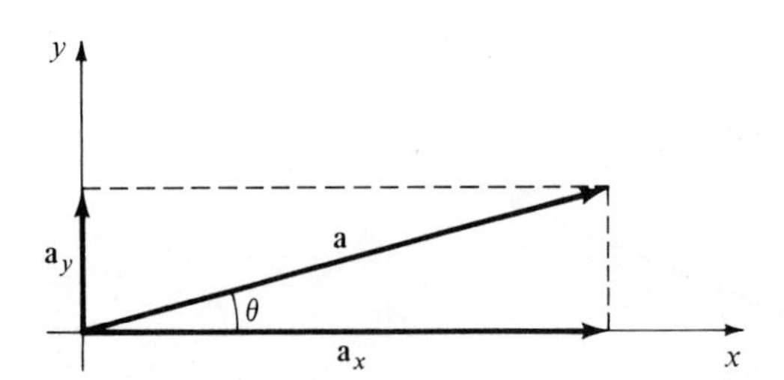

Figure 1-35 Resolving a vector into two components.

and the coordinate origin coincide (Figure 1-35). The components are the perpendicular projections of **a** onto the axes x and y.

From Figure 1-35 we deduce the useful relations

$$|\mathbf{a}_y| = |\mathbf{a}| \sin \theta \tag{1.10}$$

$$|\mathbf{a}_x| = |\mathbf{a}| \cos \theta \tag{1.11}$$

and

$$|\mathbf{a}|^2 = |\mathbf{a}_y|^2 + |\mathbf{a}_x|^2 \tag{1.12}$$

2.4 Vector multiplication

It is only reasonable to ask whether vector multiplication exists. The answer is yes, but it is more complicated than for scalar algebra.

2.4(a). Multiplication by a scalar. The simplest possibility, multiplying a vector $\mathbf{a}$ by a scalar x is not difficult to comprehend:

$$x\mathbf{a} = \mathbf{b} \quad \text{and} \quad \mathbf{b} = x\mathbf{a}$$

The result is a vector $\mathbf{b}$ with the same direction as $\mathbf{a}$ but with an absolute value enlarged by the factor x; which means that the arrow presenting $\mathbf{a}$ is stretched by the factor x.

There are two different ways to multiply one vector with another vector, each of which yields an entirely different result: scalar product and vector product.

2.4(b). Scalar product. The scalar product of vectors $\mathbf{a}$ and $\mathbf{b}$ is expressed as

$$\mathbf{a} \cdot \mathbf{b} = c \tag{1.13}$$

(read: vector a dot vector b) where

$$c = |\mathbf{a}|\,|\mathbf{b}| \cos \theta \tag{1.14}$$

and θ is the angle between $\mathbf{a}$ and $\mathbf{b}$.
Hence, for

$$\mathbf{a} \parallel \mathbf{b} \curvearrowright c = |\mathbf{a}|\,|\mathbf{b}|$$

(read: if vector a parallel vector b then . . .) and for

$$\mathbf{a} \perp \mathbf{b} \curvearrowright c = 0$$

(read: if vector a perpendicular vector b then . . .). The multiplication dot in Equation (1.13) must appear explicitly.

Note that the result of a scalar product is always a scalar, a pure number without a direction attached to it.

2.4(c). Vector product. The vector product between two vectors $\mathbf{a}$ and $\mathbf{b}$ is written

$$\mathbf{a} \times \mathbf{b} = \mathbf{d} \tag{1.15}$$

(read: vector a cross vector b) where

$$|\mathbf{d}| = |\mathbf{a}|\,|\mathbf{b}| \sin \theta \tag{1.16}$$

Hence, for

$$\mathbf{a} \parallel \mathbf{b} \smile \mathbf{d} = 0$$

and for

$$\mathbf{a} \perp \mathbf{b} \smile |\mathbf{d}| = |\mathbf{a}|\,|\mathbf{b}|$$

Note that the result of a vector product is always a vector! The direction of the resulting vector is perpendicular to the plane defined by the two original vectors (see Figure 1-36). Whether $\mathbf{d}$ points up or down with respect to the plane defined by vectors $\mathbf{a}$ and $\mathbf{b}$ is determined in the following way: Rotate the first vector (here, $\mathbf{a}$) into the second (here, $\mathbf{b}$); then $\mathbf{d} = \mathbf{a} \times \mathbf{b}$ points into the direction a right-handed screw would move under this rotation. See the left-hand

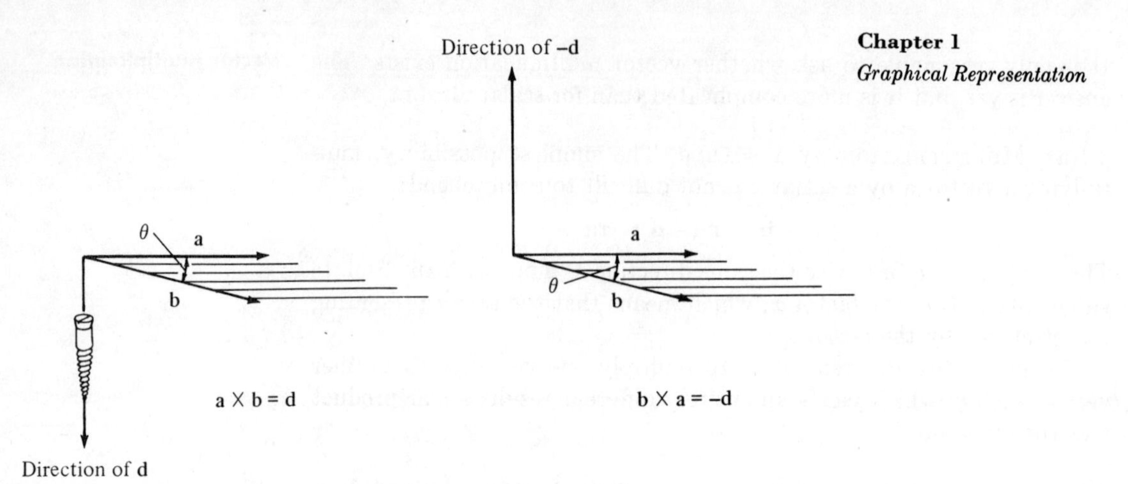

Figure 1-36 The vector product and its direction.

side in Figure 1-36. From this we realize that the vector product is noncommutative, which means that

$$\mathbf{a} \times \mathbf{b} \neq \mathbf{b} \times \mathbf{a}$$

In fact, it is

$$\mathbf{a} \times \mathbf{b} = -\mathbf{b} \times \mathbf{a}$$

3. FLOW CHARTS

It is not always feasible or even possible to mold interrelations into a mathematical formulation. If we want to find a pattern within a host of detail, display the organization of a system, analyze a logical chain, or demonstrate the interdependence of various events, we may profitably employ another form of graphical representation, the flow chart. It is mainly a child of the computer age although it was known long before the twentieth century. Have a look at Figure 1-37. It presents a rather elaborate flow chart of the primary organization of this chapter. The point of departure is obviously the central block labeled "graphical representation," since all arrows lead away from there. Arrow 1 guides the observer to "coordinate graphs," from which the path divides into branches "two-dimensional" and "three-dimensional," and so on in a self-explanatory fashion. The upper and lower halves of Figure 1-37 indicate subjects already covered in this chapter, and arrow 2 indicates what remains to be covered.

The main purpose of this and subsequent flow charts is to enable you to keep your bearings while plowing through details. This will become more obvious as we proceed through further chapters, where it is quite possible to lose sight of the forest (the encompassing physical law) by

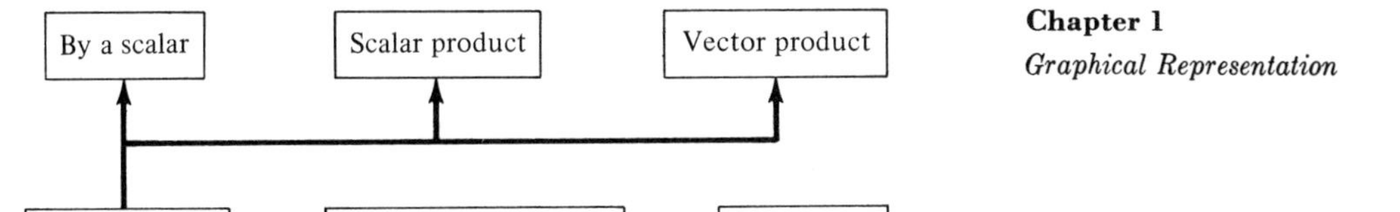

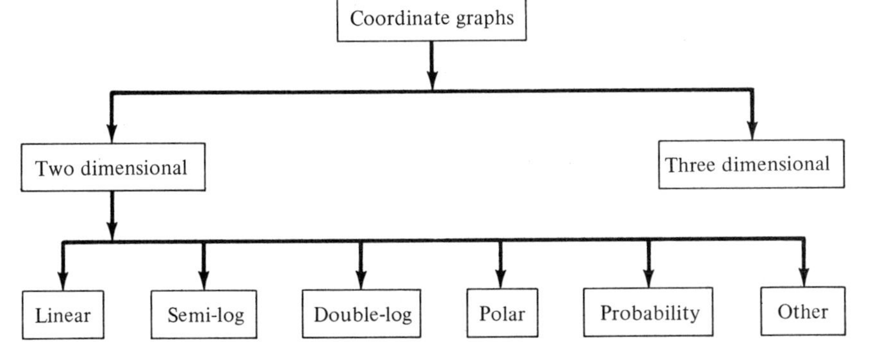

Figure 1-37 Flow chart for the contents of chapter 1.

looking at the individual trees (the detailed consequences of that law).

Another illustration of the flow chart is shown in Figure 1-38. The opening stage of a table-tennis match is analyzed in order to indicate how much information the brain must handle.

The flow chart starts at the upper left corner: Information reaches the brain via the optical and acoustical sensors. The brain reduces this information to essential input data. In parallel, an internal clock marks the time. The point of impact of the oncoming ball is calculated with the help of additional data stored in the long-term memory. After that the player determines (subconsciously, of course) whether the point of impact is on his side of the table. If it is, then the calculation continues as indicated by the arrow leading away from the right corner of the triangle; otherwise he does not swing his paddle. This flow chart demonstrates how to break down a series of complicated actions into logically linked individual steps—a prerequisite for analyzing the performance of the brain during a table-tennis game.

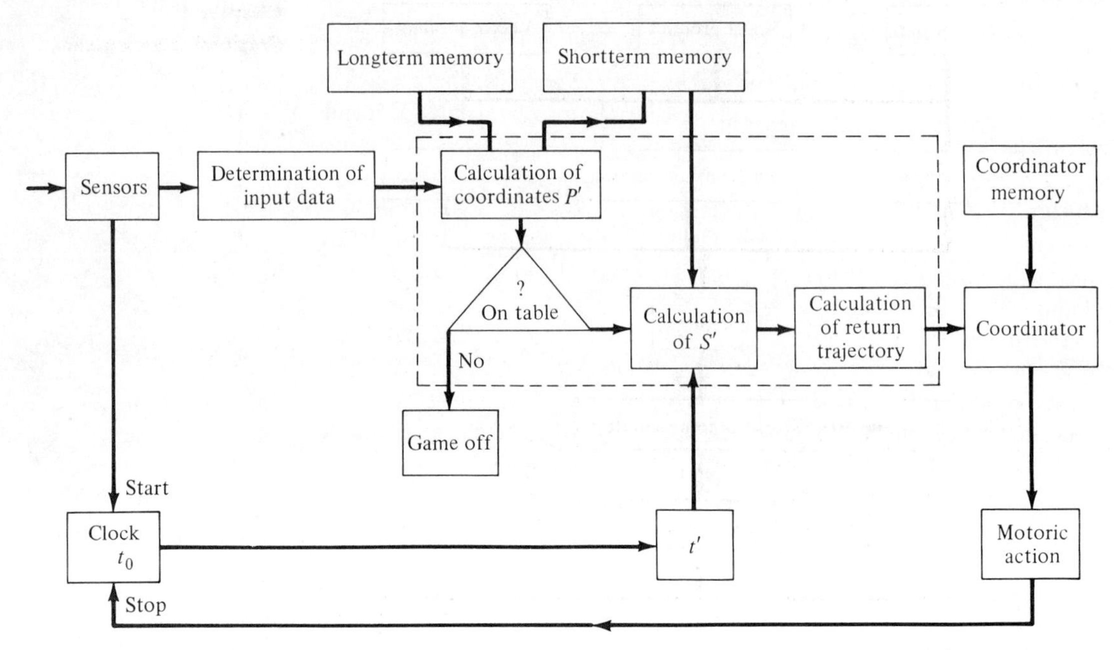

Figure 1-38 Direction of information flow during a table-tennis match. P': point of impact; S': point of interception.

A flow chart does not need to be as complex as demonstrated in the previous figure. A more straightforward one is shown in Figure 1-39, which depicts the detecting, counting, and plotting of the radiation from a radioactive source.

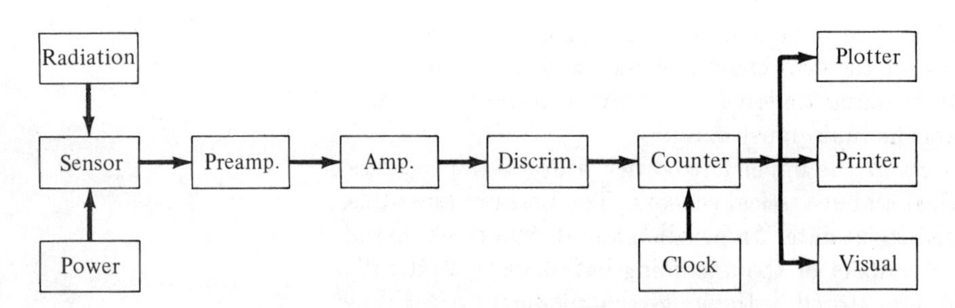

Figure 1-39 Flow chart for a radiation detection set-up.

There is an almost infinite variety in flow charts, also called block diagrams. In most cases they are self-explanatory. We shall make full use of this auxiliary device to understand the basic principles of physics.

SUMMARY

The coordinate graph of a mathematically formulated law is an efficient means of grasping the meanings and implications of that law. Depending on the relation, we can display formulas in linear, semilog, double-log, polar, or other appropriate fashion. Employing a nonlinear presentation is especially useful if we are trying to find a mathematical interrelation between experimental data. Interpreting a coordinate graph you must be aware all the time of the pitfalls in interpolation and extrapolation.

Another field where graphical representation simplifies understanding is vector algebra. A directional quantity, a vector **a**, can be represented by a pointed arrow whose length is proportional to the non-directional component, the absolute value $|\mathbf{a}|$, and whose tip points in the direction of the vector. The basic rules of vector algebra are:

Addition:

$$\mathbf{a} + \mathbf{b} = \mathbf{c}$$

or

$$|\mathbf{c}|^2 = |\mathbf{a}|^2 + |\mathbf{b}|^2 - 2\mathbf{ab}\cos\theta$$

Subtraction:

$$\mathbf{a} - \mathbf{b} = \mathbf{d} \quad \text{or} \quad \mathbf{a} + (-\mathbf{b}) = \mathbf{d}$$

Scalar product:

$$\mathbf{a}\cdot\mathbf{b} = c$$

$$c = |\mathbf{a}|\,|\mathbf{b}|\cos\theta$$

The resultant c is always a scalar.

Vector product:

$$\mathbf{a} \times \mathbf{b} = \mathbf{d}$$

$$|\mathbf{d}| = |\mathbf{a}|\,|\mathbf{b}|\sin\theta$$

The resultant **d** is always a vector.

PROBLEMS

1. Determine the decay constant λ from Figure 1-6.
2. There is really no limit to the possibilities for displaying information. One fine example is in Figure 1-40. All told, the diagram displays four variables. Name those variables. Which one is displayed twice?
3. Plot the light intensity distribution of asymmetric headlights in a polar diagram.

4. Convert the discriminating viewing power of the hawk (Figure 1-11) from a polar graph to a rectangular graph. That means, plot the viewing power versus viewing angle.
5. Draw a flow chart to describe the action you would take if your car stalled in the middle of a desert and you want to find the reason for the breakdown.
6. Produce a flow chart showing how your brain works to determine the largest one of 1000 different numbers printed at random succession on a paper strip.
7. Prove that if you add geometrically $\mathbf{a} + \mathbf{c} + \mathbf{d} + \mathbf{b}$ (see Figure 1-32) it will yield the same resultant vector $\mathbf{s}$.
8. Refer to Figure 1-33: Determine the dominant muscular force at the wrist if the muscle e.c.u. (extensor carpi ulnaris) is impaired.

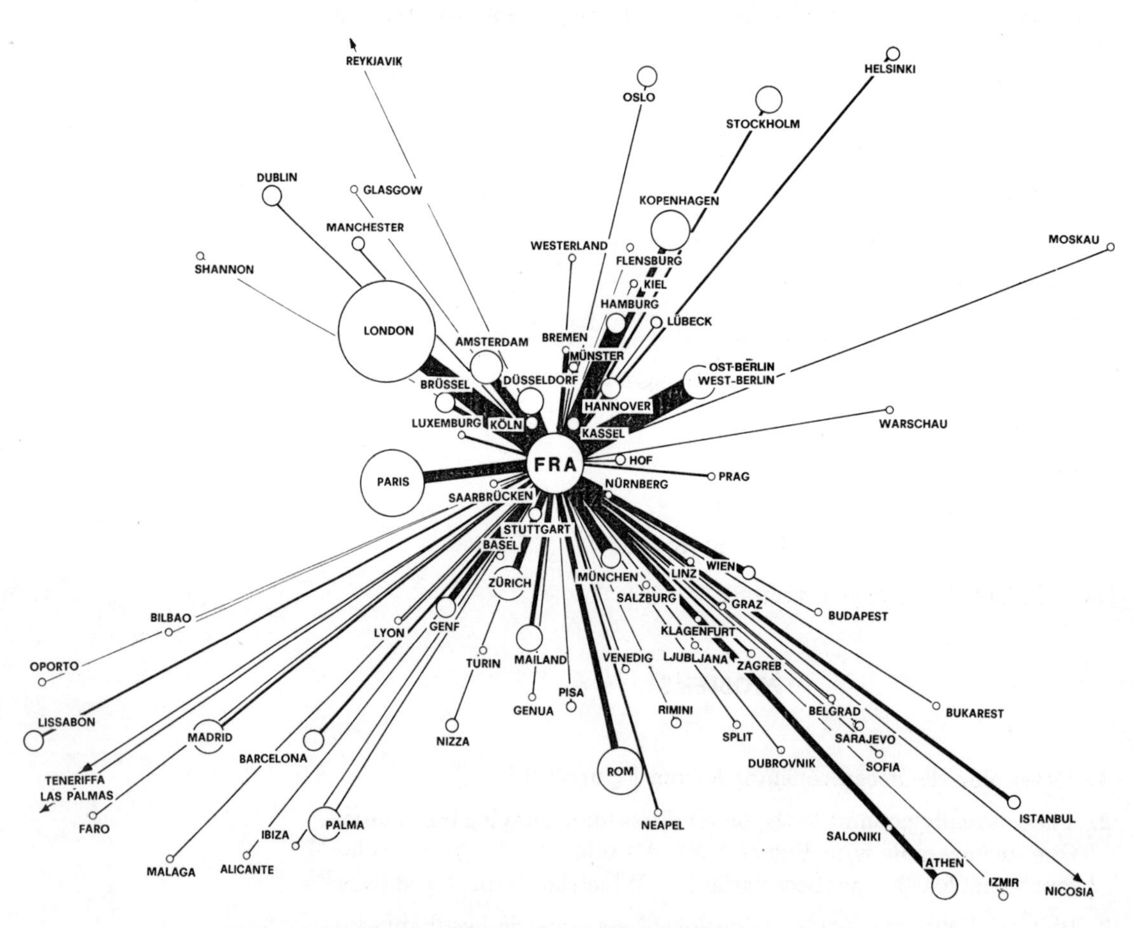

Figure 1-40 Graphical display of an international airport. It contains more information than is obvious at first glance.

9. Display Equation (5.8) from Chapter 5 for altitudes between 0 and 20 km on linear, semi-log, and double-log graph paper. Determine graphically the slope and compare with its correct value.

10. Add vectors e.c.u., a.p.l., f.d.p., and f.p.l. (see Figure 1-33). Resolve the resultant vector into a horizontal and a vertical component. Does the horizontal component point into the ulnar or into the radial direction?

FURTHER READING

A. CROW, A. CROW, *Mathematics for Biologists.* Academic Press, London, 1969.

K. MATHER, *Statistical Analysis in Biology.* Methuen, London, 1966.

S. S. STEPHENS, "Neural Events and the Psychophysical Law," *Science,* **170** (1970) 1043.

D'ARCY W. THOMPSON, *On Growth and Form.* Chapter XVII, Mac-Millan, New York, 1945.

chapter 2

TIME AND LENGTH MEASUREMENTS

In this chapter we shall move right into the mainstream of physics. But before we can look into the problems involving measurements of physical quantities, we must agree on our terminology. What do we mean by using words such as *accuracy*, *precision*, and *sensitivity?*

If we perform a measurement it is not sufficient to state a number as result. There is always an error involved, such that we really can only state within what error limits the measured quantity lies. This is of utmost importance since any measured value quoted in physics is useless unless the error limits are known. It is not always easy to estimate the errors involved but—and this cannot be emphasized enough—it must be done.

We shall see that there is also a principle barrier which limits the accuracy of all measurements.

The result of a measurement is expressed in units of the International System of Units, which encompasses the six basic units: the meter, second, kilogram, ampere, degree Kelvin, and candela. Nature provides us with many quantities which could (and did) serve as standards: The circumference of the earth, the revolution of the earth around the sun, the periodicity of the luminosity of bacteria, the frequency of a pulsar, etc. Practical considerations led to the adoption of units which are in general determined by atomic or molecular phenomena.

1. MEASUREMENTS AND THEIR EVALUATION

1.1 Terminology

The result of a measurement is a physical quantity. This physical quantity is expressed as the product of its numerical value (a pure number) and its unit, i.e.

$$\text{physical quantity} = \text{numerical value} \times \text{unit}$$

Throughout the text we shall use the International System of Units (SI system), a coherent system based on the six basic units:

Name	Symbol
meter	m
kilogram	kg
second	s
ampere	A
degree Kelvin	°K
candela	cd

The symbols of physical quantities are identical for singular and plural quantities (15 kg not 15 kgs!); they do not include a period (s not s.!).

Sometimes physical quantities are well established in units other than SI units. In those cases we shall use the established units and state the conversion.

More about symbols, units, and nomenclature is presented in the appendix.

Let us agree on the following terminology referring to the results of measurements.

Precision or reproducibility: The degree of agreement of repeated measurements of the same quantity.

EXAMPLE:

A small object is placed on a balance and repeatedly weighed. Result: 2.371 g, 2.361 g, 2.362 g, 2.377 g, 2.378 g. Average: 2.370 g. The precision is 0.4% because the largest deviation of one measurement from the average value is 0.009 g; that means 0.4%.

Accuracy: The degree of agreement between the result of a measurement and the accepted value of the quantity measured.

EXAMPLE:

The mass of the small object (example above) was calibrated at the National Bureau of Standards and certified as 2.300 g. The average of the measurement (2.370 g) deviates 0.07 g from the accepted value (2.300 g); the accuracy of the measurement is 3%.

Sensitivity: The ratio of the change in the response to the change of the quantity measured.

EXAMPLE:

The thickness of a foil is determined by transmission of radiation through it. See Figure 2-1.

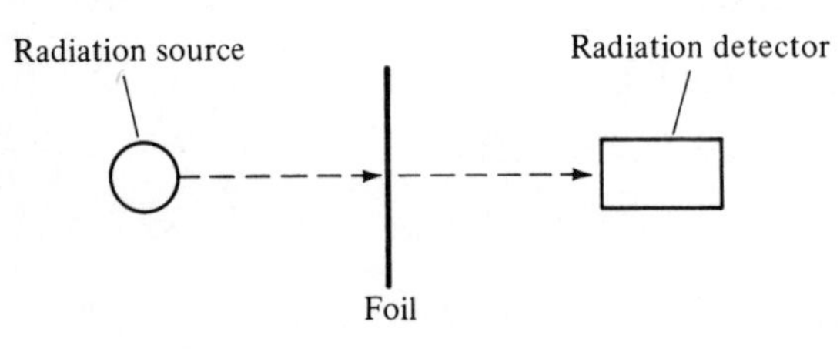

Figure 2-1

Result: For a thickness of 2 mm, the detector measured 430 events per second. For a 1-mm foil, the detector registered 5521 events per second. That means, the measured quantity (thickness of foil) changed by 100% (from 2 mm to 1 mm); the response of the measuring system changed 1300% (from 430 to 5521). Hence, the sensitivity of the measuring arrangement is (1300/100) = 13, a high sensitivity indeed.

1.2 Errors

1.2(a). ERROR SOURCES. Every measurement contains errors which are reflected in a finite accuracy of the result. Some sources of error are:

1. *Limits imposed by the measuring device:* Obviously it is impossible to determine the length of an object within one millimeter if the ruler used is subdivided into only centimeters.
2. *Disturbance of the system under investigation by the measuring device itself:* To measure the body temperature of a fly with an ordinary mercury thermometer is impossible. The internal temperature of the mercury is of greater importance than the temperature of the insect. This case calls for a thermometer with a very small heat capacity compared with that of a fly (a thermoneedle would do a fine job).

It is interesting to note that the disturbance of the system by the investigation is unavoidable in principle. In our macroscopic world it can be reduced to a tolerable level by using appropriate instruments and methods. Contrarily, in the microscopic world, on the atomic scale, this influence imposes severe limits and leads to what is known as the Heisenberg Uncertainty Principle.* More about this cornerstone of modern physics in Chapter 16.

3. *Inherent variations in the quantity measured:* A factor of special importance in the life sciences. Two consecutive measurements will not yield the same result because various quantities connected with the measurement have already changed. For example, the system to be measured grew older, the temperature changed, the barometric pressure dropped, or the oxygen supply was exhausted. It is one of the most difficult tasks in the life sciences to know which of the almost innumerable variables of a living system will

* Werner Heisenberg, 1901, physicist, Nobel laureate 1930. Major contributor in quantum theory and hydrodynamics.

significantly influence the outcome of an experiment. Otherwise, lack of reproducibility will be the inevitable result. Repetition of the measurement will improve the result to some degree.

4. *Response time:* There are many quantities which are functions of time. The concentration of Ca in bones, the acuity of the eye, the potential drop across a cell membrane, the blood pressure, and the body temperature come to mind. The measuring device for each quantity must perform the measurement during a time span that is short compared with the time variation of this quantity. The condition is easily met for the determination of Ca in bones. Modern absorption methods need a few minutes for this measurement, and the Ca content changes significantly only over years. The potential drop across a cell membrane, however, lasts only a few thousandths of a second; hence the response of the measuring instrument must be rapid. See Figure 2-2.

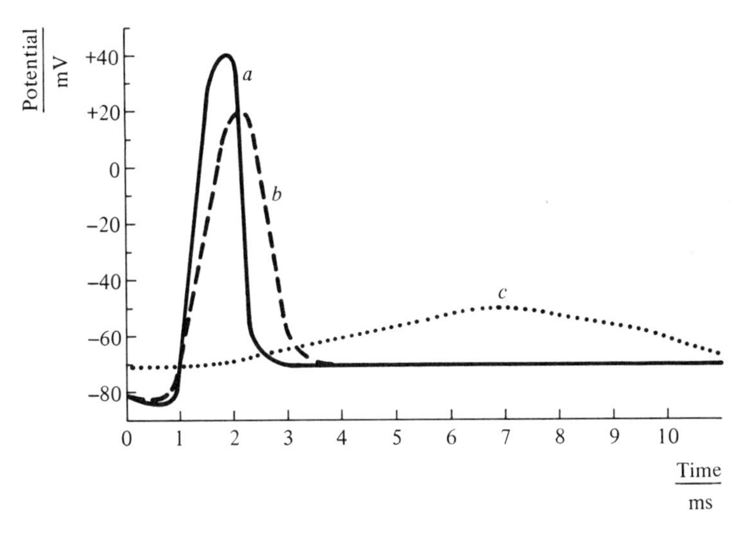

Figure 2-2 Discharge across a cell membrane as observed by instruments with different time responses: (a) time response 1 μs, (b) 1 ms, (c) 5 ms.

5. *Influence of statistical fluctuations:* For example, radioactive decay is largely determined by statistical laws; thus the number of events recorded contains an inaccuracy.
6. *Signal-to-noise ratio:* Each measurement includes a certain amount of background noise which consists mostly of random disturbances affecting the measuring device. Sitting inside a soundproof chamber, you will notice a faint murmur, the background noise produced by the Brownian motion of molecules on your eardrums. If you now try to detect a very weak sound produced outside the ear, then the intensity of this sound (the signal) should be larger

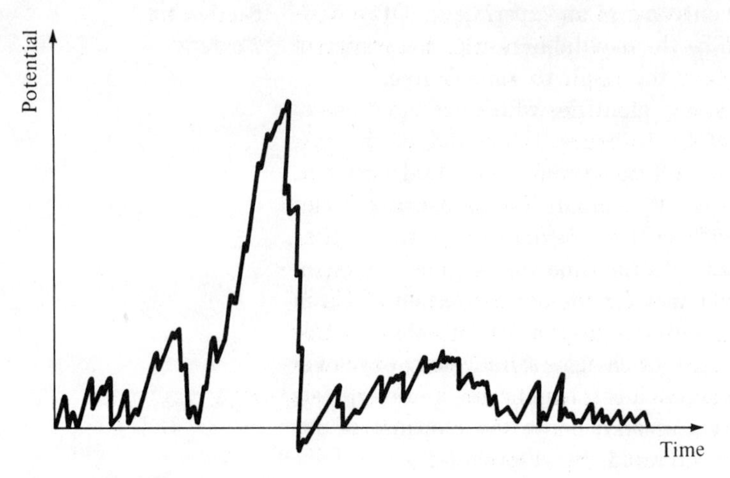

Figure 2-3 The electrocardiogram of an unborn child as observed by an oscilloscope.

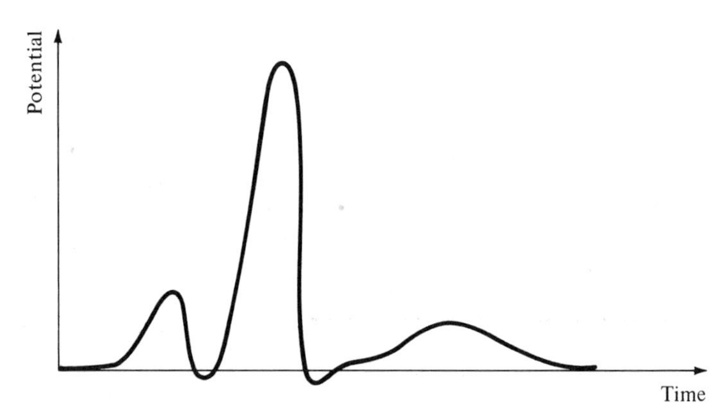

Figure 2-4 The same electrocardiogram taken by an electric signal averager. A signal averager takes advantage of the periodicity of the signal.

than the ever-present background noise. In other words, the signal must be above the noise level; the signal-to-noise ratio must be favorable.

The signal-to-noise ratio is most important for weak signals. In many measurements the uncertainty introduced by the background noise is small compared with the influence of other error sources. There are a number of tricks to better the signal-to-noise ratio; see for example Figures 2-3 and 2-4. More about signal-to-noise ratios will be presented in Chapter 14.

The two common ways to present the error of a measured quantity are best demonstrated by an example:

The mass of an object was measured to be 253.5 g and the error was

0.2 g. Thus we know that the object has a mass between 253.3 g (253.5 − 0.2) and 253.7 g (253.5 + 0.2).

The *absolute error* is

$$253.5 \pm 0.2 \text{ g}$$

The *relative error* expresses the error in percent of the measured quantity; hence

$$253.5 \text{ g} \pm 0.08\%$$

You can observe in everyday life how different the same result appears, depending on whether the absolute or the relative error is quoted. Take, for example, the outcome of the Condon report concerning unidentified flying objects (UFO): The sceptic always quotes that a mere 2% of all reported and investigated sightings could not be explained as natural phenomena by the commission—a very small fraction indeed. His adversary quotes that more than 400 cases could not be explained—quite an impressive number. They are both correct since the commission investigated more than 20 000 reports.

1.2(b). AGREEMENT WITHIN ERROR LIMITS. Two results are said to have the same value if they agree within their error limits.

EXAMPLE:

Two objects are measured to be 30.3 ± 0.3 cm and 30.1 ± 0.5 cm long, respectively.

Since the length of the first object is between 30.0 cm and 30.6 cm and the second is between 29.6 cm and 30.6 cm, we cannot determine which one is longer; hence they have the same length.

Incidentally, this is also the meaning of the equal sign (=) used in an equation. The identity sign (≡) implies that the quantities to the left and to the right of the sign are absolutely equal. In physics this is possible only in definitions.

1.2(c). SIGNIFICANT FIGURES. The value of a quantity is commonly expressed in such a way that only the last digit quoted is uncertain due to the intrinsic error. All quoted figures are then named *significant figures*.

EXAMPLE:

The above quoted mass of 253.5 ± 0.2 g is written as 253.5 g. The last digit (the 5) is uncertain, and the other three digits are not affected. Thus the value contains four significant digits.

Operations on a slide rule yield numbers with three significant digits. Throughout the text, all worked-out examples will be calculated with slide rule accuracy, that is, to three significant digits.

1.3 Error calculation

More often than not, the accuracy of a result depends on the accuracy of two or more separate measurements.

EXAMPLE:

The surface area of the human body can be calculated using the following formula:

$$S = 0.007\,184(m)^{0.425} \times (h)^{0.725} \quad (2.1)$$

where

S: surface area measured in m^2,

m: mass measured in kg,

h: height measured in cm.

The uncertainty in S will depend on the uncertainties of m and h. But how?

The rules of how to compound errors are few and straightforward.

1.3(a). ADDITION AND/OR SUBTRACTION. The result depends on quantities which are connected by addition or subtraction signs, such as

$$R = x_1 \pm x_2 \pm x_3 \quad (2.2)$$

Then

$$\Delta R = \sqrt{(\Delta x_1)^2 + (\Delta x_2)^2 + (\Delta x_3)^2} = \sqrt{\sum_{i=1}^{3} (\Delta x_i)^2} \quad (2.3)$$

where

ΔR: compounded *absolute error* of result R,
$\Delta x_1, \Delta x_2, \Delta x_3$: individual errors attached to the quantities x_1, x_2, x_3.

EXAMPLE:

Three electric cells are connected in series to form a battery. The voltage across each cell is measured: 2.12 ± 0.02 V, 2.08 ± 0.03 V, and 2.25 ± 0.02 V, respectively. The appropriate law states that the voltage across the battery is the sum of the cell voltages, hence 6.45 V. The uncertainty ΔU in this voltage is calculated according to Equation (2.3):

$$\Delta U = \sqrt{(0.02)^2 + (0.03)^2 + (0.02)^2} = 0.04 \text{ volt}$$

Thus the voltage across the battery is

$$6.45 \pm 0.04 \text{ volt}$$

Remark: To get this result we had to take the square root of 0.00017, which is 0.041 1. But since the individual errors have only one significant digit, the compounded error must be rounded off to only one digit.

1.3(b). MULTIPLICATION AND/OR DIVISION. The result depends on quantities which are connected by multiplication and/or division signs, such as

$$R = x_1 x_2 x_3 \quad (2.4)$$

Then

$$\Delta R = \sqrt{(\Delta x_1)^2 + (\Delta x_2)^2 + (\Delta x_3)^2} = \sqrt{\sum_{i=1}^{3} (\Delta x_i)^2} \quad (2.5)$$

ΔR: compounded *relative error* of the result R,
$\Delta x_1, \Delta x_2, \Delta x_3$: individual relative errors attached to the quantities x_1, x_2, x_3.

EXAMPLE:

The electric current I through a wire having a resistance R determines the potential drop U across the wire.

$$U = IR$$

If I is measured to be 12.2 ± 0.3 amperes and $R = 3.71 \pm 0.07$ ohm, then we get $U = 45.3$ volts. The uncertainty ΔU in U due to the errors contained in R and I is, according to Equation (2.5):

$$\Delta U = \sqrt{(2.5\%)^2 + (1.8\%)^2} = 3.1\%$$

therefore

$$U = 45.3 \pm 1.4 \text{ volt}$$

1.3(c). EXPONENTIATION. The result depends on one or more quantities which contain exponents, such as

$$R = x_1^{n_1} x_2^{n_2} x_3^{n_3} \qquad (2.6)$$

$$\Delta R = \sqrt{(n_1\,\Delta x_1)^2 + (n_2\,\Delta x_2)^2 + (n_3\,\Delta x_3)^2} = \sqrt{\sum_{i=1}^{3} (n_i\,\Delta x_i)^2} \qquad (2.7)$$

where

ΔR: compounded *relative error* of result R,
n_1: exponent of quantity x_1,
n_2: exponent of quantity x_2,
n_3: exponent of quantity x_3.
$\Delta x_1, \Delta x_2, \Delta x_3$: individual relative errors attached to the quantities x_1, x_2, x_3.

EXAMPLE:

Equation (2.1) stated the surface area of the human body as a function of mass m and height h. For $m = 70 \pm 1$ kg and $h = 200 \pm 2$ cm, we get

$$S = 0.007\,184(70)^{0.425} \times (200)^{0.725} = 2.03 \text{ m}^2$$

and for ΔS,

$$\Delta S = \sqrt{(0.425 \times 1.4\%)^2 + (0.725 \times 1\%)^2} = 0.94\%$$

therefore

$$S = 2.03 \pm 0.02 \text{ m}^2$$

Remark: The computed error is small, very small for a quantity of the life sciences. You should not be impressed! Although the error calculation is correct, it is highly unlikely that Equation (2.1) (empirically devised by Dubois in 1916) connects body mass and height with the surface area to better than 10%. This indicates that, in this case, the errors intrinsic to the measurements are negligible compared with the inherent error of the original formula—a situation not uncommon in the life sciences.

1.4 Powers of ten

It is convenient to express very large and very small numbers having only a few significant digits in terms of powers-of-ten. Table 1 shows some conversions from fully written numbers to equivalent numbers expressed in powers-of-ten notation.

Table 1 CONVERSION OF FULLY WRITTEN NUMBERS INTO POWERS OF TEN

1 000 000 000	10^9
100 000 000	10^8
10 000 000	10^7
1 000 000	10^6
100 000	10^5
10 000	10^4
1 000	10^3
100	10^2
10	10^1
1	10^0
0.1	10^{-1}
0.01	10^{-2}
0.001	10^{-3}
0.000 1	10^{-4}
0.000 01	10^{-5}
0.000 001	10^{-6}
0.000 000 1	10^{-7}
0.000 000 01	10^{-8}
0.000 000 001	10^{-9}

EXAMPLES:

The number of brain cells is approximately 10 billion. It would be a waste of paper and difficult to decipher to write it as 10 000 000 000. Instead we express it in powers-of-ten as 1×10^{10}.

Diameter of the Hydrogen Nucleus: Here, 1.32×10^{-15} m is preferable to 0.000 000 000 000 001 32 m.

Powers-of-ten notation is useful only for quantities having few significant digits. For example, the year 1900 was 31 556 925.975 s long. Since this number contains 11 significant digits, neither space nor time can be saved by writing it in powers-of-ten notation as $3.155\,692\,597\,5 \times 10^7$ s.

There are prefixes for some powers of ten which are used in connection with the name or symbol of a unit. See Table 2.

Table 2 PREFIXES AND THEIR SYMBOLS FOR SOME POWERS OF TEN

Number	Prefix	Symbol
10^{12}	tera	T
10^9	giga	G
10^6	mega	M
10^3	kilo	k
10^{-1}	deci	d
10^{-2}	centi	c
10^{-3}	milli	m
10^{-6}	micro	μ
10^{-9}	nano	n
10^{-12}	pico	p
10^{-15}	femto	f
10^{-18}	atto	a

Double prefixes should be avoided when single prefixes are available:

not mμs but ns

1.5 Natural and man-made units

The always repeating rhythm of day and night was probably the first time measure. So was the change of the moon phases and the position where the sun rises above the horizon. The time of the day can be determined by looking at various flowers which open at different times (Linnè's flower clock). Nowadays we could use the oscillations of a pulsar to measure time.

For reasons stated below, most natural standards are not used any more. But sometimes it gives a new insight if we convert our commonly applied units into natural units.

EXAMPLE:

The span of a generation is about 30 years; this seems to be the natural unit for history. To know that the most ancient city (probably Jericho) is 9000 years old hardly inspires our imagination. If we realize that this city was founded about 300 generations ago, the elapsed period does not seem so vast.

To measure distances there are many natural units available: a day's journey, the length of a king's foot, or a fraction of the circumference of the earth (already used in the second century B.C. during the Han dynasty in China). The natural unit for an area could be the size of a field a man can plow in one day (acre).

For scientific purposes or even for practical use, most natural units have severe drawbacks such as unreproducibility, unavailability, and variability with time. The last reason is the most important one because even the duration of the sun-year slows down (if only 10^{-5} second per year), and the best protected standard of length (like the original meter in Paris) will change (crystallization, corrosion, etc.).

Man has searched for and found standards which are believed to be unvarying. From these standards, units are derived which are applied in science and more and more in everyday life.

2. TIME MEASUREMENTS

Time is measured in seconds, one of the basic units of the SI system.

It is interesting to note that time is basically different from all other quantities measured. It has two aspects to it: First, as a coordinate, it shows the embedding of events in the universe. Second, as consecutive order, it shows the difference between past and future. More about this in: Friedrich Hund, 'Zeit als physikalischer Begriff.' Studium Generale, Vol. 23 (1970) p. 1088.

2.1 Definition

One second (abbreviated s) $\equiv$ *time required for a particular transition of the* ^{133}Cs *atom to perform* 9 192 631 770 *oscillations.*

The old definition was: 1/86 400 of the mean solar day for the year 1900.

Other units are:

$$1 \text{ day (d)} = 24 \text{ hours (h)} = 1440 \text{ minutes (min)} = 86\,400 \text{ s}$$

$$1 \text{ year (a)} = 3.155\,7 \times 10^7 \text{ s}$$

Clocks measure time. Figure 2-5 shows some time spans and the ranges of various clocks.

Any periodic process can be used to build a clock. The precision of the oscillation employed will basically limit its accuracy. Table 3 shows the accuracy (under favorable circumstances) of some clocks.

Table 3 The accuracy of some clocks

Type of Clock	Accuracy
atomic	10^{-5} second per year
quartz	10^{-3} second per year
pendulum	3 seconds per year
ship chronometer	4×10^2 seconds per year
^{14}C	50 years
magnetic	100 years
^{238}U	5×10^7 years

2.2 Various clocks

2.2(a). Biological clocks. Many processes in plants and animals have a periodic character. This is hardly surprising since they all live in a periodically changing world. Day and night come immediately to mind. Thus we can expect circadian (approximately one day) rhythms in nature. But it was discovered in the second half of this century that there are oscillatory events in living matter which are independent of outside influences. They are governed by periodic biological processes inside the organism and are named *biological clocks.*

Figure 2-6 shows an example of such a biological clock. The periodicity of the clock is approximately 25 hours.

There must be biological clocks oscillating much faster. To govern the succession of steps during orderly thinking requires an internal clock which oscillates within small fractions of a second. There are also much slower ones. Figure 2-7 shows the data for a biological clock in man with a period of about one year.

At present the processes which underlie the biological clocks are not understood.

Biological clocks are not suitable as time standards because their rhythm depends on factors such as temperature, age, availability of oxygen, and the individual. However, some of those biological clocks

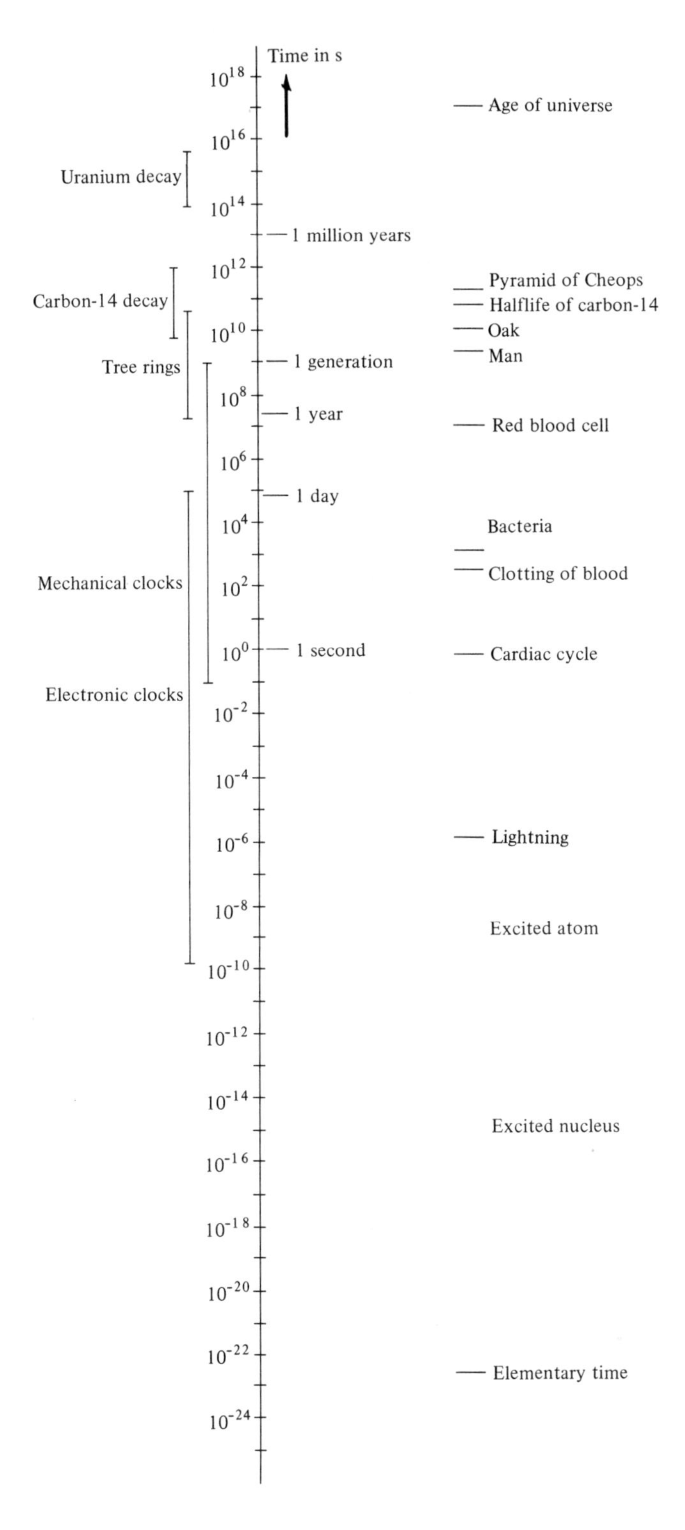

Figure 2-5 Typical time spans and the clocks used for their measurement.

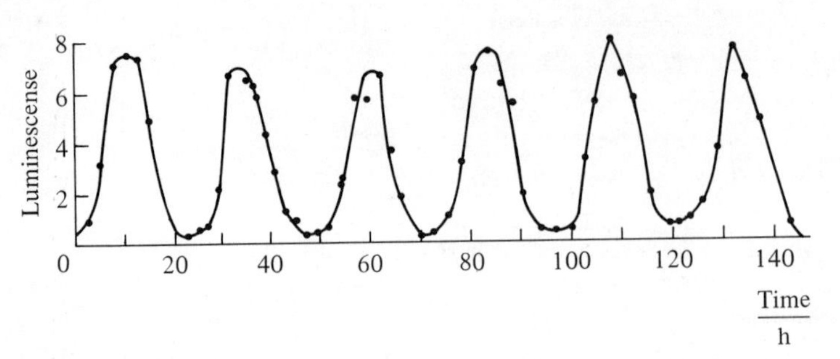

Figure 2-6 Rhythm of luminescence from cultures of Gonyaulax polyedra maintained in constant dim light and constant temperature.

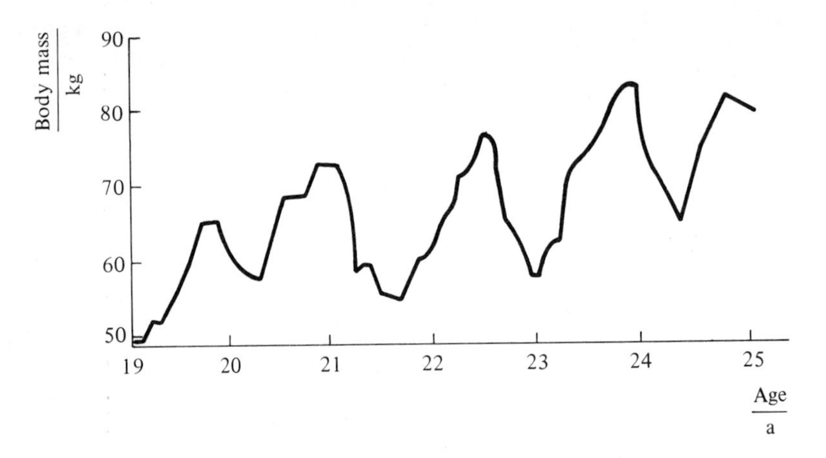

Figure 2-7 Manifestation of a very slow biological clock in man.

are known to be quite accurate. For example, the activity record of a flying squirrel was measured over a period of one month. It showed a rhythm of 1461 ± 6 min!

2.2(b). Radioactive clocks. The decay time of radioactive isotopes is known to be constant and cannot be influenced by outside events. Hence, if the decay time is known, it is possible to employ it as a time-span measuring device. Carbon dating (pioneered by Libby*) may serve as an example.

A piece of charcoal is discovered inside a hearth at a prehistoric site. If the age of the charcoal could be determined, we would know when this hearth was lit for the last time.

As long as wood is part of a living tree, it takes up CO_2 from the atmosphere via its leaves. This CO_2 contains minute amounts of the radioactive isotope ^{14}C (carbon 14). The ^{14}C is built into the structure of the wood like ordinary carbon molecules. If the tree is cut, the as-

* Willard Frank Libby, 1908, Nobel laureate 1960. Chemist and discoverer of radioactive dating.

similation ends, and the content of the radioactive carbon will disappear at a constant rate. Within 5730 years, half of it is decayed. If we measure, with suitable instruments, the radioactivity of the charcoal piece due to its remaining content of ^{14}C, we can determine its age provided we have an idea about its original carbon 14 content. If, for example, the measurement shows that it contains only half the normal activity, we conclude that the tree from which it came was cut 5730 years ago. Thus the hearth was lit about 57 centuries ago.

Remark: The long controversy about absolute dating with the help of ^{14}C has been settled. It is now an established fact that the amount of radioactive CO_2 in the atmosphere was not constant over the last few thousand years. It varied as much as 10%, which could mean an uncertainty in dating of 600–800 years! The content of radioactive CO_2 over the past 7000 years is now established by analyzing the tree rings from a Californian Pine (Pinus aristata). The tree rings serve to calibrate the radioactive clock.

Any radioactive isotope may be used as a radioactive clock under suitable circumstances. For example, ^{238}U (half-life 4.5×10^9 a) is used to determine the age of rocks; ^3H (half-life 12.3 a) is used to determine the age of wine. The range of a radioactive clock is approximately between $\frac{1}{10}$ and 10 times its half-life. Thus for carbon dating, an age between 600 and 60 000 years can be measured. Outside those limits, the derived age becomes very inaccurate. The result of radioactive dating is rarely more accurate than 5%.

2.2(c). Magnetic clocks. The magnetic poles of the earth change their position through the course of centuries. The path traveled by the magnetic poles is known. Magnetic material—if allowed to change its direction—will point toward the magnetic poles, as the compass needle demonstrates. If this magnetic material becomes fixed in position, it will stay aligned with the old position of the magnetic pole. Knowing the direction of the alignment, we can determine the time when it became fixed.

EXAMPLE:

Molten rock from a volcanic outbreak contains magnetic materials. It will align toward the poles as long as it is in its liquid state. Once hardened, it will keep its magnetic direction and hence allow a time determination of the outbreak. More about this modern method of magnetic dating in Chapter 12.

3. LENGTH MEASUREMENTS

Length is measured in meters, one of the basic units of the SI system.

3.1 Definition

One meter (abbreviated m) ≡ *length of* 1 650 763.73 *waves of the red line in the* ^{86}Kr *spectrum.*

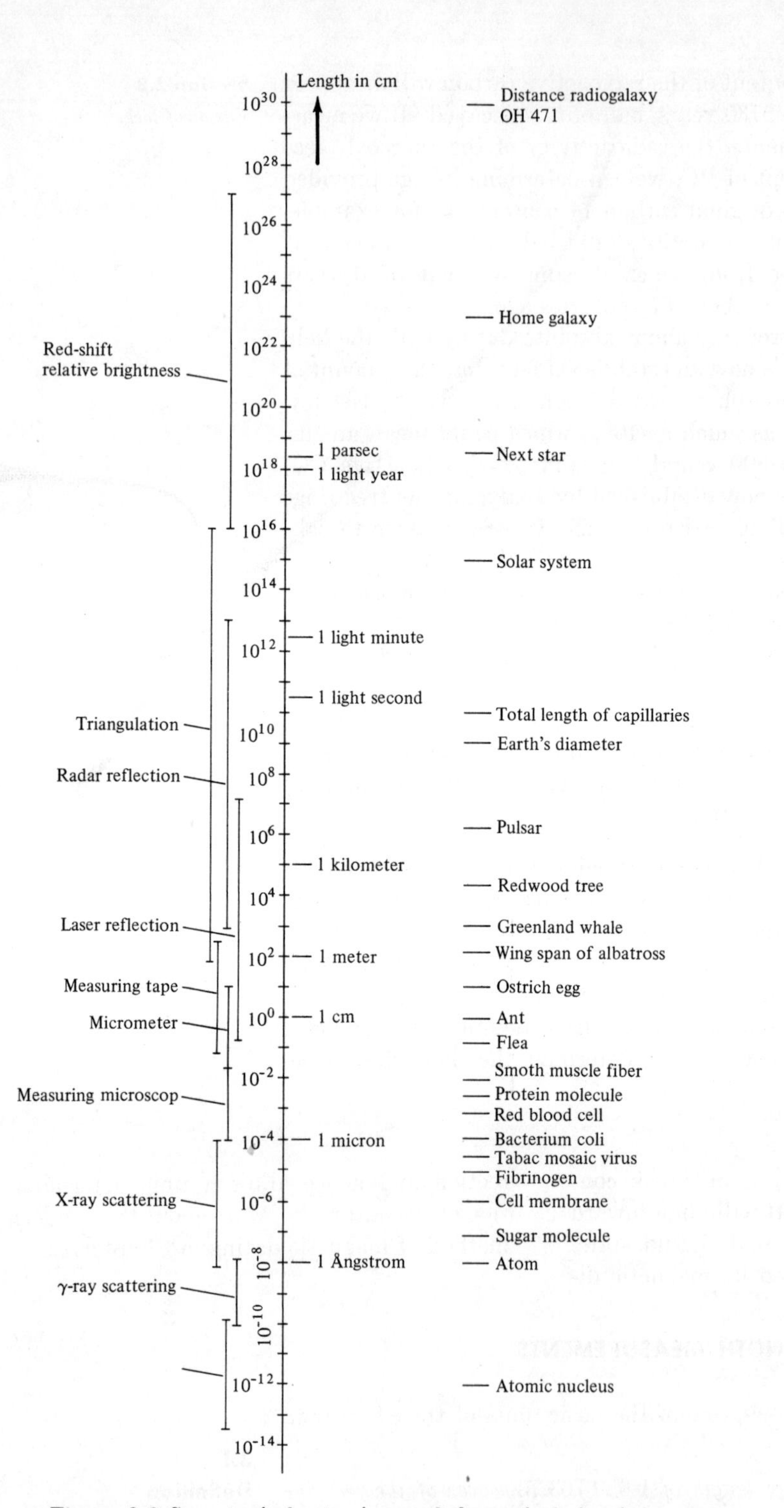

Figure 2-8 Some typical extensions and the methods for their measurement.

The old definition stated that one meter is equal to the ten-millionth part of a quadrant of a terrestrial meridian. This definition, introduced during the time of the French Revolution, is not accurate enough for scientific purposes. The original meter laid the foundation for the metric convention which is now used by 90% of the world's population.

Section 3.1
Definition

Other units of length are:

$$1 \text{ ångström (Å)} = 1 \times 10^{-10} \text{ m}$$

$$1 \text{ inch (in.)} = 0.0254 \text{ m}$$

$$1 \text{ foot (ft)} = 0.3048 \text{ m}$$

$$1 \text{ mile} = 1609 \text{ m}$$

$$1 \text{ light year} = 9.46 \times 10^{15} \text{ m}$$

The instruments used to measure length vary according to the extension to be determined. Table 4 shows some length-measuring devices, their ranges, and their accuracies.

Table 4 SOME DEVICES USED TO MEASURE LENGTH

Length Measuring Device	Range	Accuracy
ruler	3 m–1 cm	0.2%
metal scale	1 m–1 cm	0.01%
vernier caliper	50 cm–1 cm	0.01 cm
micrometer caliper	10 cm–1 mm	0.000 5 cm

In special cases it is possible to achieve an astounding accuracy in length measurements: The distance from the earth to the moon was determined with the help of a laser beam with an error of ± 0.15 cm!

Some typical extensions are presented in Figure 2-8, together with the methods employed to measure them.

3.2 Length of arc

Angles are usually measured in degrees, where a fully swept circle corresponds to 360 degrees. In physics it is often convenient to express the angle φ by the length of the corresponding arc l divided by the radius r. See Figure 2-9. The angle is then expressed in radians (symbol rad), where 360 degrees equals $2\pi = 6.28$ radians. See Figure 2-10.

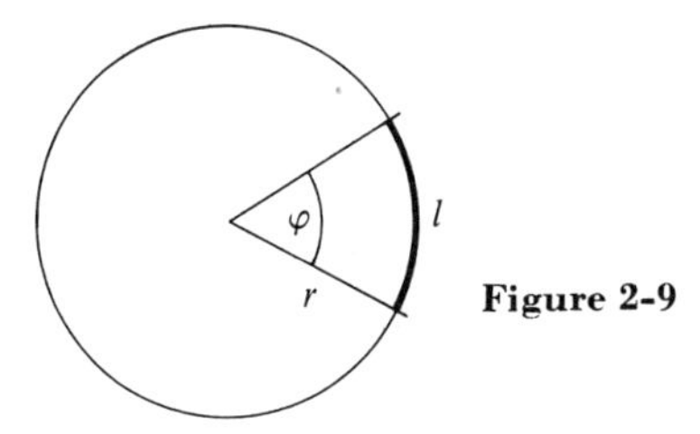

Figure 2-9

3.3 Area and volume measurements

Area is measured in square meters, symbolized m^2:

$$1 \text{ square meter} \equiv 1 \text{ m} \times 1 \text{ m}$$

$$1 \text{ m}^2 = 10^4 \text{ square centimeters, symbolized cm}^2$$

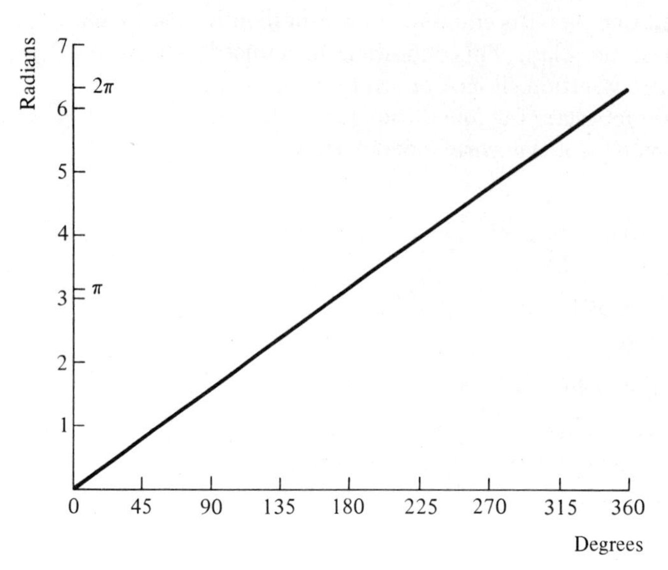

Figure 2-10

Other units of area are:

$$1 \text{ square inch (in.}^2) = 6.452 \times 10^{-4} \text{ m}^2$$

$$1 \text{ square foot (ft}^2) = 9.29 \times 10^{-2} \text{ m}^2$$

Volume is measured in cubic meters, symbolized m^3:

$$1 \text{ cubic meter} \equiv 1 \text{ m} \times 1 \text{ m} \times 1 \text{ m}$$

$$1 \text{ m}^3 = 10^3 \text{ liters (l)} = 10^6 \text{ cm}^3$$

Other units of volume are:

$$1 \text{ cubic inch (in.}^3) = 1.639 \times 10^{-5} \text{ m}^3 = 16.39 \text{ cm}^3$$

$$1 \text{ cubic foot (ft}^3) = 2.83 \times 10^{-2} \text{ m}^3 = 28.32 \text{ l}$$

$$1 \text{ U.S. gallon} = 3.785 \times 10^{-3} \text{ m}^3 = 3.785 \text{ l}$$

$$1 \text{ bushel} = 3.524 \times 10^{-2} \text{ m}^3 = 35.24 \text{ l}$$

In general, the area and volume of an object are determined by measuring characteristic extensions and employing suitable formulas. In the life sciences, however, objects are of irregular shape, and often no suitable formulas are available. Ingenious methods were developed for special cases. How to determine the surface area of plant leaves and needles may serve as an example (see Fig. 2-11). The leaves are thickly coated with pressure-sensitive adhesive; then small glass balls (diameter 0.1 mm) are poured over them. Tapping and shaking removes the loose balls. The others form a uniform layer over the entire surface. The gain in weight of the leaves is directly proportional to the area covered. The relation between surface and gained weight can be determined from a

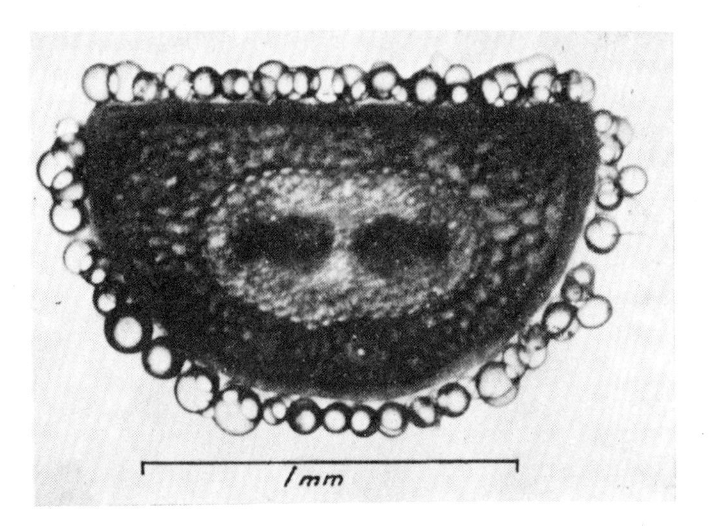

Figure 2-11 Corsican pine needle coated with Ballotini glass balls. (From *Nature*, **229,** February 1971.)

known test surface. This method is fast and proves to be accurate within 3%.

3.4 Signal travel time

Often extensions cannot be measured directly. For example, it is difficult to perform length measurements inside a living object without destroying it. Here the signal travel time is helpful. The signal (a sound impulse; a light flash) is sent into the object, and the reflected signal (from inner layers) is observed and timed in relation to the original signal. Since we know the propagation speed of the signal, we can calculate the distances between the layers from the time lapse between the original and reflected signals.

Figure 2-12 shows the echogram of the human eye. At time zero, a short ultrasonic sound impulse begins to travel through the eye. Parts of it are reflected at the various boundaries such as the cornea, lens, retina, or sclera. The reflected sound is recorded as a function of time. The speed of sound inside the various parts of the eye is known. This allows us to calculate the distances involved by observing the delay time between signal and reflection. The employed sound impulses are of extremely low intensity; therefore the eye is not harmed.

We can find a most interesting application of this method in astrophysics. Recently, blinking sources of radio waves were discovered in the sky. They are named *pulsars*. Pulsars emit regular bursts of radio waves of a duration of about 0.1 millisecond. This duration tells us something about the maximum possible diameter of the pulsar. Since the radio bursts from all parts of the pulsar are synchronized within 0.1 millisecond, information must pass from the emitting molecules of one part of the pulsar to any other part very rapidly. The upper limit for the speed of information transfer is the speed of light (3×10^8 m/s). The information to emit a radio burst travels across the pulsar as an electromagnetic wave, hence with the speed of light. In 10^{-4} s, light travels 3×10^4 m. Consequently the maximum possible diameter of a pulsar is about 30 km. This

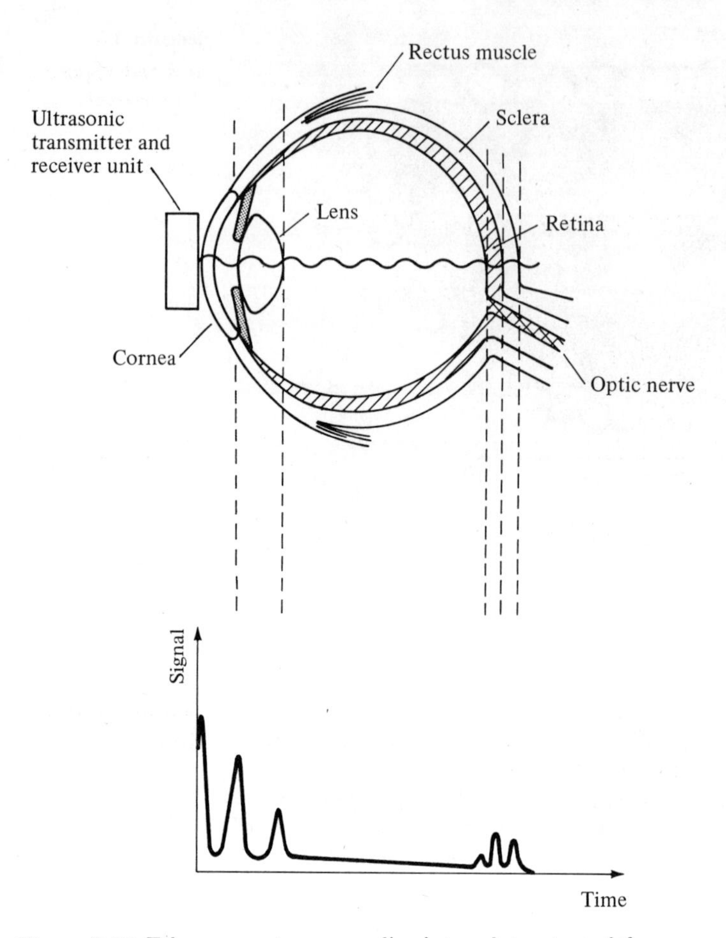

Figure 2-12 Echogram and corresponding internal structure of the eye.

is the maximum diameter because, if the information is brought across by any other means, the diameter would be even smaller due to the lower speed of information transfer. By this simple calculation we can determine the diameter of a stellar object (which still puzzles the scientists) thousands of light years away. The speed of information transfer is of outstanding importance in the living world. More about it in Chapter 14.

SUMMARY

Any measurement of a physical quantity is valid only within its error limits. The following terms are used in this connection:

Precision or *reproducibility:* The degree of agreement of repeated measurements of the same quantity.

Accuracy: The agreement between the result of a measurement and the accepted value of the quantity measured.

Physical quantities are presented using their significant digits. Only the last digit of a number may change due to the finite accuracy for the value of this quantity. To express large or small numbers having only a few significant digits, the powers-of-ten notation is convenient. For example, 303 000 may be written as 3.03×10^5; 0.000 723 may be written 7.23×10^{-4}.

Any measurement involves a comparison with a standard measure. The result is expressed in units of this standard. The standards are defined and believed to be unvarying.

In physics the International System of Units having six basic units (meter, kilogram, second, ampere, degree Kelvin, and candela) is recommended. Other units are derived from these basic units.

The unit of time is the *second* (symbol s). The modern definition links the second with a transition of the ^{133}Cs atom.

Conversion: 1 year (symbolized a) = 3.1557×10^7 s. The unit of length is the meter (symbolized m). In the original definition it is the ten-millionth part of a meridian. Since this proved to be not accurate enough, the meter is now linked to a process on the atomic scale: 1 meter (symbolized m) is the length of 1650763.73 waves of the red ^{86}Kr line.

PROBLEMS

1. Calculate the surface area of your body using Equation (2.1). How large are the relative and absolute errors?
2. Under what circumstances will a signal averager improve the signal-to-noise ratio?
3. Determine the precision of the biological clock shown in Figure 2-6.
4. What advantage can you see of the echogram over the x-ray picture?
5. What would be the maximum diameter of a pulsar if the information is spread like sound in air?
6. The blood speed *in vivo* can be measured by "tagging" water molecules and detecting the arrival of those tagged molecules further downstream. Figure 2-13 shows the signal produced by the arrival of tagged molecules. The molecules were marked 2.5 cm upstream (time 0); on the average, they arrived after 0.65 s (center of dip in Figure 2-13). Calculate the blood speed, and estimate the error.
7. Determine the time between consecutive maxima in Figure 2-6. Calculate the absolute and the relative error.
8. Express your present age in seconds and estimate its absolute error. Give only significant figures.
9. Convert 60 miles per hour in $cm \cdot s^{-1}$.
10. Calculate the distance light travels in one year (light year) to four significant figures. Express the value: (a) in centimeter and powers-of-ten notation; (b) in miles, together with the proper prefix.

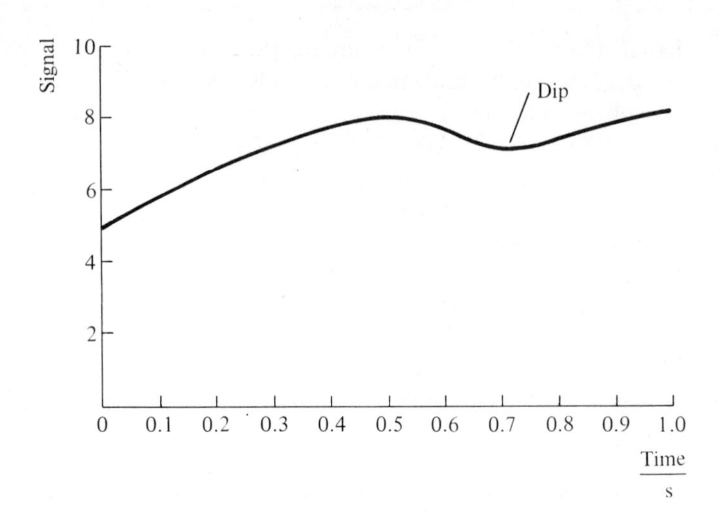

Figure 2-13 Signal produced by tagged water molecules.

11. Convert 12 square miles into m^2.

12. A standard clock emits time signals with a spacing of 1.000 minute. The time intervals are measured with a stop-watch with an accuracy of 0.2%. Calculate relative and absolute error of the measurement. Is the reaction time of the person doing the measurement of importance?

FURTHER READING

E. A. BROWN ET AL., *The Biological Clock, Two Views.* Academic Press, New York, 1970.

E. BÜNNING, *The Physiological Clock.* Academic Press, New York, 1964.

M. DANLOUS-DUMESNILD, *The Metric System.* Athlone Press, London, 1969.

J. OSTRIKER, "The Nature of Pulsars," *Scientific American* (Jan., 1971).

F. B. THOMPSON, L. LEYTON, "Method for Measuring the Leaf Surface Area of Complex Shoots," *Nature*, **229** (1971) 572.

"Metric System Status of Adoption by the United States," *Science*, **170** (1970) 1337.

"Symbols, Units, and Nomenclature in Physics," International Union of Pure and Applied Physics. S. U. N. Commission. Document U.I.P. (S.U.N.65-3) (1965).

chapter 3

MOTION AND FORCES

This chapter deals with topics usually called *kinematics* and *dynamics*. In kinematics, quantities are introduced which enable us to describe the motions of bodies. The motions may be uniform or not, on a straight line, or along a curved path. Whereas kinematics deals with how bodies move, dynamics is concerned with the problem of why they are moving. Consequently dynamics deals mainly with forces acting on bodies.

Objects of the real world are *bodies* having physical extensions and reacting to applied forces by translation, rotation, and deformation.

To simplify the formulation of the laws which govern motion and forces, it is necessary to idealize the bodies of the real world. If not stated otherwise, a body in kinematics has no physical extensions. It is treated as a *point*. The only possible motion of a point is a translation from an initial position to another position in space. Admittedly, this seems to be a rather poor approximation of real bodies. A better approximation takes into account the physical extensions of the body. Again, matters are simplified by the introduction of the rigid body.

A *rigid* body has physical extensions, that is, length, area, or volume, but those extensions are fixed. Consequently a rigid body can undergo a translation (as can the point body) and a rotation.

An even better approximation of the real body is obtained if we account for its deformation under the influence of a force. As long as these deformations are small and reversible, we can describe the body as *quasirigid*. If this is not possible, it is termed a *deformable* body. Table 1 summarizes the successive approximations toward a real body.

In this chapter we will examine new quantities such as velocity, acceleration, mass, force, torque, density, and pressure. Many of these quantities are vectors, so it is worthwhile to review first the section of Chapter 1 which describes vectors. Also, the quantities defined in this chapter are of outstanding importance to all subsequent chapters!

Table 1 SUCCESSIVE APPROXIMATIONS TOWARD A REAL BODY

Notation of Approximation	Properties	Possible Motion
point body	no physical extensions	translation
rigid body	fixed physical extensions	translation rotation
quasirigid body	physical extensions undergo small and fully reversible changes	translation rotation vibration
deformable media	real body	any motion

1. MOTION

The terminology introduced here will be used again and again. Most likely, you will forget the details, but you shall remember where to look up the individual terms.

1.1 Translatory motion

1.1(a). VELOCITY AND SPEED. Translation of a body means that the body changes its position from point A in space to another point B along a path in a given direction. See Figure 3-1.

If the translation is along a straight line, we introduce the *average velocity*, abbreviated $\bar{\mathbf{v}}$.

$$\bar{\mathbf{v}} \equiv \frac{\Delta \mathbf{s}}{\Delta t} \tag{3.1}$$

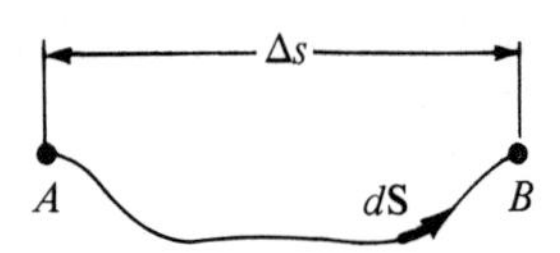

Figure 3-1

where

$\Delta\mathbf{s}$: displacement of the body moved from A to B,
$|\Delta\mathbf{s}| = \Delta s$: distance along a straight line between A and B,
Δt: elapsed time while the body moved from A to B.

The average velocity is a vector, its magnitude $\bar{v}$ is called *average speed.*

We introduce the *instantaneous velocity* (or velocity), symbolized $\mathbf{v}$.

$$\mathbf{v} \equiv \frac{d\mathbf{s}}{dt} \tag{3.2}$$

where

$d\mathbf{s}$: infinitesimal displacement of the body while moving,
dt: infinitesimal time elapsed during displacement $d\mathbf{s}$.

Velocity $\mathbf{v}$ is a vector.

The magnitude v of the instantaneous velocity $\mathbf{v}$ is called *instantaneous speed* or just *speed.*

Speed is measured in meters/second. Conversion:

$$1\ \text{m}\cdot\text{s}^{-1} = 3.60\ \text{km}\cdot\text{h}^{-1}$$

For a uniform motion along a straight line, the average and instantaneous velocities are equal.

EXAMPLE:

An airplane flying from Sacramento to Reno over the Sierra Nevada needs 1 hour and 10 minutes to complete the trip. The displacement $\Delta\mathbf{s}$ of the plane is 1.8×10^5 m eastward, and for the elapsed time of 4.2×10^3 s, the average velocity is

$$\bar{\mathbf{v}} = \frac{\Delta\mathbf{s}}{\Delta t} = \frac{1.8 \times 10^5}{4.2 \times 10^3}\ \text{m}\cdot\text{s}^{-1} = 42.7\ \text{m}\cdot\text{s}^{-1}\ \text{eastward}$$

The average speed is

$$\bar{v} = \frac{\Delta s}{\Delta t} = 42.7\ \text{m}\cdot\text{s}^{-1} = 154\ \text{km}\cdot\text{h}^{-1}$$

The instantaneous velocity $\mathbf{v}$ of the aircraft (measured at any moment by the speedometer) changed while flying between Sacramento and Reno. Certainly while climbing over the Sierra, $\mathbf{v}$ was smaller than $\bar{\mathbf{v}}$.

Addition of Velocities. If velocities are to be added or subtracted, the rules of vector addition can be applied as long as the velocities are small, i.e., less than 5% of the speed of light, 3×10^8 m/s. The reason for this restriction (which in everyday life is of no importance) is that above certain speeds relativistic effects must be taken into account.

Angular Velocity. A body orbits along a circular path (*ABCDEABC . . .*), as shown in Figure 3-2. The velocity of the body is always perpendicular

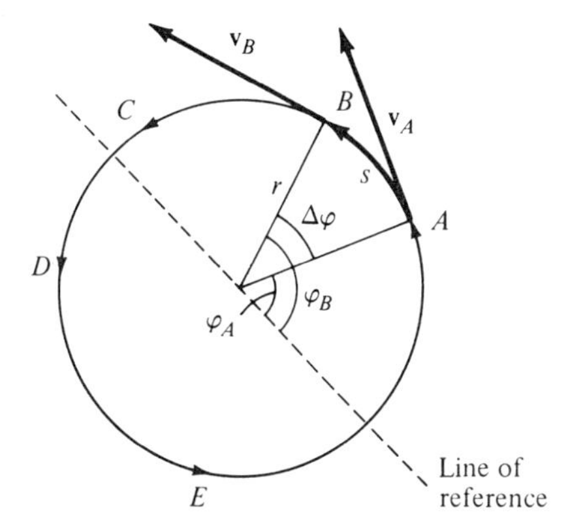

Figure 3-2 Body orbiting along a circular path. r: radius of circle; φ: central angle; s: length of arc; $\mathbf{v}$: velocity of orbiting body.

to the radius r of the circle (it is tangential), and it changes even though the speed of the body may be constant. This is due to the change in direction of $\mathbf{v}$ as the body moves from one position on the circle to another.

Remember: The velocity $\mathbf{v}$ is a vector quantity having direction and magnitude. Although the magnitude of the velocity remains constant, the direction of the velocity changes continuously as the body moves along the circular path. The magnitude of the velocity, the speed v, is a scalar (a pure number) and does not change while the body moves along the circular path.

It is convenient to express the velocity in terms of the central angle φ. If the central angle φ is expressed in radians, then the length s of the arc between A and B is

$$s = r \cdot \Delta\varphi \tag{3.3}$$

where

$\Delta\varphi = \varphi_A - \varphi_B$: change of central angle for a motion from A to B. The central angle is measured from an (arbitrary) reference line.

r: radius of circle.

An infinitesimal motion ds takes an infinitesimal amount of time dt; thus Equation (3.3) changes into

$$ds = r\,d\varphi$$

and we get for the orbital speed v

$$v = \frac{ds}{dt} = r\frac{d\varphi}{dt} \tag{3.4}$$

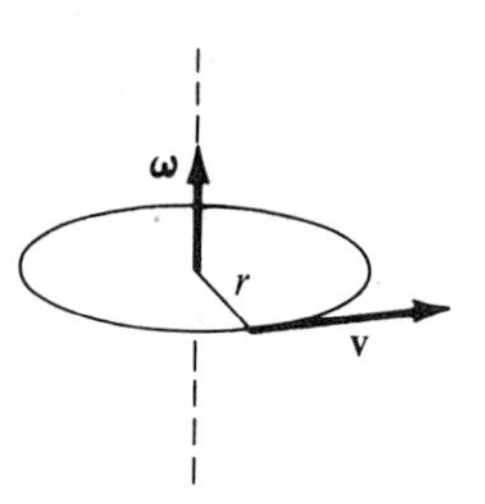

Figure 3-3

where the term $d\varphi/dt = \omega$ is called *angular speed.* ω can be expressed as a vector quantity: The angular velocity $\boldsymbol{\omega}$ is a vector pointing along the axis of orbiting. Thus it is called an *axial* vector. See Figure 3-3.

For counterclockwise motion $\boldsymbol{\omega}$ points upward. The directions of those vectors are a convention taken from vector analysis.

Period and Frequency. To describe circular motion, two new quantities are introduced, *period* and *frequency*. Period, symbolized T, is measured in seconds and is defined as:

$$T \equiv \text{elapsed time for a complete revolution}$$

Frequency, symbolized ν (or f), is defined as:

$$\nu \equiv \text{number of revolutions per unit time}$$

Frequency is measured in 1/s, and there is a special unit for it:

$$1 \text{ hertz* (symbol: Hz)} = 1 \text{ s}^{-1}$$

A useful relation is:

$$\nu = \frac{1}{T} \tag{3.5}$$

Although period and frequency have been defined for circular motion, both quantities are used to describe any periodic motion.

1.1(b). Acceleration. If the velocity of a body changes during trans-

* Heinrich Hertz (1857–1894), physicist, proved experimentally the existence of electromagnetic waves.

latory motion, it undergoes an *acceleration*. Acceleration, symbolized **a**, is defined as:

$$\mathbf{a} \equiv \frac{d\mathbf{v}}{dt} \tag{3.6}$$

where

$d\mathbf{v}$: infinitesimal change of velocity,
dt: infinitesimal time elapsed during $d\mathbf{v}$,

The magnitude of the acceleration is measured in meters/second². Sometimes it is expressed in multiples of $g = 9.81\ \mathrm{m\cdot s^{-2}}$, which is acceleration due to gravity.

EXAMPLE: ACCELERATION OF AN AUTOMOBILE

A car is advertised to accelerate from rest to 80 km/h = 22.2 m/s in 8 seconds. The magnitude of the acceleration is:

$$a = \left|\frac{d\mathbf{v}}{dt}\right| = \left|\frac{0 - 22.2}{8}\right| = 2.78\ \mathrm{m\cdot s^{-2}}$$

or expressed in multiples of g,

$$a = 0.283g$$

Remark: The above defined acceleration, Equation (3.6), is really an instantaneous acceleration. Instantaneous and average acceleration have not been introduced because the latter will rarely be encountered and there are enough new quantities without it.

Acceleration is a vector quantity; thus it can change in three different ways: by changing magnitude or direction alone or both at the same time. Consequently it is common to distinguish three types of acceleration: *linear, radial,* and *curvilinear.*

Linear Acceleration. In the case of linear acceleration $\mathbf{a}_l$, the magnitude of the velocity changes but the direction does not. The motion occurs on a straight line.

The magnitude a_l of $\mathbf{a}_l$ is

$$a_l = \left|\frac{d\mathbf{v}}{dt}\right| = \frac{d}{dt}\left(\frac{ds}{dt}\right) = \frac{d^2s}{dt^2} \tag{3.7}$$

where $|\mathbf{v}| = ds/dt$.

The direction of the linear acceleration $\mathbf{a}_l$ is parallel to the velocity $\mathbf{v}$ of the body. If $\mathbf{a}_l$ points in the same direction as $\mathbf{v}$, then the velocity will increase with time. If $\mathbf{a}_l$ points in the opposite direction from $\mathbf{v}$, then the velocity decreases. (Sometimes this is called *deceleration*).

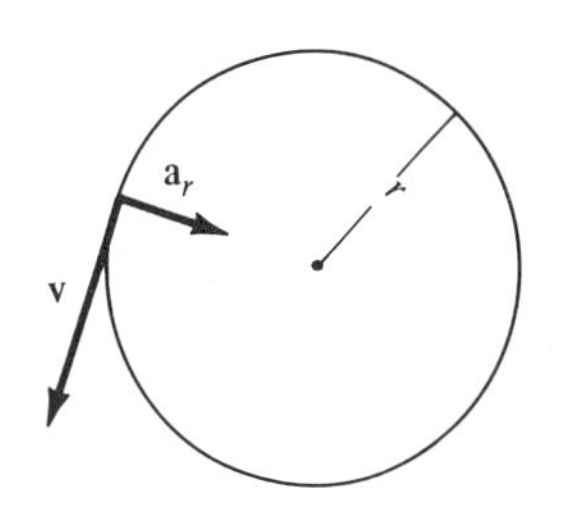

Figure 3-4

Radial Acceleration. For radial (or normal) acceleration $\mathbf{a}_r$, the direction of the velocity changes and its magnitude remains constant. The motion occurs on a circular path. See Figure 3-4.

The magnitude a_r of $\mathbf{a}_r$ is

$$a_r = \frac{v^2}{r} = \omega^2 r = 4\pi^2\nu^2 r \tag{3.8}$$

where

r: radius of orbit,
v: tangential speed of body along orbit,
ν: frequency of revolving body,
ω: $2\pi\nu$.

The radial acceleration $\mathbf{a}_r$ is a vector perpendicular to the velocity $\mathbf{v}$ of the orbiting body; it points toward the center of the orbit.

EXAMPLE: A BIRD CHANGING ITS DIRECTION IN FLIGHT

The curved part of the flight path is the arc of a circle (in good approximation). The speed of the bird remains constant. As the bird approaches position A its velocity $\mathbf{v}_A$ points toward A; see Figure 3-5. There is no acceleration because $\mathbf{v}$ is constant and it travels on a straight line. At position A, although the speed remains constant, the bird is accelerated because now the direction of the velocity changes. The radial acceleration $\mathbf{a}_r$ points inward, and its magnitude a_r is given by Equation (3.8),

$$a_r = \frac{v^2}{r}$$

For a chimney swift, you can observe:

$$v = 65\ \text{km/h} = 18\ \text{m}\cdot\text{s}^{-1}$$

$$r = 10\ \text{m}$$

hence

$$a_r = 32.4\ \text{m}\cdot\text{s}^{-2}$$

or if we express a_r in multiples of g:

$$a_r = 3.3g$$

The bird is accelerated until it reaches position B. Now the velocity is constant again, there is no acceleration.

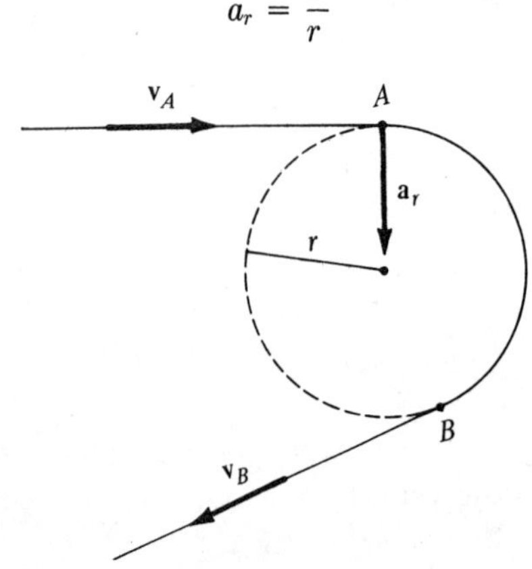

Figure 3-5 Acceleration due to a change in direction of flight. r: radius of curvature; $\mathbf{v}$: velocity; $\mathbf{a}_r$: acceleration.

EXAMPLE: THE ORBITING SPACE PLATFORM

Inside a space capsule circling the earth in a stable orbit, no acceleration due to gravity is sensed by an astronaut. This makes life and work for him difficult. He might feel more comfortable living inside a doughnut-shaped space platform which revolves around its axis. The induced radial acceleration would give his body some sense of orientation.

Curvilinear Acceleration. A curvilinear acceleration is present if both direction and magnitude of the velocity of the body in motion change. The motion takes place along a curved path, the instantaneous velocity $\mathbf{v}$ is tangential to this path. See Figure 3-6.

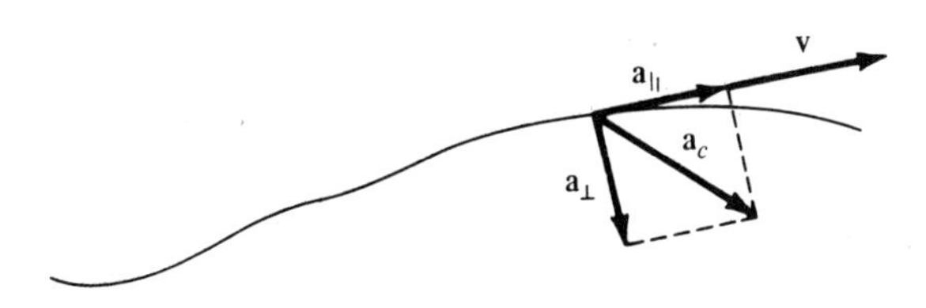

Figure 3-6

The curvilinear acceleration $\mathbf{a}_c$ points toward the inside of the curved path. To analyze $\mathbf{a}_c$, it is resolved into a component $\mathbf{a}_{||}$ parallel to $\mathbf{v}$ and into a component $\mathbf{a}_{\perp}$ perpendicular to $\mathbf{v}$.

EXAMPLE: EARTH ORBITING THE SUN

The path is an ellipse; its eccentricity is exaggerated in Figure 3-7. One of its two focal points is occupied by the sun. $\mathbf{v}$ and $\mathbf{a}_c$ are different in direction and magnitude in both displayed positions.

Winter position: $\mathbf{a}_c$ points inward, the component $\mathbf{a}_{||}$ points in the direction of $\mathbf{v}$; consequently the orbital speed of the earth increases.

Summer position: $\mathbf{a}_c$ points inward, the component $\mathbf{a}_{||}$ now points opposite to $\mathbf{v}$; hence the orbital speed of the earth decreases.

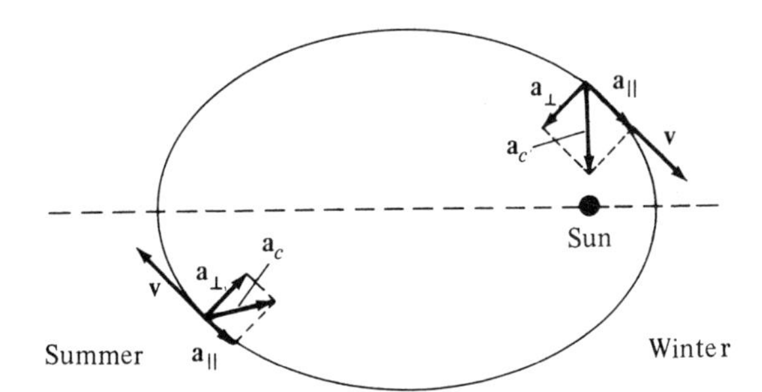

Figure 3-7 The earth orbiting the sun: an example for curvilinear acceleration. Summer and winter positions are shown.

Figure 3-8 summarizes the various types of acceleration.

1.2 Rotational motion

In this section we shall treat the extended body as rigid, referring the treatment of deformations of a body to Chapter 5.

A body is called *rigid* if the separations between its individual constituents (the molecules) do not change. The general motion of a rigid

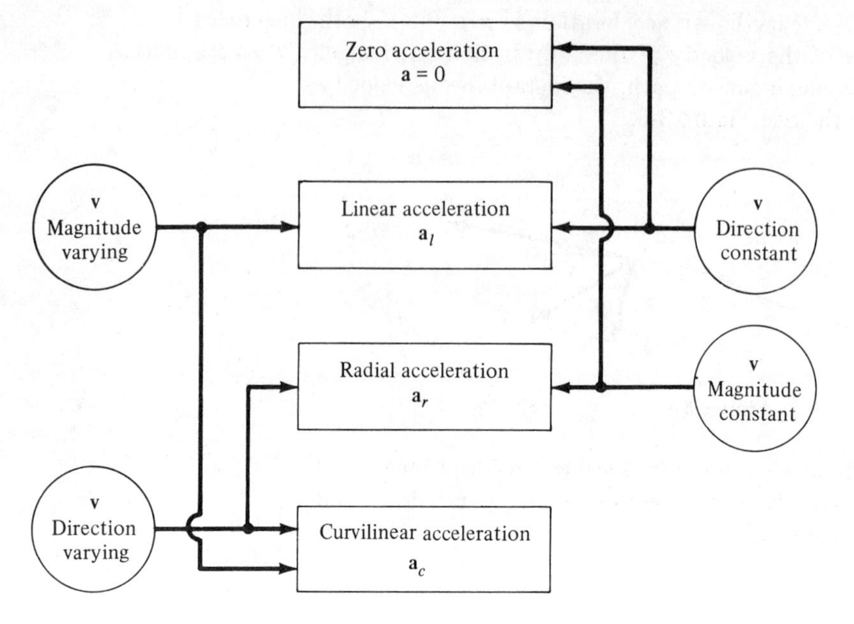

Figure 3-8 Various types of acceleration.

body is completely described by decomposing its motion into a rotation around an axis and a translation of this axis.

Figure 3-9 shows a rigid body rotating around an axis. (The position of this axis depends on the circumstances. In Figure 3-9, the tibia

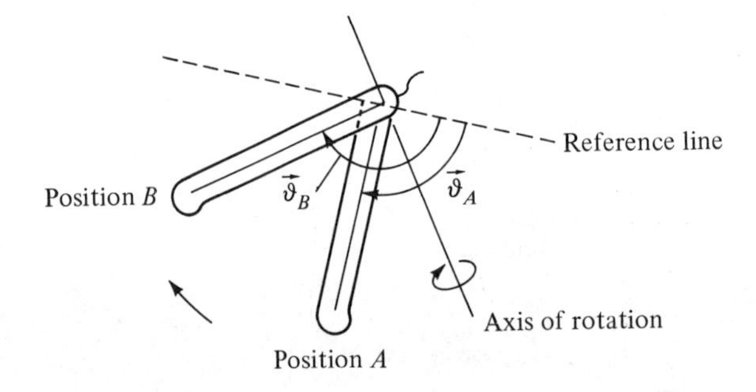

Figure 3-9 Rotational motion of a rigid body shown at the tibia.

(shin bone) is chosen to be the rigid body, and consequently the axis of rotation is perpendicular to its proximal end).

Any rotational position of the rigid body can be described by stating the angle of rotation $\boldsymbol{\vartheta}$ from an arbitrary reference line. $\boldsymbol{\vartheta}$ has vector properties because it has a magnitude (expressed in degrees or radians) and a direction (clockwise or counterclockwise).

1.2(a). VELOCITY AND ACCELERATION. Angular velocity and angular acceleration are defined in analogy to translational kinematics. *Angular velocity* (instantaneous) symbolized $\boldsymbol{\omega}$, is defined as:

$$\boldsymbol{\omega} \equiv \frac{d\boldsymbol{\vartheta}}{dt} \tag{3.9}$$

where

$d\boldsymbol{\vartheta}$: infinitesimal change of angle of rotation,
dt: infinitesimal time elapsed during change $d\boldsymbol{\vartheta}$.

The magnitude ω of $\boldsymbol{\omega}$ is measured in degrees/second or in radians/second. In engineering, ω is measured in rotations/minute (rpm).

The direction of $\boldsymbol{\omega}$ is along the axis of rotation; it points upward if the rotation is counterclockwise and downward if clockwise. See Figure 3-10.

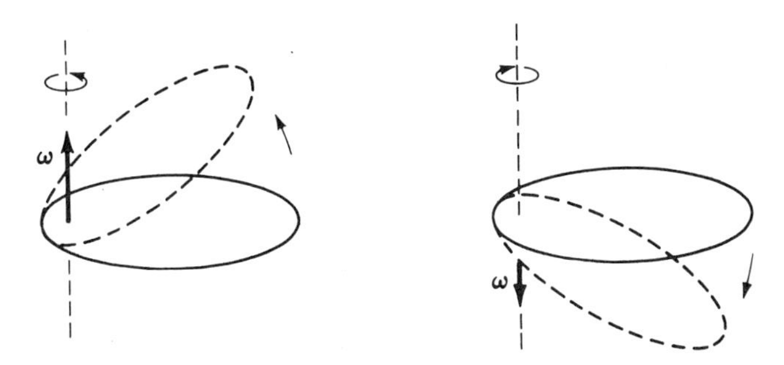

Figure 3-10

Angular Acceleration (instantaneous), symbolized $\boldsymbol{\alpha}$, is defined as:

$$\boldsymbol{\alpha} = \frac{d\boldsymbol{\omega}}{dt} \tag{3.10}$$

where

$d\boldsymbol{\omega}$: infinitesimal change of angular velocity,
dt: infinitesimal time elapsed during change $d\boldsymbol{\omega}$.

The magnitude of the angular acceleration is measured in degrees/second^2 or in radians/second^2.

The direction of $\boldsymbol{\alpha}$ is along the axis of rotation, hence, along $\boldsymbol{\omega}$. If $\boldsymbol{\alpha}$ points in the same direction as $\boldsymbol{\omega}$, then $\boldsymbol{\omega}$ increases. $\boldsymbol{\omega}$ decreases if $\boldsymbol{\alpha}$ and $\boldsymbol{\omega}$ have opposite directions.

1.2(b). Angular acceleration detector in man. The three semicircular canals in the bony labyrinth (see Figure 3-11) form an indicator for the angular accelerations experienced by the head.

Figure 3-12 shows a cut along one of those canals. If the head rotates clockwise in the plane of this canal, the endolymph fluid inside the membranous labyrinth does not follow immediately the movement of the labyrinth. Therefore the fluid obtains a relative motion counterclockwise and bends the cupula inside the ampulla. This bending indicates to the appropriate nerve center (via connecting nerve fibers) that a rotational motion began. After a few seconds of continuing rotation, the frictional forces between the walls of the membranous labyrinth and

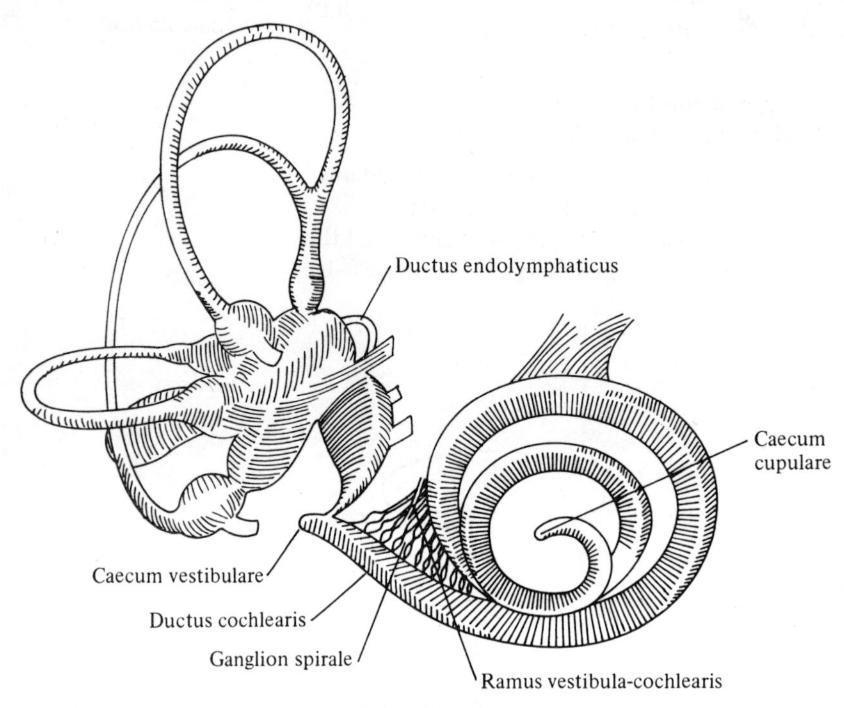

Figure 3-11 The angular acceleration detector in man. It is part of the inner ear and about the size of a pea.

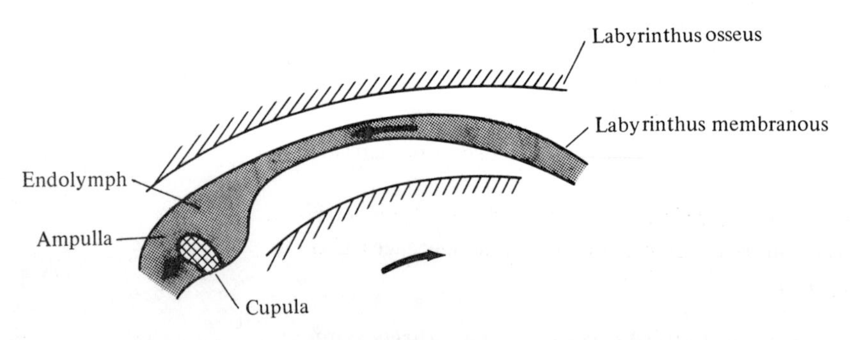

Figure 3-12 One of the semicircular canals of the inner ear. Bending of the cupula indicates angular acceleration.

the endolymph have slowed down and finally stopped the relative motion of the fluid. The elastic cupula unbends, and the sensation of rotation vanishes until the rotation stops. Now the endolymph fluid moves again with respect to the walls, but in the reverse direction. Consequently the semicircular canals signal only the beginning and end of a rotational motion, hence, the angular acceleration.

Because the three semicircular canals are mutually perpendicular, any angular acceleration is automatically resolved into three components, thus allowing a three-dimensional analysis of the undergoing motion.

The lateral line system in fish functions in the same manner; however, since the lateral canals are linear, this system responds mainly to linear accelerations.

1.3 Irregular motion

The motion of a body exhibiting an irregular path can often be analyzed by statistical methods. The Brownian motion, treated in Chapter 8, may serve as an example. It will spare you a general treatment of irregular motion.

2. FORCE

2.1 Newton's first and second Laws

In the previous section you saw how to describe the motion of a body. Now you might ask why a body moves. The answer to this question is an extrapolation and abstraction from experience: Objects move under the influence of a force.

Although most of the physical laws are formulated by abstraction from gained experience, you should realize that this abstraction is not a trivial and obvious process. That a force can act through empty space without any intermediary medium was recognized only a few hundred years ago. It is quite an intellectual achievement to realize that the force exerted by our earth on a falling stone does not depend on the size of the stone, barometric pressure, time of the day, color of the sky, locality of fall, or scattering of the measured values.

Force, symbolized **F**, is usually defined in a negative fashion by Newton's* first law: If the velocity of a body remains constant, then the resultant force acting on it is zero.

Newton's first law is an outstanding abstraction because we cannot easily perceive it from everyday experience. Not only are we groundlings subject to gravitational force, but every moving object on earth is influenced by frictional forces.

Force is a vector quantity; therefore the rules of vector addition apply. Forces can be cancelled by other forces of equal magnitude but opposite direction.

EXAMPLE: A BODY FALLING THROUGH OIL

You can observe that it moves eventually with constant speed downward. The influence of the downward-pointing gravitational force is in this instance cancelled by the opposing frictional force. The resultant force is zero.

* Isaac Newton (1642–1727), physicist and mathematician. He not only formulated the modern concepts of dynamics but made many other substantial contributions to physics. He invented independently from Leibniz the formalism of differential calculus.

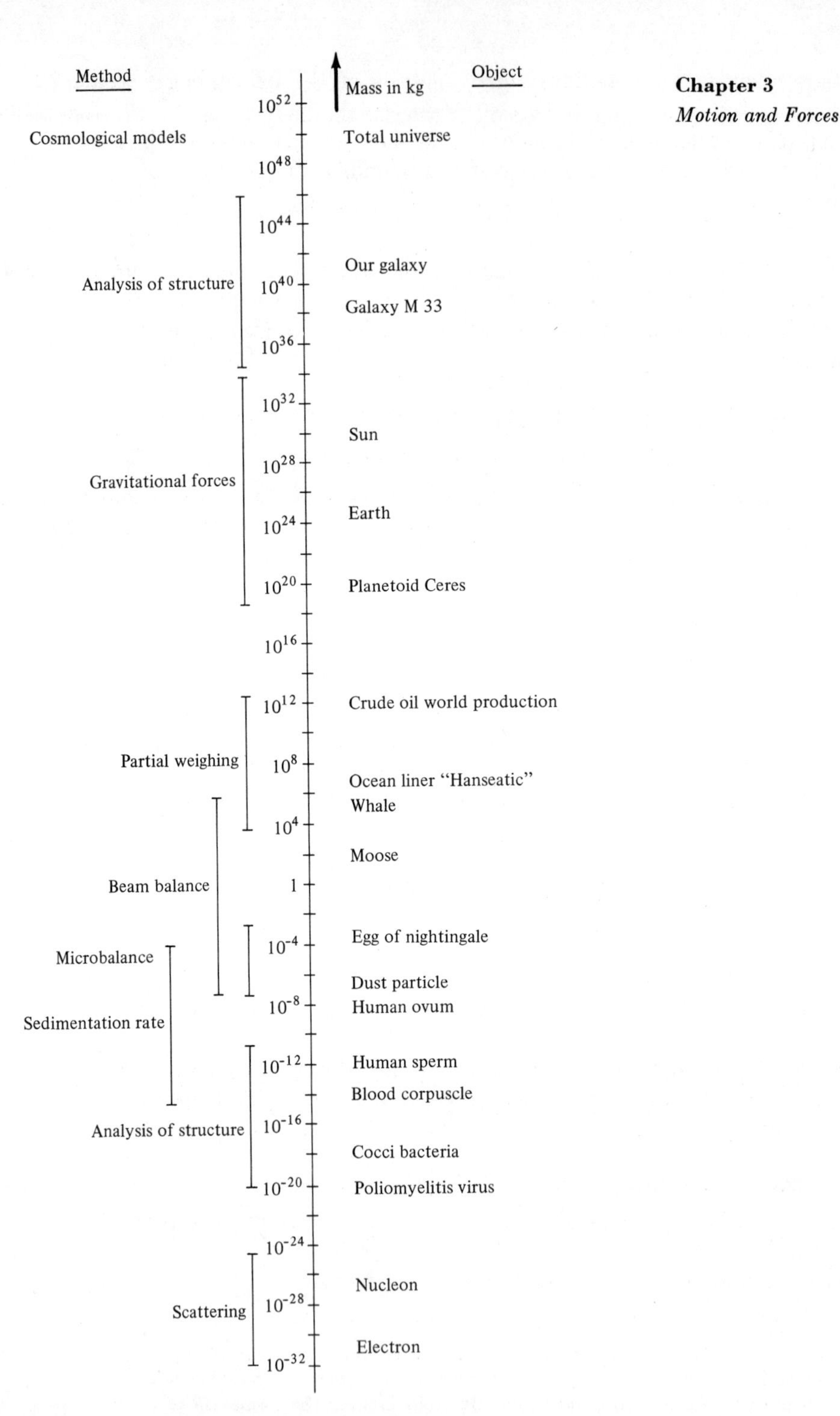

Figure 3-13 Physical objects of our universe (right) and methods used to determine individual masses (left).

Section 2.1
Newton's First and Second Laws

Newton's first law states that, if a resultant force **F** acts on a body, the velocity of this body changes. In other words, the body is accelerated. Measurements show that

$$\mathbf{F} \sim \mathbf{a}$$

where **a** is the acceleration of body under influence of **F**. Changing the proportionality into an equation, we get

$$\mathbf{F} = m\mathbf{a} \qquad (3.11)$$

where m is introduced as proportionality constant. (This is the common mathematical method of changing a proportion into an equation). The relation in Equation (3.11) is Newton's second law.

2.2 Mass

The proportionality constant m in Equation (3.11) is a scalar and is called the *mass* of the body. It is a very basic property of this body and does not change if the position of the body changes in space or time.

Mass is measured in kilograms (symbol: kg), one of the basic units of the International System of Units.

DEFINITION: 1 kg ≡ 1000 gram (g) = *mass of a standard platinum block stored at Sèvres, France*

In atomic and nuclear physics, mass is measured in atomic mass units, symbolized u:

$$1\ \text{u} \equiv \tfrac{1}{12} \text{ of the mass of the neutral carbon atom } ^{12}\text{C}$$

Conversion:

$$1\ \text{u} = 1.6604 \times 10^{-27}\ \text{kg}$$

An unknown mass is determined by comparison with a calibrated mass. The instrument used to measure mass is called a *balance*. Since balances have a limited range, indirect methods are employed to determine very large and very small masses. Figure 3-13 shows various objects and the methods used to determine their masses.

2.2(a). REMARKS CONCERNING MASS. One of the fundamental laws of physics states that the mass of a body remains constant. But there is an effect in nuclear

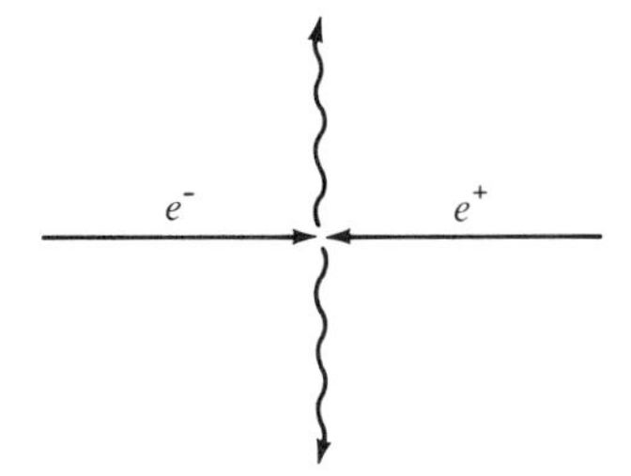

Figure 3-14

physics named *annihilation* which seems to contradict this law. See Figure 3-14.

If the two elementary particles electron (e^-) and positron (e^+), each having a mass m_e, meet, we observe that they disappear. Their masses apparently vanish with them, but at the same instant annihilation radiation (symbolized by two wavy lines in the sketch) occurs. This apparent violation of mass conservation can be explained if we realize that the mass of both elementary particles is converted into another form, into radiant energy. This indicates that the law of mass conservation is part of the more encompassing law of energy conservation. The complementary effect is also observed in nuclear physics: Conversion of radiation energy into mass, called *pair production*. See Figure 3-15. Radiation (high-energy x rays) disappears, and two elementary particles, for example an electron and a positron, are created.

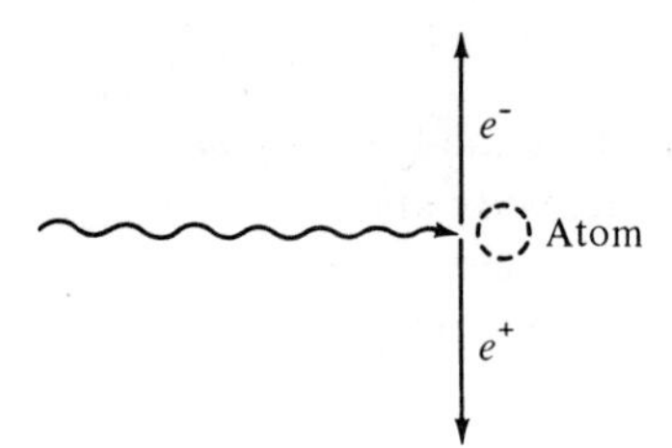

Figure 3-15

The governing equation for the conversion of mass into radiation energy, and vice versa, is

$$E = mc^2$$

where

E: energy,
m: mass,
c: speed of light.

A later chapter (Radiation) explains this relation in more detail.
Another property of mass is that it depends on its speed! If the mass observed at rest (the rest mass) is denoted by m_0, then the mass m while moving with the speed v is given by:

$$m = \frac{m_0}{\sqrt{1 - v^2/c^2}}$$

where c = speed of light = $3 \times 10^8\ \text{m} \cdot \text{s}^{-1}$.
The speeds involved in everyday life are so small compared with the speed of light that m is for all practical purposes indistinguishable from m_0. At the atomic and nuclear level, this is no longer the case. The electrons leaving an accelerator of moderate energy have an actual mass hundreds of times larger than their rest mass.

2.3 Force

Using Equation (3-11) as the definition of the force **F** we measure its magnitude in $\text{kg} \cdot \text{m} \cdot \text{s}^{-2}$. There is a derived unit for force, the newton, symbolized N:

$$1 \text{ newton} \equiv 1\ \text{kg} \cdot \text{m} \cdot \text{s}^{-2}$$

By convention, attractive forces carry a negative sign. Conversion:

$$1\ \text{N} = 10^5 \text{ dyne (dyn)} = 10^5\ \frac{\text{g} \cdot \text{cm}}{\text{s}^2}$$

Force is a vector quantity, and the addition and subtraction of forces follow the rules laid down in Chapter 1.

EXAMPLE:

Adding the forces of the posterior (p.) and anterior (a.) deltoid muscle to produce a perpendicular arm elevation. See Figure 3-16.

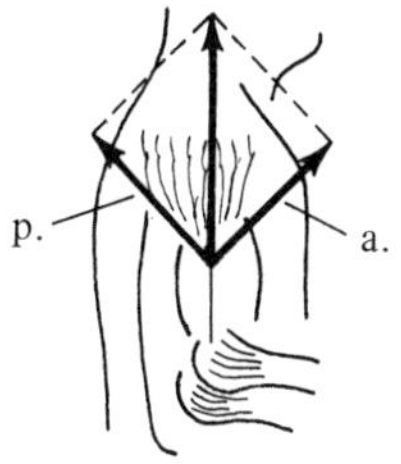

Figure 3-16

To measure a force, we take advantage of its vector properties. If we cancel the unknown force with a calibrated force of equal magnitude but opposite direction, we can measure the force. Instruments which measure force are called *dynamometers.*

There are different types of forces: gravitational, electric, magnetic, frictional, nuclear, and others. All forces observed in the life sciences can be reduced to pure physical forces. For example, muscular forces are electric forces between macromolecules.

2.4 Torque

In order to analyze rotational motion, it is convenient to introduce a new vector quantity: *Torque*, symbolized **T**, is defined as:

$$\mathbf{T} \equiv \mathbf{r} \times \mathbf{F} \tag{3.12}$$

where (see Figure 3-17)

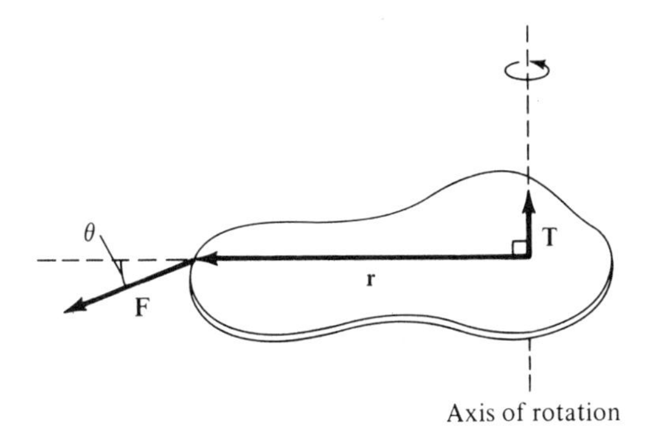

Figure 3-17

F: external force applied to body,
r: position vector, leading from the axis of rotation to the point where the external force **F** is applied.

Torque is a vector because Equation (3.12) is a vector product. **T** is perpendicular to both **r** and **F**, and its direction is given by the right-hand rule. The magnitude T of **T** is (according to the rules of vector multiplication):

$$T = rF \sin \theta$$

where

F: magnitude of external force $\mathbf{F}$,
r: distance from axis of rotation to the point where the force is applied,
θ: angle between $\mathbf{r}$ and $\mathbf{F}$.

T is measured in meter·newton.

EXAMPLE:

The performance of a car depends on the torque $\mathbf{T}$ supplied by the engine via the drive shaft. See Figure 3-18. $\mathbf{T}$ determines how much force $\mathbf{F}$ is available at the outer surface of the tire to move the vehicle. Figure 3-19 shows T for a car engine.

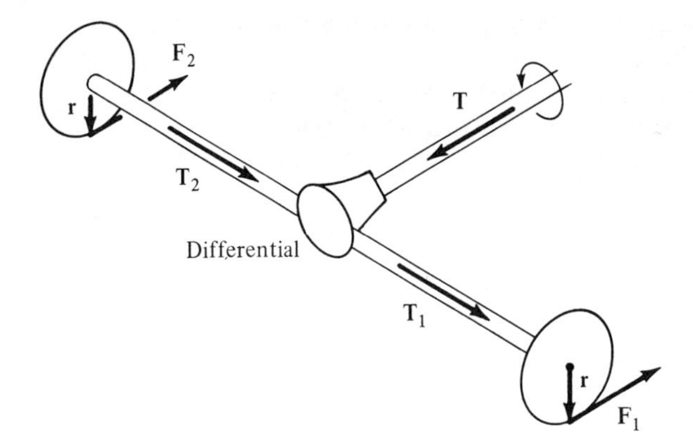

Figure 3-18

2.4(a). TORQUE AND THE DESIGN OF THE SKELETON. The locomotion of practically every large animal can be traced to a rotational motion of limbs. This means that torque plays an important role in nature's design of skeletons. The aim is to maximize torque with a minimum effort without jeopardizing other functions of the organism. Let us look at Equation (3.12) from this point of view. Flexion of the elbow serves as an example. Figure 3-20 demonstrates the torque exerted at various stages.

The angle θ changes during the partial rotation; for $\theta = 90°$ the available torque is a maximum. Since $\mathbf{T} \sim \sin\theta$, you can observe in nature that the angle of flexion, θ, varies between 30° and 150° because the torque becomes very small outside these limiting values.

$\mathbf{F}$ has an upper limit determined by the muscular cross section. $\mathbf{r}$ is fixed by the anatomical structure and the desired motion.

The optimal method for nature to increase the torque for a given force is to increase $\mathbf{r}$. The proximal end of the human femur is a fine example for this design, see Figure 3-21. $\mathbf{F}_M$ is the force exerted by the hip abductor muscles, $\mathbf{T}_1$ and $\mathbf{T}_2$ are the torques produced, and $\mathbf{T}_1 > \mathbf{T}_2$ because $\mathbf{r}_1 > \mathbf{r}_2$.

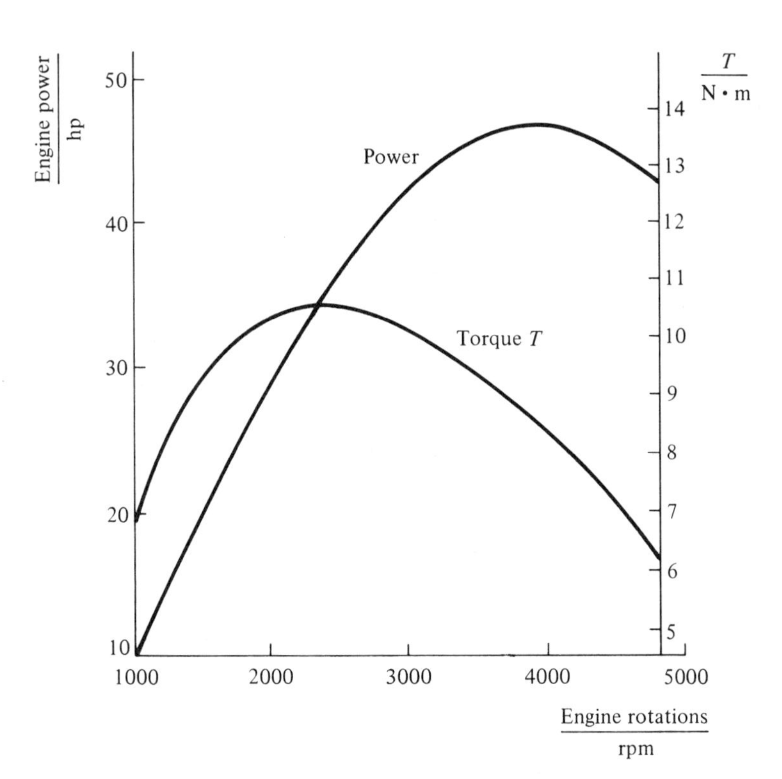

Figure 3-19 Magnitude of torque T as a function of engine rpm (VW 1600 Beetle). For comparison the power output is also shown. The car's performance depends mainly on the torque.

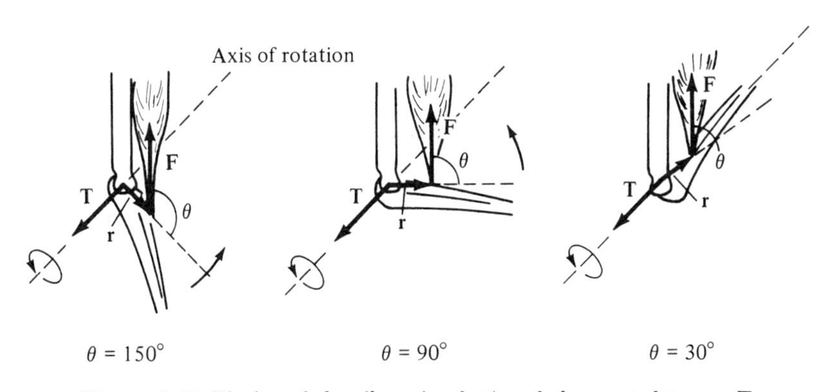

Figure 3-20 Flexion of the elbow (angle θ) and the exerted torque **T**.

2.5 Center of mass

To calculate the translatory motion of an extended body, the formalism of a point body is applicable if a new quantity is employed, the center of mass, abbreviated CM. The CM concept is a convenience for solving problems, nothing more.

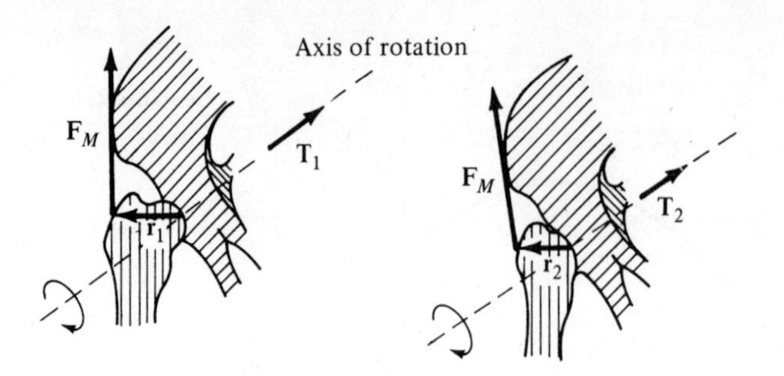

Figure 3-21 Torque **T** at a normal (left) and an abnormal (right) hip joint. Observe that the exerted force $\mathbf{F}_M$ is the same for both cases.

DEFINITION: *The center of mass of a body is a point at which forces acting on the body produce no torque in it.*

The CM depends on the physical properties of the body and is localized inside or outside. Figure 3-22 shows the CM (represented by

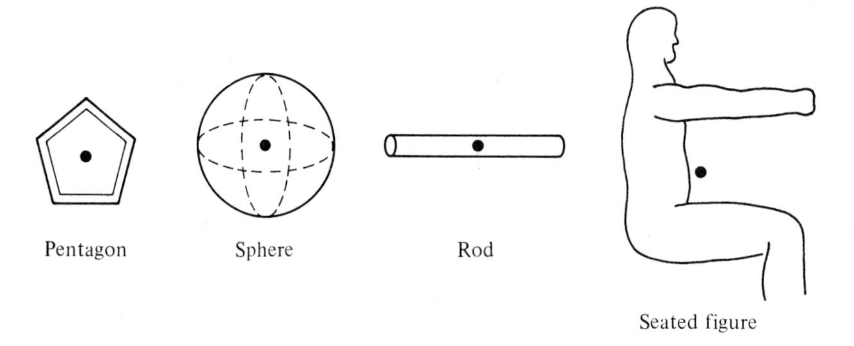

Figure 3-22

a dot) for some simple mass configurations. For homogenous geometrical configurations, formulas exist for finding the CM; however for irregular (and nonhomogenous) objects, experimental methods are employed.

2.6 Mechanical equilibrium

A body is in mechanical equilibrium if the vector sums of all forces and torques acting on it are zero:

$$\sum_i \mathbf{F}_i = 0 \tag{3.13}$$

and

$$\sum_i \mathbf{T}_i = 0 \tag{3.14}$$

$\sum_i \mathbf{F}_i$ is a short notation for $\mathbf{F}_1 + \mathbf{F}_2 + \mathbf{F}_3 + \ldots + \mathbf{F}_i$, and accordingly,

$$\sum_i \mathbf{T}_i = \mathbf{T}_1 + \mathbf{T}_2 + \mathbf{T}_3 + \ldots + \mathbf{T}_i$$

EXAMPLE: BALANCE

The instrument is said to be balanced if it is in mechanical equilibrium. See Figure 3-23.

Due to the construction of the balance, translational motions are impossible. Equation (3.13) is always fulfilled. Equation (3.14) is satisfied if

$$\mathbf{r}_1\mathbf{F}_1 \sin \theta_1 = \mathbf{r}_2\mathbf{F}_2 \sin \theta_2$$

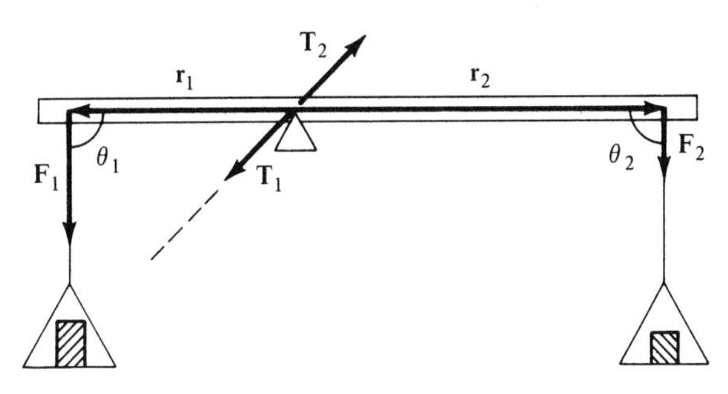

Figure 3-23

An equilibrium is called *stable equilibrium* if it possesses or develops an intrinsic resistance to change, an ability to restore the equilibrium if disturbed.

EXAMPLE: A SEAWORTHY YACHT

See Figures 3-24 and 3-25. The gravitational force acts on the CM of the vessel downward, while a buoyancy force $\mathbf{F}_b$ acts on the CM of the displaced water CB. Normally CB is above CM. Both forces are equal in magnitude but of opposite sign, and the yacht floats upright (Figure 3-24). If the yacht heels (Figure 3-25) a torque develops, forcing the boat back into the upright position. This torque restores the original balance.

If the position of CB is below CM, the torque developed during heeling will increase the angle of heel until the ship is capsized. Then a stable equilibrium is again achieved because now CM is below CB.

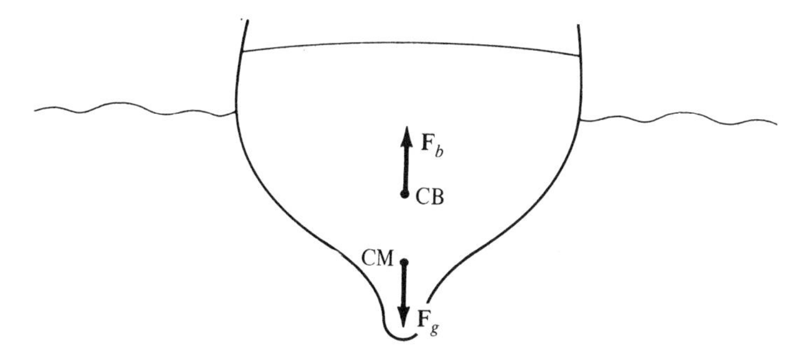

Figure 3-24

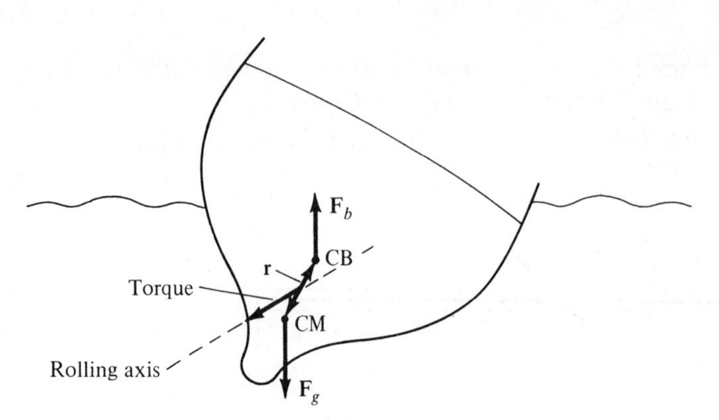

Figure 3-25

3. VARIOUS FORCES

3.1 Gravitational forces

The gravitational force $\mathbf{F}_g$ is an attractive force intrinsic to mass itself. If two masses m_1 and m_2 are separated by a distance r then the magnitude F_g of $\mathbf{F}_g$ is,

$$F_g = G\frac{m_1 m_2}{r^2} \tag{3.15}$$

where

G: universal gravitational constant, having a numerical value of $6.67 \times 10^{-11}\ \mathrm{N \cdot m^2 \cdot kg^{-2}}$.

The distance r refers to the separation of the respective centers of mass of both bodies.

Actually the gravitational force is very weak compared to forces such as electric or nuclear forces. Nevertheless, although the gravitational force is weak, it plays a dominant role in our lives because it originates in a very massive body, our earth. In addition, we are always under its influence.

3.1(a). WEIGHT. Forces are defined by Newton's second law; thus the gravitational force $\mathbf{F}_g$ acting on a body of mass m is

$$\mathbf{F}_g = m\mathbf{a}_g \tag{3.16}$$

where

$\mathbf{a}_g$: acceleration of m due to the gravitational force $\mathbf{F}_g$.

The magnitude of $\mathbf{a}_g$ for the earth's gravitational force is abbreviated g. Substituting in Equation (3.16) and using Equation (3.15) with $m = m_1$,

$$mg = G\frac{mm_2}{r^2} \tag{3.17}$$

it is easy to compute the numerical value of g at the surface of the earth (r = earth's radius, m_2 = earth's mass):

$$g = 9.81\ \text{m}\cdot\text{s}^{-2} \text{ for } 45° \text{ latitude}$$

The magnitude of $\mathbf{F}_g$ is called *weight*, symbolized W:

$$|\mathbf{F}_g| = mg = W = \text{weight of body on earth} \quad (3.18)$$

Remember: The weight of a body is a scalar quantity. It depends on the gravitational forces present. If the body changes its locality, its weight may change but its mass will not.

On the surface of the moon, the acceleration is determined by the gravitational force of the moon,

$$g_{\text{moon}} = 1.62\ \text{m/s}^2$$

Consequently a body on the moon weights only one-sixth of what it would weigh on the earth. Of course, the mass of that body is the same on the moon as on the earth!

Since all material bodies have a mass, the gravitational force is truly universal and governs the motion of all celestial bodies from satellites to planets to galaxies. On the surface of the earth, the gravitational force toward the center of the earth is always present. But can be compensated for and hence masked by other forces, for instance electric or magnetic forces or the lift produced by the profile of the wing of a bird.

3.1(b). Gravity detectors in organisms. Since the gravitational force of the earth is perpendicular to the earth's surface, it serves as a means of orientation for animals. This is especially true for those such as fish and birds which can easily move in three dimensions. Gravity detectors in animals help to establish the position of the animal with respect to the vertical. In its primitive form, a gravity detector is a rounded object called *statocyst* enclosed inside an organ with sense receptor hairs attached individually to separate nerve fibers. Figure 3-26 shows the gravity receptor of a free-swimming mollusk.

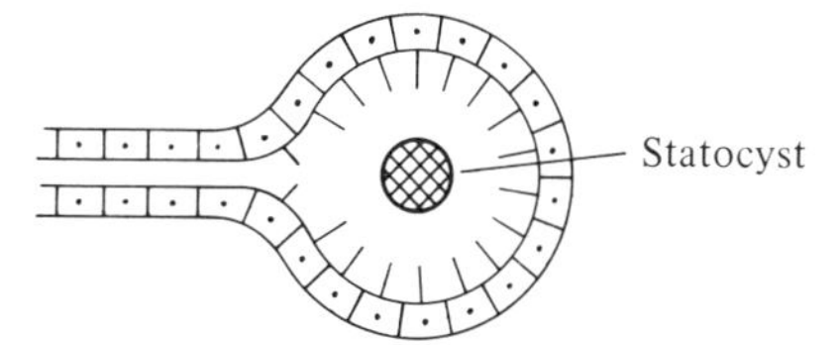

Figure 3-26

The statocyst moves freely inside, although its motion is damped because it is suspended in a fluid. It rests on the lower hairs while the mollusk is in horizontal position. Any movement out of this normal position causes the statocyst to touch other sensitive hairs and thus signals the new position.

A similar organ exists in man, albeit much more sophisticated and miniaturized. Its size is only a fraction of a millimeter, and it is part of the otic labyrinth and situated inside the utriculus. Sensitive hairs are connected to a heavy, elastically suspended membrane. The membrane changes position slightly if the head is tilted. The hairs then activate the support cells and thus the position of the head with respect to the direction of gravitation is indicated to the coordination center in the brain.

The same organ determines linear accelerations with a sensitivity of 0.12 m/s^2. That means a linear acceleration as low as 1% of g can be discerned.

Gravity Perception in Plants. The direction of growth in plants is determined mostly by gravity. Consequently there must be areas sensitive to gravity. Although the mechanism is still not really understood, it may be connected with the free-moving starch grains in the cells of extreme apical regions in roots, shoots, and leaves.

3.2 Buoyancy forces

If a body is submerged in a fluid (either a liquid or a gas), then a buoyant force acts on it. The direction of this force $\mathbf{F}_b$ is opposite to the gravitational force $\mathbf{F}_g$ acting on the submerged body. The buoyancy force acts at the center of mass of the displaced fluid (CB). See Figure 3-27.

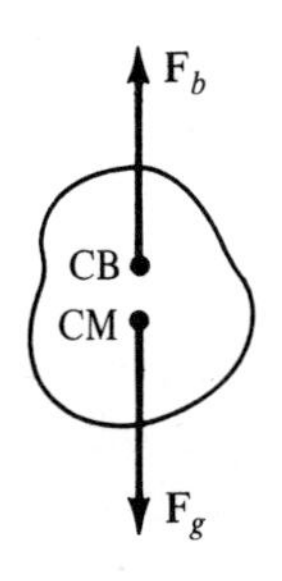

Figure 3-27

The magnitude F_b of $\mathbf{F}_b$ is

$$F_b = W_{\text{fl}} \qquad \text{Archimedes principle*} \qquad (3.19)$$

where W_{fl} is the weight of the displaced fluid.

3.2(a). Density. Now we introduce a new physical quantity, the density of a body, symbolized ρ:

$$\rho = \frac{m}{V} \qquad (3.20)$$

where

m: mass of body,
V: volume of body.

The density is measured in g/cm^3. Table 2 lists densities of various materials.

3.2(b). Density of water. The density of matter depends on its temperature and other factors. Usually the density decreases with increasing temperature. The density-temperature diagram of water, see Figure 3-28, is unique and of utmost importance to marine life. Figure 3-28 explains why open water freezes at the top first and at the bottom last.

If the water temperature decreases at the surface, the cooler top layer has a higher density than the rest and it sinks down. This downward

* Archimedes (285–212 B.C.), mathematician and physicist. Today Archimedes is known mainly for his contributions to mathematics. During his lifetime he was one of the most famous engineers. More about him in E. J. Dijksterhuis, *Archimedes*, Kopenhagen, 1956.

Table 2 DENSITIES OF MATTER

Material	ρ in g/cm^3
nucleon	10^{14}
pulsar	$10^{14} - 10^{11}$
tungsten	19.1
mercury	13.6
lead	11.3
cast iron	7.25
planet earth	5.52
moon	3.34
otokonia	2.95
nucleic acid	2.2
human bone (adult)	1.85
sugar	1.61
cotton	1.50
sun	1.41
glycerine	1.26
plexiglas	1.20
seawater	1.03
pure water	1.00
ice	0.88 – 0.92
oak	0.69
liquid hydrogen	7×10^{-2}
air (sea level)	1.3×10^{-3}
water vapor	7.7×10^{-4}
average universe	10^{-30}

water current continues until the bottom layer has a temperature of approximately 4°C. If the surface water continues to drop in temperature, it stops sinking because its density is now lower than that of the bottom water. Consequently, the water freezes at the top first and, since ice is a better insulator than water, it slows down the cooling rate of the

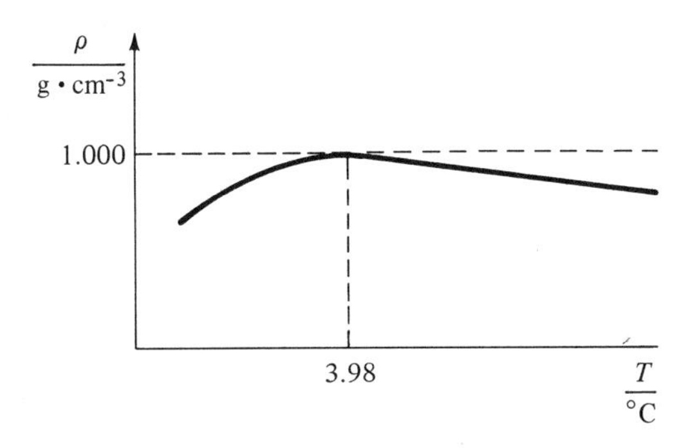

Figure 3-28 Density of water.

water body. Figure 3-29 shows the actual winter temperature distribution of a deep lake. The wind is responsible for the two distinct temperature layers.

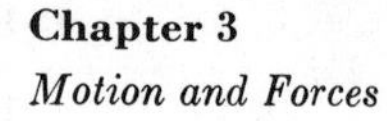

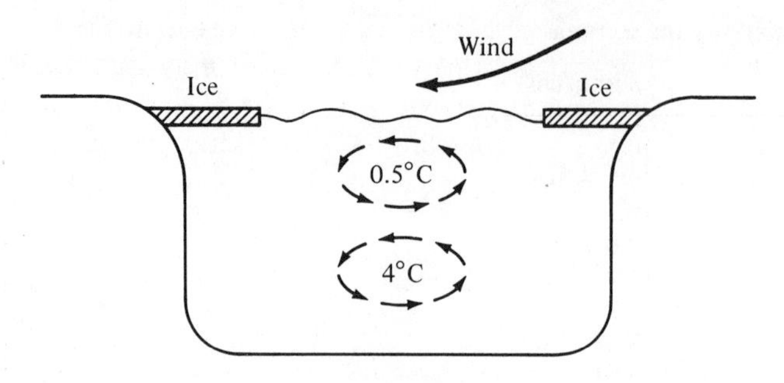

Figure 3-29

3.2(c). FLOATING. If a body is submerged in a fluid, two forces are acting on it, the gravitational force $\mathbf{F}_g$ downward and the buoyancy force $\mathbf{F}_b$ upward. The resultant force $\mathbf{F}_r$ is the vector sum of both. Since $\mathbf{F}_g$ and $\mathbf{F}_b$ act in opposite directions and by convention $\mathbf{F}_g$ carries a negative sign,

$$\mathbf{F}_r = \mathbf{F}_b - \mathbf{F}_g$$

The magnitude F_r of $\mathbf{F}_r$ is,

$$F_r = |\mathbf{F}_b - \mathbf{F}_g| = m_{fl}g - mg \tag{3.21}$$

or

$$F_r = |Vg(\rho_{fl} - \rho)| \tag{3.22}$$

where

$V = m/\rho$: volume of submerged body,
m: mass of body,
g: acceleration due to gravity,
ρ: density of body,
ρ_{fl}: density of fluid.

The direction of the resulting force $\mathbf{F}_r$:

For $\mathbf{F}_b$ larger than $\mathbf{F}_g$, then $\mathbf{F}_r$ is positive and upward, and the body swims.

For $\mathbf{F}_b$ smaller than $\mathbf{F}_g$, then $\mathbf{F}_r$ is negative and downward, and the body sinks.

For $\mathbf{F}_b$ equal to $\mathbf{F}_g$, then $\mathbf{F}_r$ is zero, and the body floats.

EXAMPLE: BUOYANCY ACTING ON MAN

Man (volume V_m) is always submerged in a fluid (density ρ_{fl})—in air. Therefore he is always under the influence of at least two forces, the gravitational force $\mathbf{F}_g$ and the buoyancy force $\mathbf{F}_b$. The ratio of these magnitudes is:

$$\frac{F_b}{F_g} = \frac{V_m \rho_{fl} g}{V_m \rho g} = \frac{\rho_{fl}}{\rho}$$

If man is submerged in air

$$\left.\frac{F_b}{F_g}\right|_{\text{air}} = \frac{1.3 \times 10^{-3}}{1.10} = 1.18 \times 10^{-3}$$

where the density of man is taken to be $\rho = 1.1 \text{ g}\cdot\text{cm}^{-3}$. In this case the buoyancy force is only 0.1% of the gravitational force.

The ratio F_b/F_g becomes very different if

man is submerged in seawater; then

$$\left.\frac{F_b}{F_g}\right|_{\text{water}} = \frac{1.03}{1.10} = 0.935$$

Now the buoyancy force is 93.5% of the gravitational force. In order to float in water, man needs an additional upward force equivalent to 6.5% of the $\mathbf{F}_g$ acting on him. He can obtain this by filling his lungs with air, using a life belt, or by swimming movements.

Due to the low density of air, the buoyancy forces do not play a significant role in the flight of birds. For fish, $\mathbf{F}_b$ is all important. But even there, in general, the buoyancy forces do not cancel the gravitational forces acting on the submerged fish. Most live fish will slowly sink to the bottom if they do not use their propulsion systems. However, with the swim bladder, they possess an organ that allows them to change their average density. This is a useful device if we recall that the density of water changes with temperature and salinity.

3.3(a). Hooke's law. A body is *quasirigid* if it changes its extensions to a small extent under the influence of forces. It assumes its original shape as soon as those forces disappear. The obvious example of a quasirigid body is the spring.

If an external force $\mathbf{F}$ acts on a spring of length l as indicated in Figure 3-30, then the spring will change its length by Δl. $\mathbf{F}$ is counteracted by

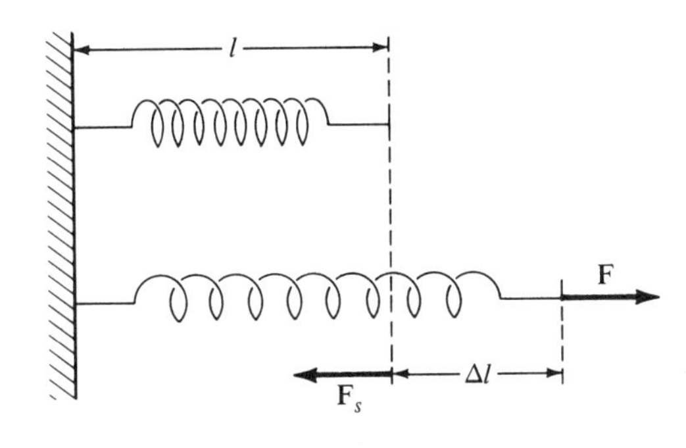

Figure 3-30

an internal force, the spring force $\mathbf{F}_s$. Experiments demonstrate that the magnitude F_s of $\mathbf{F}_s$ is

$$F_s = k\,\Delta l \qquad \text{Hooke's law*} \tag{3.23}$$

as long as $\Delta l/l$, a quantity called *linear strain* ϵ, remains small compared with 1. The proportionality constant k (the spring constant) depends on the form and material of the spring. The direction of the spring force $\mathbf{F}_s$ is opposite to the external force $\mathbf{F}$.

* Robert Hooke (1635–1703), physicist and engineer. More information in R. T. Gunther, *The Life Work of Robert Hooke*, Oxford, 1930.

Hooke's law offers a convenient way to measure forces. We let the unknown force act on a spring. It will extend the spring by Δl until it is cancelled by the opposing spring force. Equation (3.23) determines the magnitude of the unknown force.

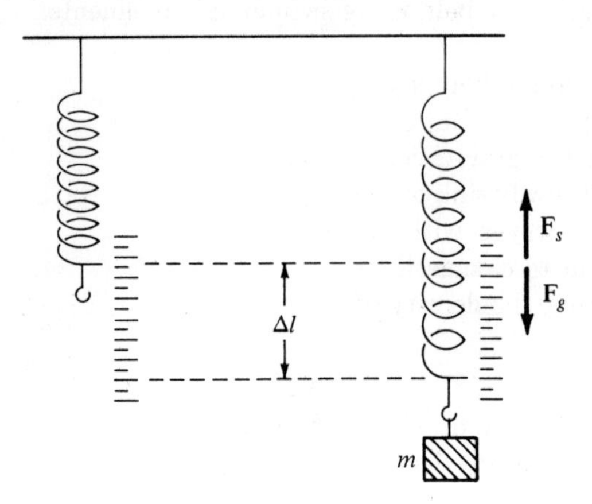

Figure 3-31

An application of Hooke's law is the spring balance. (See Figure 3-31.) The gravitational force $\mathbf{F}_g$ acting on the body stretches the spring until mechanical equilibrium is reached:

$$\mathbf{F}_g = \mathbf{F}_s$$

and

$$mg = k\,\Delta l$$

Solving for m, we obtain:

$$m = \frac{k}{g}\,\Delta l$$

The calibration constant k/g is determined by the manufacturer of the spring balance.

3.3(b). Stress, strain, and elasticity. Hooke's law can readily be extended to quasirigid bodies in general, since the internal molecular forces holding the body together act in a fashion similar to a spring force.

If two forces equal in magnitude but opposite in direction act on a body, as shown in Figure 3-32, no motion results. The body will be

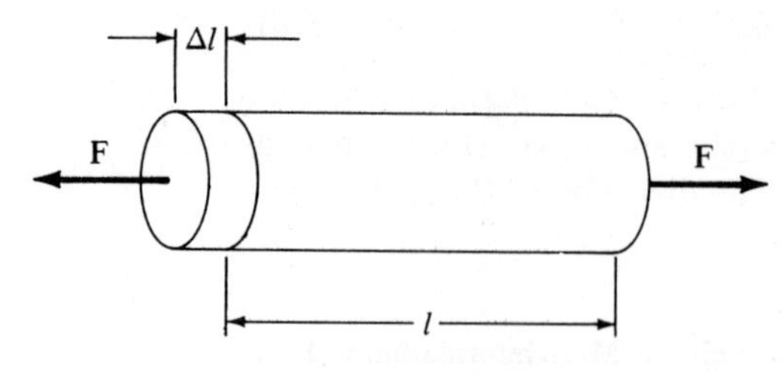

Figure 3-32

stretched by Δl. It is customary to introduce in this case the *normal tensile stress*, symbolized ρ:

$$\rho \equiv \frac{F}{A} \tag{3.24}$$

where

F: magnitude of force acting on the body,
A: cross-sectional area of that body.

Realize that ρ is a scalar quantity.

If the forces acting on the body point toward the body, then the stress developed inside is called *normal compression stress.* It is measured in newtons/meter^2. The compression stress is called *normal* because it acts perpendicular to the cross section A.

If the two forces do not act along the same line, they will exert a torque, and the developed stress is called *shear stress. Bulk stress* occurs if forces act uniformly on the body from all sides. The formal treatments of all three types of stresses are similar; therefore we shall consider only the normal (or linear) stress.

Hooke's law in its general formulation is:

$$\rho = E\epsilon \tag{3.25}$$

where

ρ: normal stress,
ϵ: linear strain,
E: modulus of elasticity (or Young's modulus).

Since the strain is dimensionless, the modulus of elasticity is measured in the same units as stress, that is, in newtons/meter^2. As long as a material obeys Equation (3.25), it is termed *elastic.*

Table 3 lists some experimentally determined values for Young's modulus.

Table 3 MODULUS OF ELASTICITY FOR SOME MATERIALS

Material	E in $\text{N} \cdot \text{m}^{-2} \times 10^{10}$
stimulated skeleton muscle	$\approx 1 \times 10^{-5}$
rubber	$\approx 1 \times 10^{-4}$
fir (parallel to fibers)	1.1
lead	1.8
bone	≈ 2
aluminum	7.1
silver	7.5
glass	≈ 9
copper	12
oak (parallel to fibers)	13
steel	20

The linear relationship between stress and strain usually holds only for small strains. If the stress exceeds a certain value (the elastic limit), then Equation (3.25) becomes invalid. After relieving the stress, the original length of the body will not be fully restored—it is irreversibly stretched or compressed.

If the exerted stress exceeds a limiting value (the breaking stress) the object will break. This behavior of an object under stress is shown in the stress-strain diagrams, Figures 3-33, 3-34, and 3-35. Observe the different scales for the strain.

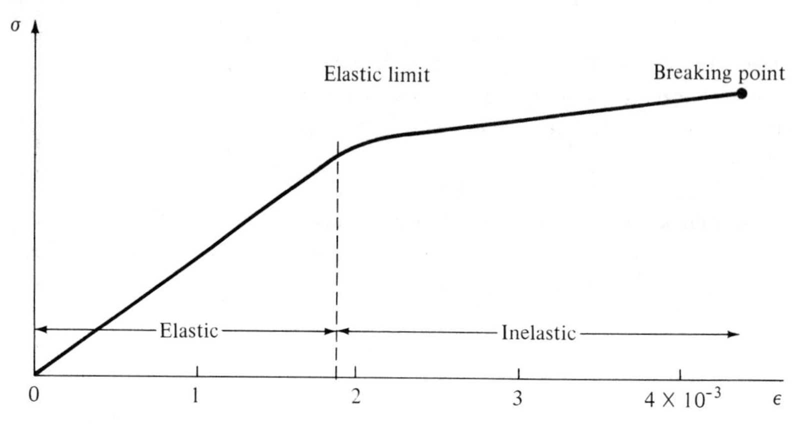

Figure 3-33 Stress-strain diagram of a metallic wire.

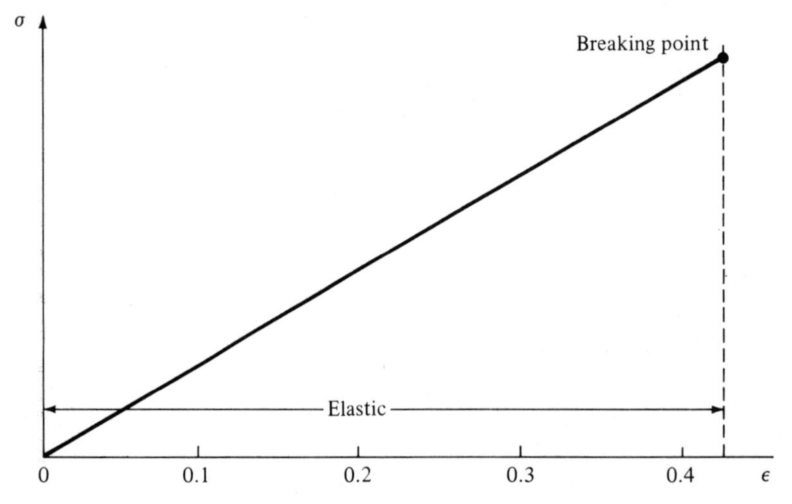

Figure 3-34 Stress-strain diagram of a bone.

The shape of the stress-strain diagram of a muscle is very different from the diagrams for wire and bone, indicating that more than one material and process are involved.

Stress detectors in man: Stress detectors inside the tendons (Golgi's tendon organ) and parallel to the muscle fibers (muscle spindle organ)

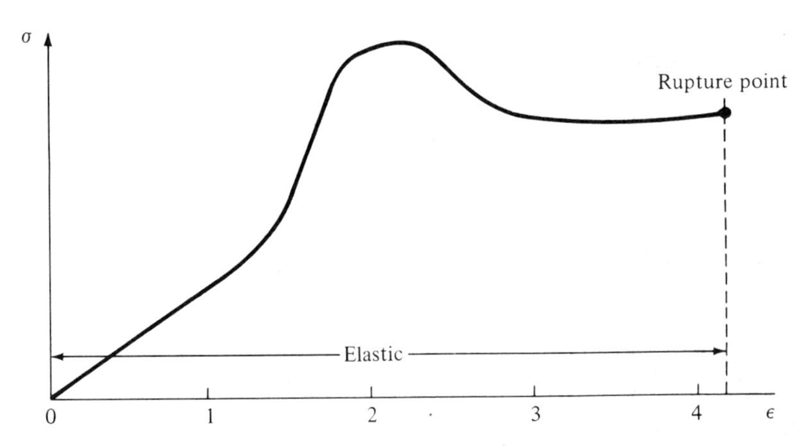

Figure 3-35 Stress-strain diagram of an activated triceps muscle.

monitor the stress developed and trigger powerful relief reflexes if the stress approaches the rupture point.

3.3(c). FATIGUE. Investigating the stress-strain properties of materials, we have not encountered any time factors up to now. It is implied that, if the stress does not exceed the elastic limit, the original length of the body will be restored when the stress vanishes. This is the definition of elastic behavior. However, if a material undergoes a change of stress repeatedly, it will develop *fatigue*. In metals this becomes evident if the material breaks at a stress far below the initial breaking point.

Biological materials also show fatigue. For example, after a frog's leg muscle contracts due to artificial stimulation about 10 000 times within three or four hours, it will fail to contract anymore. This fatigue is not caused by a lack of nutrition. The cause for muscular fatigue is not fully understood at present.

3.3(d). FINE STRUCTURE OF BONES. If a body is subject to stress, experience shows that the stress is not uniformly distributed over a given cross section inside that body. Stress occurs along certain lines, whose course depends on shape and internal structure. Engineers take advantage of that by providing only supporting material along the lines of stress and hence saving weight and cost. Nature operates in the same way.

The bones support the weight of the body; they are under compression stress. Wherever the force of a muscle acts, a tensile stress develops. The sponge-like internal structure of the hollow bones is aligned along the lines of tensile and compression stress. Figure 3-36 shows a fine example of this. The structure saves not only weight but also energy since the body needs to maintain a smaller number of living cells. (The constituents of a bone are by no means dead material.)

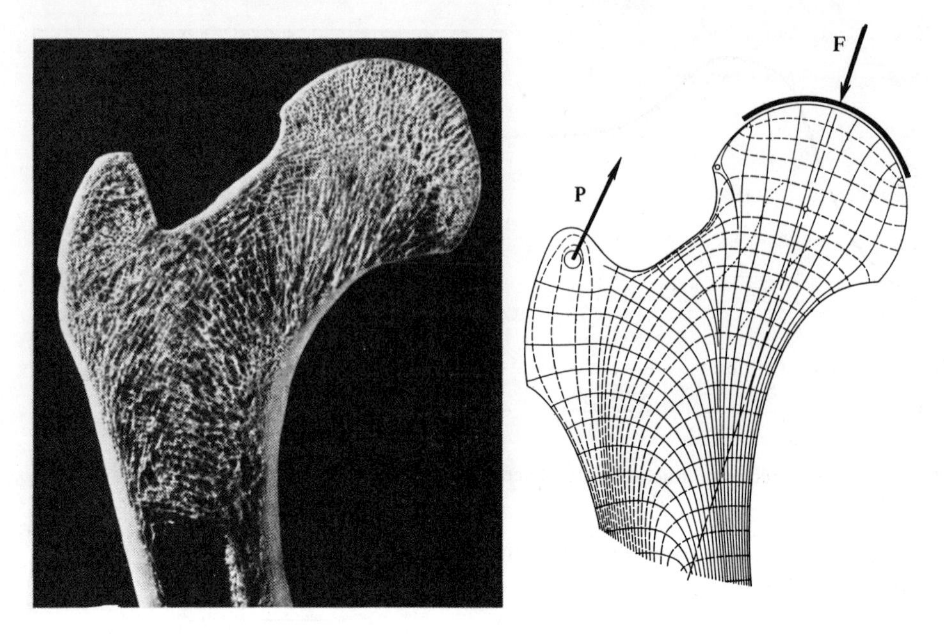

Figure 3-36 Proximal end of a human femur. The photo on the left shows a lengthwise cross section; the diagram shows the lines of tensile stress (hatched) originating from the pull **P** of the hip muscles and the lines of compression stress (solid) as produced by the combined forces **F** due to weight and muscular action.

3.4 Frictional forces

3.4(a). External friction. If two surfaces are in contact, their molecules will exert a force $\mathbf{F}_f$ which resists a relative motion between both surfaces; see Figure 3-37. A detailed description of friction is complex; however, a simplified treatment yields some insight.

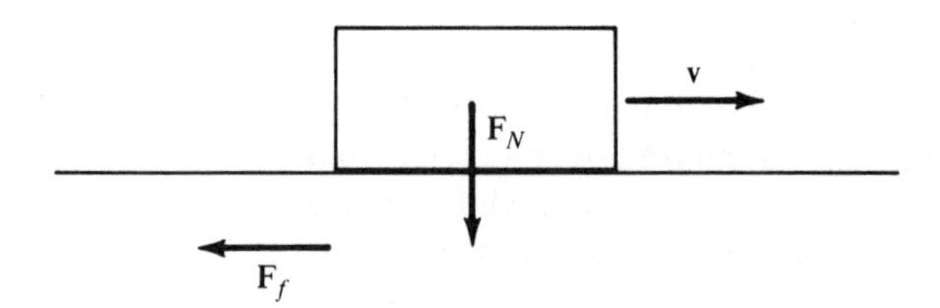

Figure 3-37

The magnitude F_f of the frictional force $\mathbf{F}_f$ is in first approximation

$$F_f = \mu F_N \tag{3.26}$$

where

μ: coefficient of friction for the surfaces in contact,
F_N: magnitude of force acting perpendicular to the surfaces in contact.

The coefficient of friction depends not only on the materials in contact, but also on whether both surfaces are at rest with respect to each other (static coefficient of friction μ_s) or in relative motion (kinetic coefficient of friction μ_k). Table 4 lists some experimentally determined coefficients of friction.

Table 4 COEFFICIENTS OF FRICTION

(μ_s: static coefficient; μ_k: kinetic coefficient)

Surface 1	Surface 2	μ_s	μ_k
wood	wood	0.50	0.34
teflon	steel	0.04	0.04
metal	oil	0.12	0.07
leather	wood	0.54	0.40
rubber	asphalt	0.60	0.40
steel	steel	0.75	0.58
glass	glass	0.94	0.40

Realize that, once a body is set in motion, the frictional forces opposing this motion will be reduced since

$$\mu_k < \mu_s$$

There is a third coefficient of friction, *rolling friction*. Its meaning is obvious and its numerical values are generally only one-tenth the kinetic friction. Ball bearings are an application of rolling friction.

Friction is not always undesirable, as illustrated in the following example:

EXAMPLE: FRICTION IN GAIT

During walking, the forward-swinging foot strikes the ground with a force **F** at an angle φ. See Figure 3-38. We can resolve **F** into a force parallel to the ground, $\mathbf{F}_p$, and a force perpendicular to the ground, $\mathbf{F}_N$. If we are not to slip, $\mathbf{F}_p$ must be cancelled by an opposing force, in this case, the frictional force $\mathbf{F}_f$ between the surface of the heel and the ground. It is

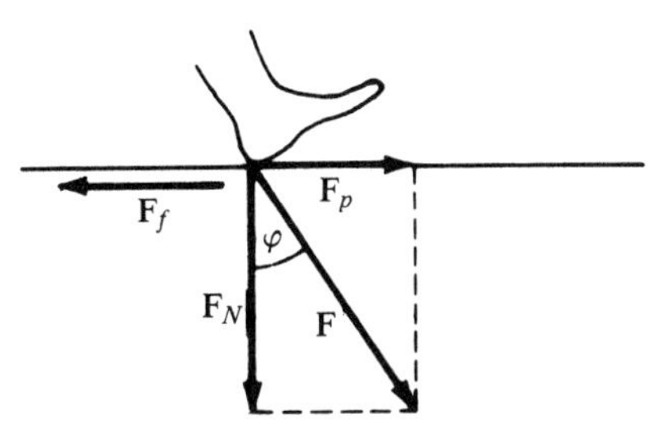

Figure 3-38

$$\mathbf{F}_p = \mathbf{F}_N \tan \varphi$$

The condition for no slip is:

$$\mathbf{F}_f = \mathbf{F}_p$$

hence

$$\mu_s \mathbf{F}_N = \mathbf{F}_N \tan \varphi$$

$$\mu_s = \tan \varphi$$

That means the no-slip condition is independent of $\mathbf{F}_N$.

For a leather heel striking a wooden floor ($\mu_s = 0.54$), we get

$$\varphi = 28°$$

The foot will not slip if the striking angle is equal to or smaller than 28°.

If the ground is covered with ice, the

coefficient of friction decreases substantially; consequently the striking angle must be reduced to prevent slip. This is automatically achieved by shortening the step length.

It is interesting to note that, if φ exceeds the limit set by the coefficient of friction ever so slightly, the heel will slip suddenly. The reason is that, as soon as the surfaces start to move with respect to each other, the coefficient for kinetic friction μ_k applies. Since $\mu_k < \mu_s$, the frictional force is suddenly reduced.

Realize that the foregoing treatment, based on Equation (3.26), is only an approximation of an intrinsically complex process.

3.4(b). INTERNAL FRICTION. If a body moves through a fluid (either a liquid or a gas), internal frictional forces $\mathbf{F}_f$ oppose this motion. The magnitude F_f of the internal frictional force $\mathbf{F}_f$ is in first approximation

$$F_f = K\eta v \tag{3.27}$$

where

v: speed of the body relative to the fluid,
K: a constant depending on the shape of the body in motion,
η: coefficient of viscosity characterizing the fluid.

As long as η is independent of the speed, the fluid is called a *Newtonian fluid.*

The coefficient of viscosity is measured in $\text{N}\cdot\text{s}\cdot\text{m}^{-2}$. Often the coefficient of viscosity is expressed in the unit poise, symbolized P.*

Conversion:

$$1\ \text{N}\cdot\text{s}\cdot\text{m}^{-2} = 10\ \text{P} = 100\ \text{cP}$$

The internal friction is due mainly to molecular interactions within the fluid. The material of the moving object is unimportant because a thin layer of the fluid coats its surface. Therefore there is no motion between the surface of the body and the immediately surrounding layer of the fluid.

Theodore von Karman, one of the fathers of hydrodynamics, illustrates internal friction with a well-chosen example: Suppose that a book containing many pages is placed on a desk and the upper cover is slowly pushed parallel to the surface of the desk. The pages slide over each other, but the lower cover sticks to the desk. Similarly, fluid particles stick to the surface of a body, so that there is no slip between fluid and solid surface. (T. von Karman, *The Wind and Beyond,* page 74.)

Equation (3.27) is valid only for low speeds; otherwise it is replaced by

$$F_f = K\eta v^n \tag{3.28}$$

where $1 \leq n \leq 2$, depending on the speed and type of fluid.

In general, the constant K of Equation (3.27) is determined experi-

* Named in honor of Jean Louis Poiseuille (1799–1869), a medical doctor who contributed to the theory of blood circulation.

mentally. For very simple geometrical configurations, formulas for K exist. For a sphere of radius R,

$$K = 6\pi R \tag{3.29}$$

introduced into Equation (3.27)

$$F_f = 6\pi R\eta v \qquad \text{Stoke's law} \tag{3.30}$$

Viscosity of Blood. The viscosity of blood is a function of the blood's temperature, and it also depends on its protein content and on the suspended particles. The viscosity of blood is substantially reduced if the red blood cells align with each other while passing through a capillary.

3.4(c). TERMINAL (OR SEDIMENTATION) SPEED. A body moves through a fluid under the influence of gravitational force $\mathbf{F}_g$, buoyancy force $\mathbf{F}_b$, and frictional force $\mathbf{F}_f$. The resultant force $\mathbf{F}_r$ is the vector sum of all three,

$$\mathbf{F}_r = \mathbf{F}_g + \mathbf{F}_b + \mathbf{F}_f$$

Since the frictional force is in the opposite direction to the velocity of the moving body, two cases are distinguished:

1. $\mathbf{F}_g > \mathbf{F}_b$. Then $\mathbf{v}$ and $\mathbf{F}_r$ will point in the direction of $\mathbf{F}_g$, that is, downward (Figure 3-39).
2. $\mathbf{F}_g < \mathbf{F}_b$. Now $\mathbf{v}$ and $\mathbf{F}_r$ point upward in the same direction as $\mathbf{F}_b$ (Figure 3-40).

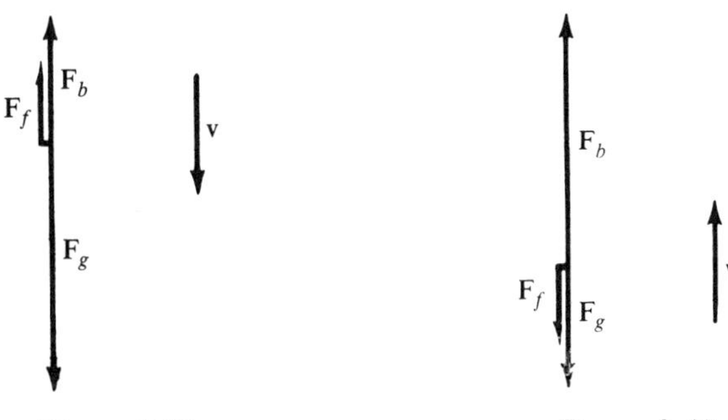

Figure 3-39 **Figure 3-40**

Only situation 1 is of interest here. The resultant force acting on the falling body is

$$\mathbf{F}_r = -\mathbf{F}_g + \mathbf{F}_b + \mathbf{F}_f \tag{3.31}$$

or

$$|m\mathbf{a}| = -V\rho g + V\rho_{fl} g + K\eta v \tag{3.32}$$

The acceleration $\mathbf{a}$ will continuously increase the speed v and hence the frictional force, which in turn decreases the resulting force $\mathbf{F}_r$. Eventually $\mathbf{F}_r$ becomes zero and the body sinks with a uniform speed. Its terminal speed v_t also called *sedimentation speed:*

$$0 = -V\rho g + V\rho_{fl} g + K\eta v_t$$

and

$$v_t = \frac{Vg(\rho - \rho_{fl})}{K\eta} \tag{3.33}$$

If the falling body is a sphere of radius R, Equation (3.33) becomes:

$$v_t = \frac{2R^2g(\rho - \rho_{fl})}{9\eta} \quad (3.34)$$

By measuring the terminal speed $\mathbf{v}_t$ of a sinking sphere, we can determine the viscosity of the fluid involved.

Note: The terminal speed for very small bodies is substantially altered by the random motion of the molecules of the fluid (Brownian motion).

Equation (3.34) is applied to ascertain the sedimentation of bodies in lakes and in the sea. In addition, other factors (e.g., water currents, wind direction) must be taken into account. It is also used to measure the viscosity of the fluid inside a cell if a suitable body can be observed. Many plant cells, for example, contain starch granuli. Their terminal speeds can be measured, and hence η for the insides of those cells can be measured.

EXAMPLE: CENTRIFUGE

A centrifuge separates substances of different densities (ρ_1 and ρ_2). The centrifugation is a sedimentation process as described above. The magnitude a_r of the radial acceleration $\mathbf{a}_r$ replaces g in Equation (3.33). Introducing Equation (3.8) we get for the terminal speed v_t,

$$v_t = \frac{4\pi V \nu^2 r(\rho_1 - \rho_2)}{K\eta} \quad (3.35)$$

where r is the average distance of the substances to be separated from the axis of rotation.

Since sedimentation is a relative motion of one substance with respect to another, it ultimately leads to a spatial separation. All the centrifuge achieves is an increase in v_t, thus making it possible to separate substances exhibiting very small density differences.

4. APPLICATIONS

4.1 Weightlessness

Since man has in the macula of the utriculus a detector for linear accelerations, he experiences a nervous sensation of his *apparent body weight*. An explanation of this detector will help you to understand the experience of weightlessness (experienced at least by a few).

Figure 3-41 shows a schematic drawing of the neuroepithelium of the macula acting as a detector for acceleration. Embedded in the otolithic membrane are tiny grains (otokina) with a high density (ρ = 2.95 g·cm^{-3}). The entire membrane acts analogously to the statocyst de-

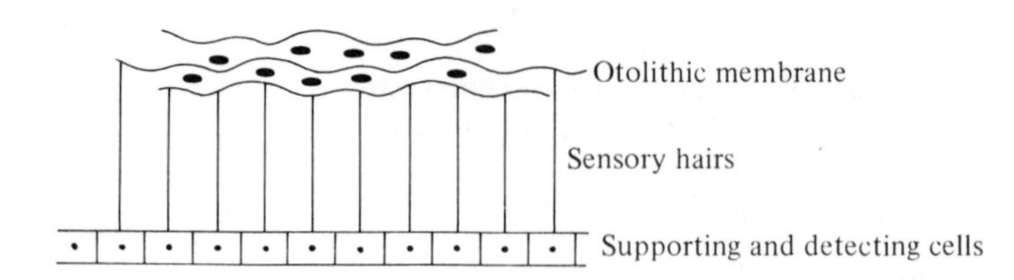

Figure 3-41

scribed earlier. The membrane can move within narrow limits with respect to the supporting and detecting cells. A sideward movement bends the sensory hairs and subsequently causes a nerve signal in the detecting cells. An accelerated up or down movement produces a stress at the sensory hairs and the detecting cells and again causes a nerve signal.

Normal Position of the Body. On the surface of the earth, the membrane (like all other things) is subject to the gravitational force of the earth. Since we do not observe any downward motion of the membrane, the gravitational force must be counteracted by a force of equal magnitude but opposite direction. This is the elastic force caused by the stress on the sensory hairs at the supporting cells. The presence of this elastic force is signaled to the appropriate brain center, causing the sensation of *apparent body weight.*

Free Fall. In a human being experiencing free fall (for example, by a paratrooper with a still-closed parachute), the otolithic membrane in the labyrinth is again under the influence of the gravitational force of the earth. But in this instance the membrane is falling under the influence of the force; hence no counteracting forces are necessary and no nervous signals occur. The free-falling person therefore experiences the sensation of no apparent body weight or *weightlessness.*

Space Platform. An astronaut on a space platform which always hovers at the same position with respect to an earthbound observer (synchronous platform) also experiences the sensation of weightlessness. The otolithic membrane is under the influence of the gravitational force of the earth and is under the compensating radial force of the orbiting platform. No elastic counterforces are present in the neuroepithelium, and the astronaut does not experience the sensation of body weight.

Actually, the nervous sensation of apparent body weight is not caused only by the nervous discharge of the supporting and detecting cells in the macula of the utriculus. Visual information and nervous impulses from various pressure detectors inside the skin and muscles are also involved.

4.2 Pressure

If a force acts over an area, it is convenient to introduce a new physical quantity, the pressure, symbolized p:

$$\text{pressure} \equiv \frac{F_{\perp}}{A}$$

where

$F_{\perp}$: magnitude of the force acting perpendicular to the area,
A: area on which the force acts.
p is a scalar and is measured in newtons/meter^2.

A number of derived units are in practical use for the pressure.

Conversion:

$$1\,\frac{\text{N}}{\text{m}^2} = 10^{-5}\text{ bar (bar)} = 9.26 \times 10^{-6}\text{ atmosphere (atm)}$$

$$1\text{ mm Hg} = 1.33\text{ N}\cdot\text{m}^{-2} = \frac{1}{760}\text{ atm}$$

Pressure-measuring instruments are called *manometers*. (Barometers are manometers used to determine atmospheric pressure.) The pressure of gases is customarily referred to standard temperature 0°C.

The standard pressure of 1 atm used to be defined as the atmospheric pressure at sea level and at a temperature of 0°C. Observing a weather chart extending over the seas will show that this obsolete definition cannot lead to reproducible values.

Nervous Sensation of Pressure. There are localized pressure detectors present in man. The best known is Meissner's corpuscle in the skin. The others are situated deeper inside, in vascular walls, in inner layers of the dermis, and in tendons. These pressure detectors react to nonuniform deformations, which means they indicate only a pressure difference between neighboring positions—a pressure gradient. We do not sense the atmospheric pressure because it is uniformly distributed over the entire body.

The minimum local pressure difference needed to cause the sensation of pressure varies widely in different areas. Most sensitive is the tip of the nose. For a pressure difference of $2 \times 10^4\ \text{N}\cdot\text{m}^{-2}$, the sensation of pressure (touch) occurs. This pressure difference is equivalent to the pressure exerted by a 2-g mass on one square millimeter.

The sensitivity of technical manometers is much higher; pressures down to $1 \times 10^{-8}\ \text{N}\cdot\text{m}^{-2}$ can be measured.

Vacuum. If the air is pumped out of a container, then we customarily say that the vessel contains a vacuum. It is impossible to remove all molecules from the container; in fact, many remain even in an ultrahigh vacuum. The remaining molecules exert a pressure on the walls of the vacuum vessel, caused by their repeated collisions with the walls as they move about.

With the help of advanced vacuum technology it is possible to reduce pressure to about 13 orders of magnitude less than atmospheric pressure. But even at this low pressure, 3×10^6 molecules per cm^3 remain inside the container.

Outside our solar system, astrophysical spectra indicate that there are still a few molecules per cm^3 present.

SUMMARY

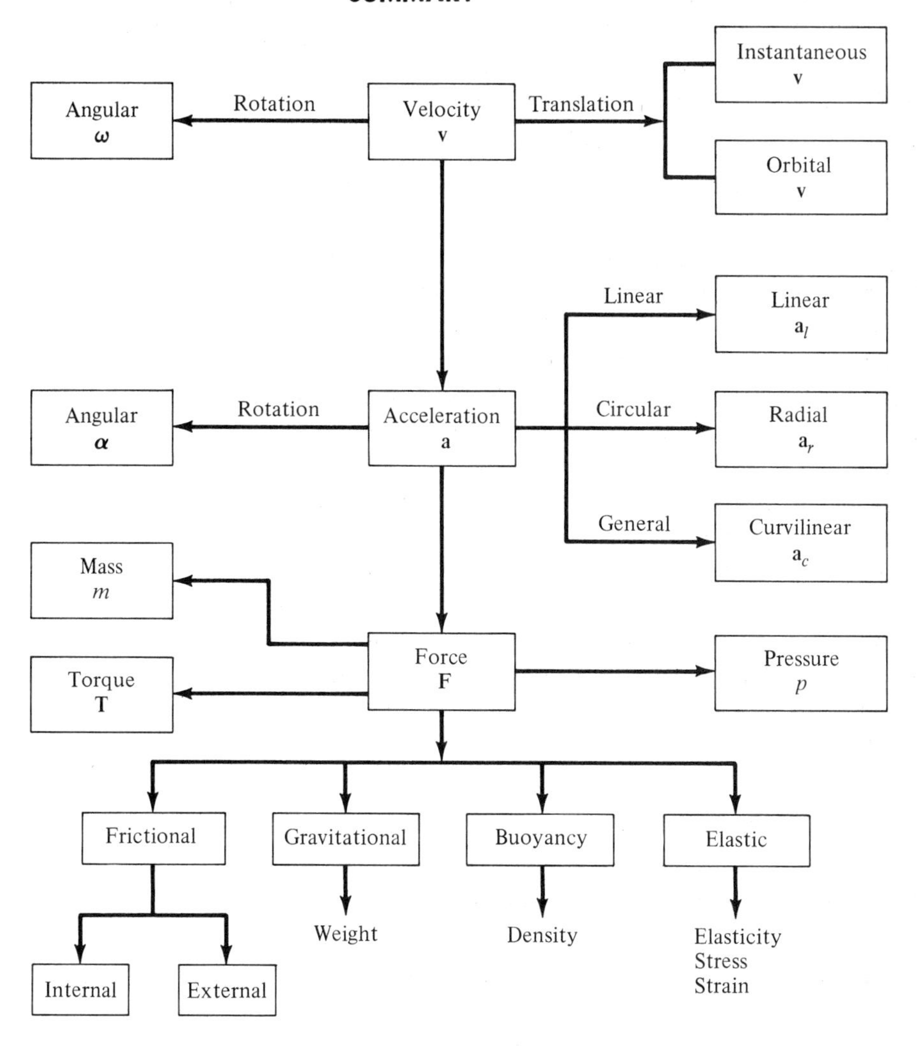

1. In critical situations an experienced car driver will pump the brakes. Why?
2. Why do you walk flat-footed on a slippery surface?
3. Calculate how fast a doughnut-shaped space platform must revolve to introduce a radial acceleration of 0.5 g. Make reasonable assumptions about the radius. How will this rotation be felt by an astronaut inside?
4. Refer to Figure 3-7. The winter position of the earth is closer to the sun than the summer position. Is there a contradiction?
5. Calculate the acceleration g_M at the surface of the planet Mars (mass = 0.11 earth's mass, $r = 3.43 \times 10^6$ m). What is the weight

and mass of a 75 kg astronaut at the surface? Convert the results into units of the English System (see Appendix).

6. Calculate the mass of the earth according to Equation (3.17) ($r = 6.36 \times 10^6$ m). Carry all units through the calculation.

7. Why is the statocyst in Figure 3-26 suspended in a liquid?

8. Demonstrate that the constant K in Equation (3.27) is measured in meters.

9. A wheel of 40 cm radius rotates with 1200 rpm. What is the orbital speed of a point situated at the surface? Calculate its angular speed. In which direction does the angular acceleration point if it is the left front wheel of a car moving forward?

10. Calculate the torque developed at the joint in Figure 3-20 if the force is lifting a mass of 9 kg. Assume $r = 4$ cm for all three positions. What happens if the force lowers the same mass?

11. A centrifuge rotates at 20 000 rpm; the distance from sample to axis of rotation is 75 cm. Calculate the relation between the sedimentation speed in the centrifuge and the sedimentation speed due to gravity.

12. Molecules having a spherical radius $R = 10^{-6}$ cm are suspended in glycerin ($\eta = 1.49 \times 10^3$ cP). The density difference between both materials is 10^{-5} g·cm^{-3}. The suspension is placed into a centrifuge ($r = 0.5$ m) and a sedimentation speed of 10^{-3} cm/s is observed. Calculate the rpm of the centrifuge.

FURTHER READING

H. E. HUXLEY, "The Contraction of Muscle," *Scientific American* (Dec., 1965) 18.

M. A. MACCONAILL, J. V. BASMAJIAN, *Muscles and Movements.* Williams & Wilkins, Baltimore, 1969.

E. W. MERRILL, "Reology of Blood," *Physiol. Rev.*, **49** (1969), 863.

R. TRAUTMAN, "Ultracentrifugation," in D. W. Newman, ed., *Instrumental Methods of Experimental Biology.* Macmillan, New York, 1969.

J. WEBER, "The Detection of Gravitational Waves," *Scientific American* (May, 1971).

M. WILLIAMS, H. R. LISSNER, *Biomechanics of Human Motion.* Saunders, Philadelphia, 1962.

HIROSHI YAMADA, *Strength of Biological Materials.* Williams & Wilkins, Baltimore, 1970.

chapter 4

CONSERVATION LAWS IN MECHANICS

This chapter deals with quantities such as work, kinetic energy, potential energy, momentum, and power. A knowledge of these quantities is necessary for the understanding of two very basic conservation laws: conservation of energy and conservation of momentum. Both laws are highly significant for the life sciences because they allow the analysis of the performance of complicated systems without demanding a detailed knowledge of either system or process. Applying these conservation laws, you will easily gain an understanding of propulsion methods as apparently different as those in rockets, fish, and birds.

The performance of a system can also be evaluated by the system's efficiency, another quantity of importance in nature.

1. WORK

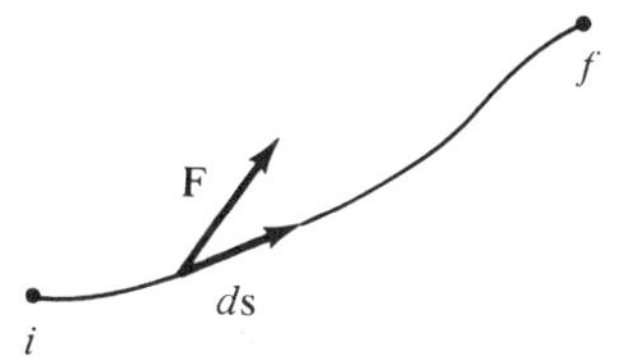

Figure 4-1

If a force **F** moves a body from an initial position i to a final position f (see Figure 4-1) then this force does *work* on the body. The definition of work, symbolized W, is:

$$W \equiv \int_i^f \mathbf{F} \cdot d\mathbf{s} \tag{4.1}$$

where

F: force acting on the body,
$d\mathbf{s}$: infinitesimal (very small) displacement of the body.
$\int$ (integral) means the sum of $\mathbf{F} \cdot d\mathbf{s}$ taken at every $d\mathbf{s}$ along the path.

Work is a scalar because it is the scalar product of the two vectors, **F** and $d\mathbf{s}$. Work is measured in newton·meter. There is a derived unit for work, the joule,* symbolized J:

* Named for James Prescott Joule (1818–1889). The owner of a brewery, Joule could afford to be an independent scholar. His interest focused on heat phenomena

$$1 \text{ joule} = 1 \text{ N}\cdot\text{m}$$

Conversion:

$$1 \text{ J} = 10^7 \text{ erg} = 1 \text{ kg}\cdot\text{m}^2\cdot\text{s}^{-2}$$

Further measures of work are (abbreviations in parentheses):

$$\begin{aligned} 1 \text{ joule (J)} &= 1 \text{ watt}\cdot\text{second (W}\cdot\text{s)} \\ &= 2.79 \times 10^{-7} \text{ kilowatt}\cdot\text{hour (kW}\cdot\text{h)} \\ &= 2.39 \times 10^{-4} \text{ kilocalorie (kcal)} \\ &= 3.73 \times 10^{-7} \text{ horsepower}\cdot\text{hour (hp}\cdot\text{h)} \end{aligned}$$

To calculate the work done, we need to know the force and the path along which the body is moving. Then we can solve the integral in Equation (4.1). Although this may be difficult, for a special case it is simple.

If **F** is constant in magnitude and direction and the body moves in a straight line, then

$$W = \mathbf{F}\cdot\mathbf{s} \tag{4.2}$$

where

$$\mathbf{s} = \int_i^f d\mathbf{s} = \text{displacement of body}$$

According to the rules for the scalar product (see page 21),

$$W = Fs\cos\theta \tag{4.3}$$

where

F: magnitude of force acting on body,
s: distance covered by the body under the influence of the force **F**,
θ: angle between **F** and s.

Notice that no work is done unless the force causes a displacement. This is contrary to the colloquial use of the expression "work." As an example, during isometric exercise (that is, pressing against an immobile object), no work in the strict physical sense is done. However, to uphold the stress, energy is dissipated and heat is produced in the muscles. But this is not included in the physical concept of work.

EXAMPLE: WEIGHTLIFTING

A weight of mass $m = 70$ kg is raised perpendicularly from the floor to a height of 2.5 m. The force $\mathbf{F}_1$ exerted by the athlete has to be at least equal and opposite to the gravitational force $\mathbf{F}_g$ always acting on the weight. See Figure 4-2.

Assuming a constant force $\mathbf{F}_1$ and perpendicular lift, we apply Equation (4.3) with $\theta = 0°$:

$$\begin{aligned} W &= mgh \\ &= 70 \times 9.81 \times 2.5 \text{ kg}\cdot\text{m}\cdot\text{s}^{-2}\cdot\text{m} \\ &= 1720 \text{ J} \end{aligned}$$

and he not only experimentally determined the quantitative relation between mechanical energy and heat, but he also discovered, together with W. Thomson, the effect which makes our refrigerators work. More about him in O. Renolds, *Memoir of J. P. Joule*, Manchester, 1892.

Figure 4-2

Carrying the same load along a horizontal path is shown in Figure 4-3. $W = 0$ because $\mathbf{F}_g \perp \mathbf{s}$; that means $\theta = 90°$ in Equation (4.3).

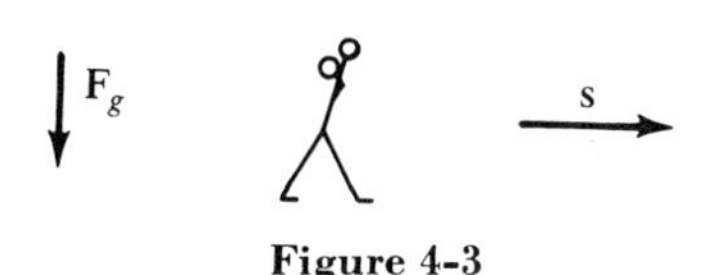

Figure 4-3

Actually this is only the simplest approximation because the load will move up and down during walking. Figure 4-4 shows the path of the center of mass (CM) of the load during walking. The displacement $\mathbf{s}_2$ in the direction of the gravitational force $\mathbf{F}_g$ has a magnitude of $h_2 = 0.15$ m per step; hence the work per step W_2 due to the load is

$$\begin{aligned} W_2 &= mgh_2 \\ &= 70 \times 9.81 \times 0.15 \text{ joule} \\ &= 103 \text{ J per step} \end{aligned}$$

This work W_2 is supplied by the muscles. During the downward motion of the load, the gravitational force supplies the work, but the skeleton muscles must again do work to slow down the motion of load and body. Consequently, the work done per step is twice W_2.

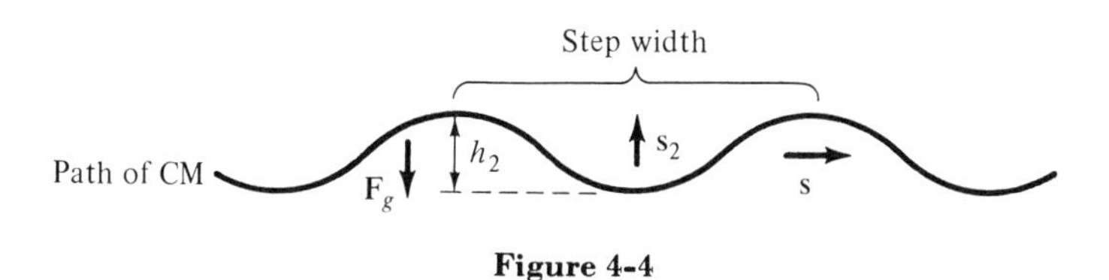

Figure 4-4

EXAMPLE: WORK DONE BY THE HEART

We can derive the force under which the heart pumps the blood out of the left ventricle into the aorta if we realize that this force has to overcome the blood pressure between systole and diastole. Taking the pressure difference Δp to be

$$\Delta p = 60 \text{ mm Hg} = 8 \times 10^3 \text{ N}\cdot\text{m}^{-2}$$

and the aortian cross section φ_a as $7 \text{ cm}^2 = 7 \times 10^{-4} \text{ m}^2$, we get for the magnitude of the force $|\mathbf{F}_v|$ that the ventricle has to exert at least

$$\begin{aligned} |\mathbf{F}_v| &= \Delta p \varphi_a \\ &= 8 \times 10^3 \times 7 \times 10^{-4} \text{ newton} \\ &= 5.6 \text{ N} \end{aligned}$$

The average stroke volume is 100 cm^3; hence we can calculate the magnitude of the displacement $|\mathbf{s}|$ of the blood along the aorta to be 14.3 cm (100 cm^3 of blood occupies a 14.3-cm length in the aorta). Since the exerted force is parallel to the displacement and assumed to be constant, we use Equation

(4.3) to calculate the work W_v done by the left ventricle:

$$W_v = 5.6 \times 0.143 \text{ N}\cdot\text{m}$$
$$= 0.8 \text{ J per beat}$$

Assuming one heart beat per second, the work done per day W is

$$W = 3.6 \times 10^3 \times 24 \times 0.8 \text{ s}\cdot\text{J}\cdot\text{s}^{-1}$$
$$= 7.8 \times 10^4 \text{ J}$$

This calculation is a crude one because the aortic pressure is not constant and hence $\mathbf{F}_v$ is not constant. For a more accurate result, we must solve the following equation

$$W_v = \int_{vi}^{vf} p\, dV \qquad (4.4)$$

where

v_i: ventricle volume at beginning of systole,
v_f: ventricle volume at beginning of diastole,
p: ventricle pressure as a function of the volume of the ventricle.

To solve this integral, we need to know the pressure-volume diagram of the ventricle. Taking the experimentally determined diagram and solving the integral in Equation (4.4) graphically (see Problem 10 at the end of this section), we get a result that is within 20% of the derived value.

In the simplified example above, we sacrificed some accuracy to gain ease of calculation. This trade-off is justified because the biological variability in the work per heart stroke from one individual to another is larger than the encountered inaccuracy of 20%.

It is recommended that you get used to orders-of-magnitude estimates. They are simple and the derived accuracy is in most instances sufficient, especially in the life sciences.

A word about simple machines such as levers, inclined planes, and pullies: Most human activity involves doing work. To achieve a maximum of work, the exerted force and the displacement should be as large as possible. The exertable force is limited by the cross section of the muscles (approximately 40 N/mm^2). The only way to increase the work done beyond the limits of the exertable force is by using a longer path, which is larger displacement. Man-powered machines are devices which achieve this goal; hence the above-mentioned devices save force, not work.

2. ENERGY

One of the most enlightening concepts in science is the concept of energy. Energy is closely linked to work and can be described as *stored work*. If work is done by a body, then its energy declines. Work done on the body increases the energy of the body. Energy is manifest in many different ways. In the life sciences we are particularly aware of chemical (bond) energy, electromagnetic (light, x rays) energy, heat, sound, and mechanical (muscular) energy. We shall now consider only two forms: kinetic energy and potential energy. The other manifestations of energy will be introduced later.

2.1 Kinetic energy

The *kinetic energy*, symbolized E_k, of a body is defined as

$$\text{kinetic energy} \equiv \frac{m}{2} v^2 \qquad (4.5)$$

where

m: mass of body,
v: speed of body.

The kinetic energy is a scalar. It is never negative, and it is measured in joules.

EXAMPLE: COLLIDING CARS

The damage done is directly proportional to the kinetic energies of the cars involved. This is so because the kinetic energies of the cars are partly or entirely converted into work which subsequently does the damage.

The kinetic energy of a Volkswagen ($m = 900$ kg) traveling at 50 km/h = 14 m/s is

$$E_{k\text{vw}} = \left(\frac{900}{2}\right) \times (14)^2\ \text{kg}\cdot\text{m}^2\cdot\text{s}^{-2}$$
$$= 8.8 \times 10^4\ \text{J}$$

For a Pontiac ($m = 2100$ kg) having the same speed,

$$E_{k\text{pon}} = \left(\frac{2100}{2}\right) \times (14)^2\ \text{joule}$$
$$= 2.06 \times 10^5\ \text{J}$$

To be on equal footing with the Pontiac, that is, having the same kinetic energy during a head-on collision, the Volkswagen must increase its speed. We calculate this speed v_{vw} by equating both kinetic energies and solving for the unknown v_{vw}:

$$2.06 \times 10^5 = \left(\frac{900}{2}\right) v_{\text{vw}}^2$$
$$v_{\text{vw}}^2 = 4.6 \times 10^2\ \text{m}^2\cdot\text{s}^{-2}$$
$$v_{\text{vw}} = 21.4\ \frac{\text{m}}{\text{s}} = 77\ \frac{\text{km}}{\text{h}}$$

The smaller car has to increase its speed by 54% to obtain the same kinetic energy as the one with 130% more mass. This is a consequence of the fact that the kinetic energy is proportional to the square of the speed.

EXAMPLE: VELOCITY FACTOR IN CARDIAC OUTPUT

During systole the heart ejects blood at a speed of approximately 14 cm/s into the aorta. About 100 g are pumped per systole. The kinetic energy of the ejected blood is, according to Equation (4.5),

$$E_k = 0.1 \times 0.5 \times (0.14)^2\ \text{kg}\cdot\text{m}^2\cdot\text{s}^{-2}$$
$$= 9.8 \times 10^{-4}\ \text{J}$$

Since this kinetic energy of the blood is supplied by the work of the left ventricle, we must add 9.8×10^{-4} J to the 0.8 J previously obtained in order to know the work done by the left ventricle per heart beat. Notice that less than 0.1% of the work done by the left ventricle is transferred into kinetic energy of the blood.

EXAMPLE: ELECTRON ORBITING THE NUCLEUS

The kinetic energy of the electron that orbits the hydrogen nucleus is

$$E_k = \left(\frac{m_{\text{el}}}{2}\right) v^2$$

where

m_{el}: electron mass $= 9.2 \times 10^{-31}$ kg
v: electron speed $= 2.2 \times 10^6$ m/s

Hence:

$$E_k = (4.6 \times 10^{-31})(2.2 \times 10^6)^2\ \text{kg}\cdot\text{m}^2\cdot\text{s}^{-2}$$
$$= 2.2 \times 10^{-18}\ \text{J}$$

The potential energy, symbolized E_p, of a body is defined as

$$\text{potential energy} \equiv -\int_i^f \mathbf{F} \cdot d\mathbf{s} \tag{4.6}$$

where

$\mathbf{F}$: conservative force acting on the body,
$d\mathbf{s}$: infinitesimal displacement of the body,
i: initial position of the body,
f: final position of the body.

The potential energy is a scalar measured in joules, and it depends only on the initial and final positions of the body. Potential energy can be either positive or negative, quite contrary to kinetic energy, which is always positive.

If the body moves under the influence of $\mathbf{F}$ from i to f, then $\mathbf{F}$ does work on the body equal to $-E_p$.

If the integral in Equation (4.6) is independent of the path of integration, i.e., the result is independent of the path the body follows to move from the initial position to the final position, then $\mathbf{F}$ is a *conservative force* by definition. Gravitational and elastic forces are conservative.

Observe that, due to the path-independent integral in Equation (4.6), no net work is done if the body returns to its initial position (closed path).

Equation (4.6) defines potential energy in physical terms. In a broader sense of the word, potential energy is stored energy which can be converted into work. This stored energy may be electrical, elastic, chemical, atomic, nuclear, or positional. Chemical (or bond) potential energy is essential for sustaining life; this form can be utilized by cells via their ATP (adenosintriphosphate).

If the force $\mathbf{F}$ is constant in magnitude and direction, then Equation (4.6) simplifies to:

$$E_p = -\mathbf{F} \cdot \mathbf{s} = -Fs \cos \theta \tag{4.7}$$

where

$\mathbf{s} = \int_i^f d\mathbf{s}$ = displacement of body,
θ: angle between $\mathbf{F}$ and $\mathbf{s}$.

Before we can calculate the potential energy of a body, we must arbitrarily define the position where the potential energy is zero. The displacement $\mathbf{s}$ is then measured from this reference position.

Convention: A body under the influence of the gravitational force has zero potential energy at the surface of the earth.

In general, a body of interest under the influence of the gravitational force is situated above the surface of the earth. It is convenient to calculate with a positive potential energy in this case; hence the unusual negative sign appears in its definition.

EXAMPLE: BODY AT ALTITUDE h

Since the gravitational force $\mathbf{F}_g$ is conservative and practically constant as long as the altitude is very small compared to the radius of the earth, Equation (4.7) applies:

$$E_p = -\mathbf{F}_g \cdot \mathbf{s} = -|\mathbf{F}_g||\mathbf{s}| \cos \theta$$

$\mathbf{F}_g$ and $\mathbf{s}$ are parallel but of opposite direction (vector $\mathbf{s}$ originates at the surface of the earth). Thus $\theta = 180°$ and $\cos \theta = -1$

$$E_p = mgh \tag{4.8}$$

because $|\mathbf{F}_g| = mg$ [see Equation (3.18)] and $|\mathbf{s}| = h$.

For $m = 10$ kg and $h = 20$ m

$$E_p = 10 \times 9.81 \times 20 \,\text{kg}\cdot\text{m}^2\cdot\text{s}^{-2} = 1.96 \times 10^3 \,\text{J}$$

The potential energy of that body at an altitude of 20 m is positive. This means that, if the body drops under the influence of $\mathbf{F}_g$ back to the surface of the earth, it will do 1960 J of work. After that, its potential energy will be zero.

EXAMPLE: AORTA DURING SYSTOLE

The walls of the aorta are elastic. When the left ventricle ejects blood, the aorta expands and in doing so stores work in the form of potential energy. This energy will be released during systole to keep the blood flowing. See Figure 4-5.

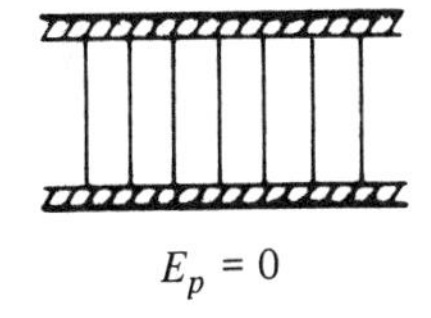

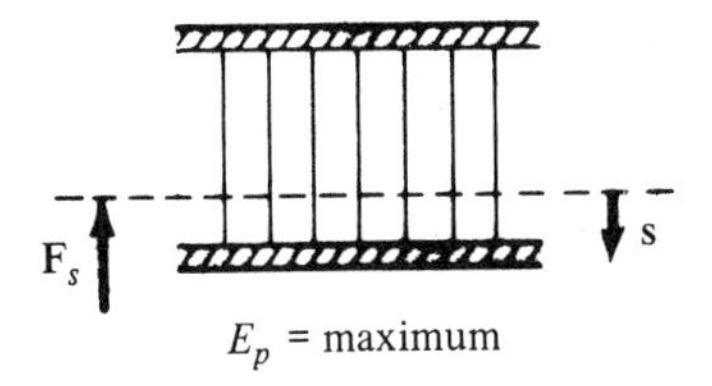

Figure 4-5

Idealizing the circumstances, we can compute the E_p of the aorta at the end of systole. Let us treat the elastic force of the aorta walls as a spring force $\mathbf{F}_s$ with a spring constant $k = 1 \times 10^5 \,\text{N}\cdot\text{m}^{-1}$. According to Equation (3.23), the magnitude F_s of $\mathbf{F}_s$ is

$$F_s = k \,\Delta l$$

where Δl is the expansion of the aorta walls measured from the position before systole.

The potential energy at the end of systole is then [see Equation (4.6)]:

$$E_p = -\int_0^{\Delta l_{max}} \mathbf{F}_s \cdot d\mathbf{s}$$

Since $\mathbf{F}_s$ and $d\mathbf{s}$ are parallel but opposite in direction ($\theta = 180°$),

$$E_p = \int_0^{\Delta l_{max}} F_s \,ds$$

$$= \int_0^{\Delta l_{max}} k \,\Delta l \,ds$$

or with $\Delta l = s$

$$E_p = \int_0^{\Delta l_{max}} ks \,ds$$

$$= \frac{k}{2} s_{max}^2$$

with the experimental values $k = 1 \times 10^5 \,\text{N}\cdot\text{m}^{-1}$ and $s_{max} = 3 \times 10^{-3}$ m:

$$E_p = \frac{10^5 \times 9 \times 10^{-6}}{2} \,\text{N}\cdot\text{m}^{-1}\cdot\text{m}^2$$

$$= 0.45 \,\text{J}$$

By comparison with the total work W_v calculated earlier, the result indicates that about half of the work done by the left ventricle during systole is stored in the form of potential energy. This potential energy will be released to the blood flow during diastole by contraction of the aorta walls.

2.3 Conservation of mechanical energy

In a system where only conservative forces are present, the total mechanical energy (symbolized E_{mech}) is conserved:

$$E_{mech} = E_p + E_k = \text{constant} \tag{4.9}$$

Formulated in another way,

$$E_{pi} + E_{ki} = E_{pf} + E_{kf} = \text{constant} \tag{4.10}$$

where the indices i and f refer to the investigated system at initial and final states, respectively.

It often happens that one or two of the quantities in the preceding equation are zero. For example, if we refer to a supported mass m at the height h, the kinetic energy of the initial system is zero. In the final system (after the support is removed and the mass is on the surface of the earth) the potential energy is zero. In this case,

$$mgh = \left(\frac{m}{2}\right) v^2$$

where v is the impact speed. Hence

$$v = \sqrt{2gh} \tag{4.11}$$

Here we encounter the free fall, first quantitatively investigated by Galilei* about four centuries ago.

EXAMPLE: A CYCLIST IN THE MOUNTAINS

Man and bicycle form a system under the influence of a conservative force, the gravitational force. All other forces such as frictional and muscular forces are neglected. If the rider is not allowed to brake or to use the pedals, conservation of mechanical energy applies. The initial system shall be the cyclist and his vehicle at rest at the top of the Continental Divide (altitude 2520 m) in Yellowstone National Park. See Figure 4-6. This means that $E_{ki} = 0$, and according to Equation (4.8)

$$E_{pi} = mgh$$

For $m = 90$ kg, follows

$$E_{pi} = 90 \times 9.81 \times 2520 \text{ kg}\cdot\text{m}\cdot\text{s}^{-2}\text{ m}$$
$$= 2.23 \times 10^6 \text{ J}$$

Now the cyclist makes a straight run for the Old Faithful Lodge (altitude 1930 m) without using the brakes. The system cyclist-bicycle will attain a speed v. This is the final state of the system.

$$E_{pf} = 90 \times 9.81 \times 1930 \text{ joules}$$
$$= 1.70 \times 10^6 \text{ J}$$

$$E_{kf} = \frac{m}{2} v^2$$

Mechanical energy is conserved and we derive for the final speed v:

$$E_{pi} + E_{ki} = E_{pf} + E_{kf}$$

$$2.23 \times 10^6 = 1.7 \times 10^6 + \frac{m}{2} v^2$$

* Galileo Galilei (1564–1642) mathematician, physicist, and astronomer. He discovered the satellites of Jupiter (too early for the Inquisition). It is worthwhile to note that he was one of the first to use his native tongue for presenting scientific results. More about him in C. L. Colino, *Galileo Reappraised*, Berkeley, 1966.

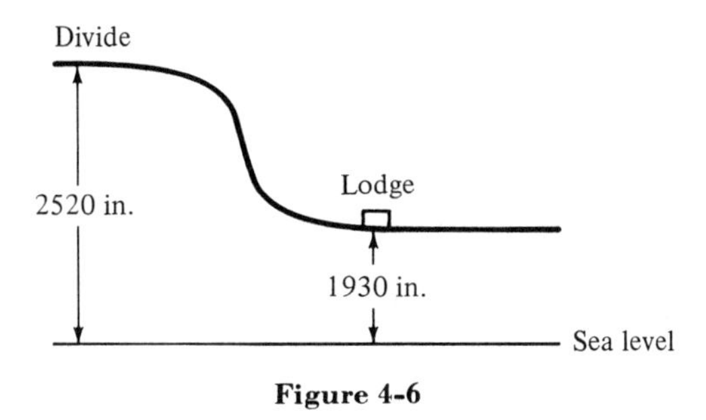

Figure 4-6

$$0.53 \times 10^6 = 45\, v^2$$

$$v^2 = 1.17 \times 10^4 \text{ m}^2\cdot\text{s}^{-2}$$

$$v = 108 \text{ m}\cdot\text{s}^{-1} = 380 \frac{\text{km}}{\text{h}}$$

A speeding cyclist indeed!

EXAMPLE: SCHRÖDINGER EQUATION

A famous equation which appears to be very complicated and seemingly impossible to comprehend is the Schrödinger* equation, the very foundation of modern physics. In one form (time-dependent, one-dimensional) it looks like

$$-\frac{\hbar}{i}\frac{\partial}{\partial t}\psi(x,t) = \frac{\hbar^2}{2m}\frac{\partial^2}{\partial x^2}\psi(x,t) + V(x,t)\psi(x,t)$$

This is really nothing but mechanical energy conservation expressed in a different formalism. The left side represents the total energy, the first term to the right of the equal sign is the kinetic energy, and the remaining part represents the potential energy.

As much as I would like to delve now or later into the beauties and consequences of Schrödinger's equation, it is impossible since it exceeds by far the scope of this presentation. Mathematically treating matter as waves (a concept introduced by Louis de Broglie†), Schrödinger achieved quantitative agreement with the experimentally determined atomic energy levels—to name only one of the successes of his theory. For the layman he started the great confusion: Does matter consist of particles or of waves?

* Erwin Schrödinger (1887–1961), Nobel laureate 1933. One of the fathers of modern physics. You will probably enjoy reading his books: *What is life ?*, University Press, Cambridge, 1944, and *Science and Humanism*, University Press, Cambridge, 1951.

† Prince Louis-Victor de Broglie, physicist, Nobel laureate 1929. Extended the scope of quantum theory and introduced the matter waves (in his dissertation!).

3. MOMENTUM

3.1 Definition

The linear momentum usually abbreviated to momentum **p** of a body is

$$\text{linear momentum} \equiv m \cdot \mathbf{v} \tag{4.12}$$

where

m: mass of body,
$\mathbf{v}$: velocity of that body.

Momentum is a vector and points in the same direction as the velocity of the body. Its magnitude is measured in $\text{kg} \cdot \text{m} \cdot \text{s}^{-1}$. There is a useful relation between the kinetic energy of a body and the magnitude of its momentum:

$$E_k = \frac{p^2}{2m} \tag{4.13}$$

where

m: mass of body,
p: magnitude of momentum p.

3.2 Momentum conservation

In an isolated (closed) system, the sum of the momenta of all bodies is conserved, that is

$$\sum_j \mathbf{p}_j = \mathbf{p}_1 + \mathbf{p}_2 + \mathbf{p}_3 + \cdots + \mathbf{p}_j = \text{constant} \tag{4.14}$$

where $\mathbf{p}_1$, $\mathbf{p}_2$, . . . are the momenta of the individual bodies in that isolated system.

This law—especially if combined with the law of conservation of mechanical energy—is one of the most powerful investigational tools in physics because, if the initial state of a system (characterized by momentum, E_p and E_k) is known, its final state is determined.

The actual transfer from initial to final state via some intermediate states may be complicated, but it is of little concern. For example, we can now investigate many aspects of the propulsion of rockets and fish and the hovering of birds without detailed knowledge of how they are actually achieved.

EXAMPLE: ROCKET PROPULSION

During takeoff from the earth, it becomes rather involved to calculate the motion of a spaceship. But once outside the earth's atmosphere the spaceship represents a closed system very well. To change the velocity of the ship, a rocket engine is temporarily ignited. This leads to the emission of gas molecules (total mass m_G) with a velocity $\mathbf{v}_G$. See Figure 4-7.

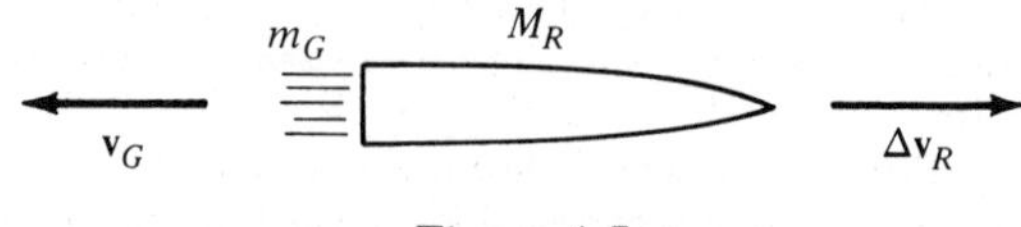

Figure 4-7

The gas carries a momentum $\mathbf{p}_G$ in the backward direction. An equal momentum but of opposite direction will be induced in the spaceship since the total momentum must be conserved; hence

$$m_G \mathbf{v}_G = M_R \Delta \mathbf{v}_R$$

where

M_R: mass of spaceship,

$\Delta \mathbf{v}_R$: additional velocity gained by the ship due to the burning of the rocket engines.

The gained (or lost, if a retrorocket was ignited) velocity of the ship is:

$$\Delta \mathbf{v}_R = \left(\frac{m_G}{M_R}\right) \mathbf{v}_G$$

Notice that the larger the mass of the ship, the smaller the gained velocity.

To achieve a significant change in velocity, the total mass of the ejected gas, m_G, or the velocity with which the burned fuel leaves the engines, or both, must be large. The mass of the spaceship remains essentially constant. Increasing $\mathbf{v}_G$ means higher temperatures in the burners, and there are technological limits to that. Increasing m_G calls for more fuel to be carried, hence increasing takeoff weight of the craft. Clearly, the most economical solution to this problem calls for higher velocities of the ejected substance. Conventional fuel (benzene, H_2 and O_2 mixtures) leads to a maximum velocity of approximately 10–20 km/s. If electrically charged particles (for example, ionized air) are used as fuel and accelerated by suitable electromagnetic fields, then the fuel molecules will reach much higher velocities. The most advanced of these ion-rocket engines ejects air molecules with a velocity of 140 km/s. Unfortunately, the number of molecules leaving the ion engine is still small; hence m_G is small.

EXAMPLE: PROPULSION OF FISH

The method of propulsion depends on the physical appearance of the fish (naturally, this sentence can be phrased the other way around). Since their shapes vary considerably, we expect different methods for fish propulsion. Nevertheless, practically all observed methods rely on momentum conservation. A good example for that is the way in which the fastest fish swim, for instance shark, tuna, and sailfish—to name a few. It is not by accident that they all have a prominent lunate tail. This is their powerful propulsion instrument. All the other fins perform predominantly balance and steering functions.

But how is the forward propulsion achieved by a sideways swing of the tail? If you observe the moving fish closely, you will see that a narrow jet of water right behind its tail moves in the backward direction. See Figure 4-8. A calculation shows that the momentum of the water stream is equal in magnitude to the momentum of the moving fish—momentum conservation at work.

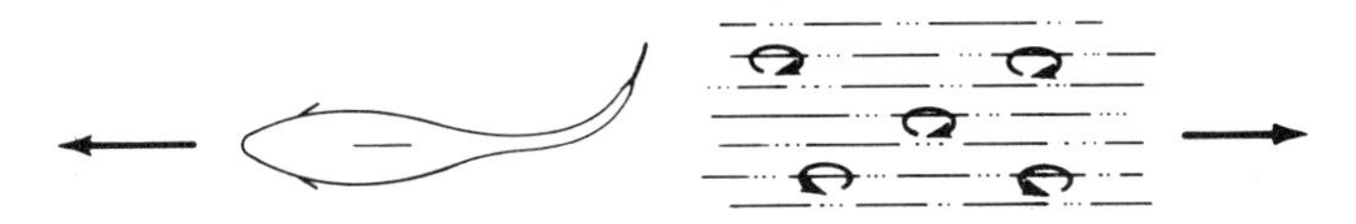

Figure 4-8

But how does the fish impart this momentum to the water? The answer to that involves some knowledge of hydrodynamics: Whenever a body moves through water it leaves a wake behind. This wake is made up of individual circular currents of the medium, *vortexes*. Aircraft designers try to avoid this wake of vortexes, called *Karman-straight.** But a fast-moving fish relies on it for propul-

* Theodore von Karman (1881–1963), physicist and pioneer of theoretical aerodynamics. Autobiography: *The Wind and Beyond*, Boston, 1967.

sion. With every movement of the lunate tail, vortexes are produced. The shed vortexes transfer their energy to the water—by way of friction—and a current in the backward direction results.

It is difficult to calculate the energy intrinsic to a vortex, but this is not necessary since we can calculate the momentum of the moving water stream. Knowing that, we can determine the speed of the fish. The remarkable speed of 20 m/s is achieved by that method.

The cephalopods have a pure jet system for locomotion. See Figure 4-9. Equation (4.14) applies directly. It is such a powerful jet propulsion system that even the giant squid can hurl itself clear out of the water. This system is rather different from the fanciful, imaginary one depicted by an artist some centuries ago. The woodcut illustrated in Figure 4-10 shows the paper nautilus employing its arms as oars and its membrane as a sail. Notice that the sail is set.

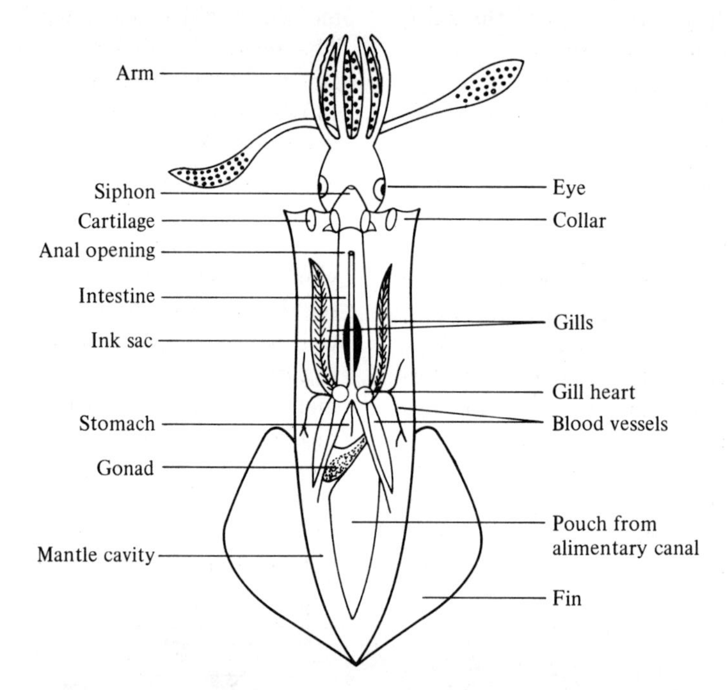

Figure 4-9 Anatomy of a cuttle fish (cephalopoda). Sea water is drawn into the mantle cavity and ejected through a muscular funnel, the siphon. Consequently, the fish moves backward (down in the figure).

EXAMPLE: HOVERING BIRD

When the humming bird hovers in front of a flower to suck out the nectar, it beats the air rapidly. Each stroke propels a small mass of air (m_{air}) downward with a velocity $\mathbf{v}_{air}$. An equal momentum of opposite direction is imparted to the bird, and consequently the animal is lifted up with a velocity $\mathbf{v}_{bird}$. Momentum conservation holds:

$$m_{bird}\,\mathbf{v}_{bird} = m_{air}\,\mathbf{v}_{air}$$

By changing the angle of attack during the upward stroke (achieved by twisting the wings), the colibri assures that now only a comparatively small momentum downward is imparted to the body. Nevertheless the bird would fall during that period, but it is elevated again during the next downbeat. The momentary position of the hovering bird thus oscillates around an average.

Figure 4-10 Medieval drawing of a paper nautilus.

A word of caution: The conservation laws [Equations (4.9) and (4.14)] are really valid only for ideal systems. Whenever you apply it to a real system, i.e., systems in the realm of nature, remember that now the conservation laws yield only approximate results. For every example, assure yourself that the investigated system does not deviate too much from the ideal system. Ask questions such as: Can the system really be approximated as a closed system? Are there forces (like frictional forces) at work which I cannot include? How large are those forces compared with the others?

4. EFFICIENCY

Work can be stored in the form of energy, and energy can be reconverted into work. In actual life, unfortunately, during that conversion, part of the cycled energy always appears in an undesirable form, for example, as heat. Take an olympic swimmer. He tries to convert the work stored in his muscles into motion, hence, into kinetic energy. But even if he uses the crawl stroke, only 2.2% of the exerted work is converted into kinetic energy of his body. That means the work he performs mainly produces heat. The 2.2% is called the *efficiency*, symbolized η, of the process involved. By definition:

$$\text{efficiency} \equiv \frac{E_{\text{out}}}{E_{\text{in}}} \qquad (4.15)$$

where

E_{in}: total energy freed in the investigated system,
E_{out}: desired energy extracted from the system.

η is a dimensionless quantity and is usually expressed in percent of E_{in}.

For economic reasons it is most desirable to make any form of energy conversion as efficient as possible. The chemical energy stored in engine fuel is converted into mechanical energy. For gasoline engines, η is 38%. Diesel engines are more economical since their efficiencies go up to 56%. The highest efficiency is not to be found in the realms of technology. A small water beetle holds the record with 84%. See Table 1 for additional values.

Table 1 EFFICIENCY η OF VARIOUS SYSTEMS

System	η Measured in Percent
man swimming	0.5–2.2
steam engine	10–15
muscle	10–22
gasoline engine	38
photosynthesis	23–43
treadmill	25–40
living cell	40–50
fuel cell (Pt in NaOH)	50
diesel engine	56
adjustable pitch propeller	82
propulsion of water beetle	84

The muscles convert metabolic energy into mechanical energy. The efficiency for that process is not high. Depending on the type of muscle, it is between 10% and 22%. It is interesting to note that the efficiency depends on the contraction speed of the muscle. This is displayed in Figure 4-11. If the muscle contracts with the maximum possible speed v_{max}, the efficiency approaches zero. If it contracts very slowly, the effi-

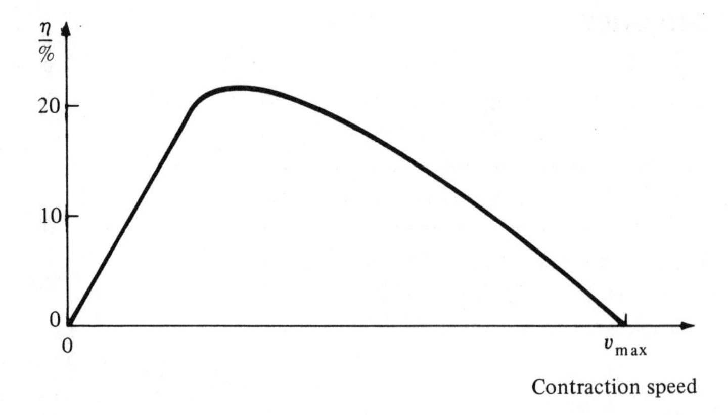

Figure 4-11

ciency also reaches zero. The highest efficiency is reached when the muscle shortens with about 20% of the maximum possible speed; but even then η is only 21% for the muscle type shown.

This efficiency curve has practical consequences. It allows us to determine the most efficient walking speed. The result is displayed in Figure 4-12. The metabolic energy of the body is converted most favor-

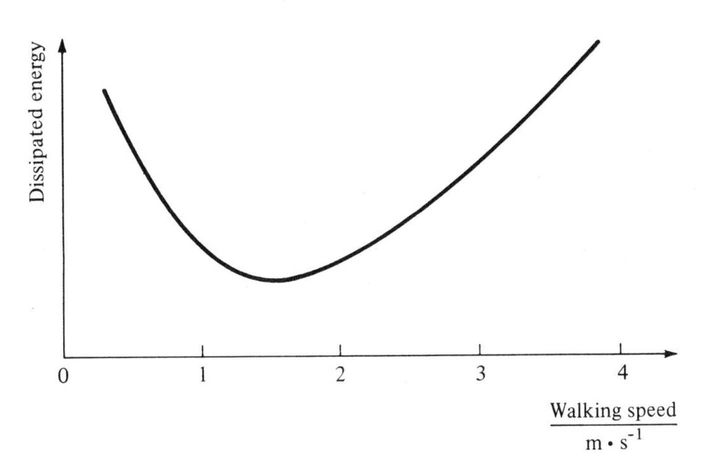

Figure 4-12

ably into motion at a walking speed of 1.5 m/s, which corresponds to 5.4 km/h or 3.4 mile/h.

The efficiency of a living cell is about 40–50%. This means that 50–60% of its received metabolic energy is converted into heat. This by-product is not wholly undesirable. The heat produced keeps the inside of the cell at an optimum temperature for the living processes.

5. POWER

5.1 Definition

Another useful quantity in mechanics is work per unit of time, called *power* (symbolized P):

$$\text{power} \equiv \frac{W}{t} \tag{4.16}$$

where

W: work done,
t: time needed in performing W.

If the work itself depends on time, then

$$P \equiv \frac{dW}{dt} \tag{4.17}$$

where

dW: infinitesimal work done,
dt: infinitesimal time needed in performing dW.

Power is a scalar quantity and is measured in joules/second. There is a derived unit for power, the *watt*, abbreviated W:*

$$1 \text{ watt} = 1 \text{ J} \cdot \text{s}^{-1}$$

A much used unit for power outside the SI system is the *horsepower*, symbolized hp. Conversion:

$$\begin{aligned} 1 \text{ W} &= 1.34 \times 10^{-3} \text{ hp} \\ &= 10^7 \text{ erg} \cdot \text{s}^{-1} \\ &= 0.39 \text{ cal} \cdot \text{s}^{-1} \end{aligned}$$

In life, power is an important factor, for in many critical situations the maximum power output of an animal determines its chance of survival. An antelope, for instance, is capable of short (some minutes) outbursts of power resulting in a high running speed. But the power it can sustain over a prolonged period of time is much smaller. The bushman of the Kalahari desert cannot outsprint the antelope; but since he is able to wear out the animal by continuously trotting, his sustained power output must be much higher.

EXAMPLE: POWER OF MAN

Man is capable of short bursts of high power. This can be observed during weightlifting. The world record in the heavyweight class for snatch stands at approximately 180 kg. The weight is lifted from the ground to a height h. The work W done by the weightlifter is equal to the increase of the potential energy E_p of the weight. According to Equation (4.8),

$$E_p = mgh = W$$

For

$$\begin{aligned} m &= 180 \text{ kg} \\ g &= 9.81 \text{ m} \cdot \text{s}^{-2} \\ h &= 2.2 \text{ m} \end{aligned}$$

we get

$$\begin{aligned} W &= 180 \times 9.81 \times 2.2 \text{ kg} \cdot \text{m} \cdot \text{s}^{-2} \cdot \text{m} \\ &= 3.9 \times 10^3 \text{ J} \end{aligned}$$

The time needed for lifting is approximately 0.5 s; hence the maximum power output P_{max} is

$$\begin{aligned} P_{max} &= \left(\frac{3.9}{2}\right) \times 10^3 \text{ J} \cdot \text{s}^{-1} \\ &= 7.8 \times 10^3 \text{ W} = 10.5 \text{ hp} \end{aligned}$$

Actually the power is even larger because not only the weight is lifted, but also the center of mass of the athlete himself. The sustained power output of man is about 75 W = 0.1 hp.

EXAMPLE: POWER OF LEFT VENTRICLE

The work W_v done by the left ventricle per stroke was calculated in the beginning of the chapter to be 0.8 joule. The duration t_v of the systole is 0.27 second; consequently the power output during systole is

$$\begin{aligned} P_V &= \frac{W_v}{t_v} = \frac{0.8}{0.27} \text{ J} \cdot \text{s}^{-1} = 2.95 \text{ W} \\ &= 3.96 \times 10^{-3} \text{ hp} \end{aligned}$$

* James Watt (1736–1819), inventor of the modern steam engine. More about him in H. W. Dickinson, *James Watt, Craftsman and Engineer*, Cambridge University Press, 1936.

Table 2 shows some interesting power outputs.

Table 2 POWER OUTPUT OF SOME SYSTEMS

System	Power Measured in watts
single neuron	10^{-9}
ant	2.5×10^{-9}
small radio	1
left ventricle	2.95
human brain	10
man walking	100
athlete, maximum	7800
small car	4.5×10^{4}
Cadillac	3×10^{5}
electric locomotive	3×10^{6}
ocean liner	2.1×10^{7}
Moon rocket	10^{8}
supernova	10^{37}

5.2 Specific power

The notion of *specific power* often used in technology is also used in the life sciences. It is the ratio between power output and some other quantity of interest. For a racing car, the total power output of its engine is clearly not as important as the power per kg of engine mass. Most desirable is an engine which produces the most horsepower for a given mass. In this case the specific power is measured in hp/kg.

Clearly, the means of transportation which needs the lowest amount of power per unit mass for propulsion at a given speed is the most economical. Look at Figure 4–13 on page 104 and you will realize how well man is made in this respect.

SUMMARY

Displacing a body by **s**, the force **F** performs the work W:

$$W = \int \mathbf{F} \cdot d\mathbf{s}$$

W is a scalar and is measured in $\text{kg} \cdot \text{m}^2 \cdot \text{s}^{-2}$ = joules.

Any body with a mass m and a speed v contains a kinetic energy E_k,

$$E_k = \left(\frac{m}{2}\right) v^2$$

Under the influence of a conservative force (spring force, gravitational force) a body has the potential energy E_p. For a gravitational force it is

$$E_p = mgh$$

where

m: mass of body,
g: acceleration due to gravity,
h: height of the body above the surface of the earth.

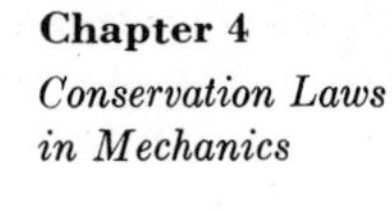

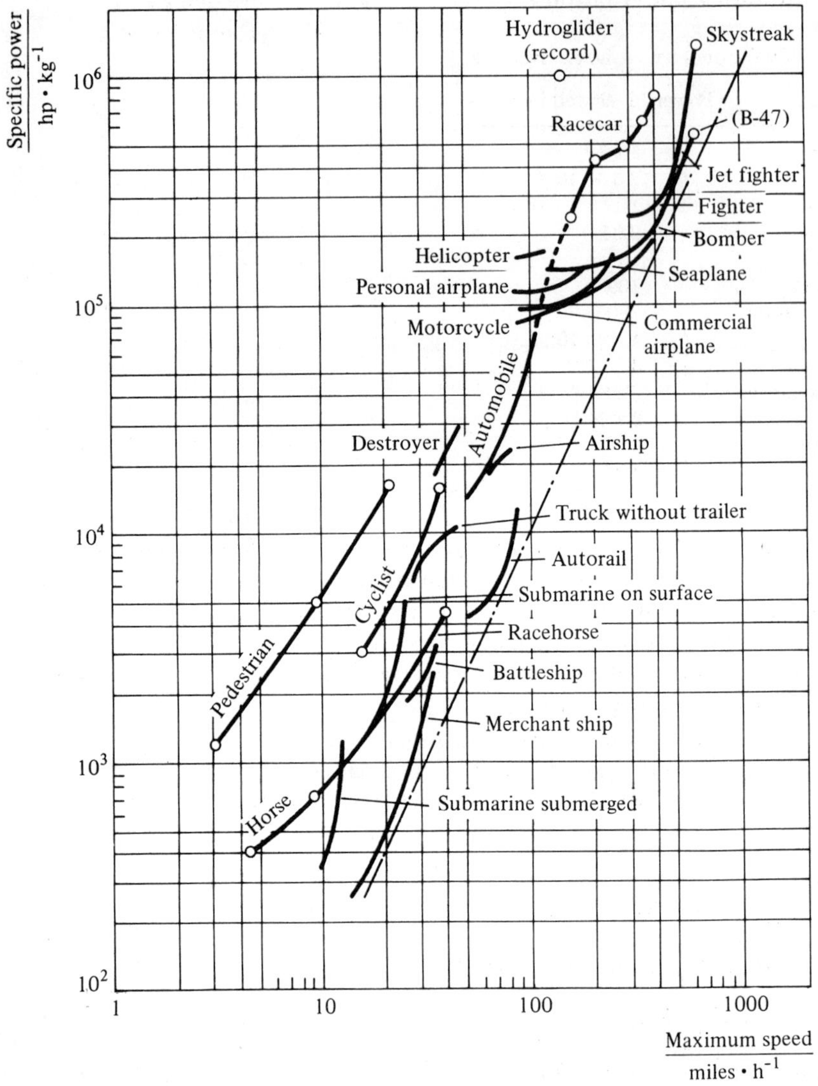

Figure 4-13 Specific power and maximum obtainable speed of various vehicles.

E_k and E_p are both measured in the same units as work.

Law of conservation of mechanical energy:

$$E_p + E_k = E_{\text{mech}} = \text{constant}$$

It is valid for bodies in a closed (isolated) system.

The momentum $\mathbf{p}$ of a body with a mass m and a velocity $\mathbf{v}$ is

$$\mathbf{p} = m\mathbf{v}$$

This momentum is conserved in a closed system.

The efficiency η is a measure of how much energy E_{in} can be converted into energy of another form E_{out}

$$\eta = E_{out}/E_{in}$$

Power is the rate of doing work:

$$P = dW/dt$$

It is measured in (joules/second) = watts.

The flow diagram in Figure 4-14 shows the interdependence of the quantities in this chapter.

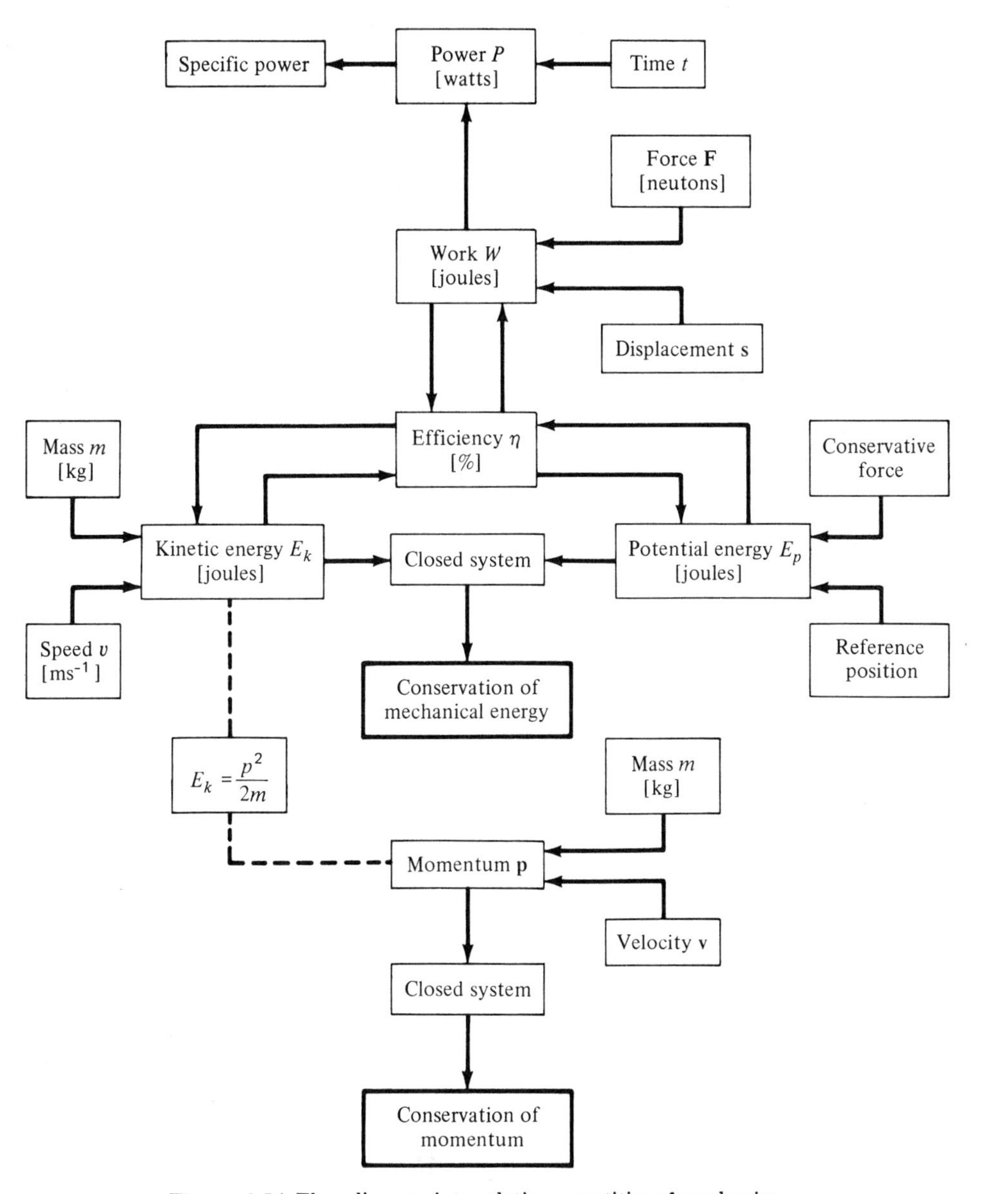

Figure 4-14 Flow diagram interrelating quantities of mechanics.

PROBLEMS

1. During an isometric exercise (e.g., pushing against a solid wall or pulling at a fastened rope), according to Equation (4.1), no work is done. But the performer nevertheless perspires, spends energy, and consequently works. Explain this apparent paradox.
2. Observing a weightlifting competition we notice that, especially in snatch, the athletes try to lift the weight as fast as possible. Why?
3. If a man is placed on a practically frictionless table, the speed of the blood leaving the ventricle during systole can be measured. How?
4. If the cyclist (see Section 2-3) does not want to exceed the speed limit (50 km/h), how much energy must he convert into heat by using the brakes until he reaches Old Faithful?
5. A body initially at rest is struck by a moving body of the same mass.

Figure 4-15

After the collision both bodies stick together and move as a single body (Figure 4-15). What is their speed?

6. A billiard play (Figure 4-16): Ball M_1 is played against ball M_2 (initially at rest). Both balls have the same mass; it is a non-head-on collision and spin effects are neglected. You may know from experience that after the collision both balls will move in different directions. The law of mechanical energy conservation applies. How large is angle φ? How does the picture change if it is a head-on collision?

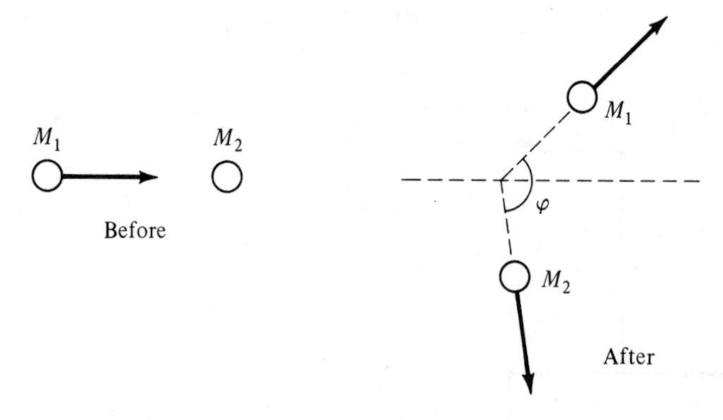

Figure 4-16

7. If a sea battle is shown in the movies, it is quite simple to discover whether or not it is a documentary or a staged representation. What will happen (even to a 42 000-t battleship) if a real broadside is fired?

8. What is the power output of your car measured in watts?

9. Take the maximum power output of your car, assume an efficiency of 22%, and calculate gasoline consumption per hour of driving. One liter of gasoline supplies 9×10^3 kcal.

10. Using the pressure-volume diagram of Figure 4-17, calculate the work done per heart beat.

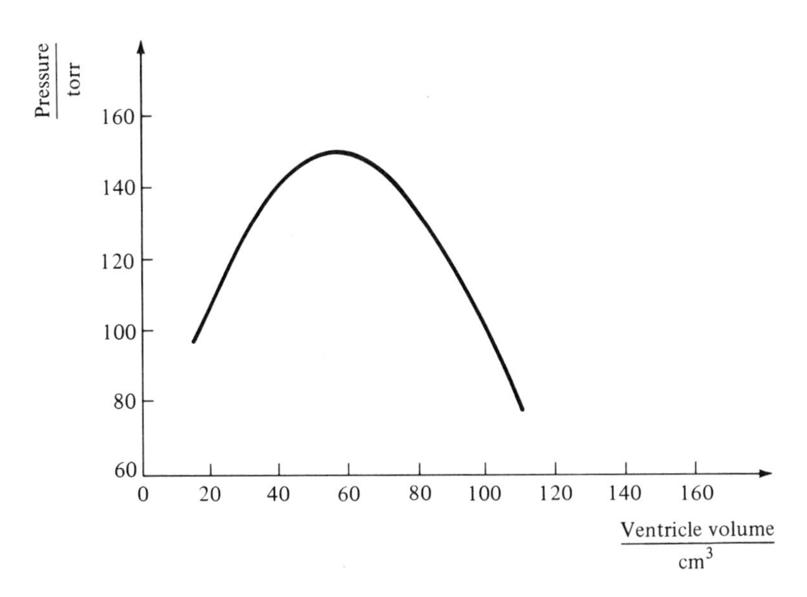

Figure 4-17

11. Larvae of chloeon (ephemeridae) and anisoptera (odonata) move by jet propulsion. They draw water into their hindgut and push it out suddenly. By using the following specifications, calculate their mass and their power output. Where do you place the larvae on Figure 4-13?
    peak velocity of larvae: 50 cm/s,
    ejection speed of water: 2.5×10^2 cm/s,
    cross section of jet: 10^{-4} cm^2,
    ejection time: 0.1 s.

12. Figure 4-18 shows the work done by the heart as a function of time. Plot the power exerted as a function of time.

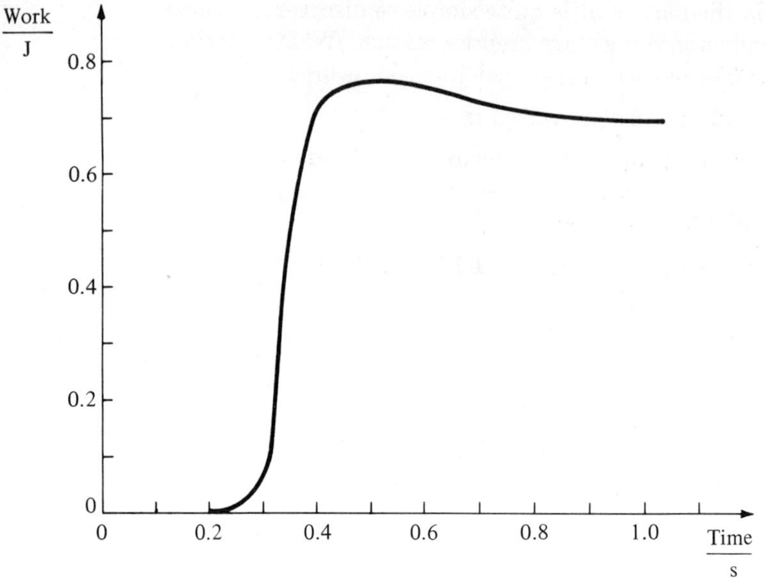

Figure 4-18

13. An elevator raises a 75-kg load to a height of 20 m within 10 s. (a) What is the power output of the lift-motor (in W and in hp)? (b) The power consumption of the lift-motor is measured to be 3 kW. What is its efficiency?

14. 5000 m^3 of water per second flows over a waterfall and drops 80 m. (a) How much work is done by gravity on the water? (b) How many kilocalories are dissipated per hour? (c) If the falling water drives a power plant at the bottom of the fall, how many horsepower will it deliver if its efficiency is 20%?

15. A ball having a mass of 0.4 kg and initially at rest drops from a height of 10 m to the ground and bounces to a height of 7 m. (a) How much work is done on the ball by gravity? (b) What is the speed of the ball striking the ground? (c) How much energy is spent during the process of bouncing? (d) What is the change in momentum during bouncing?

FURTHER READING

D. Bootzin, H. C. Muffly, eds., *Biomechanics*. Plenum Press, New York, 1969.

M. J. Lighthill, "How Do Fishes Swim?" *Endeavour*, **24** (1970) 77.

B. RICCI, *Physiological Basis of Human Performance*. Lea & Febiger, Philadelphia, 1967.

R. ROSEN, *Optimality Principles in Biology*. Plenum Press, New York, 1967.

J. MAYNARD SMITH, *Mathematical Ideas in Biology*. Cambridge University Press, New York, 1968.

A. SOMMERFELD, *Lectures in Theoretical Physics*, Volume 1, "Mechanics of Billiard." Academic Press, New York, 1950.

chapter 5

DEFORMABLE MEDIA

The static and dynamic behavior of fluids (liquids and gases) is difficult to treat due to its demanding mathematics. But those difficulties vanish if we are satisfied with a first-order approximation. This means that the quantitative agreement between theoretical elaborations and experiment is rarely better than 20–30%. The results derived under simplifying assumptions will nevertheless explain many phenomena which at first seem rather complicated. You will see, for example, that the wings of a bird not only act as lift-producing airfoils but also as propellers! A closer look at the wings will reveal a design much more sophisticated than any man-made flying machine.

Making use of hydrodynamics to understand the circulatory system of man yields useful results. In this chapter, we will see the limitations of applying physical laws to living systems.

Not only is the dynamics of fluids important in the life sciences, but its statics are also. The surface tension of water allows insects to walk on it. Reducing this tension by ejecting detergent-like secretions, some water beetles move around or even drown their pursuers.

The flow diagram on the next page shows the structure of this chapter.

1. STATICS

Fluids encompass the liquid and gaseous phases of matter. Similar laws govern both phases, so they will be treated together. The transition from one phase to the other is described in Chapter 7.

The attractive forces acting between individual molecules of a fluid are weak. Under the influence of external forces, fluids do not retain their shape (as do rigid or quasirigid bodies), but they assume the shape of their container. For liquids the internal forces are sufficient to retain a constant density (within the volume the liquid occupies) regardless of the shape of the container. For all practical purposes, the internal forces

Chapter 5

Deformable Media

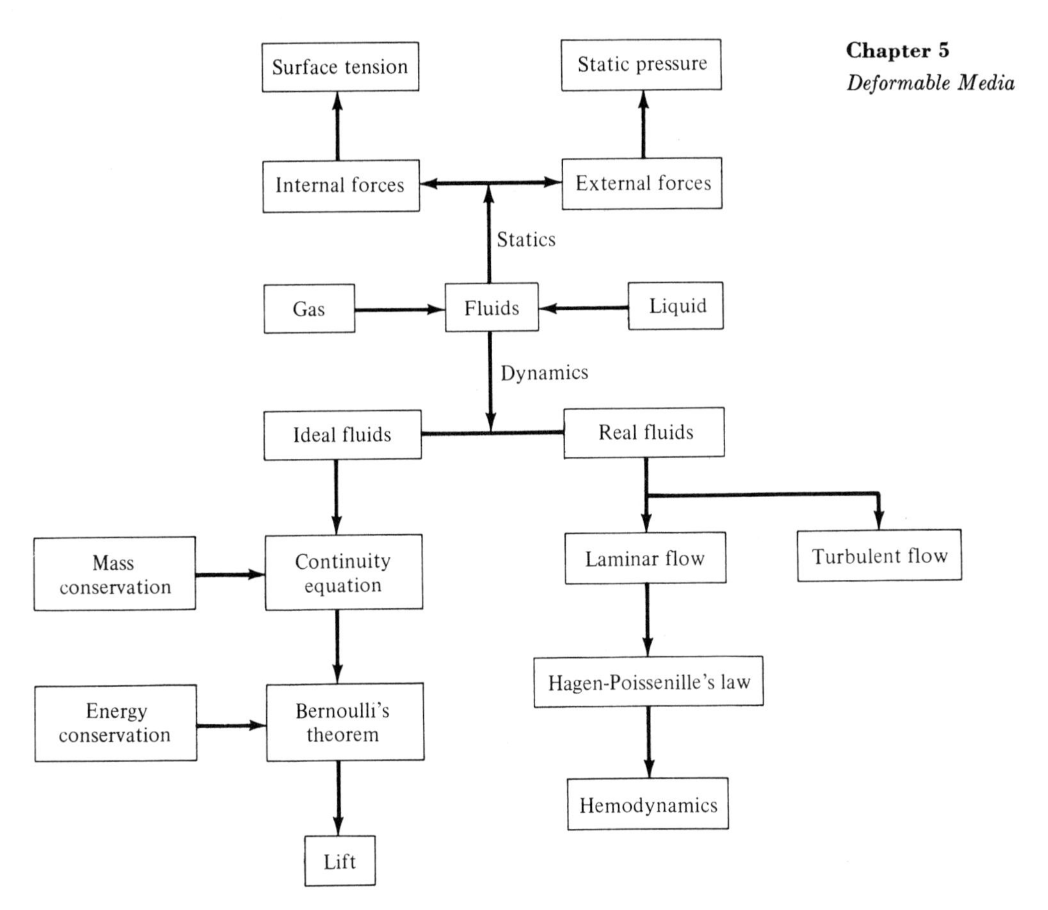

of a gas are negligible; in general, a gas will fill the entire volume of a container with uniform density.

1.1 Internal forces

1.1(a). Surface tension. The forces between the molecules inside a fluid are symmetric; there is no preferred direction. For molecules situated close to the surface, this symmetry is disturbed; therefore those internal forces have an inward resultant. See Figure 5-1. As a consequence, energy must be supplied to move a surface molecule out of the fluid.

The energy E_s required to remove all surface molecules out of range

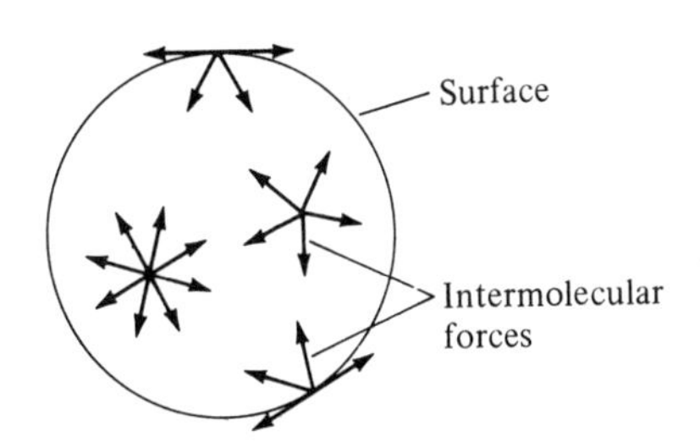

Figure 5-1

of the intermolecular forces is proportional to the surface area; hence

$$E_s = \sigma A \tag{5.1}$$

where

E_s: total energy required (surface energy),
σ: proportionality constant, called *surface tension*,
A: surface area of fluid.

The *surface tension* σ is defined as the surface energy per area:

$$\text{surface tension} \equiv \frac{E_s}{A} \tag{5.2}$$

It is measured in joules/meter^2.

Another expression for the surface tension is the magnitude of force F_σ exerted per length U of the interface

$$\sigma = \frac{F_\sigma}{U} \tag{5.3}$$

σ is expressed in newtons/meter. Since $N \cdot m^{-1} = J \cdot m^{-2}$, Equations (5.2) and (5.3) are two different expressions for the same quantity.

Conversion:

$$1\ J \cdot m^{-2} = 1\ N \cdot m^{-1} = 10^3\ dyn \cdot cm^{-1}$$

Table 1 shows some values for the surface tension for substances in contact with air.

Table 1 SURFACE TENSION

Fluid	Temperature in °C	Surface Tension in J/m^2
water	20	7.28×10^{-2}
water	100	5.9×10^{-2}
acetone	20	2.37×10^{-2}
ether	20	1.7×10^{-2}
mercury	20	0.47
platinum	2000	1.8
aluminum	700	0.84
copper	1000	1.1
lead	350	4.5

In general, the surface tension decreases with increasing temperature. Very small contaminations of substances such as lipoprotein or commercial detergents reduce the surface tension substantially.

Surface Tension of Alveoli. Each air passage inside the lungs is a closed tube limited by tiny outpouchings in the walls, the *alveoli.* Their diameters are about 20 microns (2×10^{-5} m). They are lined with an extremely thin layer of lipoprotein to achieve and keep a low surface tension which is necessary to keep the alveoli open. If the small amount of moisture present inside the alveoli determined the surface tension, its value would be about fourfold, and the alveoli would collapse during expiration.

EXAMPLE: SURFACE RIDERS

On a pond you can observe a number of insects moving right on the surface of the water. They are supported by the surface tension, and they move around like a mountaineer in climbing irons on a glacier—they dig their claws through the surface layer. If the surface tension is suddenly reduced (by a detergent for example) they must drown. A method used by some water beetles to escape their pursuer (a water spider) is to eject a liquid in the backward direction which suddenly reduces the surface tension. Unfortunately, the spiders usually launch frontal attacks!

1.1(b). CONTACT ANGLE. If a solid dips into a liquid or a liquid rests on a solid, the interfaces meet at an angle; see Figure 5-2. This contact

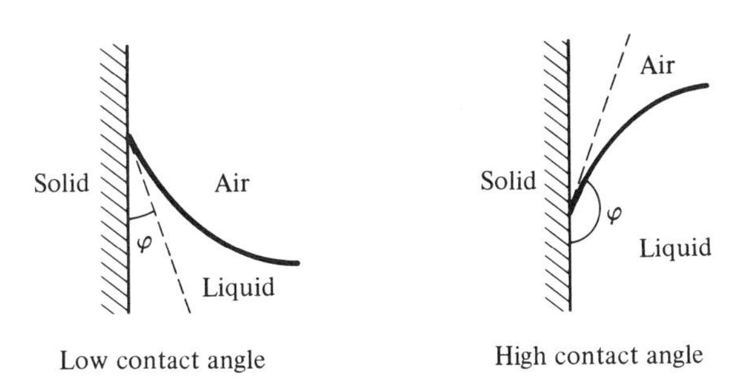

Figure 5-2

angle φ is determined by the combined intermolecular forces between solid and liquid.

A high contact angle means that the surface of the solid is wetted with difficulty. This hydrofuge property is utilized by larvae living under water with respiratory siphons penetrating the surface. See Figure 5-3. The hydrofuge property of the spiracle is caused by an oil secretion in the immediate neighborhood of the siphon. The spiracles can also be closed by hydrofuge hairs surrounding the opening. The hairs are separated by surface tension forces as long as the insect is close to the surface. Figure 5-4 shows such a natural valve.

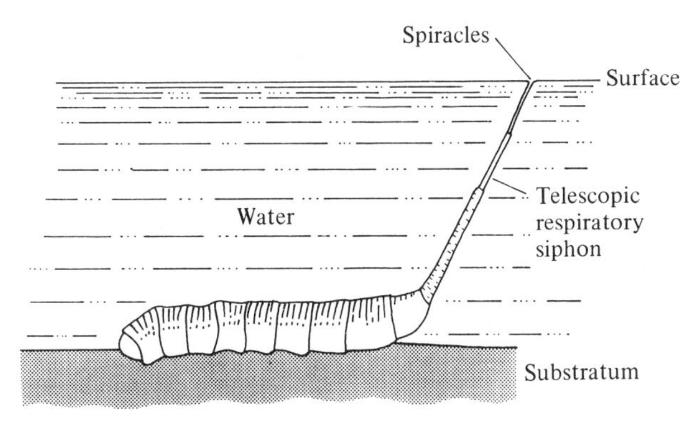

Figure 5-3 Larva of eristalis with respiratory siphon.

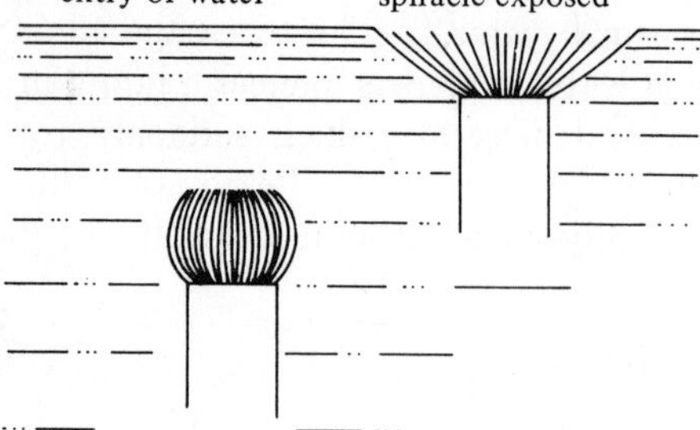

Figure 5-4 Movement of hydrofuge hairs around a spiracle of mosquito larvae. The movement of the hairs is entirely passive.

1.1(c). CAPILLARITY. A liquid will raise or sink in a tube of small cross section as a result of surface tension; this effect is called *capillarity*. Lifting of fluids by capillary action is important in nature; it is one of its means to transport fluids upward.

The magnitude of the force caused by surface tension is, according to Equation (5.3):

$$F_\sigma = \sigma U$$

where

F_σ: magnitude of force due to surface tension,
σ: surface tension,
U: length of interface.

The capillary force points upward if $\varphi > 90°$ and downward (causing capillary depression) if $\varphi < 90°$. See Figure 5-5.

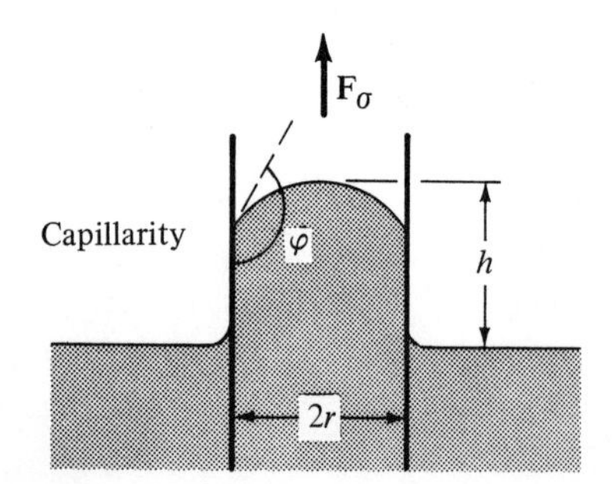

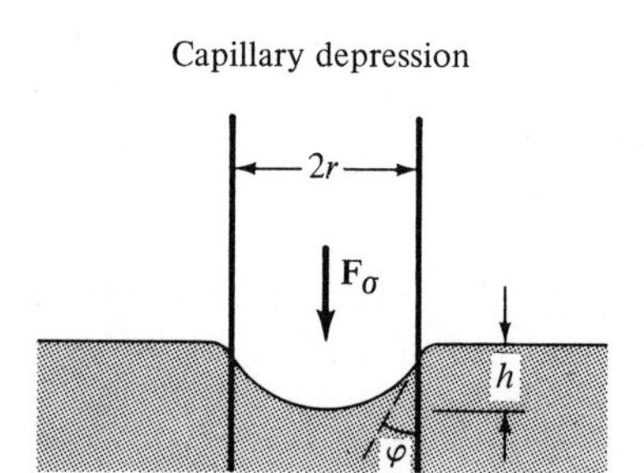

Figure 5-5

At equilibrium the capillary force $\mathbf{F}_\sigma$ has the same magnitude as the gravitational force $\mathbf{F}_g$ acting on the fluid inside the tube.

$$|\mathbf{F}_\sigma| = |\mathbf{F}_g| \tag{5.4}$$

Substituting $|\mathbf{F}_\sigma| = F_\sigma = \sigma U$

and

$$|\mathbf{F}_g| = F_g = mg$$

where

$$m = 2\pi r^2 h\rho$$

we get

$$\sigma U = mg = 2\pi r^2 h\rho g$$

Therefore

$$h = \frac{2\sigma}{\rho r g} \tag{5.5}$$

where

h: distance between the surfaces inside and outside the tube,
σ: surface tension,
ρ: density of fluid,
r: inner radius of capillary,
g: acceleration due to gravity.

There are many practical applications of surface tension. Part of the action of a detergent is to reduce the surface tension of the washing water. The resulting low contact angle allows the water to creep more easily into the minute gaps between the fibers and the dirt particles. As another example, the feathers of water fowl are covered with an oil film, resulting in a high contact angle and thus preventing wetting.

A CURIOUS METHOD OF LOCOMOTION. The surface tension of water not only allows certain bugs and the spider to walk on the water, but it is also used directly by a specialist for propulsion: The little water beetle, *Stenus*, produces a secretion in its glands at the behind. This secretion reduces the surface tension. The fluid drops on the water, the surface tension decreases, and part of the freed energy is converted into kinetic energy of the beetle. It is an emergency propulsion only, because the store of the secretion is rapidly exhausted. Later we shall elaborate on this insect. Strange as it may sound, it has something in common with relativistic electrons and supersonic airplanes.

1.2 External forces

Fluids are always under the influence of considerable internal forces. External forces, like gravity, may be present. The surface of a fluid is perpendicular to the direction of the resulting force acting on it.

EXAMPLE: ROTATING FLUID

Neglecting internal forces, the fluid shown in Figure 5-6 is under the influence of two external forces: the gravitational force $\mathbf{F}_g$ and the radial force $\mathbf{F}_r$. The resultant depends on the locality inside the fluid since $\mathbf{F}_r$ changes with distance from the axis of rotation. At the bottom there is no radial force; hence the liquid is horizontal there. The slope of the surface increases with the distance from the bottom. It will assume the shape of a paraboloid whose axis coincides with the axis of rotation.

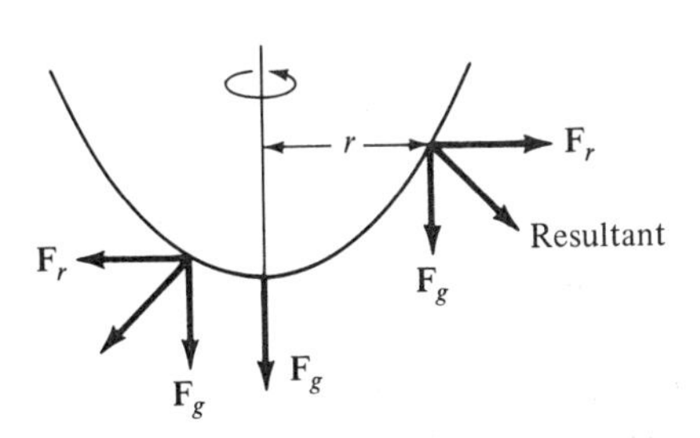

Figure 5-6

1.2(a). STATIC PRESSURE. A fluid exerts a pressure perpendicular to the walls of its enclosure. This static pressure p_s is caused by gravity and has the same magnitude at all points on the same horizontal level. It is independent of the shape of the container (Figure 5-7).

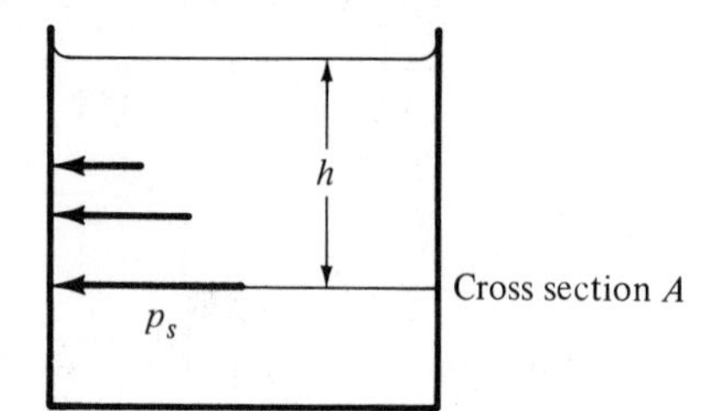

Figure 5-7

$$p_s = \frac{W}{A} \tag{5.6}$$

where

p_s: static pressure,
A: cross section of the container,
W: weight of the fluid on top of cross section A.

Since

$$W = \rho V g$$

where

g: acceleration due to gravity,
ρ: density of the fluid,
V: volume of fluid above cross section A.

The hydrostatic pressure at depth h is,

$$p_s = \frac{\rho V g}{A} = \rho h g \tag{5.7}$$

h is measured from the surface of the fluid.

EXAMPLE: BAROMETRIC PRESSURE

The weight of the earth's atmosphere causes an aerostatic pressure, called *barometric* (or atmospheric) pressure. At sea level and zero degree Celsius, the atmospheric pressure P_0 is

$$p_0 = 1.013 \times 10^5\ \mathrm{N \cdot m^{-2}} = 1.013\ \mathrm{bar} = 760\ \mathrm{mmHg}$$

Man does not notice this atmospheric pressure because his built-in sensors for detecting pressure (like Meissner's corpuscles) react to pressure differences only.

To calculate the atmospheric pressure at an altitude h, Equation (5.7) cannot be used directly because the density of air (ρ) also varies with altitude. The following relation holds

$$p_h = p_0 e^{-h/g} \tag{5.8}$$

where

p_h: barometric pressure at altitude h,
p_0: barometric pressure at sea level,
h: altitude, measured in km.

Figure 5-8 is a graphical representation of the above Equation (5.8).

A pocket barometer is sufficiently sensitive to determine the altitude of a climbed mountain.

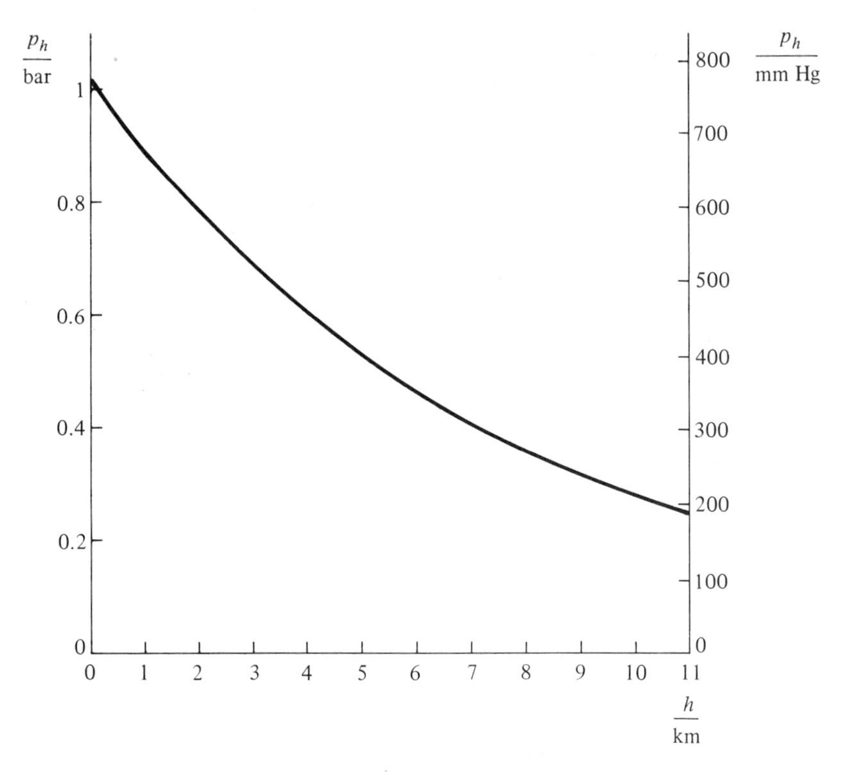

Figure 5-8 Atmospheric pressure p_h at 0°C as a function of altitude h.

EXAMPLE: HYDROSTATIC BLOOD PRESSURE

The hydrostatic component contributes substantially to the total blood pressure. In man the hydrostatic pressure in the foot artery is, using Equation (5.7),

$$p = \rho h g$$

where

ρ: density of blood ($\approx 1\ \mathrm{g/cm^3}$),
h: height of the blood column between the foot and heart (≈ 120 cm),
g: acceleration due to gravity ($= 981\ \mathrm{cm \cdot s^{-2}}$).

Hence

$$p = 1.18 \times 10^4\ \mathrm{N \cdot m^{-2}} = 88\ \mathrm{mm\ Hg}$$

This is a considerable value, about half of the total blood pressure. Adjusting for the changing hydrostatic pressure is very demanding for the regulatory systems in animals with long necks. The brain of the giraffe is about 2.5 m above the heart which consequently has to overcome a hydrostatic pressure of almost 250 mm Hg. If the animal bends down, this hydrostatic pressure drops rapidly; there must be a mechanism to prevent or compensate those rapid pressure drops.

1.2(b). Pressure-volume relations. If a fluid is subjected to an external force **F**, then we observe a uniform pressure exerted on the walls of the container. See Figure 5-9. The pressure caused by the external force **F** is equal at any location inside the container. However, the total pressure measured at the bottom of the container is largest due to the additional hydrostatic pressure caused by gravity. If we now increase the external force, the pressure will increase and the fluid compresses. Experiments show that this decrease in volume is proportional to pressure and original volume; therefore (assuming constant temperature):

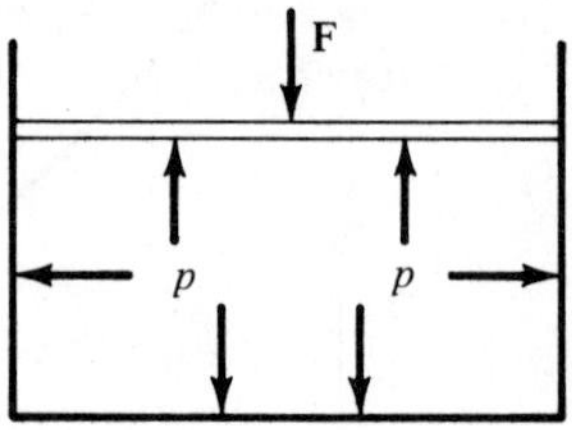

$$-\Delta V = \kappa V_0 \Delta p \tag{5.9}$$

where

$-\Delta V$: volume change of fluid (the minus sign indicates that the volume decreases),
κ: proportionality constant, called *compressibility*,
V_0: original volume of the fluid,
Δp: pressure increase caused by the increased force.

Figure 5-9

Liquids. The compressibility of liquids is very small. If the pressure is increased by a factor of 20, the volume will decrease just 0.1%. For practical applications, liquids can be assumed to be incompressible, thus making the hydraulic press possible. See Figure 5-10.

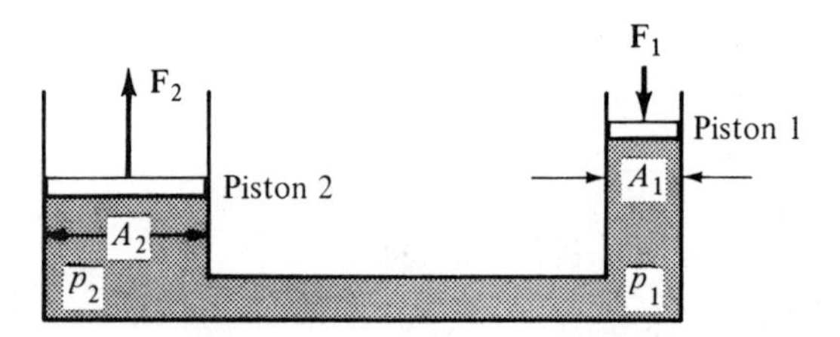

Figure 5-10

Force $\mathbf{F}_1$ with a magnitude F_1 acts on piston 1 having a cross section A_1; the pressure exerted is

$$p_1 = \frac{F_1}{A_1}$$

Since the pressure is uniform at any position inside the liquid, it has the same value inside the other container which is connected to the first one. Thus a force $\mathbf{F}_2$ is exerted on piston 2. Since

$$p_1 = p_2$$

it follows that

$$\mathbf{F}_2 = -\mathbf{F}_1 \frac{A_2}{A_1} \tag{5.10}$$

If the ratio A_2/A_1 is larger than 1, $\mathbf{F}_2$ will be larger than $\mathbf{F}_1$. Obviously, the hydraulic press is a force amplifier. Although force is gained at one side of the hydraulic press, the product of the force and the length of path traveled by the piston—the work done—is the same for both sides.

EXAMPLE: TRANSMISSION OF SOUND PRESSURE IN THE INNER EAR

Sound waves entering the ear are picked up by the eardrum. They are transferred via the ossicular chain and the oval window to the cochlear fluid; see Figure 5-11. The cochlear fluid in turn excites the basilar membrane which is the site of the auditory nerve endings.

The transmission of the sound pressure from the eardrum to the basilar membrane can be analyzed as a hydraulic press. The area of the eardrum is much larger than the area of the oval window, a gain in force of about 22 times results.

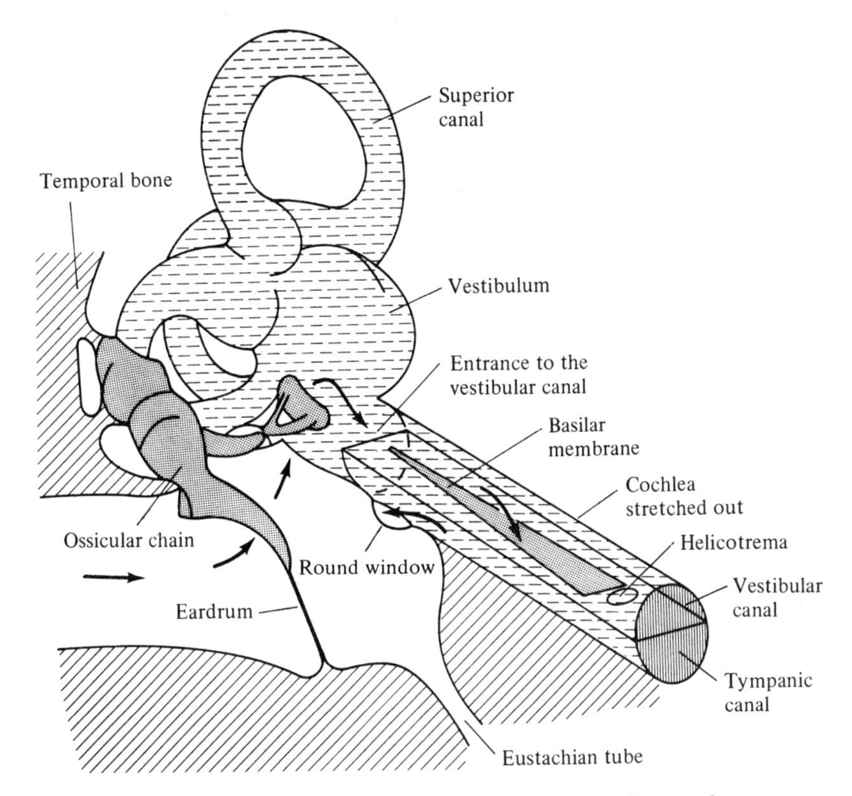

Figure 5-11 Diagram of the inner ear. Cochlea and basilar membranes are uncoiled.

EXAMPLE: PLEURAL GAP

The lungs are not fixed to the chest but are connected to the fibrous sack around it—the pleura—via the pleural gap (or pleural cavity). This is not really a gap but a thin film of liquid between the visceral layer, which adheres to the surface of the lungs, and the parietal layer which adheres to the walls of the chest and the diaphragm. Since the lungs are not fixed, they can easily move, and the thin layer of liquid in the pleural gap acts as lubricant. If the chest expands, a pull (or negative pressure) is exerted on the liquid film. It cannot expand because of its low compressibility (in this case a negative compressibility); thus it exerts a pull at the visceral layer which is fixed to the lungs. If air gets into the pleural gap, the lungs will collapse.

Gases. Compared to liquids, the compressibility of gases is large. If the temperature is constant, the relation between the volume of a gas and the pressure exerted on it, is expressed by the Boyle–Mariotte law:*

$$pV = \text{constant} \tag{5.11}$$

where

p: pressure exerted on the gas from outside,
V: volume of the gas at pressure p and constant temperature.

Ideal Gas. A gas is called an *ideal gas* if it obeys Equation (5.11). There are no ideal gases, but helium and nitrogen come rather close.

Real Gas. For real gases, Equation (5.11) was modified by van der Waals:†

$$\left(p + \frac{a}{V^2}\right)(V - b) = \text{constant} \tag{5.12}$$

where

a/V^2: a term to account for the attractive forces between molecules of a real gas,
b: a term proportional to the intrinsic volume of the gas molecules.

The transition between both laws is quite obvious:

$$\text{for} \quad \frac{a}{V^2} \to 0 \quad \text{and} \quad b \to 0$$

Equation (5.12) converts into Equation (5.11).

2. DYNAMICS

To investigate the properties of moving fluids (the fields of aerodynamics and hydrodynamics), we distinguish between real and ideal fluids. The difference is of little consequence in aerostatics and hydrostatics. Before we delve into the dynamics of fluids, we will first introduce the useful concept of the *streamline.*

2.1 Streamlines

A particle of a fluid starting at point A (see Figure 5-12) and passing through B traces out a path, the streamline. Every particle starting at A will follow this streamline and pass through B too. In general, particles following the same streamline have the same speed. The flow is then

* Robert Boyle (1627–1691), chemist and physicist. One of the founders of the Royal Society. More about him in L. T. More, *The Life and Work of the Honorable Robert Boyle*, Oxford, 1944.

Edme Mariotte (1620–1684), prior of a monastery. Investigated fluids and optics. Discoverer of the blind spot in the eye.

† J. D. van der Waals (1837–1923) physicist, Nobel laureat, 1910.

Figure 5-12

called *stationary flow*. Its velocity is indicated by arrows tangent to the streamline. The density of the streamlines (number of streamlines per cross section) is proportional to the speed of the fluid. See Figure 5-13.

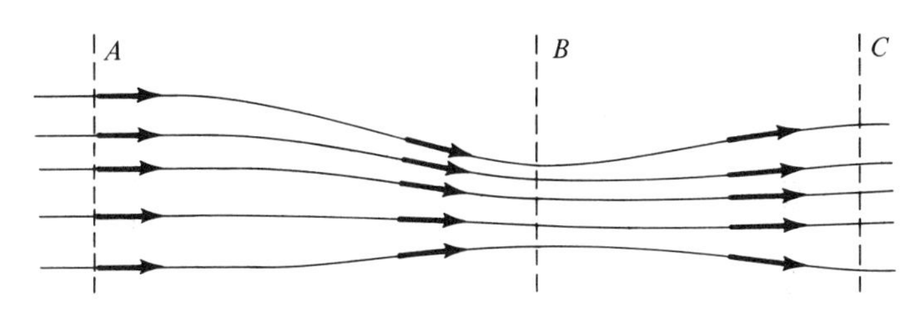

Figure 5-13

The velocity is high at B and low at A and C. The streamline concept is a familiar one from maps. Figure 5-14 displays the principal ocean currents in such a way.

Streamlines appear out of a *source* and disappear into a *sink*. A streamline closed in itself is called a *vortex* and has neither source nor sink.

Measuring the speed of a moving fluid inside nonliving systems is simple. A paddle-wheel arrangement is the most widely used method. But it is difficult to do this in the realms of life, *in vivo*. An ingenious method was developed to measure the blood speed in intact human forearms: Suitably placed (over the veins) magnets polarize water molecules in the blood. Only a small volume inside the vein is polarized. Further downstream, the arrival of those tagged molecules is detected (by means of a nuclear magnetic resonance technique). From the time difference and the separation between the tagging magnet and detector, the blood speed is determined.

2.2 Ideal fluids

An ideal fluid is characterized by the following properties:

1. No internal friction
2. Uniform density
3. Incompressibility

There are no ideal fluids. However, the laws derived for ideal fluids can be extended (with some caution) to real fluids. This extrapolation is the

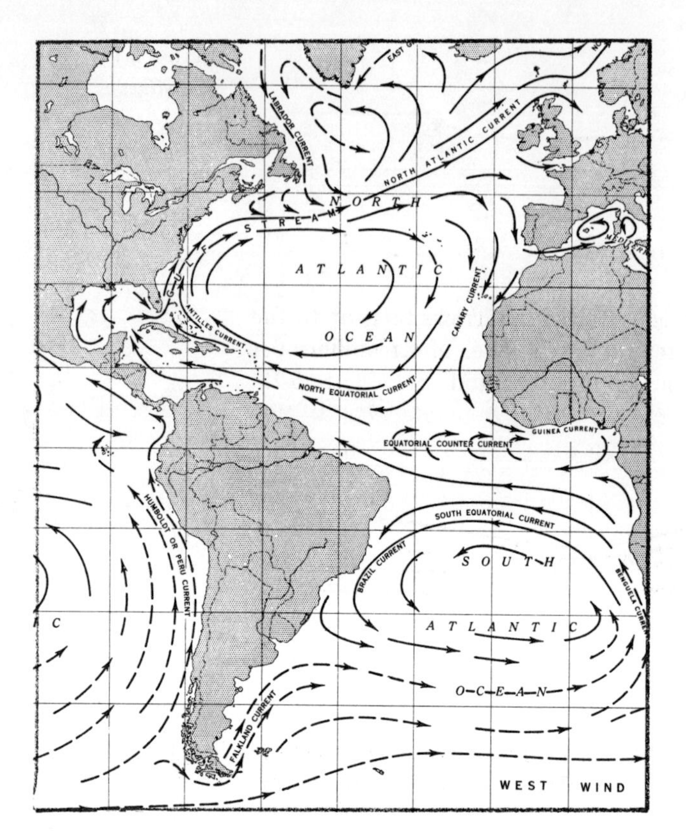

Figure 5-14 Ocean currents.

more necessary since the exact treatment of real fluids is excessively complex.

2.1(a). Equation of continuity. In an ideal fluid the same volume of fluid per time passes through any cross section. This is a consequence of the conservation of mass because no fluid will appear or disappear through the walls (no sources, no sinks). To express the same fact with different words: The number of streamlines within the containing tube is conserved.

The volume of fluid passing through a given cross section per time is

$$\frac{\text{volume}}{\text{time}} = Av$$

where

A: cross section of enclosure,
v: speed of fluid at this cross section.

The continuity equation (see Figure 5-15) is:

$$A_1v_1 = A_2v_2 = A_3v_3 = \text{constant} \tag{5.13}$$

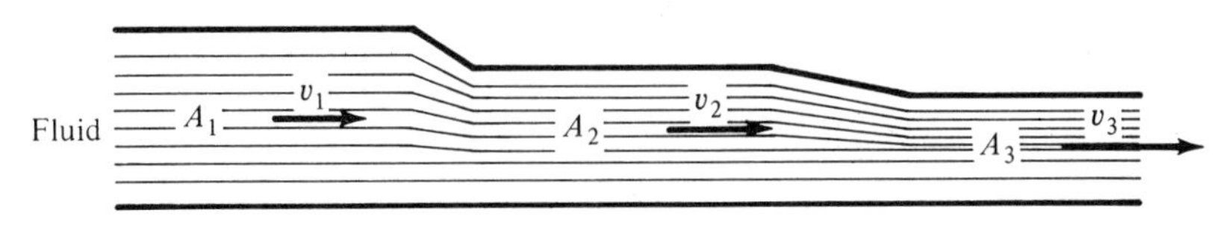

Figure 5-15

The speed of the fluid is large where the cross section is small, and vice versa.

Application: Cars moving on a road and having a uniform speed are on streamlines. In the vicinity of a road repair, the road narrows. To conserve a continuous flow, the speed of the cars should increase according to Equation (5.13). But, in general, the traffic signs enforce a lower speed, resulting in disruptive flow.

2.2(b). BERNOULLI'S THEOREM.* In ideal fluids only conservative forces are present; consequently the law of conservation of mechanical energy as formulated in Chapter 4 applies:

$$E_{\text{mech}} = E_p + E_k = \text{constant} \tag{5.14}$$

Replacing the quantities involved by their equivalents in fluid mechanics, Bernoulli's theorem results:

$$p = p_s + \frac{\rho}{2} v^2 = \text{constant} \tag{5.15}$$

where

p: total pressure of fluid,
P_s: static pressure,
ρ: density of fluid,
v: speed of fluid.

The term $\frac{\rho}{2} v^2$ is called *dynamic pressure.*

$$\text{total pressure} = \text{static pressure} + \text{dynamic pressure}$$

Deriving Equation (5.15) from Equation (5.14): According to Chapter 4

$$E_{\text{mech}} = mgh + \frac{m}{2} v^2 = \text{constant}$$

Observing that

$$m = \rho V \qquad (\rho = \text{density}, V = \text{volume})$$

and dividing both sides by V, we obtain

$$\frac{E_{\text{mech}}}{V} = \rho gh + \frac{\rho}{2} v^2$$

* Daniel Bernoulli (1700–1782), mathematician and physicist. Member of a famous family of scientists. The first to apply calculus to physical problems. He contributed mainly to hydrodynamics.

The left side of the equation is measured in $\mathrm{J \cdot m^{-3}}$, which is equal to $\mathrm{N \cdot m^{-2}}$; hence it represents a pressure. The same is valid for the right side.

Equation (5.15) is strictly valid only for fluids in a horizontal tube. For nonhorizontal flow we must take into account that the hydrostatic pressure also depends on the position of the flow with respect to the horizontal.

Since the total pressure is constant, the dynamic pressure is low where the static pressure is high, and vice versa. The streamline picture of a fluid shows that wherever the streamlines bunch together, the dynamic pressure is high and the static pressure is low.

EXAMPLE: BUNSEN BURNER

Although Bernoulli's theorem is derived for ideal fluids, it can be extended to real fluids in good approximation. The number of applications is vast. It encompasses the flight of airplanes and birds, as well as the spinning of a baseball. Later we shall demonstrate some important applications; for the moment let us be satisfied with a simple one, the Bunsen burner.* See Figure 5-16.

The total pressure of the streaming gas is constant. At the nozzle the cross section of the gas tube decreases. The streamlines bunch together and indicate a high speed in this region. At the nozzle the dynamic pressure is high (proportional to v^2) and consequently the static pressure is low. Air is sucked through holes into this region of low static pressure and supplies the needed oxygen for the burning process.

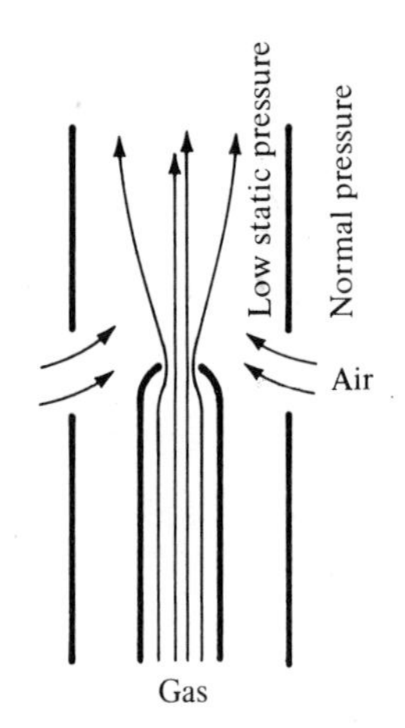

Figure 5-16 Cross section of a Bunsen burner nozzle.

2.3 Real fluids

Frictional forces are always present in fluids; fluids are also compressible to a certain extent. To treat the dynamics of real fluids with good approximation, we can neglect its compressibility but not its internal friction. For convenience, we distinguish two fundamentally different types of flow, *laminar* and *turbulent* flows.

2.3(a). LAMINAR FLOW. This flow is steady and orderly. The streamlines are parallel to the axis or wall. See Figure 5-17. The fluid molecules in contact with the walls stick to it, and friction occurs only between adjacent fluid layers. The physical characteristics of laminar flow are

* R. W. Bunsen (1811–1899), chemist. Together with R. Kirchhoff he developed spectral analysis.

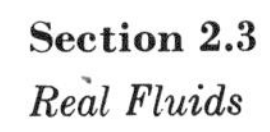

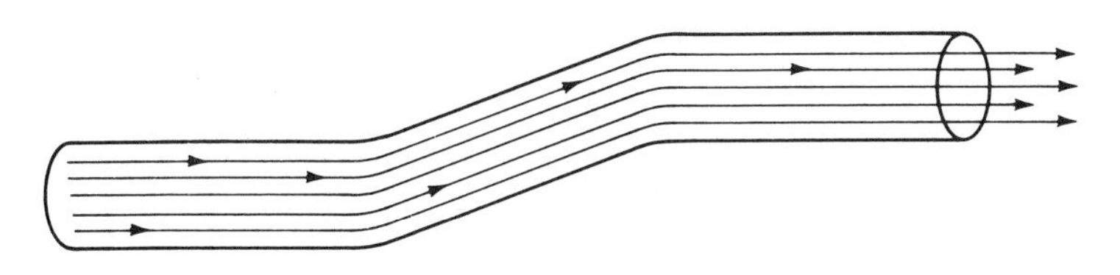

Figure 5-17

therefore determined by the internal friction of the real fluid and not by the material of the enclosure.

If the outermost layer is at rest but the fluid as a whole moves, then the velocity is highest at the center of the tube and zero at the walls. Figure 5-18 shows two ways to indicate this velocity distribution: *Left:*

Figure 5-18

The density of the streamlines is proportional to the velocity. *Right:* The arrow lengths are proportional to the velocity. They demonstrate a parabolic velocity distribution.

The flow per time passing through a rigid tube with laminar flow is given by Hagen–Poisseuille's law:*

$$i = \frac{\pi \Delta p r^4}{8\eta l} \tag{5.16}$$

where

i: volume of fluid passing through a cross section of the tube per time unit,
Δp: pressure difference between both ends of the tube,
r: radius of the tube,
η: viscosity of the fluid,
l: length of the tube.

Note that the flow is proportional to the fourth power of the radius of the tube.

Coefficient of Resistance. An obstacle in a streaming fluid offers a resistance, i.e., a counteracting force is necessary to keep the obstacle in its position. This resistance to the flow is determined experimentally be-

* Gotthilf Hagen (1797–1884), engineer of hydraulics. Discovered independently of Poisseuille the law in Equation (5.16). Investigated the transition between laminar and turbulent flow.

Poisseuille, see note on page 80.

cause, in most instances, it is too difficult to calculate. It depends on the type of fluid and on the shape and surface of the object. It does not matter whether the object moves through the fluid at rest or the fluid streams around a fixed obstacle. Always assuming laminar flow, only the *relative* motion between the fluid and obstacle matters.

The resistance is expressed as a coefficient of resistance. The lower this coefficient, the less force is necessary to move the object in the fluid. Many animals can change their coefficient of resistance, like a bird bracing its wings and tail to slow down its flight. The frontal part of swimming water beetles is well streamlined. If they want to stop quickly, they turn 90°, thus increasing their resistance tenfold.

2.3(b). TURBULENT FLOW. This flow is irregular. The streamlines are in general not parallel to a wall or axis; they may turn backward or spiral. Figure 5-19 shows the velocity distribution of turbulent flow. *Left:* Streamline picture. *Right:* The length of an arrow is proportional to the velocity at its position.

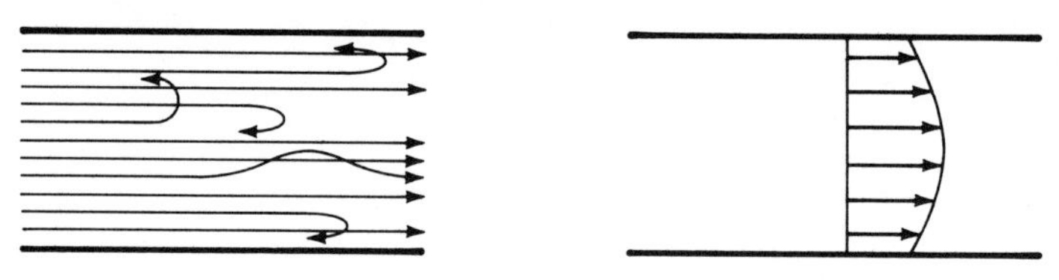

Figure 5-19

The velocity distribution near the center is more uniform; consequently the velocity gradient close to the wall is higher and the frictional losses are greater than in laminar flow. Examples of turbulent flow are the wake of a ship or airplane, air in the tracheo-bronchial tree during breathing, the flow around obstacles in rivers, and the flow inside a sharp bend of a water pipe.

EXAMPLE: DIAMETER OF THE AORTA

Presence of turbulence increases the energy necessary to maintain the flow. Nature designs systems with minimum energy requirements; consequently the blood flow is laminar. From this principle of optimal design, the diameter of the aorta can be calculated. It will be as small as compatible with nonturbulence. Computations agree within 20% of measured diameters of the aortas in various animals.

EXAMPLE: MEASURING THE BLOOD PRESSURE

A convenient and widely used method for measuring the arterial blood pressure in man is to close an artery by applying external pressure. This is performed at the upper arm with the help of an inflatable sleeve. The pressure is then slowly reduced until the arterial pressure exceeds the external pressure and the blood begins to flow again. At the moment when the artery opens, its diameter is very small. The resulting turbulent flow causes a gurgling noise which is observed by a stethoscope. Laminar flow is silent.

2.3(c). TRANSITIONS BETWEEN LAMINAR AND TURBULENT FLOW. The transition of turbulent into laminar flow is rare; we will not examine it. A laminar flow becomes turbulent as soon as a speed-dependent quantity called Reynolds'* number (symbolized R) exceeds a critical value.

This transition can be made visible in a simple way: One or more fine nozzles introduce a dye into the fluid. At laminar flow the dye forms a thin straight thread parallel to the axis of flow, the picture of a streamline. If the speed is increased step by step, we observe at a certain speed a sudden change. The thread becomes agitated, and the dye spreads over the entire fluid. Now the flow has become turbulent.

$$R = k\frac{v\rho r}{\eta} \tag{5.17}$$

where

R: Reynolds' number,
k: proportionality constant,
v: speed of the fluid,
ρ: density of the fluid,
r: radius of the enclosing tube,
η: viscosity of the fluid.

The proportionality constant k depends on the shape and size of the container. The exact value of R is usually determined experimentally.

Reynolds' number is of utmost importance in applied hydrodynamics. If we know its value, we can draw conclusions from scale models; it actually determines the scale.

EXAMPLE: DAMMING A RIVER

A dam will alter the flow of the river above and below. It is impossible to theoretically predict the changes this obstacle will cause; thus it is necessary to build a model of the project and study the hydrodynamic behavior of the planned outlay. The model is built to a scale that has the same Reynolds' number as the real project. This is possible because R depends on shape and dimensions (besides other quantities). With this scale model (the same scale is not used for all physical features) engineers study the hydrodynamic behavior of the changed river; water will actually flow. The model is altered at locations where unfavorable situations occur. This change is transferred to the real site in such a way that Reynolds' number for this feature is conserved.

The same technique is used in wind tunnel tests. The scale of the model airplane is not arbitrary but is determined by Reynolds' number.

3. APPLICATIONS

There are important applications of fluid dynamics. The ones in physics and engineering are obvious, so let us concentrate on some in the life

* Osborne Reynolds (1842–1912), physicist.

sciences. In this field, you will be surprised how well physical laws can be used to explain quantitatively features in the realms of life. However, although the formulas look impressive, the computed results in general contain an error of at least 20%.

3.1 Hemodynamics

Hemodynamics is the application of hydrodynamics to the circulatory system in man (or animal).

The blood flow is laminar through tubes of diameters ranging from 2.5 cm for the aorta to 10^{-3} cm for the capillaries. Hagen-Poisseuille's law [Equation (5.16)] applies, at least in the first approximation. The driving force is supplied by the contracting (systolic phase) and relaxing (diastolic phase) heart. The average blood pressure drops from 170 mmHg in the aorta to a few mmHg in the vena cava. Static and dynamic blood pressures are approximately equal except in the aorta, where the relative high speed of the blood leads to an increased dynamic contribution.

We shall consider only one aspect of hemodynamics, the flow speed of blood. Other aspects (e.g., energy and work done) have already been treated in Chapter 4.

Speed of Blood. The speed varies considerably in the different parts of the circulatory system. Rearranging Equation (5.16), we obtain the average speed of blood in veins:

$$\bar{v} = \frac{\Delta p A}{8\eta l} \tag{5.18}$$

where

$\bar{v}$: average speed of blood,
Δp: pressure difference between the beginning and end of the vein,
A: cross section of the vein,
η: viscosity of blood,
l: length of the vein.

Equation (5.18) yields only the correct order of magnitude because the assumptions which led to Equations (5.16) and (5.18) are not strictly fulfilled by veins. Let us look at those assumptions and how well they are fulfilled by the circulatory system. Here we encounter an enlightening example of the limitations in the application of physics to the life sciences:

1. *Steady flow* (the speed of fluid particles passing any given point does not change with time): Due to the pulsating cardiac action, the speed of blood changes with time. This is most pronounced in the aorta. Only at the arterioles and capillaries does the flow become steady.
2. *Laminar flow* (velocity distribution across the vessel is parabolic): The rhythmic pressure change leads to pressure waves in the walls

of the arteries closest to the heart. These pressure waves propagate faster than the blood. The net result is that the velocity distribution across the vessel is more uniform.

3. *Rigid tubes* (the diameter of the fluid-carrying vessel is time-independent): The circulatory veins are elastic. They expand to some degree with increasing pressure and contract with decreasing pressure. Figure 5-20 shows the speed of blood in rigid tubes and

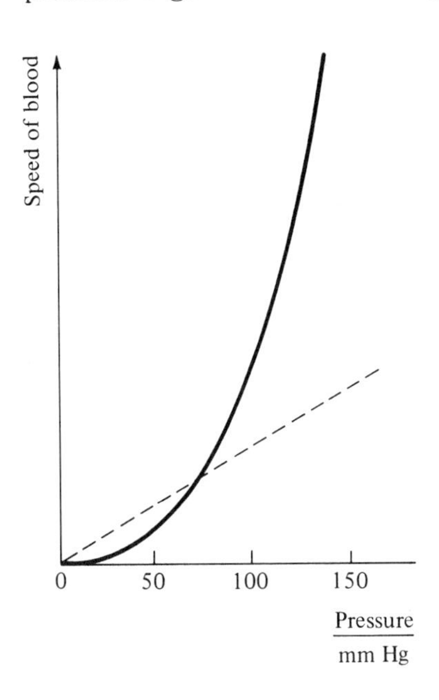

Figure 5-20 Speed of blood in rigid tubes (dashed line) and in elastic veins (solid line) as a function of pressure.

in elastic veins as a function of pressure. Obviously, the blood flow in an elastic vessel is not linear as Equation (5.18) demands. The elasticity of a vein can be determined by measuring the speed of the pulse wave.

Veins are not only elastic, but their diameters also depend on external influences. The peripheral veins (arterioles and capillaries) change their diameters according to the needs of the surrounding tissue. A small enlargement of the diameter will increase the blood flow considerably because blood flow is proportional to the fourth power of the radius. Therefore, if the diameter doubles, the flow will increase by a factor of 16 ($2^4 = 16$).

4. *Constant viscosity:* The viscosity of blood changes with its speed because blood is not a uniform fluid—particles are suspended in it. According to Equation (5.18), the pressure difference measured is not sufficient to drive blood through the capillaries if the normal value of the viscosity (5×10^{-3} centipoise at 20°C) is assumed. Actually the viscosity is much lower in the capillary region since the blood particles line up in single file.

Even with all these limitations, Hagen-Poisseuille's law is a useful tool for analyzing the circulatory system—if you realize its restricted applicability.

3.2 Lift and propulsion

Lift. An obstacle in a moving fluid causes a change in the direction and density of the neighboring streamlines. Consequently, static and dynamic pressures lead to a force acting on the obstacle. The magnitude of this force not only depends on the type and speed of the fluid, but also is critically dependent on the shape of the object. Figure 5-21 shows the

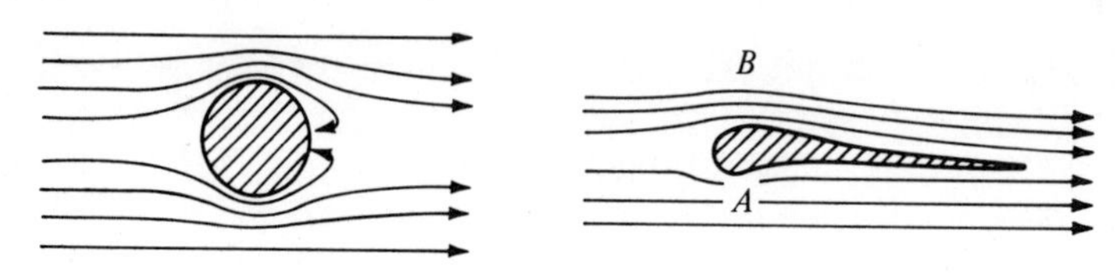

Figure 5-21

streamlines of a real fluid around two objects, a cylinder (left) and an airfoil (right). The streamlines above the airfoil (region B) are denser than below (region A), indicating a faster flow along the upper surface. This is a consequence of the continuity theorem, Equation (5.13), because the airfoil is shaped in such a way that the path on top is longer than the one below. To maintain continuous flow, the fluid particles above must flow faster than the ones below the profile.

Although Bernoulli's theorem is derived for ideal fluids, it is still valid in the first approximation for real fluids. Thus B is a region of low and A of high static pressure. The net result is an upward force, the lift. Figure 5-22 shows the lifting force at various positions on an airfoil.

The fluid layer (boundary layer) adjacent to the airfoil is of utmost importance for the magnitude of the lift. This boundary layer peels off

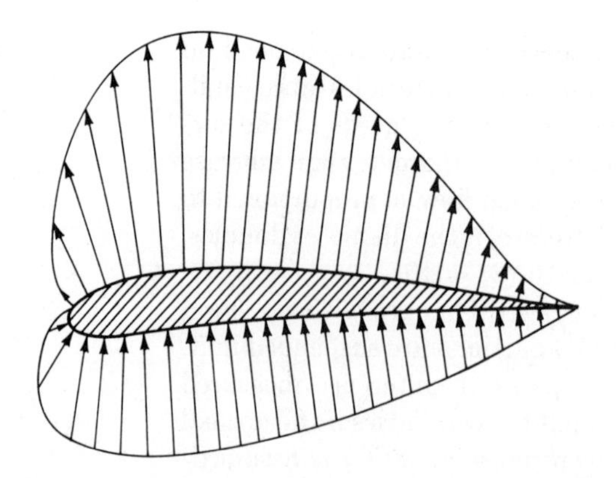

Figure 5-22 Contributions to lift from various parts of an airfoil. Arrow lengths are proportional to the lift produced at those positions.

toward the rear of the airfoil, and the streamlines form vortexes. The closer to the rear this separation occurs, the greater the lift. A turbulent boundary layer that sticks better to the surface than a laminar layer is thus highly desirable for a wing profile. There are a number of ways (high-lift devices) to produce a turbulent boundary layer. For example, a slot may be placed in the front part of the wing, parallel to the wing axis. The detailed theory of lift is extremely complicated, so much in fact that new airfoils are designed experimentally in a wind tunnel.

Propulsion. Propellers are used to propel airplanes and ships, objects moving in a fluid. The locomotive force produced by a propeller can be computed in two ways. We can treat it as a device which, due to the pitch of its blades, increases the speed of a certain amount of fluid and thus increases its momentum. According to the law of momentum conservation [Equation (4.14)] an equal momentum but of opposite direction is imparted to the object attached to the propeller. The other propeller theory treats it as an airfoil shaped in such a fashion that a low static pressure develops in front and a high static pressure develops behind the blades. In this way, a lift forward is produced. Of course, both treatments yield about the same result.

3.3 Flight of birds

Obviously the wings of a bird are designed as airfoils and produce lift as long as air flows past them. But this design is more sophisticated than appears at first sight. For example, lift can be increased by increasing the angle of attack (angle between the direction of flow and the airfoil). For small angles (at the stalling angle) the boundary layer separates from the wing surface, and this is accompanied by a sudden decrease of lift. The maximum possible angle of attack is increased by slots in the front and rear parts of the wing. Figure 5-23 shows those high-lift devices in bird and airplane wings.

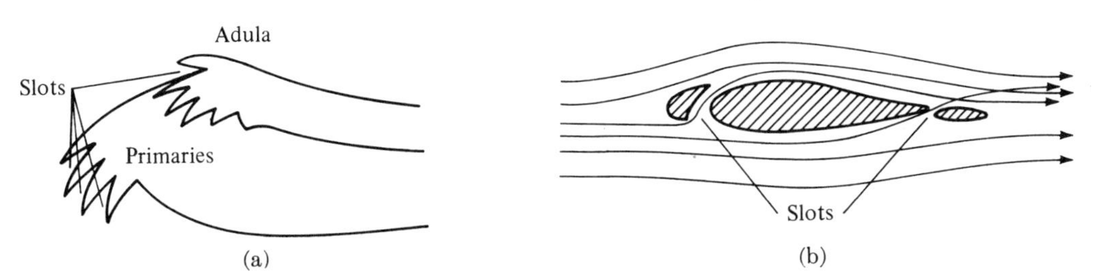

Figure 5-23 High-lift devices in (a) bird and (b) airplane wings. During ordinary flight, the feathers of the adula are held against the leading edge of the wing. At the stalling angle, the adula is spread forward to form a slot. By spreading the first few primaries apart, slotted wingtips are introduced.

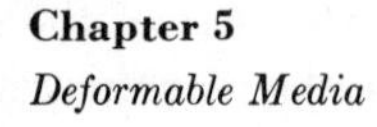

The forward propulsion of birds is achieved by means of a propeller! This is a very sophisticated propeller because it is built right into the wings; it is the hand section. The wingtips move through an arc, and they move faster than the arm section. High-speed photography reveals that during a downstroke the wingtips act as pulling propellers and during a recovery stroke they act as pushing propellers. Figure 5-24 gives an impression of the complex motion of the wings of a bird.

Figure 5-24 The hand part of the wing acts as a propeller, while the arm section maintains lift.

As in a modern airplane, a bird can adjust the pitch of its "propeller." Compared with a bird, the propeller-driven airplane is of crude design.

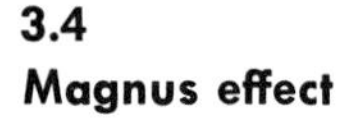

3.4 Magnus effect

If a sphere inside a moving fluid rotates around an axis which is not parallel to the direction of flow, a sideward force appears. This was discovered in the early days of gunnery when it happened that a ball fired from a howitzer hit its own gun! Magnus* used Bernoulli's theorem to explain this strange effect (and won a prize competition). See Figure 5-25.

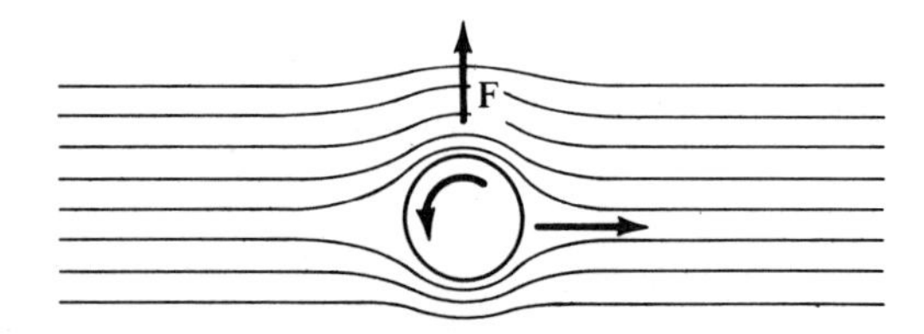

Figure 5-25

The speed between the fluid and the boundary layer is greater on top of the ball than under it. Since the total pressure is conserved, there is a low pressure area on top and a high pressure area below. A force results which is perpendicular to the direction of flow and the axis of rotation. The sphere will be driven upward. If the axis of rotation is suitable, the ball may actually return.

The deviation of the path of a "cut" ball in tennis or baseball is simply explained by the Magnus effect. The boomerang works on the same principle.

* Heinrich Gustav Magnus (1802–1870), chemist and physicist.

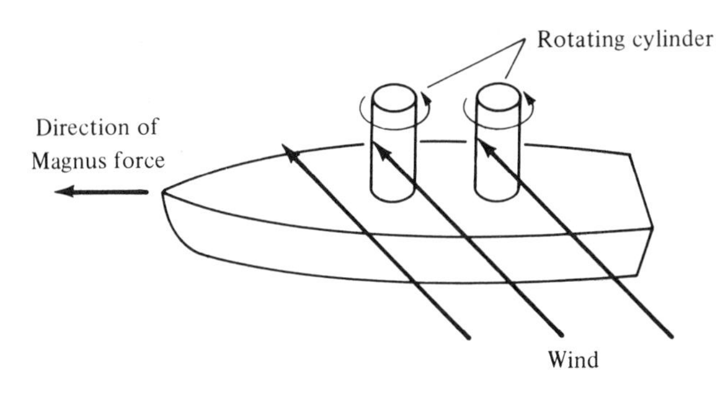

Figure 5-26

In the late 1920s, the Magnus effect was used to propel an experimental ship. See Figure 5-26. The rotors were driven by small auxiliary engines. It worked well, but the expensive ball bearings for the rotors made the construction economically unfeasible.

SUMMARY

Fluids encompass the liquid and gaseous phases of matter. If external forces are absent, the weak internal forces (mainly intermolecular forces) will cause a spherical shape of the fluid. At the boundary, the intermolecular forces are manifest as surface tension. This produces an effect similar to a fragile elastic film. Capillary action is a consequence of surface tension.

External forces such as gravity are the reason for static pressure exerted perpendicular to the walls of a container filled with a liquid. The aerostatic or hydrostatic pressure p_s caused by gravity only is

$$p_s = \rho h g$$

where

ρ: density of the fluid,
h: height of the fluid,
g: acceleration due to gravity.

Fluids change volume under external pressure. Gases have a high compressibility, and liquids have low compressibility. In aerodynamics and hydrodynamics, we distinguish between ideal and real fluids. Ideal fluids exhibit no friction, are incompressible, and are of uniform density. If we apply the law of mass conservation, the continuity equation results:

$$\text{cross section of container} \times \text{fluid speed at this cross section} = \text{constant}$$

As a consequence, a fluid inside a tube moves faster at a constriction and slows down at an enlargement.

The conservation of mechanical energy yields Bernoulli's theorem:

$$p = p_s + \frac{\rho}{2} v^2 = \text{constant}$$

where

p: total pressure of the fluid,
p_s: static pressure,
ρ: density of fluid,
$\frac{\rho}{2} v^2$: dynamic pressure of fluid.

Therefore, wherever the dynamic pressure is high (regions of high speed), the static pressure inside the fluid is low, and vice versa.

Real fluids are characterized mainly by internal friction. This causes a velocity distribution across the fluid. The velocity is zero for the layer adjacent to the walls and highest at the center. Laminar flow is steady and orderly; turbulent flow is irregular and agitated. The point of transition between both types of flow is determined by Reynolds' number.

PROBLEMS

1. Calculate the altitude for which the barometric pressure decreases to half its value at sea level.
2. Calculate the average density of air between sea level and 1000 m altitude.
3. What is the speed of water running out of a small hole on the side of a water tank? (Consider energy conservation.)
4. To calculate the hydrostatic pressure in the foot artery, we used the distance from heart to foot as height. Why not the distance from head to foot?
5. How large was the hydrostatic pressure on Beebe's diving sphere after he reached 923 m?
6. If a person rests on his head, how much does he increase the hydrostatic pressure in the brain?
7. Explain the equality of levels in commuting tubes.
8. How high can a water column be lifted with a siphon?
9. Two hollow semispheres of 1 m radius are placed together to form an airtight enclosure. It is evacuated to one-tenth of the atmospheric pressure. Calculate (a) the force sufficient to separate both spheres; (b) the same force if the sphere is submerged at a water-depth of 22 m.
10. Calculate the capillary depression of Hg in a tube with $r = 0.1$ mm.
11. What will happen to the arrangement in Figure 5-27 if a fluid passes through it?

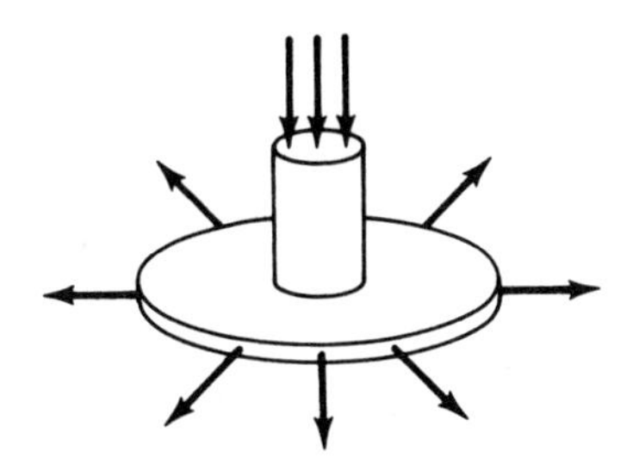

Figure 5-27

12. Calculate the speed of blood in the capillaries (radius = 0.004 mm, length = 0.05 cm, p = 10 mmHg).

13. Calculate the pressure drop over the length of the aorta (v = 40 $cm \cdot s^{-1}$, A = 5 cm^2, length = 30 cm).

14. A square (side length 2 cm) is shaped out of thin wire and submerged vertically into a liquid. If lifted out slowly a thin film will stretch across the square. The surface tension of the film opposes the lifting force. Calculate the minimum force to rupture the film for (a) water, (b) ether, (c) mercury. Note that the film has two surfaces.

15. A leaf does not fall edge downward, but approximately horizontal. Why?

FURTHER READING

C. H. BEST, N. B. TAYLOR, 8th ed. *The Physiological Basis of Medical Practice,* Chapter 36. Williams & Wilkins, Baltimore, 1966.

LOIS AND LOUIS DARLING, *Bird,* chapter "Flight." Houghton Mifflin Company, Boston, 1962.

O. C. MORSE, J. R. SINGER, "Blood Velocity Measurements in Intact Subjects," *Science,* **170** (1970) 440.

W. NACHTIGAL, "Hydrodynamics of Insects" in M. Rockstein, *The Physiology of Insects,* Vol. 2. Academic Press, London, 1965.

N. RASHEVSKY, *Mathematical Biophysics,* 3rd ed., Chapter XXVII. Dover, New York, 1960.

R. W. STACY, D. T. WILLIAMS, R. E. WORDEM, AND R. O. MCMORRIS, *Essentials of Biological and Medical Physics,* Chapter 30. McGraw-Hill, New York, 1955.

chapter 6

TEMPERATURE, HEAT AND THERMAL ENERGY TRANSFER

1. TEMPERATURE

Temperature is a nervous sensation. Our skin contains a large number of minute temperature receptors, the organs of Krause (cold) and the organs of Ruffini (warmth). They occur in all parts of the skin and are especially dense around the mouth and within the mouth cavity.

Within certain limits we can distinguish between hotter or cooler objects and order them in succession. For reproducibility we need instruments, called thermometers, to measure the temperature. Thermometers not only transfer the subjective nervous sensation of hot or cold into reproducible numbers, but they also extend the concept of temperature into ranges far outside the temperature receptors of the body. Thus we do not need to rely on a subjective nervous sensation to compare objects of various temperatures.

1.1 Temperature and thermometric scales

To measure the temperature, any adequate temperature-dependent property of matter may be used. It is convenient to choose the thermal expansion of fluids, especially alcohol and mercury. For special purposes other effects are employed, as we shall see in due course. The terminology *hot* and *cold* is dropped in physics; a number characterizes the temperature.

1.1(a). Customary temperature scales. Most thermometers rely on the thermal expansion of mercury. A thin thread of mercury inside a glass capillary extends out of a storage container. The expansion or contraction of the mercury due to the changing temperature is reflected in the height of the visible thread. A graded scale subdivides the capillary into convenient intervals. The subdivision is arbitrary. For practicability the difference between two (again arbitrarily chosen) fixed points is subdivided into intervals.

For a temperature interval, the name *degree* (abbreviated deg) is used. The customary temperature is expressed in *degree Celsius** (symbol: °C).

The two fixed points are:

temperature of melting ice: 0°C

temperature of boiling water: +100°C

Both points are determined at atmospheric pressure, i.e., at 760 mm Hg = 1.01×10^5 N/m².

The term *degree Celsius* was adopted in 1948 by international agreement and should be used instead of the expression *centigrade*.

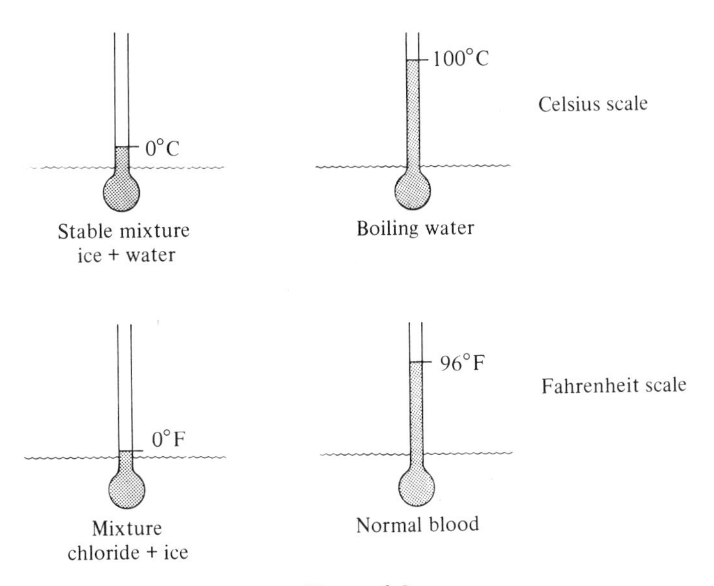

Figure 6-1

Figure 6-1 illustrates the two fixed points for the Celsius scale. Since the Fahrenheit† thermometric scale is still in use, its fixed points are

* Anders Celsius (1701–1744), astronomer. One of the first to subdivide the scale between the two fixed points into one hundred parts. He named 0 degree the boiling point of water and 100 degrees the melting point of ice.

† Daniel Gabriel Fahrenheit (1686–1736), physicist. The first to build usable thermometers. The reason he subdivided the difference between both fixed points into 96 intervals is that he measured with a duodecimal system and consequently divided the difference into $8 \times 12 = 96$ parts.

included. The scale is extended above and below the two fixed points, thus leading to positive (greater than 0°C) and negative (less than 0°C) temperatures. Where no mix-up is possible, the plus sign is dropped.

The bottom is reached at −273.15°C. A space traveler will encounter slightly higher values, since the lowest possible temperature is present only at some very remote places of the universe. There seems to be no upper limit for the temperature. Organic life is restricted to a tiny temperature interval of a few hundred degrees. Unaided, man can survive even smaller ranges. Table 1 shows some representative temperatures.

Although international agreement has abolished the degree Fahrenheit (symbol °F), it is still widely used. Conversion:

$$0°\text{C} = +32°\text{F}$$

$$100°\text{C} = +212°\text{F}$$

To calibrate thermometers, the boiling and melting points of materials such as sulfur, antimony, silver, and gold are used by the manufacturers. The procedures are specified by the National Bureau of Standards and are only of interest to the specialist.

1.1(b). THERMODYNAMIC (OR ABSOLUTE) TEMPERATURE SCALE. For scientific purposes it is convenient (formulas describing temperature-dependent effects are more clearly arranged) to choose as the reference point for the temperature scale the lowest possible temperature, the so-called *absolute zero.*

The thermodynamic temperature is measured in the basic unit *degree Kelvin* (symbol °K).* Its fixed points are:

$$0°\text{K} \equiv -273.15°\text{C}$$

and

$$273.15°\text{K} \equiv 0°\text{C}$$

The conversion between customary and thermodynamic temperature is:

$$t = T - T_0$$

where

t: customary temperature
T: thermodynamic temperature
$T_0 = 273.15°\text{K}$

1.2 Temperature-dependent properties

Practically everything in the universe, may it be animated or not, is influenced by changes in its temperature. Only the observed temperature range determines whether or not this temperature dependence can be easily observed and thus employed as a thermometric indicator.

* William Thomson, Lord Kelvin (1824–1907), physicist. Founded in 1854 the thermodynamic temperature scale. He also investigated electric oscillations.

Table 1 Some representative temperatures

(All values are in °C)

absolute zero	−273.15
melting point of hydrogen	−253
melting point of nitrogen	−190
freezing point of gasoline	−150
dry ice	−78
melting point of mercury	−38.9
melting point of ice, definition	0.00
triple point of H_2O	0.0075
hybernating ground squirrel	1.7
melting point of butter	32
average body temperature of man	37
average body temperature of birds	42
boiling point of water, definition	100.00
melting point of sugar	160
campfire	800
melting point of gold	1 063
gas-stove fire	1 600
filament of an electric bulb	2 300
melting point of tantalum	3 955
surface of sun	7 300
center of earth	14 000
exploding wire (electric discharge)	20 000
critical temperature for D-T reaction	10^7
critical temperature for D-D reaction	10^8

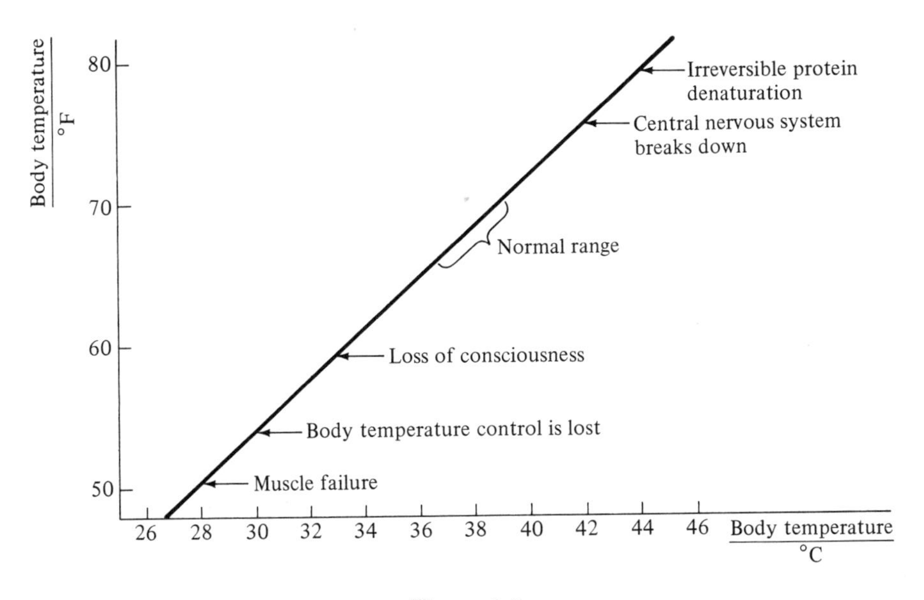

Figure 6-2

EXAMPLES:

The chemical reaction speed is a function of temperature and thus it is not surprising that enzyme activity is too. (See Figure 6-2.)

Fireflies are not observed at temperatures around zero degree Celsius. As tropical insects they show the light of their lanterns only at higher temperatures. The light intensity curve of the underlying bioluminescent system (Figure 6-3) could be used as a thermometer. It is not an unequivocal one.

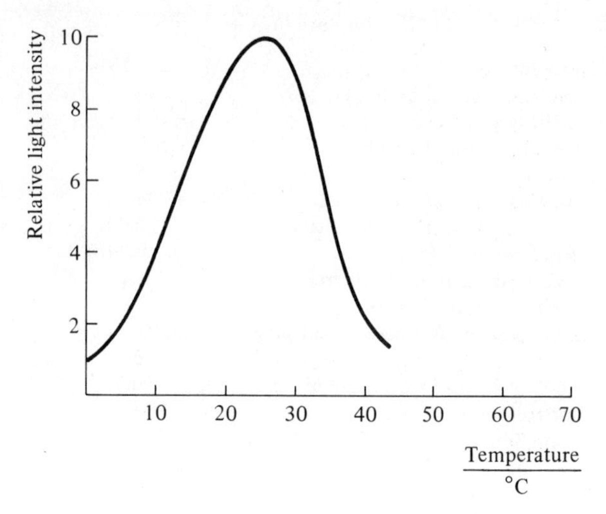

Figure 6-3 Light intensity of the firefly.

These are examples of thermometric effects within familiar temperature ranges; they are easily comprehended. But look at Figure 6-4, which displays the temperature dependence of the electric resistance at very low temperatures. As a rule, the electric resistance of metals is

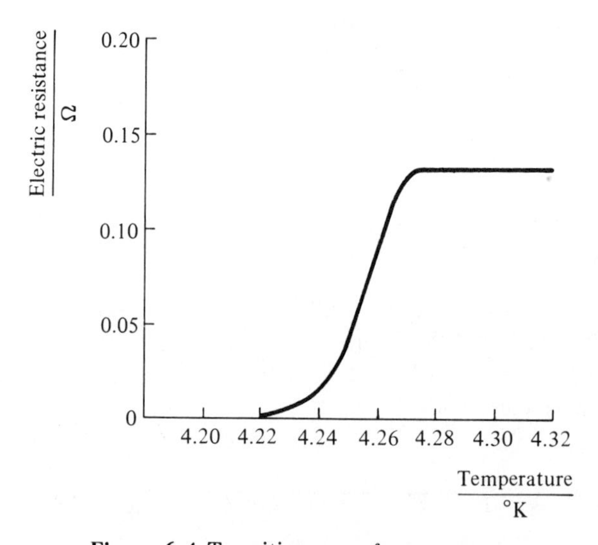

Figure 6-4 Transition curve for mercury.

proportional to its temperature. But at very low temperatures the resistance drops suddenly to exactly zero—a phenomenon called *superconductivity*. This sudden transition, which begins at a few degrees Kelvin, was discovered as late as 1911.

Technological development points to using this superconductivity. The electric resistance and its resulting energy loss is of importance if very large electric currents are conducted. Power generators and fusion reactors come to mind. Unfortunately, the cost of keeping electric conductors at the required low temperature is formidable. Maybe there is a way out: The theory of superconductivity seems to be well understood; the Nobel Prize for physics in 1972 was awarded for that to Bardeen, Cooper, and Schriefer. It might be possible to synthesize materials which are super-conductive at room temperature!

1.2(a). THERMAL EXPANSION OF MATTER. In the first approximation, temperature and volume of matter are related by

$$V = V_0(1 + \gamma t) \tag{6.1}$$

where

V_0: volume at 0°C,
V: volume,
γ: cubic expansion coefficient expressed in deg^{-1},
t: temperature expressed in °C.

The pressure—of significance only for the description of thermal expansion of gases—is assumed to be constant. In general, the expansion coefficient γ is positive.

The temperature-dependent expansion (or contraction) of all matter—in its gaseous, liquid, or solid state—takes place in all three dimensions. Thus an expansion coefficient refers to expansion in volume. But in many cases the expansion is restricted (e.g., in a rigid capillary containing a thermometric fluid) to one or sometimes two dimensions. This restriction leads to the (linear) expansion coefficient α. It is about one-third the value of the cubic expansion coefficient.

For special applications, an area expansion coefficient is introduced; its value is two-thirds the value of γ.

For the same material, it is sometimes necessary to state a different expansion coefficient for each of the three spatial directions. Crystallized materials are an example.

Gases. A gas trapped within a container with expandable walls (to achieve a constant pressure) will follow the volume-temperature relation [Equation (6.1)] well. This relation was first established by Guy-Lussac* and consequently bears his name.

For ideal gases (and most real gases) experiments show that

$$\gamma \approx 0.003\,7 \text{ deg}^{-1}$$

* L. J. Guy-Lussac (1778–1850), chemist and physicist. He generalized various experiments to formulate the law for the expansion of gases. He also investigated the high atmosphere (4000–7000 m) by balloon and wrote the first monography on a chemical element (Iodine). More about him in E. Farber, *Great Chemists*, New York, 1961.

which we may rewrite as

$$\gamma \approx \frac{1}{273 \text{ deg}}$$

Substituting this into the formula of Guy-Lussac [Equation (6.1)] we get

$$V = V_0\left(1 + \frac{t}{273 \text{ deg}}\right) \tag{6.2}$$

(It is permissible to change from an approximation ($\approx$) to an equation because this is a first approximation only.)

According to Equation (6.2), the volume of a gas vanishes as soon as the temperature reaches absolute zero (0°K). Naturally this cannot be correct since atoms and molecules have an intrinsic volume. The only possible conclusion is that Guy-Lussac's law is invalid for extremely low temperatures. This is another example of the fact that all laws are valid only within a certain range. Applying a law without knowing its limits of validity may lead to absurd results.

Knowing the relation between pressure p and volume V of a gas [Boyle-Mariotte's law, Equation (5.11)], it is simple to derive the law which governs the pressure as a function of temperature for constant volume.

$$p = p_0(1 + \gamma t) \tag{6.3}$$

where

p: pressure of gas at temperature t,
p_0: pressure of gas at temperature 0°C,
γ: cubic expansion coefficient expressed in deg^{-1},
t: temperature expressed in °C.

Equation (6.3) may also be written:

$$p = p_0\left(1 + \frac{t}{273 \text{ deg}}\right)$$

Liquids. The numerical value for the thermal expansion coefficient of liquids is, in general, smaller than the one for gases but larger than the one for solids. Table 2 presents a few examples. For comparison, γ for ideal gases and a representative solid are included.

Table 2 CUBIC THERMAL EXPANSION COEFFICIENT FOR LIQUIDS

(All values are in deg^{-1})

ideal gas	0.003 7
acetone	0.001 5
ether	0.001 7
glycerine	0.000 5
mercury	0.000 18
water (about 20°C)	0.000 21
iron (0°C to 100°C)	0.000 036

Of utmost interest in the life sciences is the thermal expansion coefficient of water. Only because this coefficient shows an abnormal variation with temperature, life in ponds and lakes is possible. We can even say that such life depends on this slight anomaly!

In studying density in Chapter 3, you already encountered a graph for the density of water (Figure 3-28). From this figure it is easily understood why a lake freezes from the top on downward and not the other way around. This remarkable variation of the density of water also leads to the formation of a *thermocline* in an open water body. This is a discontinuity layer in which the temperature decreases rapidly with depth. Figure 6-5 shows such a thermocline.

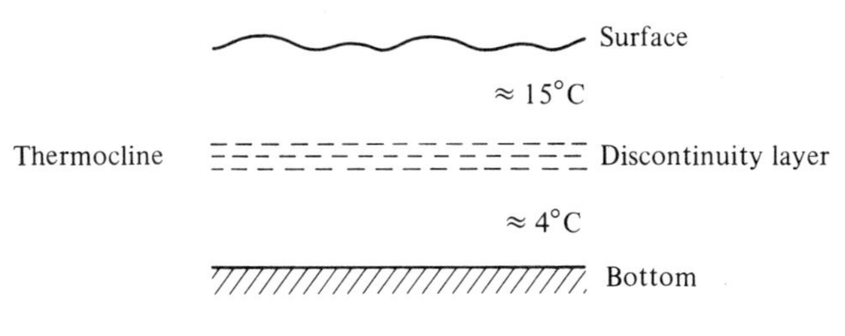

Figure 6-5

There is very little exchange of water or solutes through a thermocline. The actual position and thickness of this discontinuity layer depends mainly on the amount of heat coming from the surface. If this comparatively steep temperature gradient over a limited range didn't exist, the temperature of the entire lake would be low, thus preventing many forms of life.

Another example of a density anomaly is the alloy known as type metal (58% Pb, 26% Sn, 15% Sb, 1% Cu). Contrary to the general behavior, its solid phase has a lower density than its melted (liquid) phase. Thus, if it is cast, it will expand while cooling and fill the mold precisely.

Solids. The thermal expansion coefficient for solids is much smaller than the coefficients for gases and liquids. Table 3 shows a few examples. For comparison, γ for a gas and a liquid are included. Realize that the thermal expansion coefficients themselves depend on temperature.

In technology a thermal expansion is constantly kept in mind, since the forces exerted by expanding materials are formidable. Take iron for example: a cube of 1 cm^3 heated from room temperature to 100°C will, if prevented from thermal expansion, exert a pressure of about a thousand atmospheres.

Bimetal Strip. In most instances the thermal expansion (or contraction) of solids leads to undesirable consequences; the bimetal strip, however, uses it to advantage.

Two metal strips are riveted together. The thermal expansion coefficients are different, thus the length of each strip will be different if the temperature changes. Consequently the strip bends and the degree

Table 3 CUBIC THERMAL EXPANSION COEFFICIENT FOR SOLIDS AT TEMPERATURES AROUND 50°C

(All values are in deg^{-1} and should be multiplied by the factor 10^{-6})

ideal gas	3700
water	210
asphalt	600
bismuth	40
diamond	3.5
Jena glass	25
ice	112
marble	40
porcelain	11
quartz	20
rock salt	120
silver	58
tin	69
rock crystal:	
parallel to axis	25
perpendicular to axis	450
fir:	
parallel to fiber direction	20
perpendicular to fiber direction	100
oak:	
parallel to fiber direction	160
perpendicular to fiber direction	22

of bending is a function of temperature. In the bimetal strip in Figure 6-6, the expansion coefficient for material *A* is larger than the one for *B*.

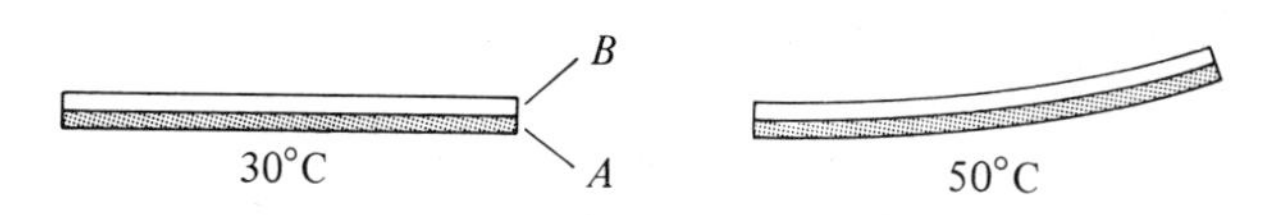

Figure 6-6

These strips are reliable and enduring thermoswitches used, for example, in refrigerators, air conditioners, and temperature-controlled ovens.

1.3 Thermometer

Any effect which depends on temperature can be used as a thermometer effect after proper calibration. Before we consider a few thermometers together with their ranges, an important but often overlooked factor must be considered: time.

Time Factors. It takes time until the thermometer substance is in thermal equilibrium with the object to be measured. This delay depends mainly on the amount of thermometric substance. As you will learn in Chapter 7, the thermometric mass must be very small compared with the object's mass. In most cases, the temperature to be determined is not time-independent; thus the response time of the thermometer must be short compared with the temperature variation. The slowly responding room thermometer is sufficient for a proper reading inside but will not indicate a warm puff of air passing by from a fireplace. It is neither sensitive nor fast enough. A thermocouple would detect the warm air.

There are three conditions a working thermometer meets:

1. Sufficient sensitivity over the working range
2. $(\text{Mass})_{\text{thermometer}} \ll (\text{mass})_{\text{object}}$
3. $(\text{Response time})_{\text{thermometer}} \ll (\text{time change})_{\text{object}}$

Table 4 shows some typical ranges and response times for thermal indicators.

Table 4 SOME TEMPERATURE-DETECTING DEVICES

Device	Approximate Range in °C	Response Time in s
receptor of pit vipers	$20 \rightarrow 40$	0.1–0.3
expansion thermometer	$-50 \rightarrow +1000$	10–1000
bimetal strip	$-100 \rightarrow +500$	10–100
mercury-thermometer; other liquids	$-50 \rightarrow +300$	10–100
thermocouple	$-20 \rightarrow +400$	10^{-2}
thermoneedle	$-20 \rightarrow +400$	10^{-3}
$^{16}O/^{18}O$	$4 \rightarrow 40$	million years
infrared cell	$0 \rightarrow 2000$	10^{-4}
clay cones in kiln	$700 \rightarrow 1400$	10–100

1.3(a). A BIOLOGICAL TEMPERATURE RECEPTOR. A unique organ developed in vipers, which puzzled naturalists for many centuries. It is so conspicuous that it appeared even in ancient Egyptian drawings. Vipers have between their eyes and nose a deep hole closed by a thin (about one-hundredths of a millimeter thick) membrane which is interwoven by a large number of fine nerve endings. This organ reacts to temperature changes of as small as 3×10^{-3} °C.

With the help of this organ, the viper can discriminate a small warm-blooded animal like a mouse in the dark. Experiments show that it can perceive a warm spot over a distance of one meter with sufficient accuracy to strike the prey. The detailed functioning of this most interesting organ still awaits discovery.

1.3(b). PALEOTEMPERATURES. There are three stable isotopes of oxygen: ^{16}O, ^{17}O, ^{18}O, the first has an overwhelming abundance of 99.78%. If any molecule containing oxygen (e.g., CO_2, H_2O, O_2) reacts with other molecules or atoms, the formation speed of new molecules depends on many factors. Here we are only interested in its dependence on temperature and mass.

Some sedimentary rocks such as marble and limestone are formed if Ca^{++} and CO_3^{--} ions react to $CaCO_3$. This takes place in the sea. Other factors remaining unaltered, the reaction speed will depend on the temperature of the sea. It also depends on whether the carbonate partner contains ^{16}O or ^{18}O. With the isotope ^{18}O, the reaction speed will be a trifle slower than with ^{16}O. This difference becomes more marked as the temperature increases.

It seems to be well established that the ratio of ^{16}O to ^{18}O (99.78% to 0.2%) is time-independent. Consequently, if we can determine the relative amount of ^{16}O to ^{18}O in the same piece of sedimentary rock, we are able to calculate the temperature at the time of rock formation.

All that was worked out in detail by Urey* in 1947. The analysis of sedimentary rock proves that over the last 150 million years the average temperature dropped by 5–10°C. Comparatively short temperature oscillations are superimposed—the ice ages.

2. HEAT AND ASSOCIATED QUANTITIES

2.1 Quantities

2.1(a). HEAT. Heat is thermal energy being transported from one place to another because of a temperature difference. The net transfer of heat ends as soon as all matter has equal temperature throughout. This state is called *thermal equilibrium.*

In the life sciences, the quantity of heat is measured in calories, abbreviated cal.

DEFINITION: *One calorie raises the temperature of* 1 *gram of water from* 14.5°C *to* 15.5°C.

The unit calorie was chosen out of practical considerations and is still widely used. In principle it is not necessary to use a special unit since heat is a form of energy—thermal energy—and the appropriate metric unit is the joule. (Recall that energy is interpreted as stored work and is thus measured in units of work. See Chapter 4.)

Conversions:

$$\begin{aligned} 1 \text{ calorie (cal)} &= 4.186 \text{ joules (J)} = 4.186 \times 10^7 \text{ ergs} \\ &= 4.186 \text{ newton}\cdot\text{meters (N}\cdot\text{m)} \\ 1 \text{ joule (J)} &= 0.239 \text{ cal} \end{aligned}$$

Although electrical units are introduced later, here are the conversions:

$$\begin{aligned} 1 \text{ cal} &= 4.186 \text{ watt}\cdot\text{second (W}\cdot\text{s)} \\ &= 1.16 \times 10^{-6} \text{ kilowatt}\cdot\text{hour (kW}\cdot\text{h)} \end{aligned}$$

2.1(b). SPECIFIC HEAT CAPACITY. The *specific heat capacity* is the quantity of heat required to raise the temperature of 1 gram of matter by 1°C. It is abbreviated *c* and measured in calories per gram and degree temperature difference.

* Harold Clayton Urey (born 1893), chemist. Nobel Prize winner, 1934.

Conversion:

$$1 \frac{\text{cal}}{\text{g} \cdot \text{deg}} = 4.186 \frac{\text{J}}{\text{g} \cdot \text{deg}}$$

There is no special unit for the specific heat capacity.

Note: The specific heat capacity depends on the temperature. As the temperature approaches zero degree Kelvin (or $-273°\text{C}$), the specific

Table 5 NUMERICAL VALUES FOR THE SPECIFIC HEAT CAPACITY

(All values in cal/(g·deg))

water	1.000
aluminum	0.22
copper	0.092
silver	0.055
mercury	0.033
diamond	0.12
air	0.23
heavy water (D_2O)	1.02
ice (at $-10°\text{C}$)	0.53
wax	0.69
concrete	0.7
steam (at 100°C)	0.48
ether	0.52
methyl alcohol	0.57
asbestos	0.19
cellulose	0.37
glass	0.20
leather	0.36
porcelain	0.26
synthetic rubber	0.45
wood	0.42
tissue of man	0.85

heat capacity also approaches zero. However, at room temperature c may be regarded as constant to a good approximation.

c_p and c_v:

Transferring thermal energy to matter usually leads to an increase in temperature, which in turn may cause an expansion or a rise in pressure. A different amount of energy is required for either effect. Thus the specific heat capacity will have different numerical values, depending on whether pressure or volume is kept constant.

Since solids and liquids are highly incompressible, the experiments yield only slight variations for their values of c. For gases this is different and we must distinguish between specific heat capacity at constant volume (abbreviated c_v) and specific heat capacity at constant pressure (abbreviated c_p). c_p is always larger than c_v.

2.1(c). MOLAR SPECIFIC HEAT CAPACITY. The *molar specific heat capacity* (abbreviated C) is the thermal energy (heat) required to raise the temperature of one mole of matter by one degree. There is no special unit for the molar specific heat capacity.

The following relation holds for ideal gases:

$$C_p - C_v = R \qquad (6.4)$$

where

C_p: molar specific heat capacity at constant pressure,
C_v: molar specific heat capacity at constant volume,
R: molar gas constant $= 8.314\ \text{J}\ {}^\circ\text{K}^{-1}\cdot\text{mol}^{-1}$
$= 8.314 \times 10^7\ \text{erg}\cdot{}^\circ\text{K}^{-1}\cdot\text{mol}^{-1}$
$= 1.987\ \text{cal}\cdot\text{mol}^{-1}\cdot\text{deg}^{-1}$

2.1(d). LATENT HEAT. *Latent heat* is the quantity of heat required to change the phase of matter while keeping its temperature constant. Latent heat is abbreviated L and measured in calories per gram. There is no special unit for it.

We distinguish:

Transition from solid to liquid: latent heat of fusion (L_f)
Transition from solid to gas: latent heat of sublimation (L_s)
Transition from liquid to gas: latent heat of vaporization (L_v)

Table 6 LATENT HEAT OF FUSION L_f AND LATENT HEAT OF VAPORIZATION L_v FOR SOME MATERIALS

(All values in cal/g)

	L_f	L_v
benzene	30.3	94.3
water	79.7	596.0
bismuth	12.6	
nitrogen	6.1	47.6
mercury	2.8	70.6
oxygen	3.3	50.9

During transition,

$$\text{liquid} \rightarrow \text{solid}$$
$$\text{gas} \rightarrow \text{liquid}$$
$$\text{gas} \rightarrow \text{solid}$$

thermal energy (heat) is released. If, for example, ice forms in a lake, the water temperature decreases more slowly than before. More details about phase transitions are introduced in Chapter 7.

2.1(e). HEAT CAPACITY. Up to now physics seems to be so well defined in terms and quantities that it could serve as an example worthy of imitation for other sciences. Investigating heat capacity, however (which would be better named *thermal energy capacity*), we find that there are various explanations for it. We shall examine one explanation. A worthwhile exercise is to list other explanations and form your own opinion. Since we shall not use the term *heat capacity* outside this chapter, no confusion is possible. But keep in mind that other terms such as *specific heat capacity* are internationally agreed upon and well defined.

The heat capacity of a substance is its thermal energy content and should consequently be measured in the appropriate unit, joules. Since 1 joule = 0.239 calorie, we can express the thermal energy in calories as well.

As long as the temperature of this substance is greater than 0°K, its heat capacity is greater than zero. The numerical value is the product of specific heat capacity, mass, and temperature (measured in °K). Because the specific heat capacity is temperature-dependent, this leads to an integration:

$$\text{heat capacity} = m \int_0^T c\, dT \qquad (6.5)$$

where

m: mass of the substance,
T: temperature, expressed in degree Kelvin,
c: specific heat capacity.

The temperature dependence of c is determined experimentally; thus the integration has to be a graphical one. The numerical result will give the energy of the substance which is stored in the form of thermal energy. In principle, this energy can be released and transferred.

For practical purposes we are satisfied with the thermal energy released by sudden temperature changes, not approaching the neighborhood of absolute zero. Then we can rewrite Equation (6.5):

$$\text{thermal energy released} \equiv m \int_{T_i}^{T_f} c\, dT \qquad (6.6)$$

where

T_i: initial temperature, expressed in °K,
T_f: final temperature.

For $\Delta T = T_i - T_f$ not too large, the specific heat capacity c is constant and we get:

$$\text{thermal energy released} = mc\, \Delta T \qquad (6.7)$$

expressed in calorie (or joule).

EXAMPLE:

The released heat capacity of a 1000-g copper block if cooled from 350°K (77°C) to 300°K (27°C) is

$$1000 \times 0.39 \times 50 = 1.95 \times 10^4 \text{ joules}$$

or

$$4.56 \times 10^3 \text{ calories}$$

This thermal energy will raise the temperature of 1 liter (1000 g) of water by

$$\Delta T = \frac{1.95 \times 10^4}{1000 \times 4.19} = 4.6 \text{ degrees}$$

Note that it is more difficult to change the temperature of water than of other nongaseous substances. This fact can be read off the table for the specific heat capacities (Table 5).

This was a lengthy excursion. As was stated before, we shall not employ the term *heat capacity* anymore.

2.2 Calorimetry

2.2(a). General. The amount of thermal energy (heat) which moves from a warmer substance to a cooler one until both systems are in thermal equilibrium can be calculated if the specific heat capacities, the masses, and the initial temperature differences are known. By this method the specific heat capacity of an unknown substance is determined in a mixing calorimeter. Also, if a substance is burned within an isolated container, the thermal energy produced can be determined. The experimental method is straightforward as the following example shows:

EXAMPLE: MIXING CALORIMETER

m_1 grams of the unknown substance with a temperature T_1 is dropped into a calorimeter (an insulated vessel) containing m_2 grams of water. The initial temperature of the water is T_2, and a final temperature T_f at thermal equilibrium is observed. What is the specific heat capacity c_1 of the substance?

Energy conservation yields:

$$m_1 c_1 \, \Delta T_1 = m_2 c_2 \, \Delta T_2 \qquad (6.8)$$

with $\Delta T_1 = T_1 - T_f$

$\Delta T_2 = T_2 - T_f$

c_2 = specific heat capacity of water

An example with

$m_1 = 700$ g

$m_2 = 1000$ g

$T_1 = 500°\text{C}$ $(= 773°\text{K})$

$T_2 = 30°\text{C}$ $(= 303°\text{K})$

$T_f = 94°\text{C}$ $(= 367°\text{K})$

yields $c_1 = 0.226$ cal/(g·deg).

A precise calorimetric experiment is not easy to perform. The entire system must be insulated against any thermal exchange with the outside, and the thermal equilibrium for all parts of the final mixture must be assured. Ideally, time factors do not enter, but in practice it may take a long time until thermal equilibrium is reached.

Cooking by Prehistoric Man. The mixing method described above was employed by prehistoric man for cooking. In the very early days he had neither pot nor pan to heat water over the campfire. So he stuck to roasted meat until he learned to dig a pit in waterproof soil. He filled the pit with water and dropped heated stones into it. Thermal energy transferred from the hot stones to the cold water. If the temperature of the stones was sufficiently high and they were not too few, the water would boil for a while. The specific heat capacity of water is three to four times larger than that of most stones, indicating that he had to heat up many stones beforehand.

Ice or Water to Make Coffee? As an example of applied calorimetry, in winter the outdoors man gets water from a hole in the ice to make his coffee instead of the seemingly easier task of melting ice. To heat one liter (1000 cm^3) of water to 100°C, 10^5 calories are needed. To melt the same amount of ice, the latent heat of fusion must be supplied before the temperature of the water starts to rise. In this case an additional 7.97×10^4 calories are needed, indicating that the man must wait almost twice as long to get his refreshment.

2.2(b). Food calories. The energy content of food is expressed as thermal energy and listed in calories. Since the life-sustaining processes depend predominantly on oxidation, this makes sense.

To find the thermal energy content of food, it is burned in a calorimeter under controlled laboratory conditions. From the measured values and the actual intake of food, the energy consumption of man can be calculated. Figure 6-7 shows the arrangement of a bomb calorimeter on the left; on the right is an actual calorimeter.

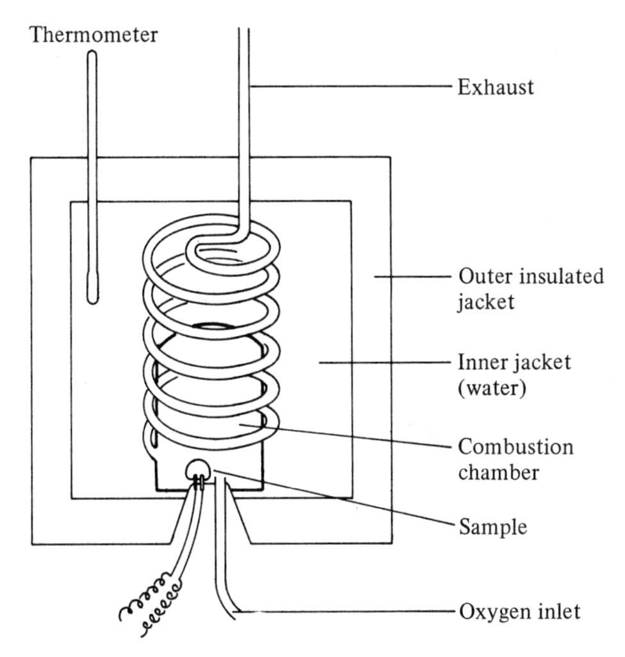

Figure 6-7 Arrangement of a bomb calorimeter (left) and an actual calorimeter (right).

There is a more direct way to determine the energy turnover of living objects—the *calorimetric chamber:* A person lives inside a thermally controlled area. All input (food, liquids, air) and output (waste, liquids, air) are carefully monitored. The thermal energy transferred from the test object to the enclosing walls (and outgoing substances) is measured. As you can imagine, those chambers are complicated, uncomfortable, and expensive.

In general, indirect methods—like the oxygen consumption method—are employed to measure the caloric intake. Both methods agree within 3%. This relatively high accuracy is necessary, because even a 10% increase of the caloric turnover in man indicates a serious malfunctioning of his body.

Man is a very economic working system. His daily needs are only about 2.5×10^6 calories. The average car will run merely 2000–3000 m per day if fed with as many calories in the form of gasoline.

Energy Output of the Left Ventricle. In Chapter 4 (Conservation laws) we calculated the work done (or energy output) per day by the left ventricle to be 7.8×10^4 J. Converting this into the thermal energy unit calorie, we get 1.88×10^4 cal. This energy is the thermal energy output of the appropriate heart muscle. Again, the energy consumption of an automobile is monstrous in comparison. Supplied with those 2×10^4 cal, it would run about 20 m!

2.3 Heat production

Sources of heat are the sun, geothermal heat, frictional heat, and chemical, electric, and nuclear reactions. In the life sciences, the most important heat sources are the metabolic processes, mainly oxidation. The amount of heat produced does not depend on whether this is a rapid (like fire) or a slow process (like food consumption). Table 7 shows the amount of heat produced by the oxidation of various substances. Only recently have nuclear reactions been used by man. Their incredible amount of heat production per gram of fuel is shown for comparison.

Table 7 Some examples of thermal energy production

(All values in cal per gram of fuel)

dry wood	3000–4000
coal	7500
city gas	up to 12 000
hydrogen	34 000
bread	2400
apple	600
pork meat	3600
trout	1000
egg	1600
nuclear reactor (uranium)	5×10^8
fusion reactor (hydrogen)	2.5×10^{10}

During muscular activity heat is generated because the efficiency of a muscle is approximately 25%. This means about one quarter of the metabolic energy supplied to the contracting fibers is converted into mechanical energy; the remaining three quarters are transformed into unrecoverable heat. This heat production is a function of time. During actual contraction, more heat is observed than during recovery. Figure 6-8 shows as an example the frog's muscle. The heat production of the muscle is displayed during various stages of muscular activity.

2.4 Heat pollution

2.4(a). Water. Since all forms of energy are ultimately converted into heat, man must be aware of the rises in temperature of water and atmosphere due to energy production in highly industrialized areas. A seem-

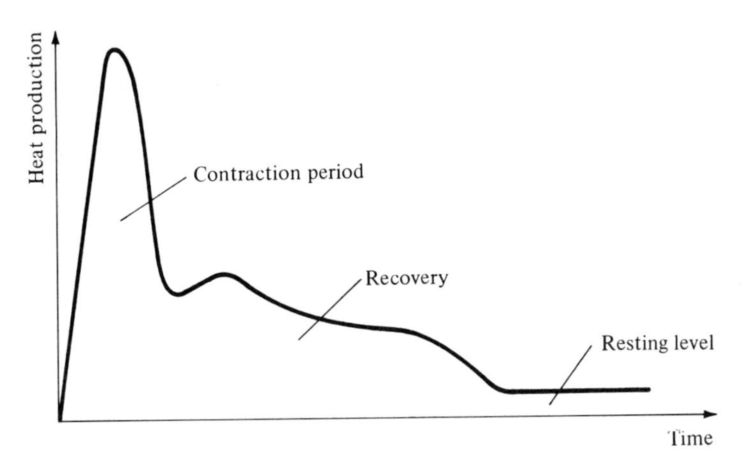

Figure 6-8

ingly small temperature increase in a lake or river can lead to significant consequences. Let us briefly look at one of those consequences: The solubility of oxygen in water is a function of temperature. With rising temperatures, less oxygen becomes available, thus microorganisms and fish are deprived of their supply. Figure 6-9 shows the drop in oxygen content with rising freshwater temperature.

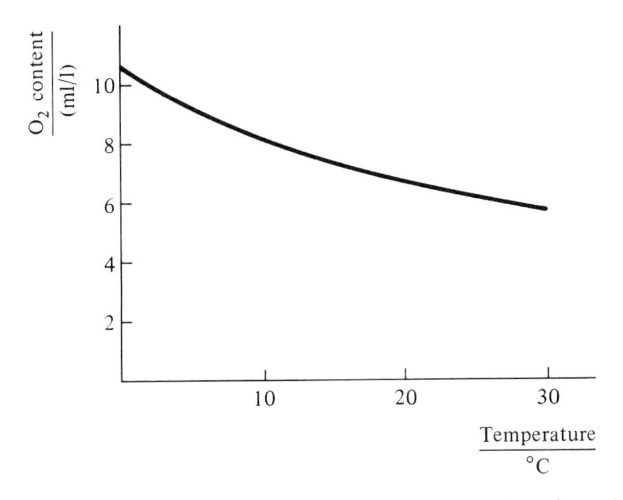

Figure 6-9 Oxygen content of fresh water as a function of its temperature.

If we take the current world production of electric energy, about 4×10^{15} W·h $= 4 \times 10^{18}$ cal, and assume that it is finally converted into heat and used to raise the average water temperature of all water on earth (approximately 7×10^{24} cm³), this temperature increase will be less than 10^{-9} °C. However, for limited water bodies, the calculation looks quite different. To cool the big new 1000 megawatt turbines of electric power stations, 50 m^3 of water is needed per second. The temperature difference between inflowing and outflowing water is 10°C, indicating that 5×10^8 cal is released continuously to the cooling water.

In summertime when the water level is low, a big river like the Rhine carries only 500 m^3/s water. If the entire Rhine is used to cool one of these turbines, the water temperature would still rise by one degree. And this is only for one turbine.

In the long run, even the mighty Mississippi, carrying 20 000 m^3 water per second to the sea, may not be sufficient for cooling purposes. One way out is to cool the heated water in gigantic towers and thus release the thermal energy into the atmosphere.

Unfortunately, nuclear and fusion power reactors offer no better solutions, since up to now they have been used only as boilers to heat the turbine water. The direct conversion (in magneto-hydrodynamic generators) from nuclear or fusion energy into electric energy is still in its infancy.

2.4(b). Fever. Fever is heat pollution of the body. During this period, the average oral temperature can rise from 98.4°F to 107.8°F. Fever may be caused by infection, drugs, or other reasons and is always accompanied by an increased metabolism leading to an increased thermal energy production. See Figure 6-10. Various powerful temperature-

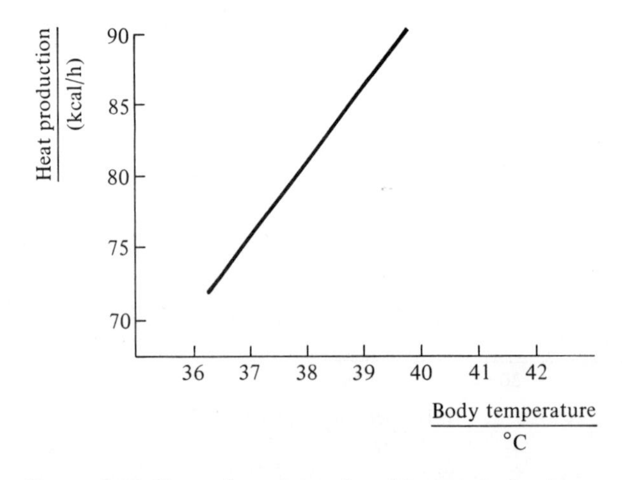

Figure 6-10 Thermal energy produced in man during fever.

regulating mechanisms are at work to keep the body temperature constant. If they cannot achieve this, heat pollution will swamp the body and cause failure of the heart muscle.

3. THERMAL ENERGY TRANSFER

As long as temperature differences exist, thermal energy (heat) is transferred from places of higher temperature to those of lower temperature until thermal equilibrium is achieved. A position of higher temperature is called a *heat source*, and one of lower temperature is called a *heat sink*.

3.1 Temperature gradient

The average temperature gradient is:

$$\frac{T_1 - T_2}{x_1 - x_2} = \frac{\Delta T}{\Delta x} \tag{6.9}$$

where

T_1: temperature at position x_1,
T_2: temperature at position x_2,
ΔT: temperature difference between T_1 and T_2,
Δx: distance between x_1 and x_2.

The average temperature gradient is expressed in deg/cm. Table 8

Table 8 SOME AVERAGE TEMPERATURE GRADIENTS

(All values in deg/cm)

sun	10^{-6}
troposphere (0–17 km height)	-6.5×10^{-5}
thermocline in lake	-0.11
human body	0.4
earth	3×10^{-4}

presents some average temperature gradients. The reference point is always the surface. For the troposphere, the point of reference is the earth's surface.

The average temperature gradient does not take into account that the temperature between heat source and heat sink may not vary uniformly. If the temperature drop is not uniform, a differential expression is introduced:

$$\text{temperature gradient} = \frac{dT}{dx} \tag{6.10}$$

where

dT: infinitesimal change of temperature over the infinitesimal distance dx.

Remark: By convention, the gradient is negative if the temperature drops with increasing distance from a reference position.

The following example demonstrates the difference between average temperature gradient ($\Delta T/\Delta x$) and temperature gradient (dT/dx). This difference can be of utmost importance.

EXAMPLE: TEMPERATURE STRATIFICATION IN A LAKE

Previously we studied the thermal expansion of water and found that, in a lake, a thermocline will develop. We observe that within the thermocline the water temperature drops rapidly. See Figure 6-11.

$$\text{average temperature gradient} = \frac{15 - 4}{0 - 20}$$

$$= -0.55 \frac{\text{deg}}{\text{m}}$$

$$= -5.5 \times 10^{-3} \frac{\text{deg}}{\text{cm}}$$

This value is misleading since it hides the presence of the thermocline. The temperature gradient itself yields more information. See Figure 6-12.

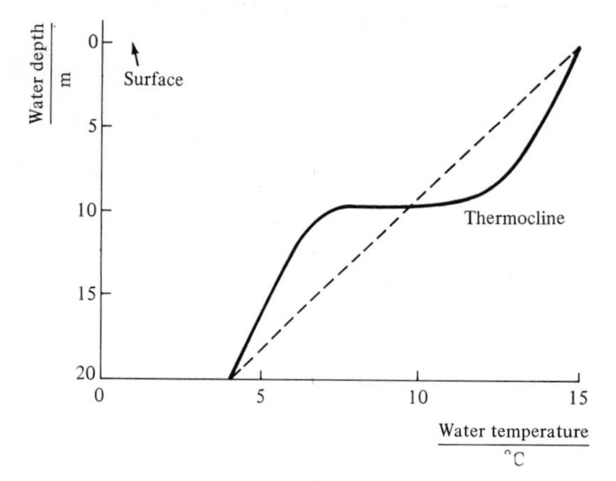

Figure 6-11 Temperature distribution in a lake.

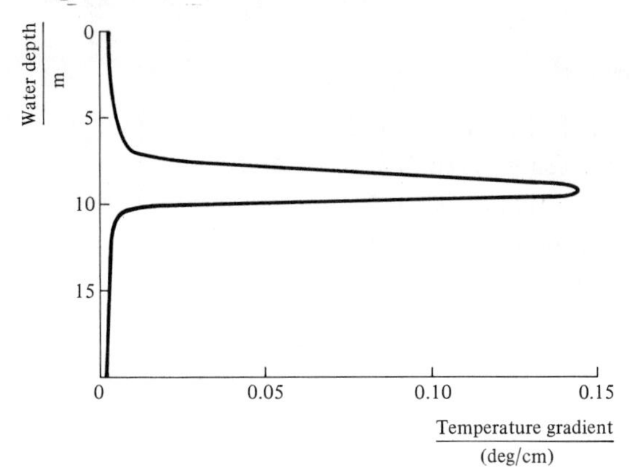

Figure 6-12 Temperature gradient of the same lake as in Figure 6-11.

3.2 Temperature rate

The temperature gradient should not be confused with the temperature rate, which in turn describes the time rate of change of temperature at the same locality.

$$\text{average temperature rate} = \frac{\Delta T}{\Delta t} \qquad (6.11)$$

where

ΔT: temperature difference between two coordinates in time,
Δt: difference between those two time coordinates.

The temperature rate is measured in deg/s.

EXAMPLE: SKIN TEMPERATURE AND ENVIRONMENT

See Figure 6-13. After a time delay of 13 minutes, the skin temperature drops due to increased sweat production. For the following 14 minutes, the skin temperature decreases at the constant rate:

$$\text{temperature rate} = \frac{\Delta T}{\Delta t}$$

$$= 3.6 \times 10^{-4} \frac{\text{deg}}{\text{s}}$$

Twenty-seven minutes after entering the warm environment, the skin temperature has achieved a constant value (adaption), and consequently the temperature rate is zero.

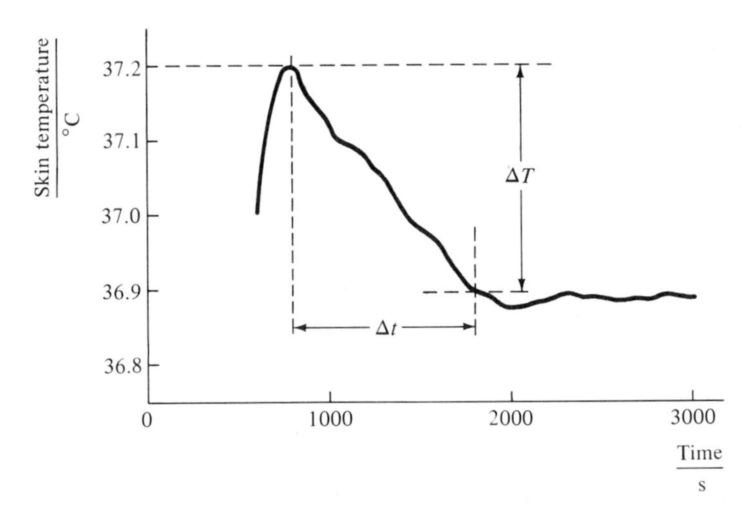

Figure 6-13 Skin temperature of a person entering a warm environment.

3.3 Means of thermal energy transfer

3.3(a). THERMAL CONDUCTION. If the heat source and sink are connected by matter (either in its solid, liquid, or gaseous state), thermal energy—a heat current—will flow from source to sink. As long as no bulk movement of matter takes place, this energy transfer occurs by thermal conduction.

The theory of thermal conduction is difficult because, on the molecular and atomic levels, more than one process contributes to thermal conduction. On the macroscopic level, it is straightforward. See Figure 6-14.

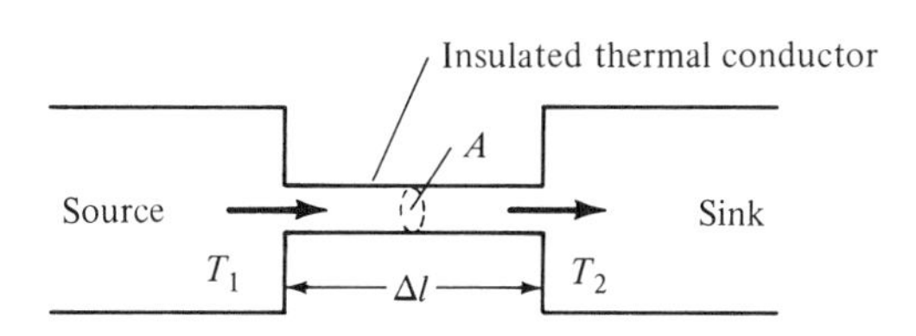

Figure 6-14

Thermal energy flows from source to sink. Both are connected by matter (for simplicity, this is indicated by a rod having a cross section A). The connecting matter is insulated against energy loss during transfer. The amount of energy transported, the thermal energy current, is obviously

$$\Phi \sim A \frac{\Delta T}{\Delta l} \tag{6.12}$$

where

Φ: thermal energy current
$\Delta T = T_1 - T_2$ = temperature difference between source and sink,

Δl: distance from source to sink,
A: cross section of the heat conductor.

Introducing the proportionality constant λ, we get

$$\Phi = \lambda A \frac{\Delta T}{\Delta l} \tag{6.13}$$

where

λ: thermal conductivity expressed in cal/(s·cm·deg) or W/(cm·deg),
Φ: thermal energy current expressed in cal/s or W,
$\frac{\Delta T}{\Delta l}$: average temperature gradient expressed in deg/cm.

The temperature gradient is assumed to be constant (stationary system); otherwise time factors come in and the formulas for the thermal energy current become involved. The thermal conductivity λ is characteristic for each material and closely related to the electrical conductivity. Table 9 lists λ for some materials.

Table 9 THERMAL CONDUCTIVITY FOR SOME MATERIALS

All values are in cal/(cm·s·deg)

Material	λ
silver	1.000
aluminum	0.48
ice	0.005
wood	0.003 5
porcelain	0.002
human skin	0.002
water	0.001 6
brick wall	0.001 4
snow	0.001 0
body fat	0.000 4
dry beaver fur	0.000 13
cork	0.000 1
caribou fur	0.000 09
eiderdown	0.000 06
air	0.000 057

Metals are good thermal energy and electric conductors. This is not by accident. Both processes depend on the number of free electrons inside the material. The similarity between thermal conductivity and electric conductivity is utilized by designing heat conducting systems in the same manner as an electrical system.

As an example, in experimental nuclear physics, some detector crystals operate best at low temperature. They are cooled by heat conduction down to liquid air temperature. For experimental reasons, the heat source (crystal) is often away from the heat sink (cooling unit). The cooling rate (most important for crystals) and final temperature are then calculated with the help of Equation (6.13).

3.3(b). THERMAL CONVECTION. The development of a thermal energy current between source and sink is not limited to thermal conduction but is also established by *thermal convection:* bulk movement of matter.

Thermal convection is restricted to liquids and gases. If the heat current is not forced (a fan could do that), thermal convection is caused by the temperature-dependent density of gases and liquids. The temperature stratification in a lake is one example, or the thermal motion inside a container (see Figure 6-15).

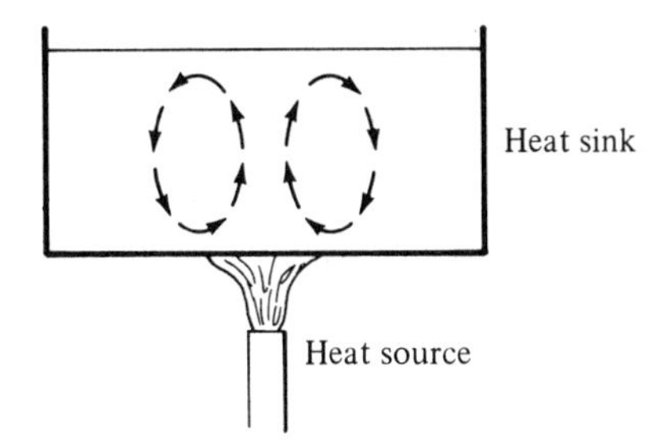

Figure 6-15

In the figure, a pot with water is heated at the bottom. The warmer water has a lower density and thus rises to the surface. The cooler water forms a countercurrent. This is the working principle of a warm-water central heating system. No pump is needed; thermal convection moves the water around. The heat source is the boiler in the basement of the building, and the rooms connected to the system act as heat sinks.

To calculate in detail the thermal energy transfer caused by thermal convection is far too complicated; instead engineers use semiempirical formulas and scale models.

EXAMPLE OF THERMAL CONVECTION

You may have wondered why it takes so long for soup to cool, since the heat flow rate from soup to air should be quite large. The answer is that the temperature gradient between soup and air is small due to a stagnant layer of warm air above the surface. This layer is a poor heat conductor. Blowing gently, you start forced convection, remove this layer, and thus appreciably increase the thermal energy current.

*Dewar Container:** To store material having a large temperature difference with its surroundings, we thermally isolate it in a double-walled container. See Figure 6-16. The space between the walls is kept at low

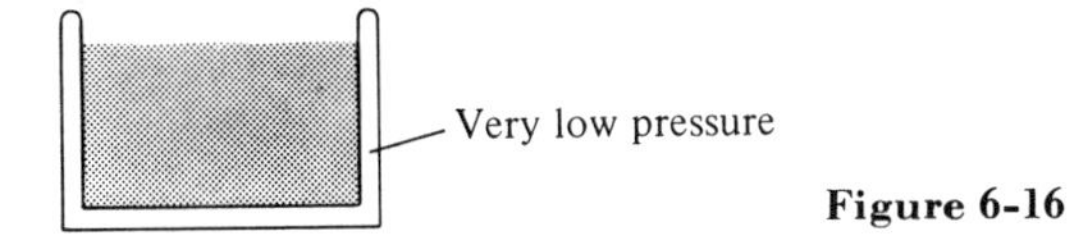

Figure 6-16

* James Dewar (1842–1923), chemist, known for his work with benzene and explosives. To store liquid air, he invented the *Dewar container*.

air pressure (low conduction, no convection). The connection between both walls is as thin as possible (low conduction). In everyday life, this container is called a *Thermos bottle.*

The in-between space can be filled (for stability reasons) with modern heat insulators such as styrofoam. These are synthetics consisting mainly of air bubbles kept in place (Figure 6-17). Conduction is very

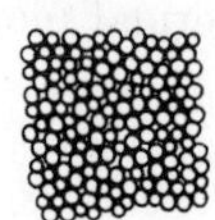

Figure 6-17

poor (λ for air is small and only a little material surrounds the air bubbles, resulting in a low cross section A), and convection does not take place since the enclosed air cannot move.

Touching styrofoam for a few seconds gives a sensation of warmth: the thermal energy current from the warmer skin to the outside is prevented, thus causing a local rise in temperature.

The Pelt of Animals. The heat loss of a warm-blooded animal would be disastrously high even in a moderate climate, were it not for the animal's pelt. The hairs prevent the free flow of air close to the body surface, thus forming a semistagnant boundary layer of poorly conducting air. It helps if the undercoat is thicker than the remaining fur. Human clothing serves the same purpose.

Convection Inside Centrifuge Containers. The term *convection* is not limited to thermal energy transfer processes. It extends to any bulk movement of fluid within a fluid. Inside a centrifuge probe container having parallel walls, the density of the solute becomes slightly larger at the walls than at the center. Consequently a radial motion called *convection* results. See Figure 6-18. The arrows

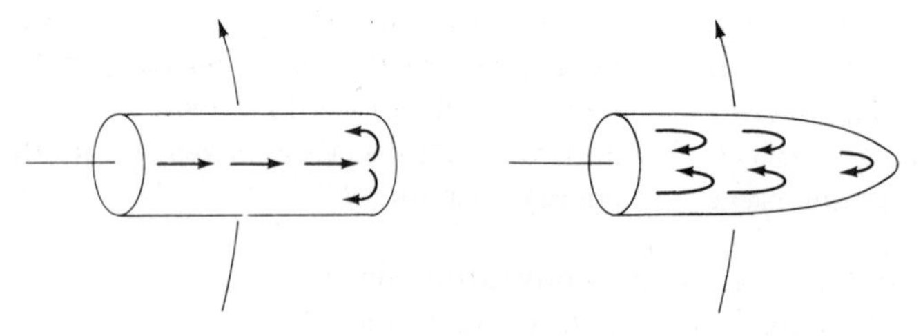

Figure 6-18

show the sedimentation of particles with respect to solvent. This effect can be prevented by conically shaped probe containers or by nonuniform solution densities.

3.3(c). Thermal radiation. Thermal energy transfer by means of thermal radiation is very distinct from thermal energy transfer via conduction and convection. As opposed to conduction and convection, thermal energy transfer by radiation occurs also if:

1. The temperature gradient is negative, positive, or zero.
2. A transferring or conducting medium does not exist.

Most of the thermal energy is transferred by the red and infrared (also named ultrared) part of the electromagnetic spectrum. Chapter 15 is devoted to radiation; there we shall present thermal radiation in its proper and more enlightening context. Note, however, two interesting facts about heat transfer via thermal radiation:

The emission of electromagnetic radiation (which encompasses thermal radiation) is proportional to the fourth power of the temperature. This means that, if the temperature of a body is doubled (always measuring on the absolute temperature scale in °K), its transfer of thermal energy by radiation is sixteenfold. For low temperatures the energy loss by radiation is only a small part of the total.

A good thermal radiator is also a good absorber. It does not matter whether a tent is black or white. The heat-absorbing black tent will also emit thermal energy very well. This is applied by the Sahara nomads. They live in black, woolen tents. There is no need to bleach the wool of their black sheep before weaving it into tent material!

3.4 Thermal energy transfer during change of phase

Up to now, we have always assumed that thermal energy transfer indicates a transfer of energy from one coordinate in space or time to another. We may extend the concept of thermal energy transfer to the energy difference between various phases of matter. The *latent heat* discussed earlier is the amount transferred. This transfer is of importance in nature; it harmonizes harsh differences. Again, a lake may illustrate this. See Figure 6-19.

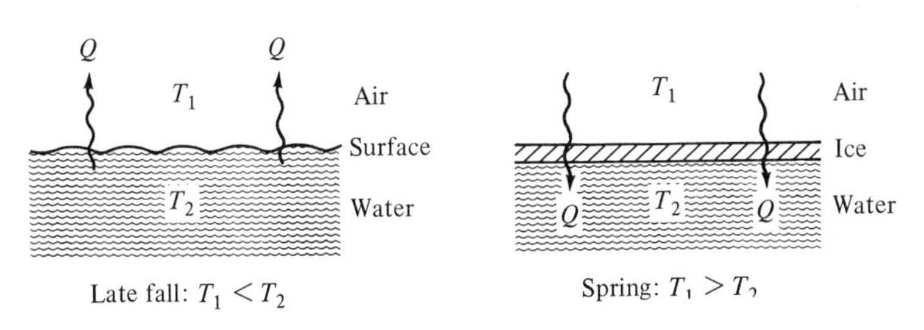

Figure 6-19

In the fall, the air temperature T_1 is lower than the water temperature T_2. Consequently, thermal energy Q is transferred from water to air at a certain rate. As soon as the water temperature drops low enough to form ice at the lake's surface, the cooling rate declines because heat is set free due to the formation of ice—latent heat of solidification. In the spring we observe the opposite effect: now T_1 is greater than T_2 but a rapid temperature change of the water is prevented because thermal energy is needed to change ice into water—latent heat of fusion. The overall effect is that temperature changes of the lake surface are slow, thus allowing biological material plenty of time to adapt.

A similar temperature harmonizing system is built into the skin of man. But how sweat production regulates the surface temperature, you may work out yourself.

SUMMARY

Temperature is measured with a thermometer, and all thermometers are calibrated in degrees. The two most important scales are the Celsius and Kelvin scales.

The Celsius temperature scale has two fixed points:

$$0°\text{C} = \text{melting point of ice}$$
$$+100°\text{C} = \text{boiling point of water at normal pressure}$$

The thermodynamic temperature scale starts at the lowest possible temperature, 0°K. The intervals are the same as for the Celsius scale.

$$0°\text{K} \equiv -273.15°\text{C}$$

Any reproducible temperature-dependent property of matter may be used to build a thermometer; for example:

expansion of matter
electric resistance
cooling rate of a heat source
growth rate of bacteria, etc.

Expansion of matter is given by (in first approximation):

$$V = V_0(1 + \gamma t)$$

where

$$V_0 = \text{volume at } 0°\text{C}$$
$$V = \text{volume at temperature } t$$
$$\gamma = \text{expansion coefficient}$$
$$t = \text{temperature}$$

The orders of magnitude for the expansion coefficients are:

$$\text{gases} \approx 10^{-3}\ \text{deg}^{-1}$$
$$\text{liquids} \approx 10^{-4}\ \text{deg}^{-1}$$
$$\text{solids} \approx 10^{-5}\ \text{deg}^{-1}$$

In general, the thermal expansion coefficients are temperature-dependent and may change their signs over limited regions (example: water).

Heat is another manifestation of energy. It is called *thermal energy* and is consequently measured in energy units, in joules.

In the life sciences and in everyday life, thermal energy is expressed in calories. Conversion:

$$1\ \text{cal} = 4.186\ \text{joules}$$
$$1\ \text{J} = 0.239\ \text{calorie}$$

Specific heat capacity (abbreviated c) is the quantity of thermal energy required to raise the temperature of 1 gram of matter by 1 degree. Typical values for c are between 1 and 0.1 cal/(g·deg).

Latent heat (abbreviated L) is the quantity of thermal energy required to change the phase of matter while keeping its temperature constant. L is expressed in cal/g. Latent heat is stored for transitions solid $\rightarrow$ liquid and liquid $\rightarrow$ gas. Latent heat is freed for phase transitions gas $\rightarrow$ liquid and liquid $\rightarrow$ solid. The numerical values for L are between a few and a hundred cal/g.

Thermal energy transfer between heat source and heat sink takes place by conduction, convection, and radiation. Thermal conduction:

$$\Phi = \lambda A \frac{\Delta T}{\Delta l}$$

where

Φ: heat flow rate expressed in cal/s,
λ: thermal conductivity, expressed in cal/(s·cm·deg),
A: cross section of heat conductor,
$\frac{\Delta T}{\Delta l}$: average temperature difference ΔT between heat source and heat sink, divided by their separation Δl.

The numerical values for λ vary between 10^{-5} and 1 cal/(s·cm·deg).

Thermal convection takes place by bulk movement of matter. Its calculation is difficult and usually replaced by measurements at scale models.

Thermal radiation is the thermal energy transfer by means of the red and infrared (ultrared) part of the electromagnetic spectrum.

Time effects are neglected in thermal energy transfer. Thermal equilibrium is always assumed, thus making the thermal energy transfer instantaneous. Nonequilibrium considerations are beyond the scope of the text.

PROBLEMS

1. Calculate how much force is needed to prevent solid matter from thermal expansion. Which material resists most?
2. How many degrees Fahrenheit is 1 degree Celsius? Derive a conversion formula to convert degrees Celsius into degrees Fahrenheit. Express absolute zero in degrees Fahrenheit.
3. A wire may be lengthened by thermal expansion and also by a force pulling at it. Derive a relation which couples both effects and discuss its range of validity.
4. Superconductivity, as shown in Figure 6-4, could be employed to measure temperature: (a) Range? (b) Maximum sensitivity?
5. A flame inside an orbiting satellite does not burn brightly, although enough oxygen is supplied. Why?

6. A water drop rides for a short while on a hot surface. Explain this Leidenfrost phenomenon.
7. The dog has no sweat glands. How does it regulate its body temperature?
8. The thermoreceptors do not regulate the skin temperature directly. Why?
9. Convert your daily food intake from calories into kW·h. Calculate the equivalent cost if this thermal energy is supplied by electricity.
10. A silver ring of 1.50 cm diameter at room temperature is to be fitted over a rod of 1.55 cm diameter. Calculate the temperature needed for fitting (a) in degrees Celsius, (b) in degrees Fahrenheit.
11. Out of a 90 liter cylinder containing compressed air at 42 atm pressure, 2950 liters are consumed. During the release the temperature inside the cylinder dropped by 12 deg. Calculate the remaining pressure.
12. A silver cube of 1 cubic centimeter volume is heated from 0°C to 500°C. The expansion is restricted to one direction. Calculate the linear expansion and the counterpressure necessary to prevent this expansion.
13. How many additional food calories are needed if the body temperature of a 70 kg man rises from 36.9°C to 40.5°C?
14. A 0.7 liter bottle of champagne is wrapped in cloth and exposed to a blowing fan. Water drips at a rate of 2 cm^3/min on the cloth and evaporates. How long will it take to cool the bottle from room temperature (22°C) to the suitable temperature of 10°C? (Neglect the influence of the bottle itself and assume that all water evaporates).
15. How much more time will it take to obtain boiling water from ice rather than from liquid water having the same temperature?
16. A 75-kg man is placed inside a thermally insulated chamber having a volume of 9 m^3 and an initial temperature of 20°C. How long will it take until the air temperature reaches 100°C?

FURTHER READING

T. H. BENZINGER, "Homeostasis of Central Temperature in Man," *Physiolog. Rev.*, **49** (1969), 671.

C. EMILIANI, "Isotopic Paleotemperatures," *Science*, **154** (1966), 851.

D. M. GATES, "Heat Transfer in Plants," *Scientific American*, Dec., 1965, 77.

chapter 7

PHASES OF MATTER

1. INTRODUCTION AND NOTATION

In this chapter we shall look at the various phases or states of matter. There are only four phases (solid, liquid, gaseous, and plasma), but in general two or three phases may coexist and transitions between the phases occur frequently. It is worthwhile to demonstrate the interrelation in graphical form before proceeding into details.

1.1 Interrelating flow chart

Phases of matter and phase transitions are always defined for the same material, either a pure element (e.g., hydrogen, copper, or iron) or a compound (e.g., water, methane, or methyl alcohol). For many materials some phases are not stable or do not exist at all. Iodine, for example, has no stable liquid phase. Compounds have no plasma phase because this phase develops at very high temperatures ($> 3000°K$) only, and the positive and negative components of the atoms and molecules separate before this temperature is reached.

Figure 7-1 shows the four states (or phases) of matter. The connecting arrows show the process for transitions with negative latent heat L. The notation is not always clearly defined. For example, it is not customary to call the energy required to transfer a gas into a plasma

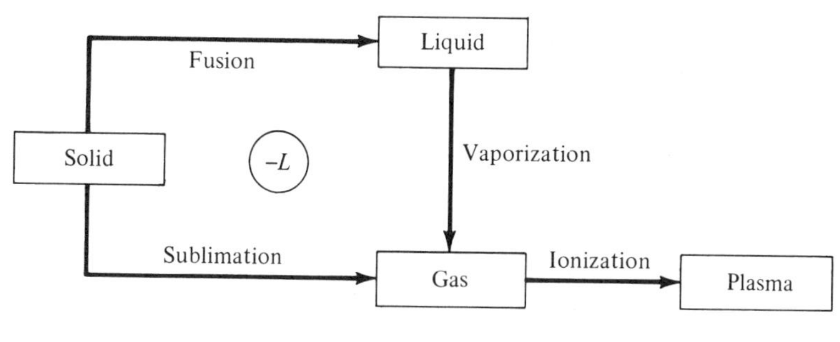

Figure 7-1

latent heat of ionization. Instead it is named *ionization energy* and measured in electron volts, abbreviated eV.

Some calculations: The ionization energy for hydrogen is 13.5 eV per atom.

$$13.5\ \text{eV} = 2.15 \times 10^{-18}\ \text{J} = 5.13 \times 10^{-19}\ \text{cal}$$

One gram of hydrogen contains 1.66×10^{24} atoms; consequently the latent heat of ionization is

$$5.13 \times 10^{-19}\ \text{cal} \times 1.66 \times 10^{24}\ \text{g}^{-1} = 8.55 \times 10^{5}\ \frac{\text{cal}}{\text{g}}$$

This value is four orders of magnitude larger than the latent heat of fusion and still three orders of magnitude larger than the latent heat of vaporization! This calculation assumes a completely ionized plasma; this means the positive (the nucleus) and the negative (the electrons) components of all hydrogen atoms are separated. The name *plasma* applies even if only a small percentage of the atoms are ionized. Gas and plasma blend into each other.

Figure 7-2 shows phases and phase transitions where thermal energy is set free, that is, transitions with positive latent heat L.

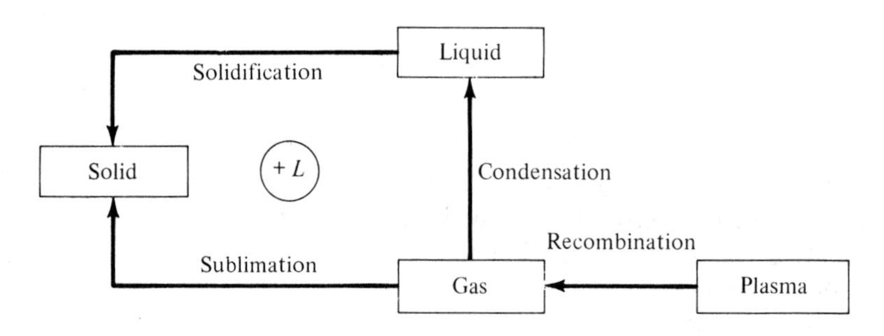

Figure 7-2

1.2 Phase diagram

The state in which matter happens to be—solid, liquid, gaseous, or plasma—depends on its temperature and the pressure exerted on it. The phase diagram in Figure 7-3 presents the interrelations.

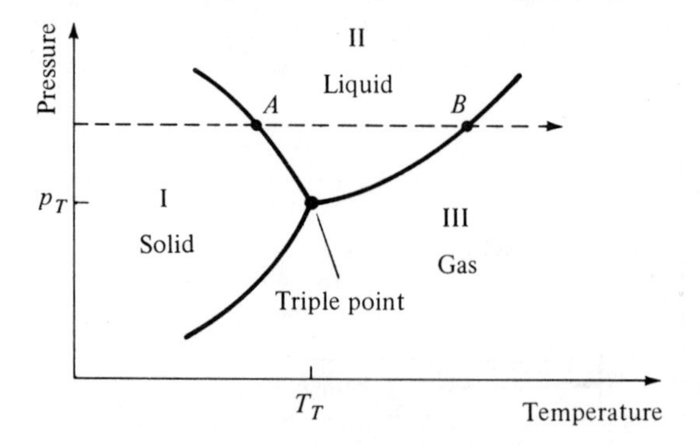

Figure 7-3

Triple Point. All three phases coexist at this pressure (p_T) and temperature (T_T).

EXAMPLE:

The triple point of water is at $p_T = 4.58$ mm Hg and $T_T = 0.0075°C$. This triple point is actually used as a fixed point for the Celsius temperature scale.

Some materials such as carbon have more than one triple point, indicating modifications of their solid phases. For carbon, those modifications are graphite and diamond.

How to Read a Phase Diagram. The path along the arrow in Figure 7-3 follows a line of constant pressure. It enters the graph at area I, denoting the solid phase. Proceeding along the hatched line to the right (the direction of increasing temperature), we encounter the boundary which separates the solid and liquid phases. There is the melting point (A) for the chosen pressure. Still raising the temperature, we pass through the liquid state and reach area III, the gaseous state. We crossed the line separating areas II and III at the boiling point (B) for the appropriate pressure. If we shift the horizontal arrow toward lower pressure, the temperature difference between the melting point and the boiling point becomes smaller, until it vanishes at the triple point. Shifting the arrow towards still lower pressure, we notice that for this pressure no liquid state exists. The solid phase converts directly into the gaseous phase, a process called *sublimation.*

Work out what will happen if the arrow is set for constant temperature, that is, parallel to the pressure axis.

1.2(a). APPLICATION: ICE SKATING. In general, the coefficient of friction between two surfaces increases with pressure. The system *steel on ice* is different, as you can observe during ice skating. Obviously, friction is minimal here. A look at Figure 7-4 and at the phase diagram will explain why this is true.

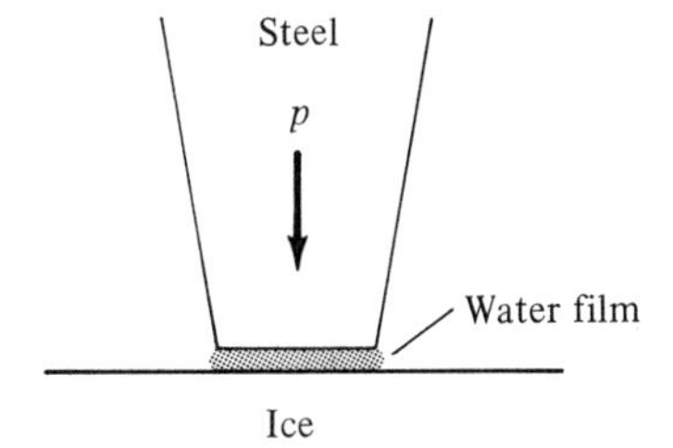

Figure 7-4

The runners are very narrow and therefore exert a high pressure on the supporting ice. According to the phase diagram, ice will melt under the pressure and the runners will glide on a thin film of water. The coefficient of friction is drastically reduced. However, if the ice temperature is very low, the exerted pressure of ordinary skates may not be sufficient to melt the ice. Narrower runners will solve this problem.

1.2(b). Principle of Le Chatelier.* During the last century, a broad-ranging principle was formulated by Le Chatelier, which can be applied here. The Principle of Least Constraint. If stress is brought to bear on a system in equilibrium, a change occurs such that the equilibrium is displaced in a direction which tries to cancel the effect of this stress.
Applying this principle to ice-skating: The pressure from the runner tries to compress the ice, that is, reduce its volume.
The reaction of the system: Solid water (ice) occupies a larger volume than liquid water at the same temperature. Consequently solid water melts under pressure, thus reducing its volume and the stress.

2. PHASES OF MATTER

Listing the characteristics of the various phases of matter, you should realize that matter usually does not exist in only one phase. In the following paragraphs, we shall idealize and neglect all but the one state of matter under discussion. Modern physical chemistry has also revealed some phases of matter which fit poorly into our conventional headings. Fibers with elastic and extremely long molecules are one example. Glass occupies a phase somewhere between solid and liquid, while most crystals hardly fit the description of a solid at all. Or, how would you classify a film of material?

2.1 Solid phase

In the *ideal solid phase*, the particles (atoms or molecules) constituting the solid body have neither translational nor rotational motions. This is true only at 0°K.

In the *real solid phase*, the particles (atoms or molecules) constituting the real solid body oscillate and rotate around an average position in space. It keeps a constant volume and shape.

Compressibility is approximately 10^{-6} atm^{-1}. This means that, if the pressure on a solid cube of 1 cm^3 is increased by one atmosphere, its volume will decrease by 10^{-6} cm^3 = 10^{-3} mm^3. For all practical purposes, the solid phase is incompressible.

Solid bodies are also characterized by properties such as hardness, elasticity, surface structure, internal structure, and thermal expansion.

2.2 Liquid phase

In the *real liquid phase*, the particles (atoms or molecules) of a real liquid can easily be shifted in position, but they keep a constant average separation. A liquid has practically a constant volume, but it changes its

* Henri Louis Le Chatelier (1850–1936), chemist. Worked in a field now called physical chemistry and formulated his universal principle in 1888. More about him in R. E. Oesper, *Journal of Chemical Education,* Vol. 8, 1931.

shape easily. Except for special purposes, it does not make sense to define an ideal liquid body because it is difficult to separate it clearly from solid and gas bodies.

Compressibility is approximately 10^{-4} atm^{-1}. This means that, if the pressure on one cubic centimeter of liquid is increased by one atmosphere, its volume will decrease by 10^{-4} cm^3 = 0.1 mm^3.

Liquids are also characterized by properties such as viscosity, surface tension, and local variations of density.

2.2(a). GLASS. Glass displays some very special features. Actually it is an undercooled liquid of extremely high viscosity. A pane of ordinary window glass fixed at only one side in a horizontal position will bend within a few weeks. *Undercooled* means that, if the molten glass is cooled below its melting temperature, the molecules will remain unordered. They will not achieve the regular internal structure of a solid. This unorderliness is the reason why glass is transparent.

2.3 Gaseous phase

In the *ideal gaseous phase,* the particles (atoms or molecules) of an ideal gas move about freely with high speed. Their speed depends on temperature and mass. For example, oxygen molecules at 0°C have an average speed of 461 m/s. Gas occupies any available space uniformly.

Gases obey Boyle-Mariotte's law (see Chapter 5):

$$pV = \text{constant} \tag{7.1}$$

where

p: pressure exerted on gas,
V: volume of gas at pressure p and constant temperature.

The intrinsic volume of the constituting particles is neglected.

In the *real gaseous phase,* the particles (atoms or molecules) interact with each other and have an intrinsic volume. A real gas obeys van der Waal's law (see Chapter 5):

$$\left(p + \frac{a}{V^2}\right)(V - b) = \text{constant} \tag{7.2}$$

where

$\frac{a}{V^2}$: term to account for the attractive interparticle forces,
b: term proportional to the intrinsic volume of the particles.

Caution: van der Waal's law is invalid if two phases coexist: for example, if a gas begins to liquify.

2.3(a). VAPOR. This is the gaseous phase of a material just before its transition into a liquid. (The notion is not clearly agreed upon: vapor can also be the gas in thermodynamic equilibrium with its liquid phase. Thermodynamic equilibrium means that as many particles move from the liquid phase into the gaseous phase as from the gaseous phase into the liquid phase. Sometimes the gaseous phase of a material that is a solid at room temperature is also named vapor.)

2.4 Plasma

The particles constituting a plasma are not all electrically neutral, as in solid, liquid, or gaseous phase. The bonds between the positive atomic nucleus and its surrounding negative electrons are broken to some extent. Consequently, a plasma consists of freely moving electrons (negative), positive ions, and electrically neutral (intact) atoms. Observed from the outside, a plasma is electrically neutral because the number of positive and negative charges is equal.

Plasma occurs only at temperatures beyond 3000°C, there is no upper limit for its temperature. However, the temperature concept is most difficult to apply to a plasma.

The number of freely moving negative charges varies between 10^8 per cm^3 (thin plasma) and 10^{14} per cm^3 (dense plasma). Plasma can be produced by heating, electrical discharges, or by very high pressure. Most matter of the universe is in its plasma phase. Examples of plasma are:

Lightning. In its bright path, a plasma is formed and sustained for 10^{-5} to 10^{-6} second.

Sun. All elements in the sun (and most other stars) are in their plasma phase.

Applications: Propulsion systems for spacecraft; production of artificial (and still very small) diamonds; direct transformation of thermal energy into electrical energy (magneto-hydrodynamic generators); thermonuclear reactors. Most of these applications are still in the experimental stage. However, there is one widely used application, the plasma-burner.

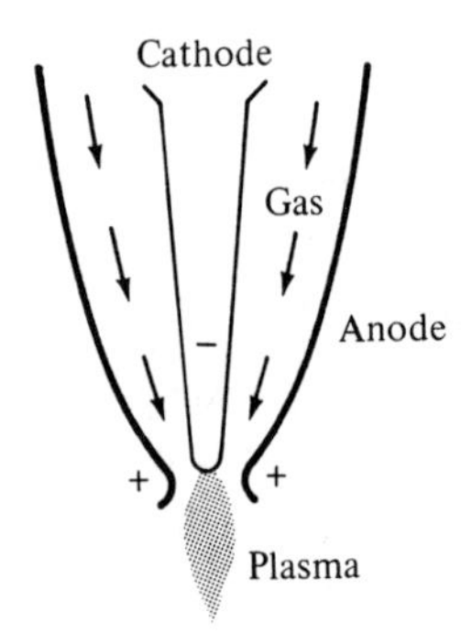

Figure 7-5

Plasma-Burner. Figure 7-5 shows a cross section through a modern plasma-burner. An electrical discharge is produced between the central cathode (negative) and the opening in the surrounding anode (positive). The discharge causes a plasma which is deflected through the nozzle with the help of a fast-flowing gas current entering the burner from the top. The temperature achieved (up to 6000°C) will melt any material.

3. TEMPERATURES AND TRANSITION ENERGIES

3.1 Temperature ranges

The transition temperature is not the only number used to characterize a transition between phases of matter, but the most evident one. Instead of listing the temperatures for the transitions between solid, liquid, gaseous, and plasma phases, Figure 7-6 presents the appropriate temperature ranges over which the four phases exist. Notice that the scale is logarithmic. The elements occupying the extremes for the vari-

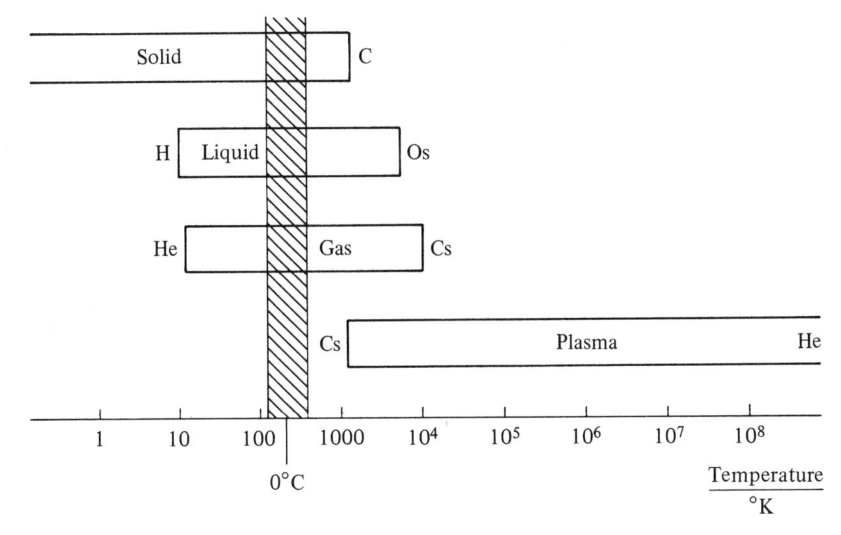

Figure 7-6 Temperature ranges for various phases. The hatched area shows the life supporting interval.

ous phases are also noted. For example, hydrogen is the last element to convert from a liquid into a solid if cooled. On the other extreme, osmium is the last element to transform from its solid into its liquid phase.

3.1(a). SUPERCOOLING. We can observe that very pure water freezes not at 0°C, but a few degrees lower. This supercooling occurs because the solidification needs a nucleus to initiate the phase transition. A tiny dust particle is sufficient. This is a most important feature for the survival of hybernating insects. If the intercellular water in their bodies does not contain a source of solidification, they may survive temperatures a few degrees below the freezing point. The (fatal) freezing can be introduced by a shock or if a crystal enters the cell.

The northern lights, vapor trails of high flying airplanes, and the traces in a cloud chamber are further examples for supercooling effects. There the gaseous phase cannot transform into its liquid state for lack of condensation nuclei. Ionizing particles or combustion products will serve as growth centers for tiny droplets.

3.1(b). SURVIVAL OF LIFE. The hatched area in Figure 7-6 shows the limits for the existence of life in any form. It looks like a large range, but this is due to the logarithmic scale. Displaying the temperature on a linear scale will demonstrate how tiny the life-supporting temperature range is. Also, this range occupies a narrow region at the lower end of the temperature scale. (We are talking about highly organized living matter, as we know it. There may be other highly organized forms in ranges beyond. But those speculations belong in the realms of science fiction. The astronomer Fred Hoyle made an interesting contribution: *The Dark Cloud, Penguin Books,* 1960).

3.2 Transition energies

If a phase changes into another phase, the internal energy of the system involved changes. The liquid phase has a higher internal energy than the solid phase for the same material because energy—in this case thermal energy—is absorbed by a solid before it starts to melt. This energy is called *latent heat* (see Chapter 6).

The numerical value for the latent heat increases during a transition from a state of higher density to a state of lower density. Figure 7-7 indicates the transition energies for hydrogen; all values are in cal/g. Since energy is conserved, the appropriate latent heat is set free if the phase transition is from lower to higher density.

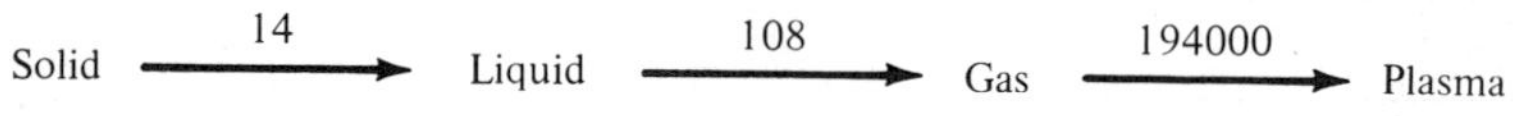

$$\text{Solid} \xrightarrow{14} \text{Liquid} \xrightarrow{108} \text{Gas} \xrightarrow{194000} \text{Plasma}$$

Figure 7-7

3.2(a). COOLING OF HUMAN BODY. We can calculate that the temperature of a normal person would increase to 40°C within one day if the heat produced inside by the consumed food warms up the body and no cooling system exists.

Thermal energy is transported by the blood from the inside of the body to the skin. There, a temperature difference between the surrounding air and the skin leads to a heat transfer via thermal convection. However, the main cooling process is the evaporation of water. As we already know, 596 cal is needed to evaporate 1 cm^3 of water. The vapor leaves the body through the lungs, the sweat glands, and directly through the skin. Most powerful is the cooling by evaporation of sweat. The temperature regulation itself is governed by a center in the brain (hypothalamus).

Again notice the exceptional qualities of water: The latent heat of vaporization for water is 596 cal; most other fluids have only 10–20% of that value. This means that five to ten times as much liquid would be needed for the same degree of cooling by other liquids. By the way: plants cool their leaves; they evaporate water through tiny openings.

4. APPLICATIONS

4.1 Freeze-drying

In the phase diagram in Figure 7.3, notice that under favorable circumstances every material may sublimate, that is, transfer from the solid phase directly into the gaseous phase. This characteristic is used in a process called *freeze-drying* or *drying by sublimation.*

In freeze-drying, the sample (a microtome cut, blood plasma, food, etc.) is placed inside a closed container (see Figure 7-8) and cooled below the triple point. See Figure 7-9. When the sample reaches the temperature T_s, the container is connected to a vacuum pump and the pressure is reduced from p_s to p_f. Now the sample dries by sublimation. The vaporized water leaves through the pump or is absorbed by the desiccant, for example, sulfuric acid.

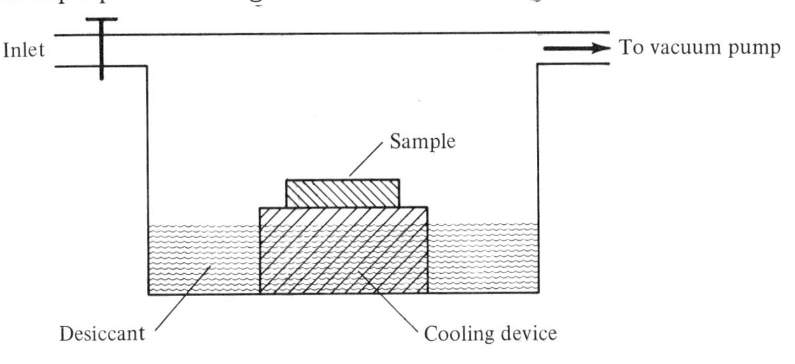

Figure 7-8 Principal arrangement for freeze-drying.

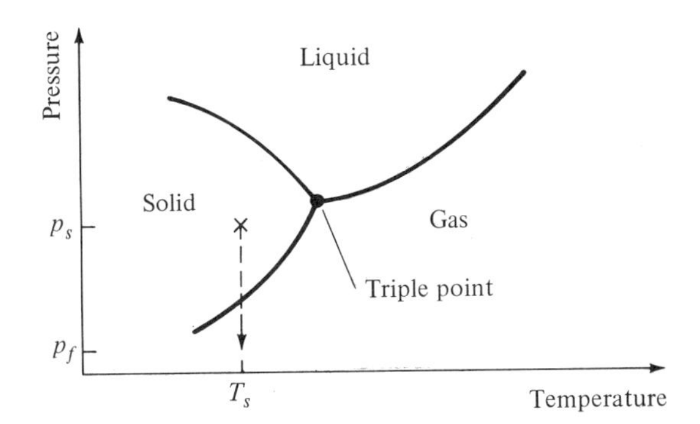

Figure 7-9 Freeze-drying. The dashed line shows the direction of the process.

One of the most attractive features of this method is the preserved texture of the dried material. The food industry already uses this process on a large scale. Not only is the texture and nutritional value of the foodstuff preserved, but also the taste of it. Modern freeze-dried instant coffee is a typical example.

Bacteriologists and histologists make extensive use of freeze-drying for the storage and later analysis of delicate samples. Because freeze-dried materials have an enormous surface, they must be kept sealed. Oxidation would be fast and destructive. There are many techniques involved in freeze-drying applications. For more detailed information see the reference at the end of this chapter.

4.2 How to make diamonds

Lavoisier* discovered in 1788 that diamonds are a modification of the solid phase of carbon; the other two modifications are graphite and soot. Of course, many tried to convert the inexpensive modification graphite into diamonds. The density of diamond is 3.5 g/cm^3; the density of graphite is 2.2 g/cm^3. Obviously, high pressure is needed for the transition. Although the efforts were as great as the stakes, success came only after Bridgman† determined the complicated phase diagram of carbon. See Figure 7-10.

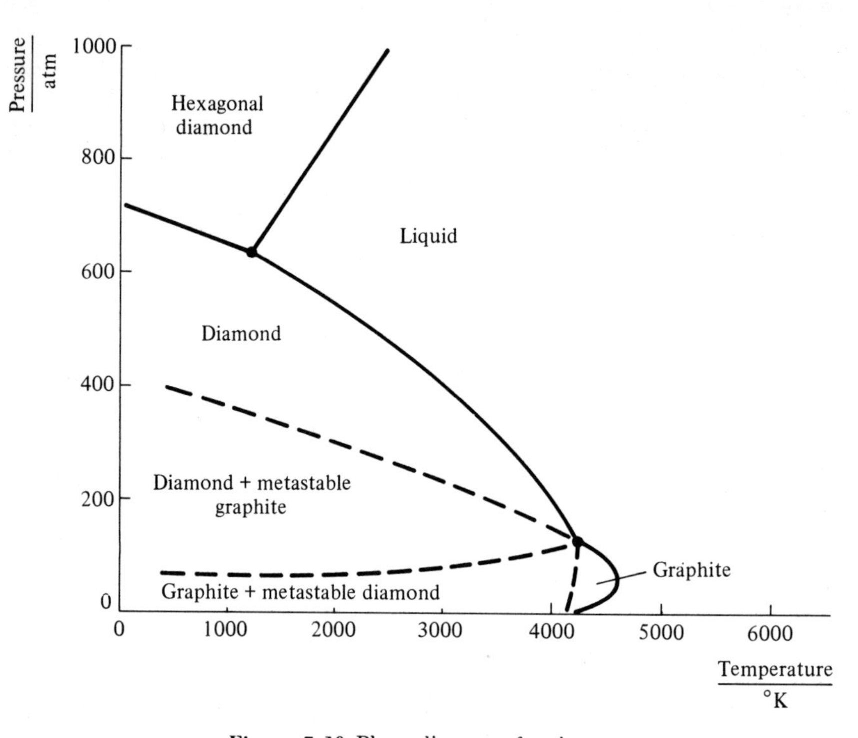

Figure 7-10 Phase diagram of carbon.

To make diamonds, four conditions must be met:

1. High pressure, above 50 000 atm.
2. High temperature, above 2000°C.
3. Sufficient growing time, some hours.
4. Suitable catalysts such as traces of Ni, Ta, or Cr.

Condition 3 in particular explains the failure of previous tries: the crystal needs time to grow. The first synthetic diamonds were produced

* Antoine Laurent Lavoisier (1743–1794), father of modern chemistry. He formulated the law of mass conservation and explained combustion as an oxidation process. More about him in D. McKie, *Antoine Lavoisier*, London, 1952.

† Percy Williams Bridgman (1882–1961), physicist. Investigated phase transitions at extremely high pressures. Nobel laureate, 1946.

in 1953, at 3100°C, 70 000 atm, and 16 hours. They are small and are used in industry as cutting tools.

Incidentally, looking at the phase diagram in Figure 7-10, you will discover that, at room temperature and atmospheric pressure, stable diamonds cannot exist! This apparently odd conclusion is correct; diamonds are metastable and convert into the more stable modification graphite. Fortunately for the ladies, the conversion speed is infinitesimally small.

4.3 Liquifying gases

The pressure-volume relation of gases is described by van der Waal's law, Equation (7.2). For

$$\text{interparticle forces} \rightarrow 0$$

and

$$\text{intrinsic volume} \rightarrow 0$$

Van der Waal's law for real gases converts into the Boyle-Mariotte law for ideal gases. See Equation (7.1).

In Figure 7-11 these relations are displayed for five different temperatures. T_1 is a hyperbola, reflecting Boyle-Mariotte's law. Moving to-

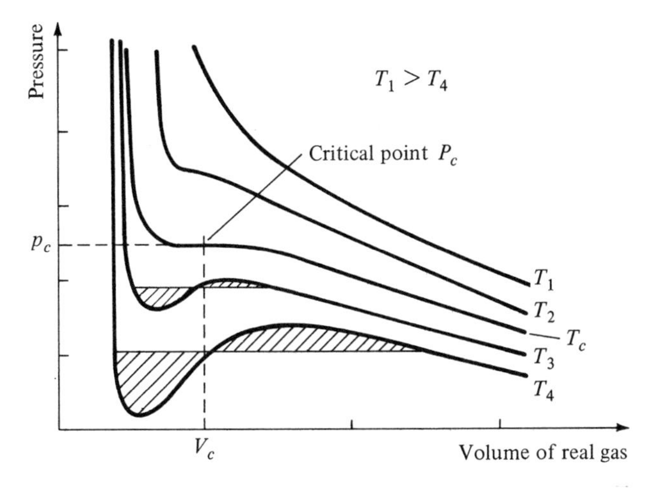

Figure 7-11 Pressure-volume relation for real gases at different temperatures. (V_c = critical volume.)

ward lower temperatures, we notice for T_c that, below the point P_c (the *critical point*), the volume of the gases decreases without a rise in pressure. The only possible interpretation is that the gas liquifies. In the hatched region, both phases (liquid and gaseous) coexist. Note the following definitions:

Critical temperature, T_c, is the temperature above which no liquification can occur regardless of pressure.

Critical pressure, p_c, is the minimum pressure needed to liquify a gas at the critical temperature.

Table 1 displays T_c and p_c for some gases.

Table 1 CRITICAL TEMPERATURE T_c AND CRITICAL PRESSURE p_c FOR SOME GASES

	T_c in °C	p_c in atm
helium	−267.9	2.26
hydrogen	−239.9	12.8
nitrogen	−147.1	34
oxygen	−118.8	48
CO_2	31.1	73
acetylene	36	62
ammonia	132.4	111.5
water vapor	374.0	217.7

Even with the highest pressure it is not possible to liquify most gases at room temperature. Carbon dioxide is an exception. For temperatures below 31°C it converts into its liquid phase if the exerted pressure exceeds 73 atm. To liquify the other gases, they are first cooled below T_c and then subjected to a pressure higher than the critical pressure p_c.

4.4 Hygrometry

Water vapor in the atmosphere is essential to life. Note the following definitions:

Water vapor: water in its gaseous phase below its critical temperature (+374°C).

Saturated water vapor: gaseous and liquid phases of water coexist.

Relative humidity: (water vapor pressure)/(saturated water vapor pressure), expressed in percent.

Dew point: temperature at which the air is saturated with water vapor, that is, the relative humidity is 100%.

The relative humidity is a function of temperature. A high value may be achieved by lowering the temperature or by increasing the water

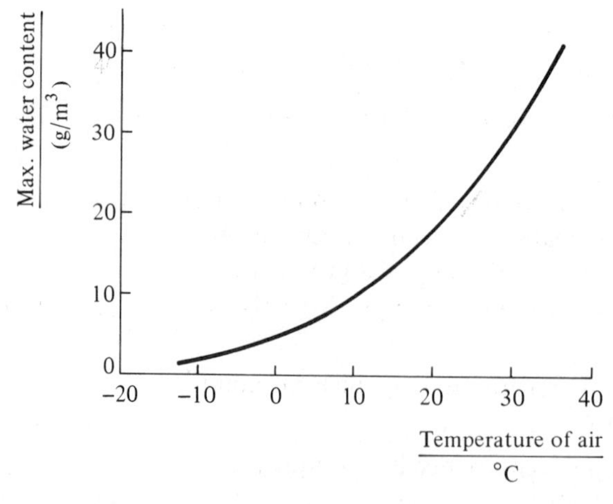

Figure 7-12 The water content of air as a function of its temperature.

content of the air. Phenomena such as fog, rain, dew, and clouds are manifestations of various degrees of humidity. Figure 7-12 shows the maximum possible water content (kept as saturated water vapor) of air as a function of temperature.

The humidity of air can be measured with a hygrometer. This device uses a human hair, which expands proportionally to the amount of water vapor in the air, to measure humidity.

4.5 Mixtures of phases and materials

In reality, two or even three phases of the same material coexist. A well-known example is water vapor (mixture of liquid and gaseous H_2O). But much more interesting are the mixtures of different phases of different materials. The possible combinations are countless. Table 2 presents

Table 2

	Solid II	Liquid II	Gas II
solid I	alloy		
liquid I	solid emulsion gel, sol suspension	emulsion	
gas I	solid foam dust SMOG smoke	fog foam aerosol	gas

the names of some stable mixtures for two different materials (I and II) and three phases (solid, liquid, and gas)

EXAMPLES:

Aerosol. Droplets of a liquid are dispersed in a gas. This is not merely a convenience (as in a perfume atomizer); it is also useful because the large total surface of the dispersed droplets is of importance in medicine (inhalators).

Emulsion. Droplets of one liquid are suspended in another liquid. Milk (fat, casein, and proteins in water) is a well-known example.

Gel, sol. Tiny particles in a liquid form, called *gel* if the result has a pudding-like consistency. It is named *sol* if the mixture is liquid. Modern no-drip paints exploit the easy transfer gel → sol and sol → gel. The paint is stored as a gel. The energy supplied by stirring transforms it into the liquid sol. Right after brushing, the paint returns to its gel state and thus cannot run. It thus has time to dry in the intended position.

Solid foam. In general, gas bubbles are enclosed by synthetic materials. If the bubbles are completely enclosed, having no connection with each other, it is an excellent thermal insulator (e.g., styrofoam). If the gas bubbles are interconnected, the solid foam has good sound-absorbing properties.

Smog. A new word synthesized from *sm*oke (solid particles dispersed in a gas) and f*og* (droplets dispersed in air). Three materials, each in a different phase, form a stable mixture: dust particles, droplets (in which many other compounds may be dissolved or attached to its surface), and air.

SUMMARY

Elements and most compounds appear in four phases: solid, liquid, gas, and plasma. Phases change along the lines of Figure 7-13.

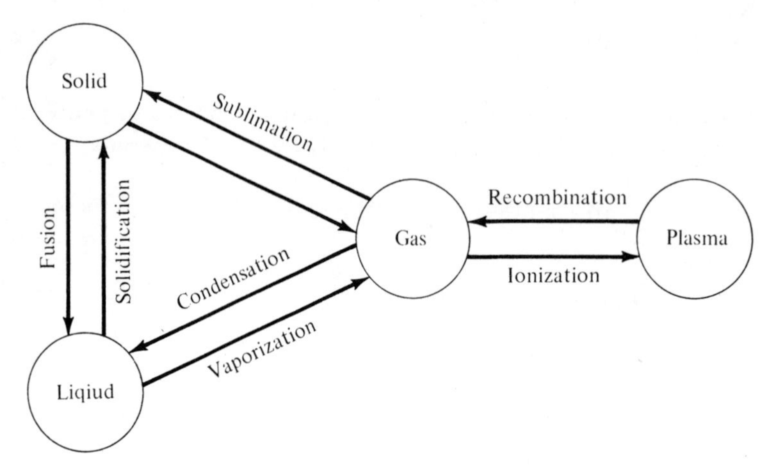

Figure 7-13

Plasma—freely moving electrons (negative electric charge) and ions (positive electric charge)—occupies an exceptional position. It can only be reached by a transition from the gaseous phase. Since the chemical bonds are broken, compounds have no plasma phase.

The transition from one phase to another is characterized by transition temperature and transition energy. This energy is positive (set free) or negative (absorbed), depending on the direction of transition.

Throughout the living world the vaporization energy is of utmost importance because it makes cooling of inner and outer surfaces possible. Using the thermal energy gained by oxidation, animals and plants can regulate their intrinsic temperatures and thus stay within the narrow temperature range determined for life.

PROBLEMS

1. State the triple point of water in degrees Fahrenheit and in atmospheres.
2. Extract the numerical values for the two critical points of carbon from Figure 7-10. Give the error limits for each value.
3. An indispensable tool of nuclear physics is the bubble chamber. It is a reversed cloud chamber. Explain its action referring to the phase diagram.

4. The heat produced by the digestion of food would raise the body temperature into the fever range if the cooling system fails. Make reasonable assumptions to explain the temperature rise within 24 hours.
5. To liquify graphite a pressure of 70 atm is needed. What is the minimum temperature to achieve the transition?
6. The triple point of CO_2 is at $p = 5.4$ atm and $T = -56.6°C$. What happens to solid CO_2 (dry ice) at normal pressure if its temperature increases?
7. For a substance in its liquid phase there are two ways of transfer into a gas: either by heating at constant pressure or by reducing the pressure at constant temperature. Explain the equivalence of both ways.
8. Describe what happens to a real gas if its volume is decreased along curve T_1 and along T_4 in Figure 7-11.
9. Fever may lead to an increase in produced thermal energy of 20%. How much additional sweat is produced per hour to achieve the necessary additional cooling?
10. If the relative humidity is 80% at 30°C and the temperature drops, at what temperature will it begin to rain? See Figure 7-12.
11. If at 20°C the relative humidity measures 80%, compute the additional pressure caused by this water content of the air.
12. Devise methods to separate (a) emulsions, (b) smoke, and (c) smog.
13. Calculate the surface area of 2 gram water-aerosol in square meters. Assume a radius of the droplets of 10^{-5} cm.

FURTHER READING

R. J. HARRIS, ed., *Biological Applications of Freezing and Drying*. Academic Press, New York, 1954.

DAVID W. NEWMAN, ed., *Instrumental Methods of Experimental Biology*, chapter 6. Macmillan, New York, 1969.

chapter 8

TRANSPORT PHENOMENA

1. INTRODUCTION

In the most general sense of the word, transport involves practically all sciences. At least it is possible to summarize most fields under this heading. Consequently, we must clearly understand what we want to cover in this section.

Some aspects of transport have already been introduced in Chapter 6 under the subheading Thermal Energy Transfer. Later in Chapter 14, we will encounter one of the most important transport phenomena, the transfer of information. In this chapter, we will study phenomena such as Brownian motion, diffusion, osmosis, and electrophoresis. These may be called transport in the narrowest sense of the word.

We will examine both passive and active transport. During active transport, energy must be supplied to the transporting system from the outside. Passive transport does not require an external energy source.

EXAMPLE:

The circulatory system is an example of active transport, where the heart supplies the energy to transport blood through arteries, capillaries, and veins. Blood corpuscles in turn carry oxygen, carbon dioxide, and other supplies. The subsequent transport of oxygen through cell membranes is partly active (via electric forces) and partly passive (via diffusion).

1.1 Stationary and nonstationary transport

In *stationary transport*, the same amount of a substance is transported per unit time through the same cross section. In *nonstationary transport*, the amount transported changes per unit time.

EXAMPLE: VOLUMES OF BLOOD PUMPED BY THE HUMAN HEART

(See Figure 8-1.)
Between midnight and 8 a.m. the same amount of blood per unit time (100 cm^3/s) is transported by the heart. During the active

part of the day—between 9 a.m. and 10 p.m. —the transport is still stationary but on a higher level (150 cm^3/s). In the transition phases between 8 a.m. and 9 a.m., and between 10 p.m. and midnight, a nonstationary transport is indicated: the amount of blood pumped per unit time changes. This example is highly idealized, but it still shows the difference between stationary and nonstationary transport.

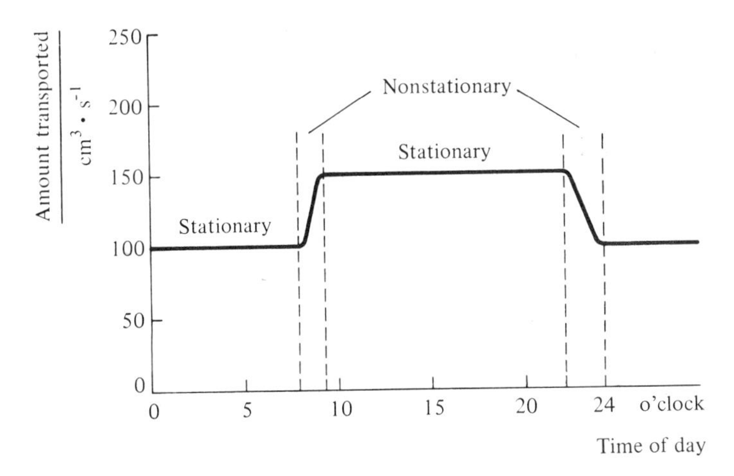

Figure 8-1 Volume of blood transported by the human heart as a function of the time of day (idealized). Regions of stationary and nonstationary transport are indicated.

In mathematical terms, if $n(t)$ is the amount of substance transported per unit time, then (we exclude the irrelevant case where $n(t) = 0$, that is, no transport at all):

Stationary transport,

$$n(t) = \text{constant}$$

or

$$\frac{d}{dt}(n(t)) = 0 \tag{8.1}$$

Nonstationary transport,

$$n(t) = \text{variable}$$

or

$$\frac{d}{dt}(n(t)) \neq 0 \tag{8.2}$$

1.2 Gradient

Whether or not a transport is stationary, a gradient will exist to initiate and sustain the transport. The term *gradient*, symbolized by Δ, is best explained by an example.

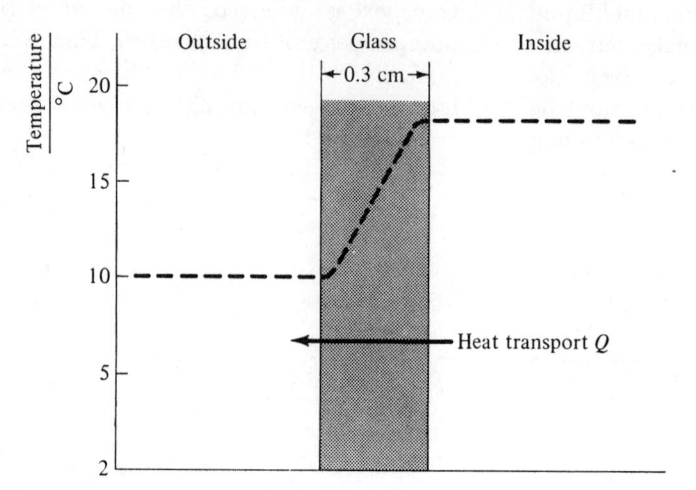

Figure 8-2 Temperature distribution in the immediate neighborhood of a window.

EXAMPLES:

Temperature gradient. The inside and outside of a window are at different temperatures, e.g., 10°C outside and 18°C inside. See Figure 8-2. A temperature gradient exists across the window.

$$\Delta\vartheta = \text{temperature gradient}$$

$$= \frac{\text{temperature difference}}{\text{spatial separation}}$$

$$\Delta\vartheta = \frac{18 - 10}{0.3} \text{ deg}\cdot\text{cm}^{-1} = 2.7 \frac{\text{deg}}{\text{cm}}$$

As long as $\Delta\vartheta \neq 0$, a thermal energy transfer takes place between the inside and the outside.

A gradient may be positive or negative; the direction of the transport involved is indicated by the sign.

Concentration gradient between two solutions of the same substance separated by a penetrable wall. See Figure 8-3. A net transport through the wall takes place until the concentrations on both sides are equal.

Pressure gradient between the inside and outside of a fresh soap bubble. See Figure 8-4. The

Initial System

Wall thickness 0.2 cm

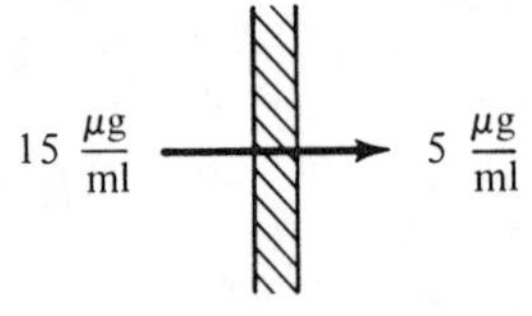

$$\text{Gradient} = \frac{15-5}{0.2}\ \mu\text{g}\cdot\text{ml}^{-1}\cdot\text{cm}^{-1} = 50\ \frac{\mu\text{g}}{\text{ml}\cdot\text{cm}}$$

Final System

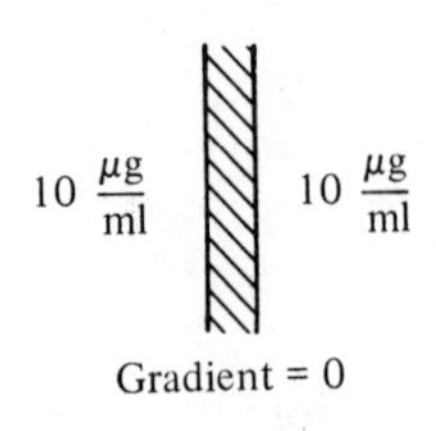

Figure 8-3

bubble will expand or contract until the pressure gradient Δp is zero. (Like all examples, this one is not perfect. The surface tension of the bubble must be taken into account; thus the gaseous pressures inside and outside are not equal. Which one is higher?)

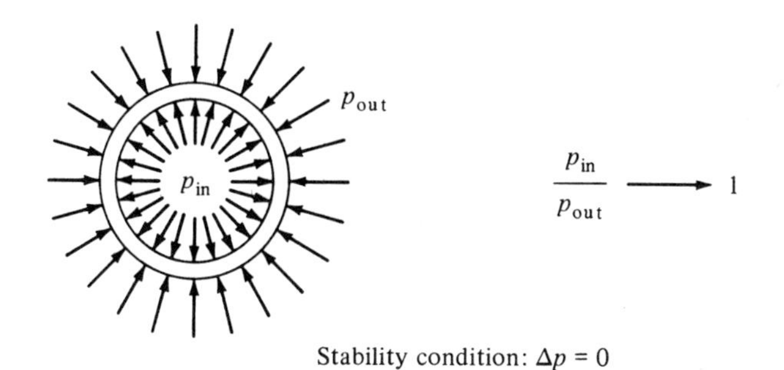

Stability condition: $\Delta p = 0$

Figure 8-4

Summary: Transport phenomena are governed in two ways by the gradient involved. The sign of the gradient indicates direction, and the magnitude of the gradient indicates the amount transported. As long as the gradient is constant, the transport is a stationary transport.

1.3 Net transport

Looking into the details of a transport, we shall often discover that it is not always a one-way street. The transport in one direction overlays a countertransport. Thus it is convenient to introduce a *net transport.* An example of net transport is the transport of waste products from tissue into the blood stream. As shown in Figure 8-5, the wall of a capillary

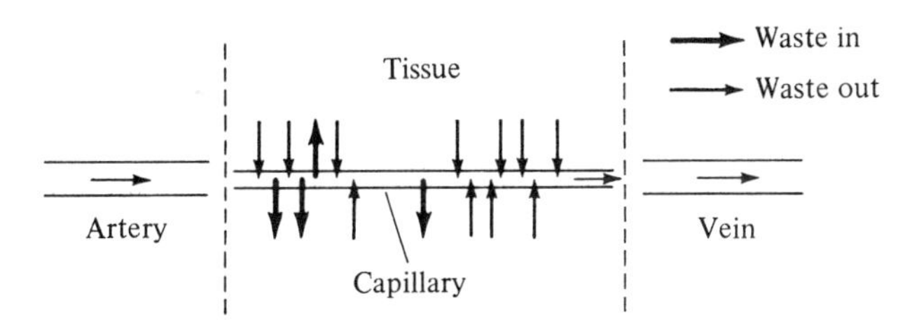

Figure 8-5

is permeable to waste products. The osmotic pressure (we shall see later what this is) p_w of the surrounding tissue drives the metabolic waste products into the blood stream. The hydrostatic pressure of the blood, p_s, counteracts this transport by driving those products back into the tissue. Towards the venous end of the capillary, p_w is about three times larger than p_s and therefore more waste products are transported across the walls from the tissue into the blood stream. Some waste products are still driven back into the tissue; however, the net transport leads into the blood.

2. PASSIVE TRANSPORT

Energy is necessary to sustain a constant gradient. If this energy is not available, the gradient will decrease and the net transport through any given cross section is not only passive but also nonstationary. This makes sense. In nature all differences show the tendency to be equalized, or expressed in other terms: the gradients tend to be zero. This is a consequence of the most general law in nature: The intrinsic energy differences of any system will be minimized in the long run. At least this is the case in classical physics. All processes run down like a wound-up clock and must finally come to an end due to the lack of energy to sustain gradients. This was a much discussed topic at the turn of the century. But nobody needs to be scared; the ultimate *Wärmetod* lies far ahead in the future. In the chapter on thermodynamics, we shall dwell longer on this interesting topic.

2.1 Diffusion

Atoms and molecules are in eternal motion. The speed of this motion depends on temperature, the type of particle, and the phase of matter. The direction of motion is not ordered. The particles undergo collisions with each other, changing their directions all the time.

The speeds of the individual constituents are not uniform, but display a characteristic distribution. Theoretical considerations, which agree well with experiments, lead to the *Maxwell* distribution* for the speed of gas molecules; see Figure 8-6. Due to this intrinsic motion, gas molecules will move into any available space until this space is occupied uniformly, i.e., until the gas's density is uniform over the available space.

The phenomenon just described is called *diffusion*. In the narrowest sense of the word, diffusion takes place if one gas moves into a space already occupied by another gas. The term diffusion also applies to the movement of a gas into empty space, the movement of a gas into a fluid or a solid, and the intermixing of fluids with nearly equal densities.

Let us return to passive transport. Since molecules have an intrinsic speed, a passive transport of those molecules takes place without energy supplied from an outside source. As long as the density is not homogeneous, this passive transport via diffusion is also a net transport. Finally, the net transport terminates: the number of molecules moving from any given direction is equal to the number of molecules moving into the opposite direction. The available space is then occupied with uniform density. (Remember that density is a macroscopic quantity. It does not make sense to talk about density if the volume is very small.)

* James Clerk Maxwell (1831–1879), physicist. Investigated theoretically the interrelation between magnetism and electricity (Maxwell relations). Formulated mathematically the kinetic theory of gases. He also made contributions to physiology, especially in color perception. More about him in C. Domb, *J. C. Maxwell and Modern Science*, London, 1963.

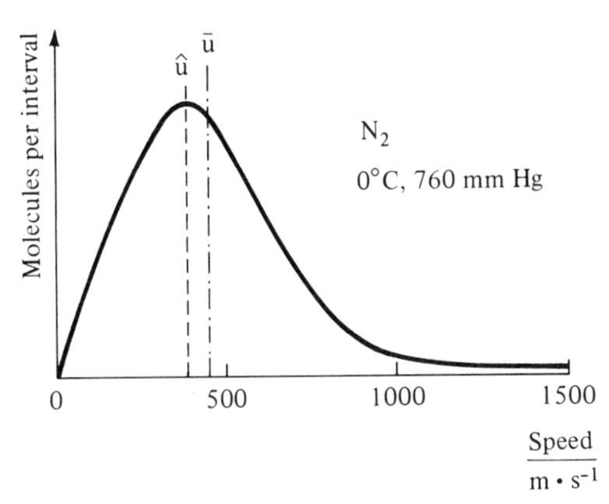

Figure 8-6 Maxwell's speed distribution of nitrogen molecules at 0°C. The most probable speed ($\hat{u}$) is 402 $m \cdot s^{-1}$; the average speed ($\bar{u}$) of the molecules is 455 $m \cdot s^{-1}$.

Imagine a separating area within the space available for diffusion. See Figure 8-7. The diffusion across this area is described by Fick's law:*

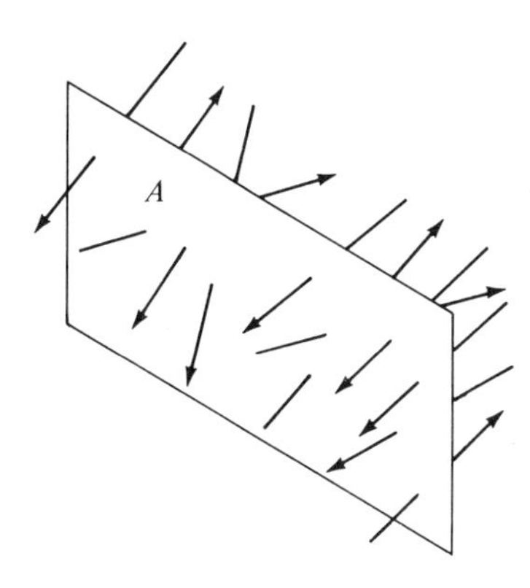

Figure 8-7

$$\frac{m}{t} = -D \cdot A \cdot \Delta \tag{8.3}$$

where

m: net amount of substance diffusing through area A during time t,
D: diffusion coefficient, measured in $cm^2 \cdot s^{-1}$,
A: area of penetration,
Δ: concentration gradient across area A.

Δ is measured in $g \cdot cm^{-3}$ per cm, hence in $g \cdot cm^{-4}$. Table 1 shows some typical diffusion systems and the corresponding values for D. The diffusion coefficient increases with increasing temperature.

* Adolf Fick (1829–1901), physiologist. Formulated many physiological phenomena in exact terms, applied thermodynamics to muscular activity, and introduced the first (and still used) exact method to measure the time-volume relation of the heart.

Table 1 DIFFUSION COEFFICIENT FOR VARIOUS SYSTEMS

Diffusion of	Into	Temperature (in °C)	Diffusion Coefficient (in cm^2/s)
CO_2	air	0	0.14
H_2	air	0	0.634
water vapor	air	8	0.24
glycerine	water	10	4×10^{-6} or 0.34 cm^2/day
NaCl	water	15	1.1×10^{-5}
sugar	water	12	3×10^{-6}
urea	water	15	1.1×10^{-5}

Application: Brownian motion.* Early in 1827, the botanist Robert Brown accidentally made a discovery which for the first time allowed the direct observation of a molecular phenomenon. The term *direct* refers to those sceptics of the past century who did not believe that matter is made up of tiny particles, of molecules. They wanted to see it with their own eyes before they would believe it. Brown did not actually observe the molecules directly in his microscope—this was reserved for the electron microscope of the twentieth century. What he did see was an erratic, irregular motion of tiny particles. Looking through his microscope at soot particles suspended in a fluid, he observed each in restless motion on a zigzag path. See Figure 8-8.

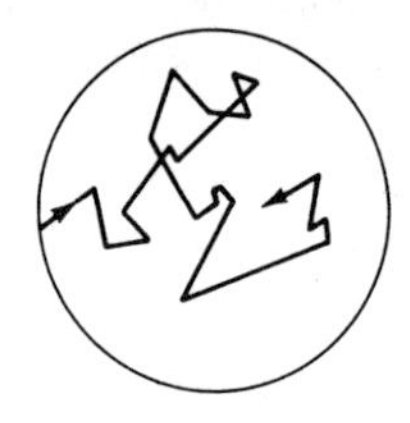

Figure 8-8

The momentary motion occurred in a straight line until the particle abruptly changed its direction. There was but one explanation: those particles, having a mass of about 10^{-18} g and moving at about 0.1 cm/s, changed their direction of motion whenever they were hit by a molecule of the fluid. An elementary calculation shows that this is a reasonable explanation: During a collision, the direction of motion of two colliding particles will change markedly only if both have comparable momenta.

The magnitude p is given by $|\mathbf{p}| = m \cdot v$, where m denotes mass and v is speed. (Refer to Chapter 4.) Some approximate values of momentum are as follows.

p of a water molecule at room temperature:

$$p_{\text{water}} = 3 \times 10^{-23} \times 5 \times 10^{3}\ \text{g} \cdot \text{cm} \cdot \text{s}^{-1} = 1.5 \times 10^{-19}\ \text{g} \cdot \text{cm/s}$$

p of a suspended soot particle at room temperature:

$$p_{\text{soot}} = 10^{-18} \times 0.1\ \text{g} \cdot \text{cm} \cdot \text{s}^{-1} = 10^{-19}\ \text{g} \cdot \text{cm} \cdot \text{s}^{-1}$$

The momenta of the colliding partners are of the same order of magnitude.

The microscope reveals to the observing eye only the movement of one partner—a sufficient proof for the existence of the other.

This is a convenient place to introduce two quantities widely used in molecular physics:

* R. Brown (1773–1858), botanist. Observed in 1827, through a microscope, the irregular, thermal motion of tiny suspended particles.

Mean free path (symbol l) is the average distance a molecule moves between collisions. Its order of magnitude is 10^{-5} to 10^{-6} cm.

Mean collision frequency (symbol Γ) is the number of collisions of the same molecule per second. Its order of magnitude is $10^9\ s^{-1}$.

The above quantities contain the term *mean* because the distance traveled by the molecule between collisions varies around a mean value. It is a statistical quantity. The same consideration applies to collision frequency. The mean free path is given by the relation:

$$l = \frac{\bar{v}}{\Gamma} = \sqrt{\frac{3kT}{m\Gamma^2}} = \frac{1}{\sqrt{2}\,\pi n d^2} \tag{8.4}$$

where

l: mean free path,
$\bar{v}$: mean molecular speed,
Γ: mean collision frequency,
T: absolute temperature,
m: mass of molecule,
n: number of molecules per cm^3,
d: molecular diameter,

and

k: Boltzmann's constant* $= 1.38 \times 10^{16} \frac{\text{erg}}{°\text{K}}$.

Molecules in motion represent a transport phenomena. l and Γ allow us to calculate [with the help of Equation 8.4] how fast this transport on the molecular level occurs. It is a passive transport. As long as the density of the molecules remains constant, there is no net transport. A net transport occurs only if the mean free path is larger in one direction than in any other. A density gradient produces a preferred direction. In this preferred direction, the collision frequency will be smaller and the molecules on the average will travel in this direction. Consequently, there is a connection between diffusion coefficient and mean free path,

$$D = \frac{1}{3}\bar{v}\cdot l \tag{8.5}$$

where

D: diffusion coefficient,
$\bar{v}$: mean molecular speed,
l: mean free path.

2.2 Osmosis

A special case of diffusion of utmost importance in nature is osmosis. *Osmosis* is the diffusion of a solvent through a membrane in response to a concentration gradient. (In the life sciences, osmosis refers to the transport of water only.)

* Ludwig Boltzmann (1844–1906), physicist. Main contributions in statistical thermodynamics and black-body radiation.

Figure 8-9 illustrates osmosis. A sugar solution inside a tube dips into water. The lower end is closed with parchment, and the upper end is open. Observing the system for a while, you will discover that the fluid column within the tube rises until it reaches a final height above the original water level. Further analysis shows that a net transport of water (the solvent) occurred through the parchment (the membrane) into the tube. This transport is in response to the concentration gradient of the sugar solution across the membrane. The function of the membrane is to sustain the concentration gradient. In general, the membrane will be semipermeable, that is, it allows the solvent to pass, but not the solute. It is actually a one-way passage open to the solvent only. The net transport ends as soon as the hydrostatic pressure of the raised sugar solution column forces as much water back through the membrane as is entering.

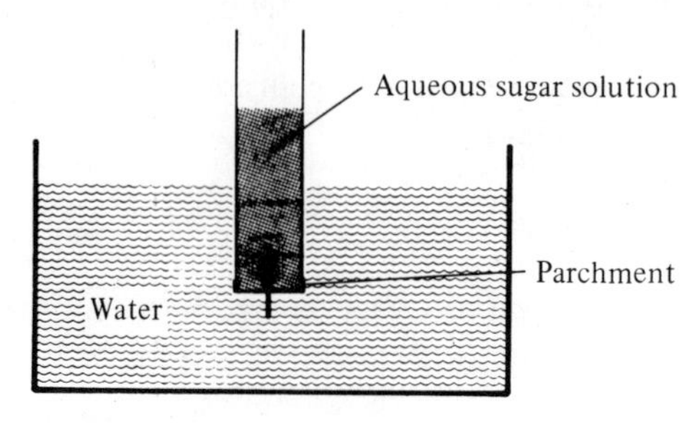

Figure 8-9

Another example is shown in Figure 8-10. In a plant cell the protoplast is normally held firmly against the cell wall, giving the plant a certain rigidity. If the cell is placed into a concentrated aqueous salt solution, water molecules will leave the cell. The protoplast contracts, and the cell becomes plasmolyzed. (See Figure 8-11.) This process is reversed if the salt solution is replaced by fresh water.

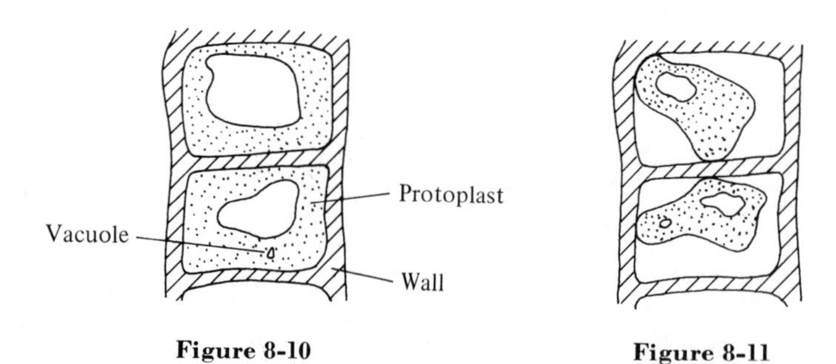

Figure 8-10 **Figure 8-11**

This is a convenient opportunity to introduce the osmotic pressure (symbolized Π): The *osmotic pressure* (caused by a concentration gradient across a semipermeable membrane) forces the solvent into the solution. An arrangement such as the one shown in Figure 8-9 may easily serve to measure the osmotic pressure—to act as an osmometer. Π is equal to the hydrostatic pressure as soon as the net transport across the membrane ends. This is fine as long as the amount of fluid is sufficient. But in the life sciences, the quantities involved are generally very small. Here indirect methods are called for. In observing the freezing point, for instance, tiny amounts of a substance are drawn into a capillary tube and frozen. The melting is observed under a microscope and compared with standard solutions.

A living cell may also serve as an osmometer. It will shrink or swell according to the osmotic pressure of the solution to be measured. However, a word of caution: The movement of water across the cell membrane is in response to the concentration of osmotically active molecules only. Whether or not the molecules are osmotically active must be determined experimentally. This makes sense if you realize that the term *semipermeable* for a living cell membrane should really be replaced by *selectively permeable.*

The law of osmosis was experimentally determined by van't Hoffs:*

$$\Pi \cdot V = \nu \cdot R \cdot T \qquad (8.6)$$

where

Π: osmotic pressure of the solution,
V: volume of solution,
ν: number of moles of the solute present in the solution,
R: molar gas constant $= 8.31 \times 10^7$ erg$\cdot$°K$^{-1}\cdot$mol^{-1},
T: temperature.

The law is valid only for weak solutions. Observe that the formula is independent of the type of solvent. It looks very much like the general gas law (as formulated in Chapter 5), and therefore the osmotic pressure may be interpreted as the gaseous pressure exerted by the solute if the solvent is removed.

There is an important difference between gas pressure and osmotic pressure. The osmotic pressure is limited to systems separated by a semipermeable wall.

The effect which forces the solvent across the membrane into a solution should not really be called pressure! Although it is measured in pressure units and interpreted as an analogy to gaseous pressure, the notion is still misleading. Values of hundreds of atmospheres for the osmotic pressure are by no means extremely large, but you may find it difficult to understanding how those fragile cell membranes will survive this enormous "pressure." The best way out of this confusion may be to think of osmosis as a transport mechanism that is sustained by concentration gradients, and leave it at that.

* J. H. van't Hoffs (1852–1911), chemist.

Living systems rely on osmosis for their supplies. Their cells are separated by semipermeable walls which permit the selection of only those molecules the cell needs. These subdividing walls are not only semipermeable, but quite often the semipermeability is variable.

EXAMPLES:

The sap moves into the leaves of even the highest trees solely due to osmotic action. It passes through the cells of the xylem on the inside of the cambium. The osmotic pressure reaches values up to 20–30 atmospheres, leading to sap speeds up to one meter per hour.

Changing the permeability of their membranes, cells may shrink or swell. This is the method by which plants fold blossoms and leaves or change the rigidity of their uppermost tips. See Figure 8-12.

Certain stages of malaria are accompanied by a pathologic change in permeability. In response, the red blood cells swell until rupture and release their hemoglobin to the plasma. Drinking distilled water in sizable quantities will produce a similar effect. The large concentration gradient across the stomach walls forces the water into the surface cells until they rupture. Internal bleeding is the consequence.

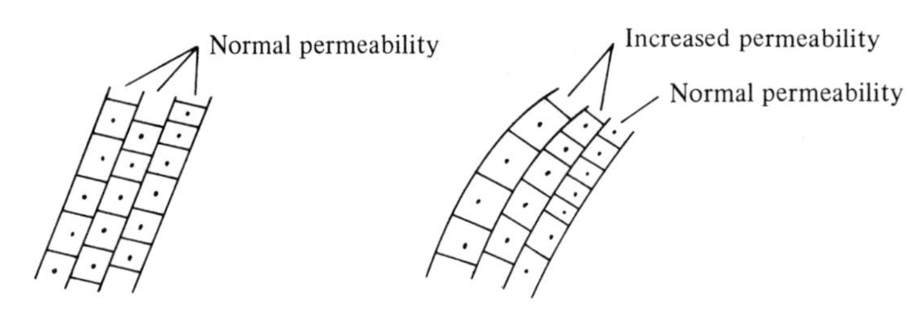

Figure 8-12

2.3 Application: separating uranium isotopes

Most chemical elements have some isotopes. The isotopes of an element are chemically indistinguishable; only their masses differ. A well-known pair of isotopes is uranium 238 and uranium 235. The numbers indicate how many nucleons (protons and neutrons) make up an individual atomic nucleus of uranium.

As the explosive for atomic bombs (Hiroshima-type) or—more desirable—as fuel for atomic power reactors, the isotope U-235 occupies a key position. But how do we separate the lighter isotope from the natural admixture? Chemical methods fail because isotopes of the same element are chemically identical. The small mass difference 238 to 235, that means 1.3%, offers a solution. Both isotopes are separated by diffusion, a process depending on mass. But first the natural uranium is converted into a gas, UF_6, to make it suitable for gas diffusion.

The average kinetic energies of both isotopes are equal:

$$\frac{m_{235}}{2} v_{235}^2 = \frac{m_{238}}{2} v_{238}^2$$

where

m: mass,
v: average speed.

The indices refer to the individual isotopes. Solved for the speeds,

$$\frac{v_{235}}{v_{238}} = \sqrt{\frac{238}{235}} = 1.006\ 5$$

This means that the average speeds of both isotopes differ by only 0.65%. Nevertheless, U-235 diffuses faster than U-238 and thus allows the separation in a diffusion column.

UF_6 containing the normal isotopic mixture—which by the way is equal for all members of our solar system—is fed into a diffusion pipe with walls acting as semipermeable membranes. These walls are not really semipermeable, but if the size of the pores is suitably chosen, the lighter compound (here $(U\text{-}235)F_6$) will pass more easily.

Figure 8-13 shows such a diffusion tube. The gas enters from the left. Some of the molecules diffuse through the semipermeable walls into the surrounding mantle. Since the lighter component diffuses faster, the gas

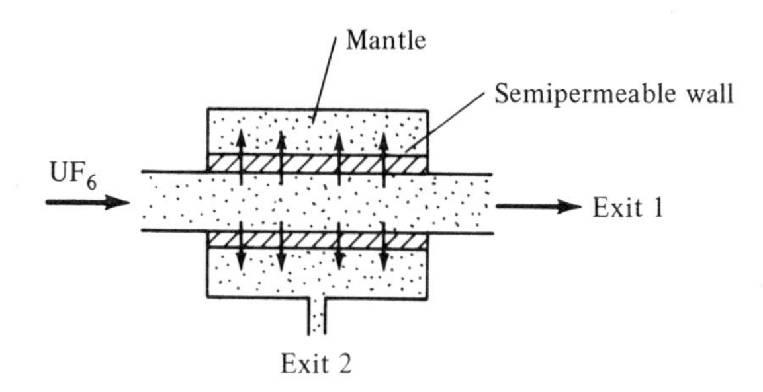

Figure 8-13

within the mantle will be slightly enriched by $(U\text{-}235)F_6$. At exit 1 the heavy compound is collected, and at exit 2, the lighter one. This very small difference in mixture is enhanced by coupling both exits to the next diffusion tubes to form a cascade. See Figure 8-14.

With each additional stage, the downward-moving part will contain more of the lighter compound. Actually, the system in Figure 8-14 is by no means complete. Each stage, for example, is again subdivided to return a larger fraction of the heavy compound in the upward direction. The separating effect of the entire cascade depends on the number of stages. A cascade of forty stages is sufficient to separate the neon isotopes (Ne-20 and Ne-22) completely. Here the speed relation

$$\frac{v_{\text{Ne-22}}}{v_{\text{Ne-20}}} = \sqrt{\frac{22}{20}} = 1.049$$

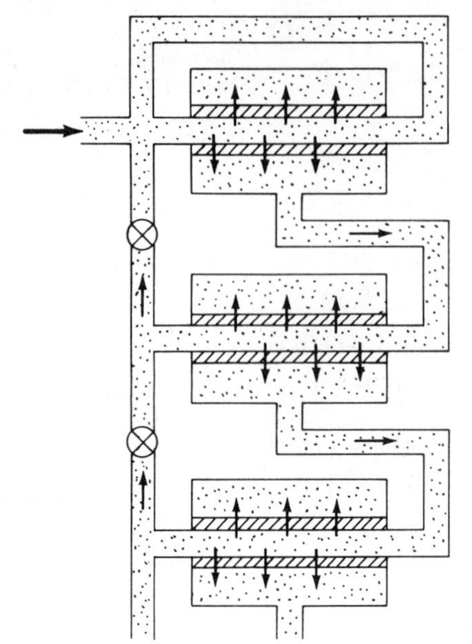

Figure 8-14

is very favorable compared with the uranium isotopes. A uranium diffusion cascade has a few thousand stages! To gain separated isotopes by the kilogram—a must for bombs and reactors—not only are a few thousand stages necessary, but also each stage must enclose a volume of some cubic meters.

Recently, other methods for isotope separation—such as the gas ultracentrifuge—have been employed to avoid the tremendous technological and financial undertaking of a gaseous diffusion plant.

3. ACTIVE TRANSPORT

During active transport, energy is required from the outside for a net transport.

3.1 Transport by mechanical pumps

A pump produces a pressure gradient and in this way makes the medium (it may be a gas, liquid, or powder) flow against an outside force. Pumping water upward from a well is achieved against the gravitational force. As soon as the pump stops, the water will assume a level determined by gravity. Most pumps belong to one of the three following types:

Piston Pump. In Figure 8-15(a) the piston moves to the right. As a result, a low pressure develops inside the chamber, the lower valve is drawn, and liquid is sucked in. In Figure 8-15(b), the piston moves to the left, producing an overpressure inside the chamber, which in turn opens the upper valve and presses the liquid upward and out.

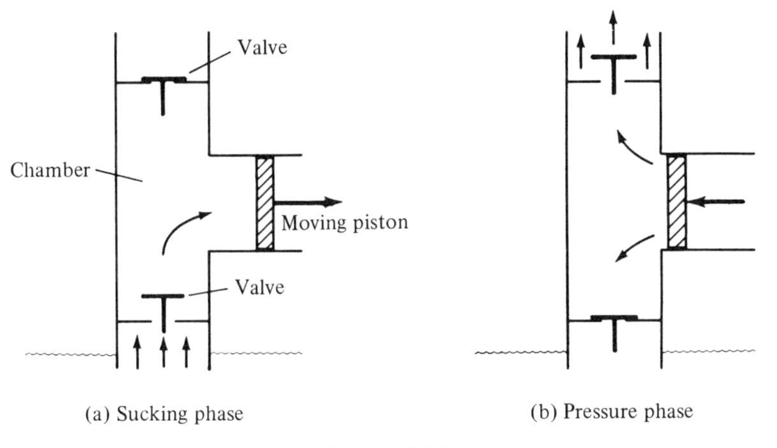

Figure 8-15

Whether the piston is replaced by an elastic membrane or rotates inside a suitable chamber, the pumping principle remains the same. An outside energy source drives the piston and facilitates the transport.

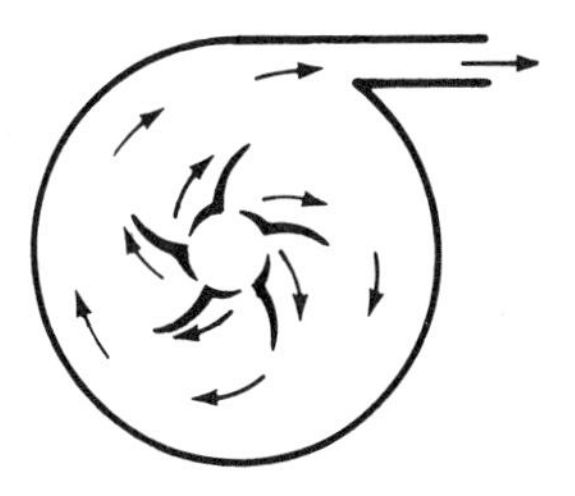

Figure 8-16

Centrifugal Pump. The liquid to be pumped enters a chamber with turning blades axially. It undergoes a radial acceleration, and the radial force drives it out of the chamber. See Figure 8-16.

Jet Pump. Here Bernoulli's theorem (see Chapter 5), which states that the sum of static pressure p_s and dynamic pressure p_d is constant, is employed. Figure 8-17 demonstrates the working principle.

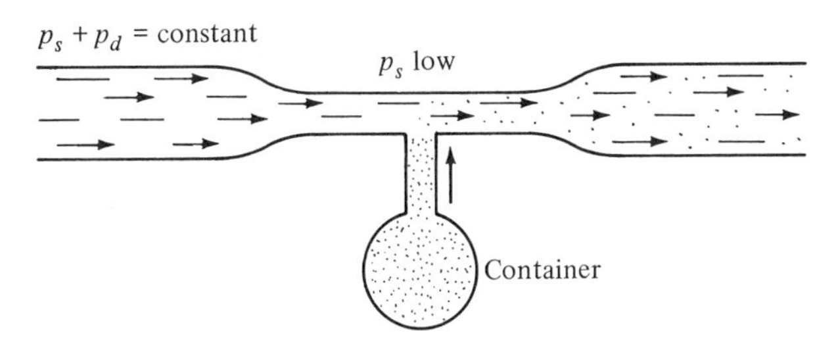

Figure 8-17

At the position where the cross section of the horizontal tube narrows, the static pressure drops. The medium to be pumped out of the container is sucked into the horizontal stream and transported away. Figure 8-18 shows a technical application of this pumping principle, the mercury vacuum pump.

There are many variations and combinations of the foregoing three types of pumps. A particular type is selected according to the desired application: maximum achievable pumping rate (measured in liters/

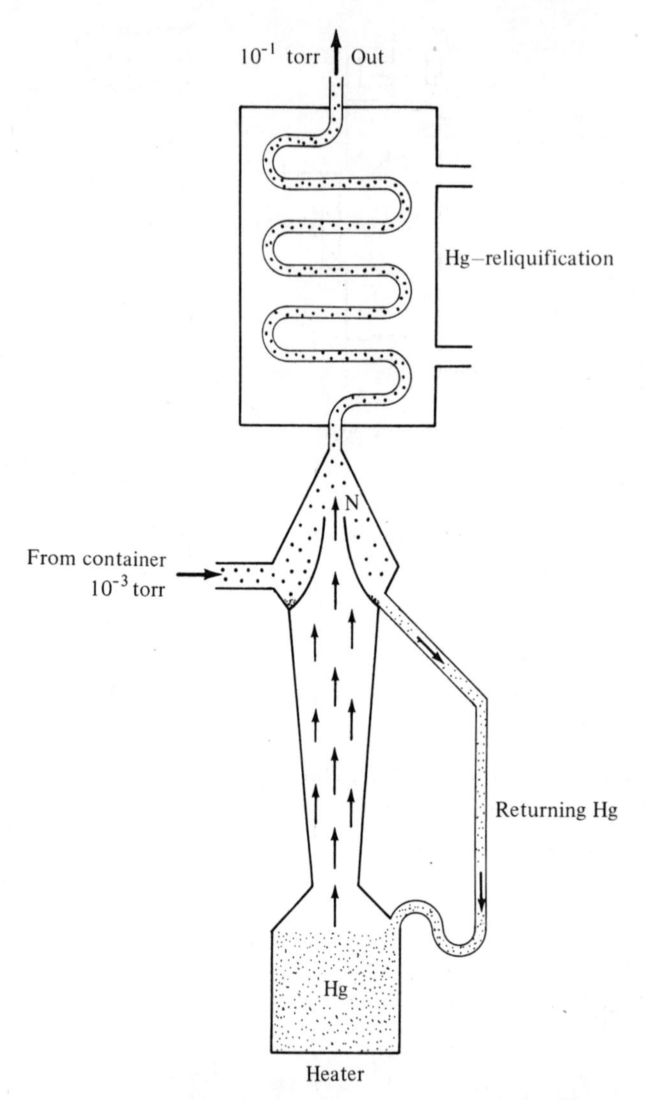

Figure 8-18 Mercury vacuum pump. A beam of fast Hg molecules leaves the nozzle N, causing a low-pressure area. Air molecules are sucked out of the container C and transported upward. The Hg steam is reliquified and flows back into the Hg container to be heated up again. Vacua down to 10^{-3} torr are achievable with this simple apparatus.

second), maximum obtainable pressure difference between intake and output (important for vacuum pumps), or uniform pumping rate (here the piston pump would not be the optimum solution).

3.2 Transport by electrophoresis

Although we have not yet discussed electrical forces, we shall nevertheless examine an active transport mechanism which is important for analyzing and separating small amounts of substances. Recall from high school that electrically charged particles will migrate if placed between electric poles. A positive particle moves to the negative pole, and a negative one moves to the positive pole. If brought into suspension or solution (with a suitable solvent or *buffer*), molecules will obtain an electric charge. If then placed between two electrodes, those charged molecules will migrate toward the respective poles. The speed of migration is in general determined by the net charge, and its sign dictates the direction of motion (Fig. 8-19).

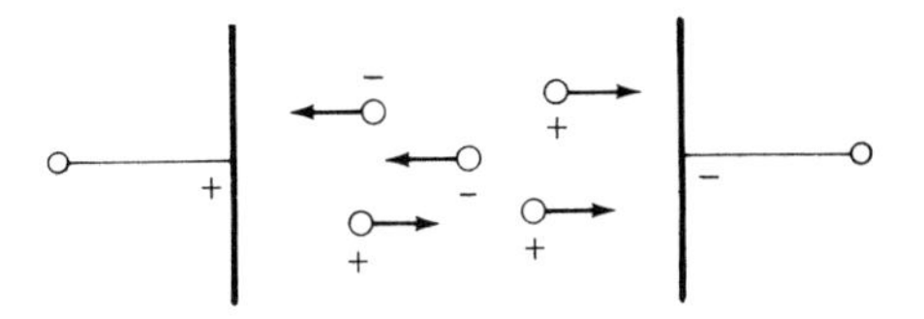

Figure 8-19

The movement of a solid phase with respect to a liquid at rest under the influence of an external electrical field is called *electrophoresis*. In general this word is reserved for the movement of coloid particles and macromolecules; otherwise, it is named *ionophoresis*.

Although other methods are possible, the favorite one is to use support media such as paper or a gel for stabilizing the electrophoretic system. Figure 8-20

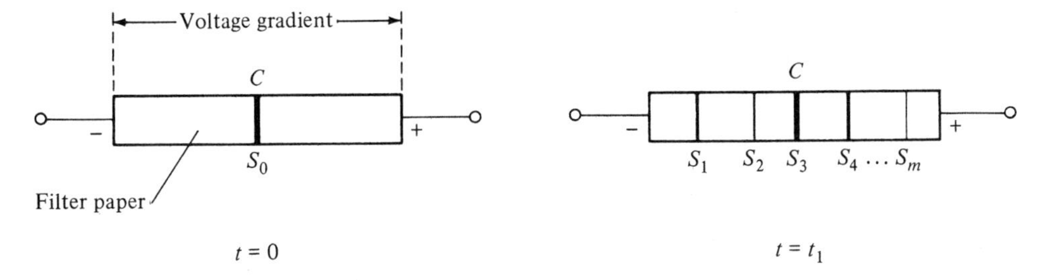

Figure 8-20

shows how it works for *paper electrophoresis*. A strip of filter paper bathes in the buffer. Both ends are connected to electric poles. The substance (S_0) to be analyzed is dropped onto the center of the strip (C) at time $t = 0$ and starts to move. Different components of the substance will migrate with different velocities. The migratory speed differs and sometimes also the direction of migration (either to the left or to the right). At time $t = t_1$, the various compounds (S_1 to S_n) are transported to different positions. The paper is taken out of the buffer and dried. Then the separated substances are analyzed.

The voltage gradients used are between a few and hundreds of volts/cm. The migratory speeds vary from a few cm/min to some cm/hour. The flow rates are small, rarely more than a few ml/hour. With suitable modifications, a stationary transport may be achieved.

The energy for this active transport is supplied by the electrical power sustaining the voltage gradient. The method just described looks like an esoteric transport system (which, by the way, has innumerable pitfalls and involves 1001 tricks), but it is indispensable for separating and analyzing the tiny amounts of protein and amino acids which are necessary for life.

3.3 Carrier-mediated transport

To sustain life, substances such as amino acids, sugar, and peptides are transported across biological membranes. In most cases the transport is achieved by osmosis, which means that the internal structure of the separating membranes supports a concentration gradient (semipermeable membranes). But there is another transport mechanism: Look at the concentration of K^+ and Na^+ ions inside and outside a cell. Figure 8-21 shows an erythrocyte (red blood cell) suspended in blood plasma.

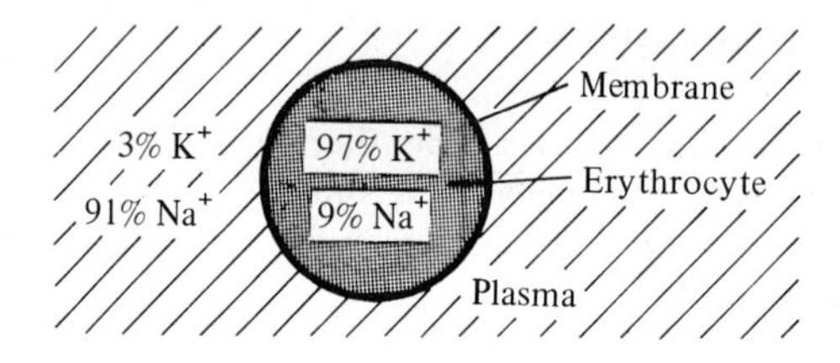

Figure 8-21

The concentration of K^+ ions in the plasma is 3% of that inside the erythrocyte, while the concentration of Na^+ ions outside the cell is tenfold compared with the inside. Analyzing the permeability of the membrane which separates inter- and extracellular fluid, we see that an equilibrium should be achieved within a few hours by osmosis. But it is not! The only feasible explanation is that an active transport mechanism is at work. The mechanism is named K-pump or Na-pump, and it is assumed to be a carrier-mediated transport: The K^+ or Na^+ ions combine with a special carrier molecule at the surface of the membrane. The formed complex passes somehow to the other side and breaks apart. The carrier is confined to the membrane; it acts merely as a shuttle. Passive transport by osmosis counteracts this motion, preventing a complete depletion on one side. Carrier-mediated transport is applied in modern medical treatment. Some drugs are coupled to carriers (they travel in disguise) and are thus able to penetrate into areas which have a protective system against those drugs in pure form.

SUMMARY

Without transport of matter and energy, nature is inconceivable. To describe the transport of matter, an overall description is sufficient; for an understanding we search at the molecular level.

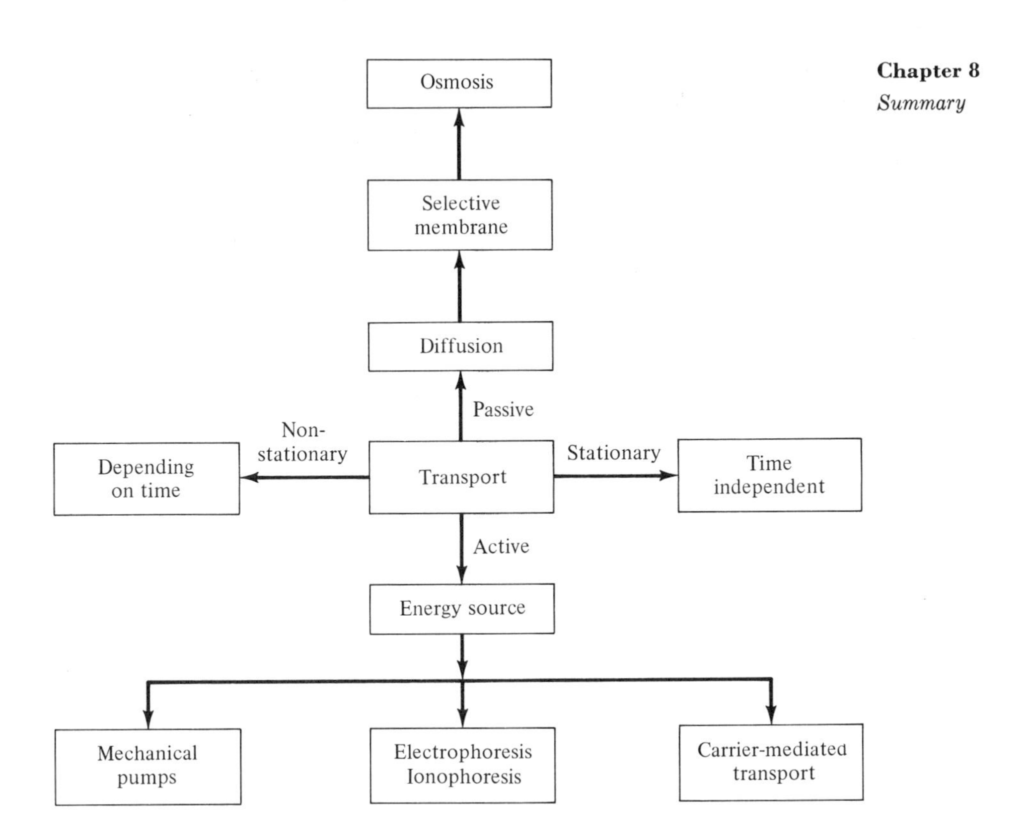

On both levels, stationary (amount transported per unit time is constant) and nonstationary (amount transported per unit time varies with time) transport occurs. A net transport in one direction may partly screen a countertransport, observable only at the molecular level.

Active transport requires an obvious energy source. It may be a man-made mechanical pump (piston, centrifugal, jet) or a more complex device like the heart (the source of active transport of blood, which in turn actively transports the blood cells). Electrophoresis belongs in the same category: electrical energy drives dissolved or suspended matter directly through a buffer medium.

A passive transport occurs in response to a *gradient*, a spatial difference of quantities such as temperature, concentration, density, electric field, or pressure. To sustain this gradient, energy is required, although the source of energy is often not obvious. *Diffusion*, the spontaneous intermixing of gaseous or liquid substances, is a passive transport phenomena. The same applies for *osmosis*, the spontaneous transport across a selective membrane.

Molecules of a gas or a liquid move eternally and in a mostly unordered way. They collide among themselves with the *mean collision frequency* after traveling a *mean free path*. The term *mean* is necessary because each molecule is an individual with an unpredictable future. Only the response of a large number of individual molecules can be described exactly. This intrinsic movement of molecules explains *diffusion:* molecules of one type move into the space between molecules of another type. Thus osmosis is a special case of diffusion: various molecules are separated by a (semipermeable) membrane, which allows only one kind to penetrate.

PROBLEMS

1. Determine the two gradients for the nonstationary transport of blood as shown in Figure 8-1.
2. In general, a rotary pump needs priming. Why?
3. Design an apparatus for continuous separation by electrophoresis.
4. The erythrocytes do not shrink or swell if placed in a 0.9% NaCl solution. Calculate the osmotic pressure inside the cell. Keep in mind that NaCl is practically completely ionized.
5. A gas diffusion plant is fed with UF_6, a highly toxic compound. Why isn't a less dangerous gaseous compound of uranium used?
6. Describe mathematically the drop in temperature across the window glass (see Figure 8-2).
7. Brownian motion: What do you observe if the momenta of the colliding partners are different by some orders of magnitude?
8. Estimate from Figure 8-6 how many molecules of the gas move with a speed twice $\hat{\mathbf{u}}$.
9. See Figure 8-3. Assume that the initial system represents the diffusion of NaCl into water. Calculate the amount of NaCl diffusing through an area A per second. ($A = 1\ \text{cm}^2$).
10. What is the approximate osmotic pressure necessary to drive the sap to the top of a redwood tree that is 130 m tall?
11. What is the maximum depth from which a piston pump can raise water?
12. (a) Calculate the mean free path, mean collision frequency, and diffusion coefficient for N_2-molecules as shown in Figure 8-6. (b) Calculate the mean free path of those N_2-molecules at high vacuum (10^{-6} mm Hg). How often will the molecules collide with each other and with the walls of their cylindrical vacuum container (diameter 12 cm)?

FURTHER READING

C. H. BEST, N. B. TAYLOR, *The Physiological Basis of Medical Practice*, Chapter 1, "The Cell." Williams & Wilkins, Baltimore, 1966.

A. CROWE, A. CROWE, *Mathematics for Biologists*. For the mathematically minded student: Chapter 3.5, "Passage of Substances Across a Biological Membrane." Academic Press, London, 1969.

A. C. GIESE, *Cell Physiology*, Chapter 12, "Movement of Water Across Cell Membranes." Saunders, Philadelphia, 1968.

D. W. NEWMAN, ed., *Instrumental Methods of Experimental Biology*, Chapter 3, "Zone Electrophoresis," and Chapter 13, "Osmotic Pressure Measuring Devices." Macmillan, New York, 1969.

R. ROSEN, *Optimality Principles in Biology*, Chapter 8, "Transport Across Membranes." Plenum Press, New York, 1967.

chapter 9

OSCILLATIONS, WAVES, AND SPECTRA

1. INTRODUCTION

Physics is a supporting science for the vast majority of students of the life sciences. A purely mathematical approach is not difficult, but it is tedious and may be boring. Avoiding the mathematics altogether is irresponsible because too many effects and applications in acoustics, optics, information, and submicroscopic physics rely on it. Therefore, you will find a compromise: about half of this chapter is devoted to the mathematical treatment, which in turn is illustrated by examples from acoustics. In this way I hope to serve two purposes; the chapter will be arranged clearly enough to serve as a reference, and we will spare ourselves an additional chapter on acoustics.

Since there are so many new quantities introduced within this chapter, it does not make sense to conclude it by a summary. The detailed table of contents will better serve this purpose.

1.1 Oscillations

An oscillation is the periodic change in time of an observable effect. The oscillation originates at a *source* (transmitter), transfers through a *medium* (or empty space), and is observed by a *detector* (receiver). See Figure 9-1. The transferring phenomenon is called a *wave*. A wave depends on time and space.

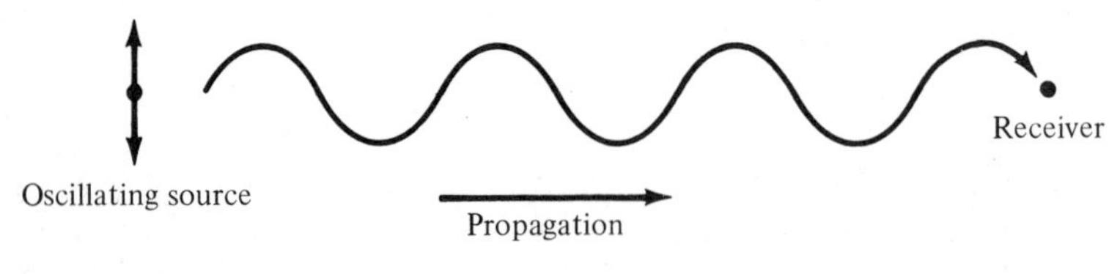

Figure 9-1

EXAMPLE:

Man produces a sound by a vibration of the vocal chord (source of sound) in the larynx. The oscillation of the chord leads to a periodic density change of the surrounding air which propagates through the air and is detected at the ear (receiver). The periodic density change of the air is called a *sound wave.*

We may represent the oscillation as a periodic displacement around an average position. Figure 9-2 shows a widely used and convenient notation. At time $t = t_0$ the oscillation starts at displacement $d = d_0$. At $t = t_1$ it reaches its maximum upward displacement d_{max} from the

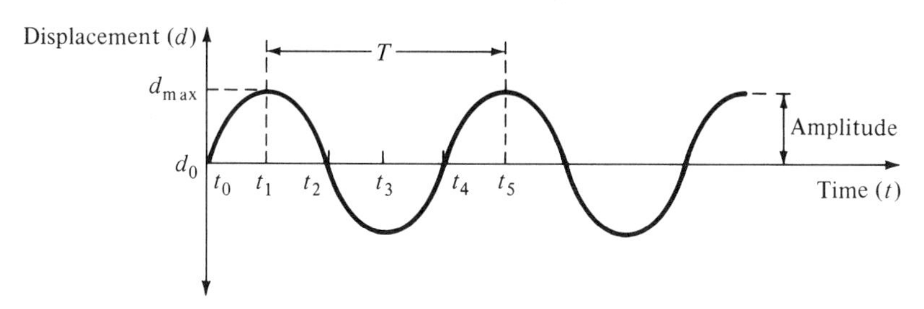

Figure 9-2

rest position. The distance $d_{max} - d_0$ is called the *amplitude* of the oscillation. Now the direction of oscillation reverses, passes through zero displacement at t_2, and again reaches a maximum displacement at t_3. Once more the direction of oscillation reverses and the displacement becomes zero at t_4. From here on, the oscillation follows the same pattern as from $t = t_0$. At every instant the oscillation is characterized by its displacement (the distance from the average position) and its direction (toward the average position or away from it). Both quantities together are called the *phase* of the oscillation. Figure 9-2 demonstrates that the oscillation repeats its phase periodically. For example, t_0 and t_4 have the same phase; t_1 and t_5 are also equal in phase.

1.2 Period and frequency

Period, symbolized T, is the time span between neighboring positions having the same phase. *Frequency,* symbolized ν, is the number of periods per unit time, that is

$$\nu = \frac{1}{T} \tag{9.1}$$

Frequency is measured in s^{-1}. There is a special unit for frequency, the hertz (abbreviated Hz),

$$1 \text{ Hz} = 1 \text{ s}^{-1}$$

(The hertz was already introduced in Chapter 3.) Figure 9-2 showed the dependence of the wave on time; Figure 9-3 is a space-dependent display.

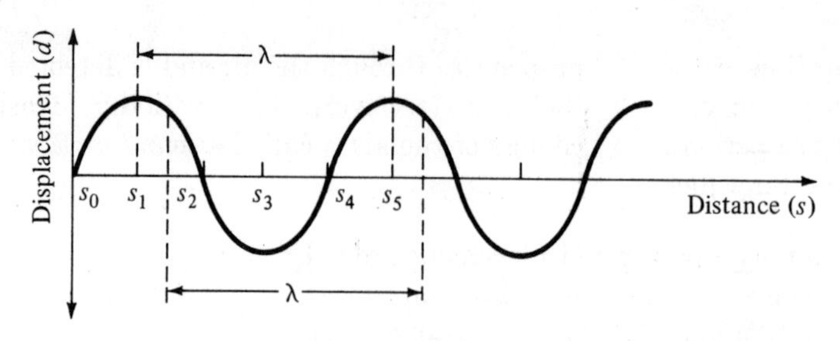

Figure 9-3

Again, the oscillation repeats its phase periodically. For example, positions s_1 and s_5 have the same phase, as do s_0 and s_4.

1.3 Wavelength and wave number

Wavelength, symbolized λ, is the distance between the nearest positions having the same phase (Figure 9-3). The wavelength is measured in cm. *Wave number*, symbolized σ, is the number of wavelengths per unit distance, that is

$$\sigma = \frac{1}{\lambda} \tag{9.2}$$

There is no special unit for the wave number; it is mainly used in spectroscopy and in atomic and nuclear physics.

The product of wavelength and frequency is called *phase velocity*, symbolized v:

$$v = \lambda \nu \tag{9.3}$$

Realize that v is the velocity of a phase which propagates from source to receiver. This is of utmost importance if a medium connects source and receiver. The particles making up the connecting medium do *not* move with v. It is the phase of the wave which moves with the phase velocity v.

EXAMPLE:

An audible source emits sound waves with a frequency between 20 and 20 000 Hz. The upper and lower limits of audibility vary from person to person and with age. The corresponding wavelengths are 1700 cm and 1.7

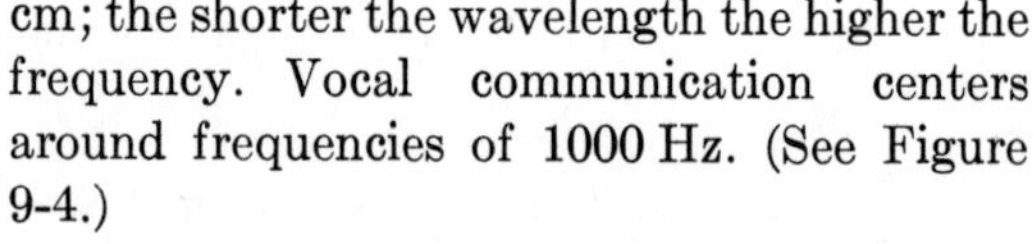
cm; the shorter the wavelength the higher the frequency. Vocal communication centers around frequencies of 1000 Hz. (See Figure 9-4.)

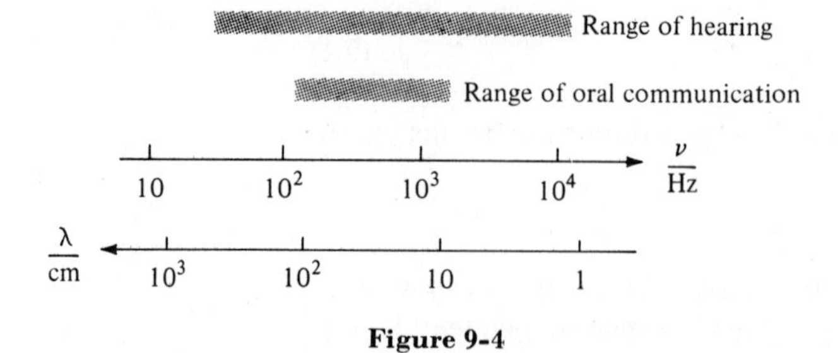

Figure 9-4

Sound propagates through air with a speed of about 330 m/s. The transferring air molecules do not leave their average positions at all.

1.4 Transverse and longitudinal waves

The direction of displacement with respect to the direction of the propagating wave leads to a discrimination:

Transverse wave: The displacement occurs perpendicular to the wave's direction (Figure 9-5). Figure 9-6 shows the range of some well-known transverse waves.

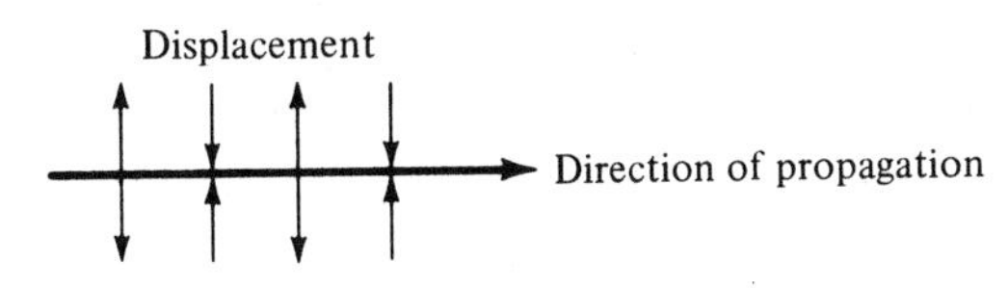

Figure 9-5

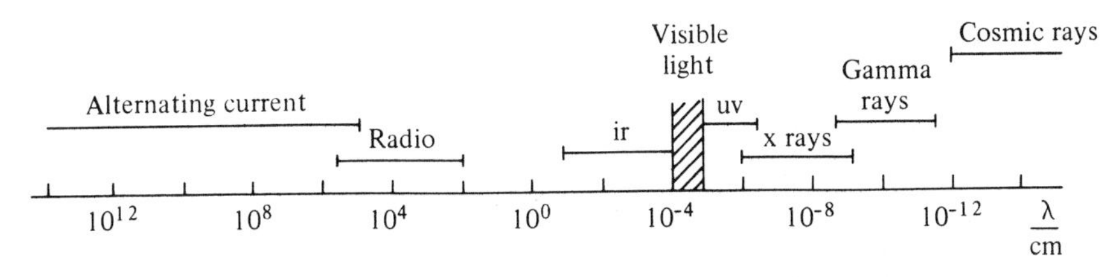

Figure 9-6

Longitudinal wave: The displacement occurs parallel to the wave's direction (Figure 9-7). Figure 9-8 shows the range of some well-known longitudinal waves.

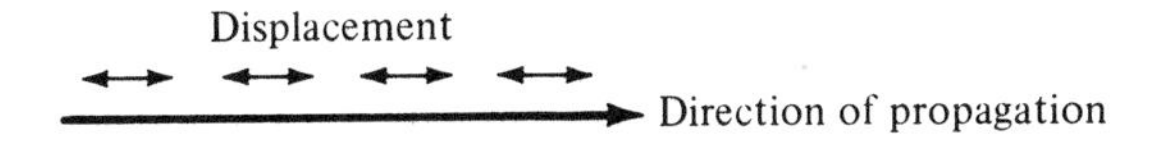

Figure 9-7

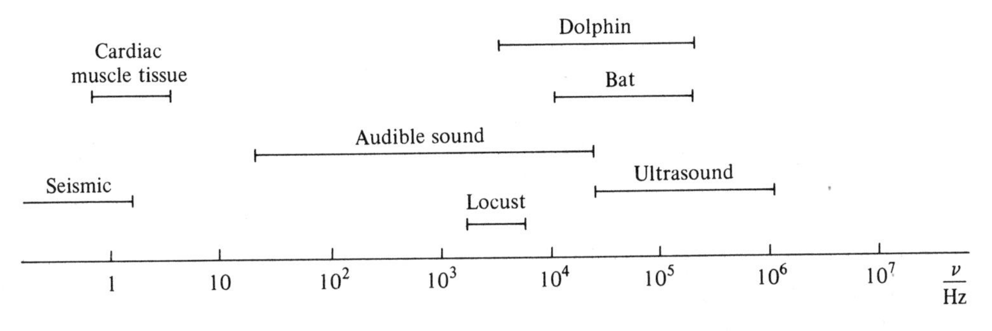

Figure 9-8

Most waves as observed in nature are conducted through a medium. Friction will occur and consequently the wave will be neither purely longitudinal nor purely transverse.

EXAMPLES:

The swell of the open sea is a transverse wave. The water molecules swing up and down, perpendicular to the horizon. The wave itself, its phase, propagates parallel to the surface of the sea.

Inside the human ear, at the *basilar membrane* a transverse wave develops whenever a sound reaches the ear. Figure 9-9 shows an idealized picture of the uncoiled basilar membrane. The phase velocity of the transverse wave at the basilar membrane varies between 150 and 1 m/s, depending on position and frequency. The actual amplitude is very small, about a few molecular diameters. The sound wave itself, which propagates through air, is a longitudinal wave. The air molecules swing parallel to the direction of sound propagation, therefore, the density of the air changes periodically along the path of the sound. Figure 9-10 shows an instantaneous representation of this local density change.

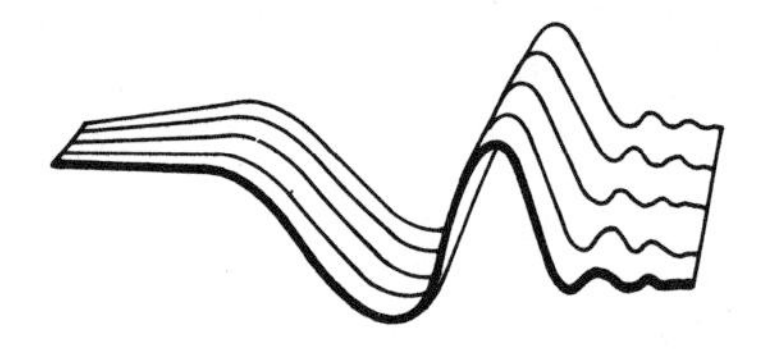

Figure 9-9

A good example of a real wave is a wind-blown field of wheat; Figure 9-11 shows an instantaneous picture.

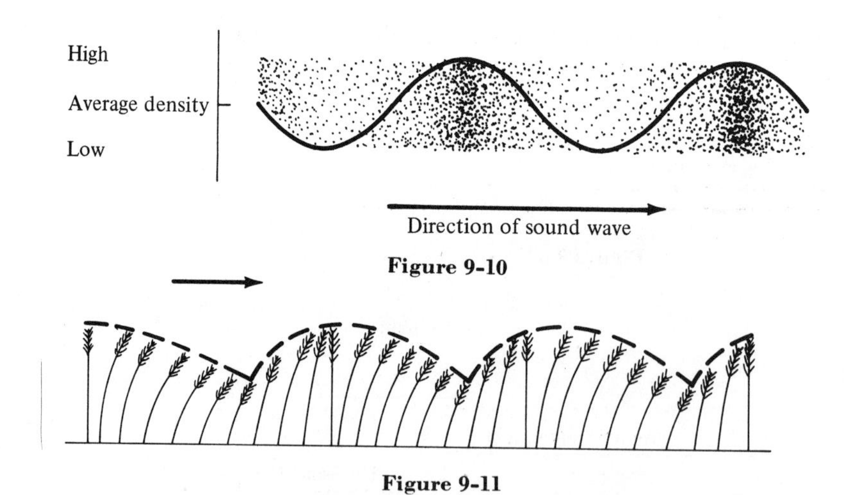

Figure 9-10

Figure 9-11

1.5 Polarization

Transverse waves may be *polarized*. In a polarized wave, the oscillations are not only perpendicular to the direction of propagation (the definition of a transverse wave), but also confined to one plane in space—the plane

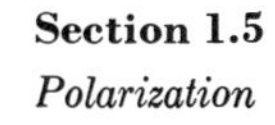

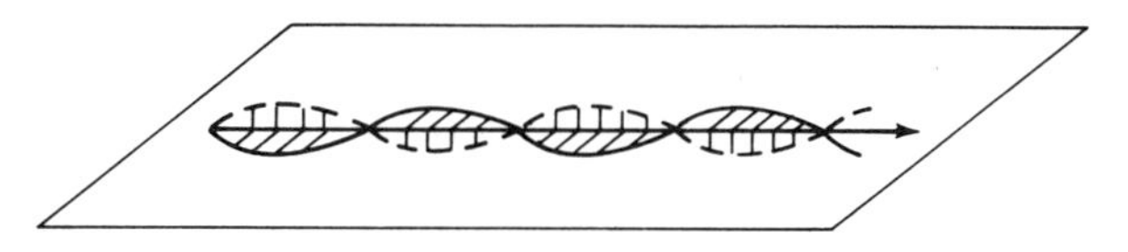

Figure 9-12

of polarization. See Figure 9-12. The dashed wave is polarized perpendicular (90°) to the reference plane, while the solid one is polarized parallel (0°) with respect to the reference plane.

2. HARMONIC OSCILLATIONS AND WAVES

An oscillation which can be described by either of the following equations:

$$x = x_0 \cos \omega t = x_0 \cos (2\pi\nu t) \tag{9.4}$$

or

$$x = x_0 \sin \omega t = x_0 \sin (2\pi\nu t) \tag{9.5}$$

where

x: displacement from the mean position,
x_0: amplitude,
$\omega = 2\pi\nu$: (angular frequency),
ν: frequency,
t: time

is called a *harmonic oscillation;* it is accompanied by a harmonic wave. See Figure 9-13.

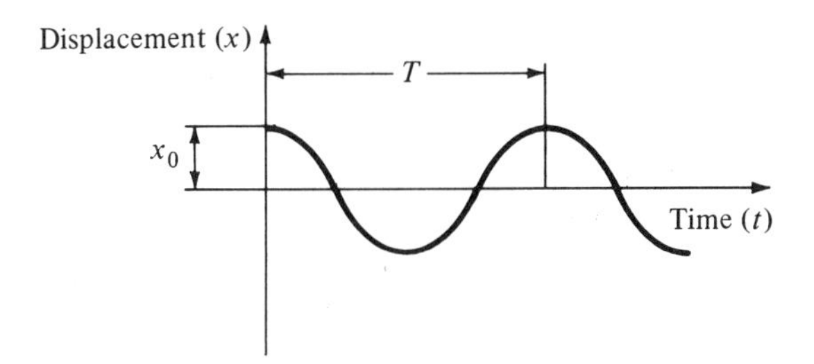

Figure 9-13

Frequency ν, wavelength λ, period T, and speed of propagation v are related by

$$\nu = \frac{1}{T} = \frac{v}{\lambda} = \frac{\omega}{2\pi} \tag{9.6}$$

Remark: Any periodic oscillation can be analyzed as a sum of harmonic oscillations—a method called *Fourier analysis.*

EXAMPLE:

Pitch is a single harmonic oscillation. The concert-pitch A, for example, has a frequency of 440 Hz. A *tone* is made up of two synchronous harmonic oscillations. If the relation between both frequencies is 1:2, an octave results; for a relation 3:4 we get a major fourth.

2.1 Free and damped oscillations

For a free oscillation, amplitude and frequency do not change. No forces interfere. However, free oscillations are rare. Since frictional forces affect most oscillations, a damped oscillation results. See Figure 9-14. In damped oscillation, the maximum displacement decreases from one oscillation to the next.

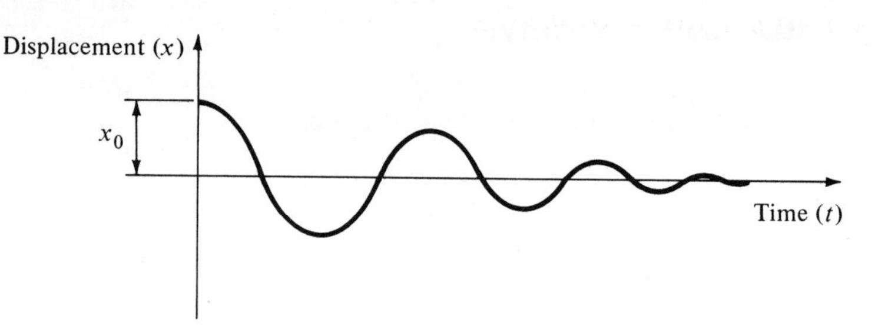

Figure 9-14

EXAMPLE:

Air leaving the lungs and passing the vocal chords stimulates oscillations. The frequency is determined by the speed of the air and the tension of the vocal chords; this tension may be adjusted by muscular action. Normally, the produced oscillations are damped, and no pure pitch occurs. With suitable training, an opera singer, for example, can compensate for the effect of the damping friction. Cutting off a pitch means applying a strong damping rapidly, so that the displacement of the vocal chords decreases toward zero.

2.2 Energy and intensity

Thus far we have dealt with the geometrical features of waves. We should not lose sight of the fact that what happens from a physics point of view is a propagation of energy. It is convenient not to deal with the energy itself but with the intensity (symbolized I) of a wave. *Intensity*, I, is the energy transmitted through an area per unit time. In other words, it is the flow rate of energy through an area at right angles to the direction of wave motion.

$$I \sim c\nu^2\xi_0^2 \tag{9.7}$$

where

I: intensity of wave, measured in $\text{erg}\cdot\text{cm}^{-2}\cdot\text{s}^{-1} = 10^{-7}\ \text{watt}\cdot\text{cm}^{-2}$,
c: velocity of wave propagation (phase velocity),
ν: frequency,
ξ_0: amplitude of wave.

At the moment there is no need to convert the above proportionality into an equation because it would be different for various types of waves. In general, it is sufficient to know that:

1. The intensity is proportional to the square of the frequency and thus inversely proportional to the square of the wavelength (because $\nu \sim 1/\lambda$).
2. The intensity is an energy passing through a cross section in a given time.

EXAMPLE: INTENSITY OF A LONGITUDINAL WAVE

The proportionality factor in Equation (9.7) is $2\pi^2\rho$; hence

$$I_1 = 2\pi^2\rho c\nu^2\xi_0^2 \tag{9.8}$$

where

I_1: intensity of longitudinal wave,
ρ: average density of transferring medium,
c: phase velocity,
ν: frequency,
ξ_0: amplitude of the oscillating molecules of the medium.

The auditory system of man acts as a detector for longitudinal waves. The human ear is most sensitive for frequencies around $\nu = 1$ kHz. An intensity of $I = 10^{-9}$ erg·s^{-1}·cm^{-2} = 10^{-16} watt·cm^{-2} is just audible. This corresponds to an amplitude ξ_0 of the oscillating molecules of

$$\xi_0^2 = \frac{I}{2\pi^2\rho c\nu^2}$$

$$= \frac{10^{-9}}{19.8 \times 1.3 \times 10^{-3} \times 3.3 \times 10^4 \times 10^6}$$

$$\xi_0 = 1.2 \times 10^{-9} \text{ cm}$$

This means that air molecules oscillating with an amplitude only one-tenth of their diameter will cause an acoustical impression.

2.3 Forced oscillations

In general, all oscillatory systems display damped oscillations due to the ever-present frictional forces. The damping may be compensated for by applying an outside oscillatory force having the same frequency. We distinguish two cases:

1. Magnitude of outside force = magnitude of damping force. The energy transferred to the oscillating system is equal to the energy taken away by the damping. *Result:* The system oscillates with a constant amplitude.
2. Outside force > damping force. The energy supplied to the system is greater than the energy taken away by damping. *Result:* The amplitude of the resulting oscillation will increase with time.

2.3(a). EIGENFREQUENCY. Any system capable of oscillation has one or more preferred frequencies—the *eigenfrequencies.* The numerical value of an eigenfrequency depends on factors such as mass, elasticity, physical extensions, and internal structure.

EXAMPLES:

The eigenfrequencies of a string are its harmonics.

The eigenfrequency of a mathematical pendulum (a heavy body attached to a string) is given by:

$$\nu_{ei} = \frac{1}{2\pi}\sqrt{\frac{g}{l}}$$

where

g: acceleration due to gravity,

l: length of the string.

The eigenfrequencies of the hydrogen atom are given by

$$\nu_H = R\left(\frac{1}{m^2} - \frac{1}{n^2}\right)$$

where

R: Rydberg's constant $= 3.29 \times 10^{15}\ s^{-1}$,
m: integral numbers 1, 2, 3, . . . ,
n: integral numbers, with $m > n$.

For more complex structures such as bridges, cars, airplanes, radio receivers, and cells the eigenfrequency is not a single frequency but covers a limited continuous range, known as the *eigenfrequency bandwidth.*

2.3(b). RESONANCE. A system suffering a forced oscillation will—after an initial adaption time—have the same frequency as the exciting force. If the frequency of the exciting force is equal to an eigenfrequency, the system is said to *resonate*. The energy needed to excite a system is a minimum at resonance. Thus an oscillation at an eigenfrequency leads to large amplitudes. The smaller the damping, the larger the resulting amplitude.

In engineering, resonance is of utmost importance. If a bridge or an airplane's wing is not carefully designed, its eigenfrequency may be excited by wind forces. The structure begins to resonate and catastrophic mechanical vibrations may occur.

3. SUPERPOSITION

3.1 General

Oscillations occurring simultaneously in space and time superimpose and produce a resulting oscillation. The resulting oscillation is the sum of the original ones. To simplify display and formulas, the following treatment is restricted to the superposition of two harmonic waves. Each is characterized by its amplitude, frequency, and phase difference (with respect to the other oscillation). See Figure 9-15.

$$x = x_1 + x_2 = x_{01} \cos (2\pi\nu_1 t) + x_{02} \cos [(2\pi + \theta)\nu_2 t]$$

The resulting oscillation (see Figure 9-16) is generally not periodic. There are three interesting special cases: beat, resonance, and cancellation.

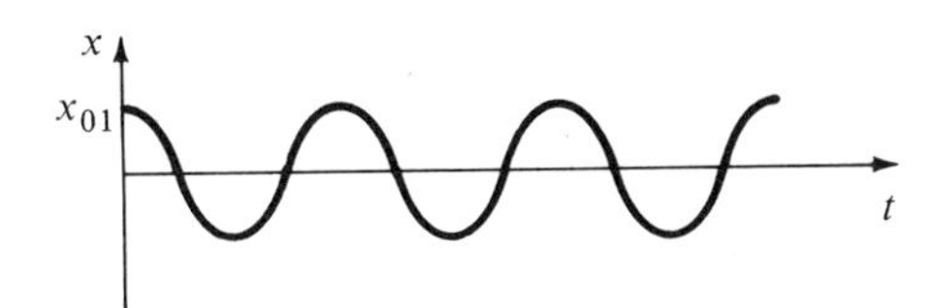

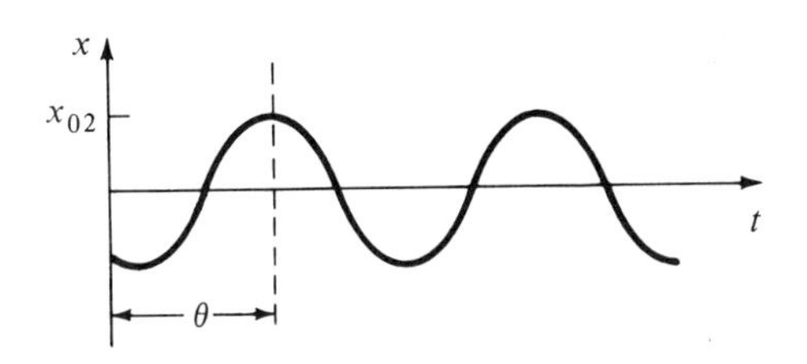

θ: phase difference

Figure 9-15

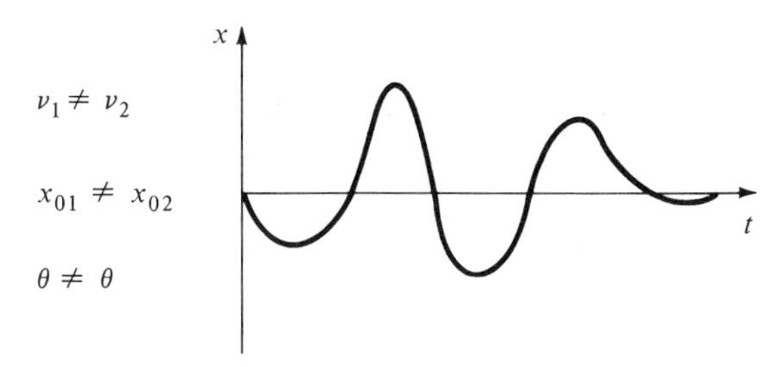

Figure 9-16

3.1(a). BEAT

$$\nu_1 \neq \nu_2 \quad \text{but} \quad |\nu_1 - \nu_2| = \Delta\nu \text{ is very small compared with } \nu_1 \text{ and } \nu_2$$

$$x_{01} = x_{02} = x_0, \text{ phase difference } \theta = 0$$

$$x = 2x_0 \cos(\Delta\nu\pi t) \cos(2\pi\nu t) \tag{9.9}$$

where

$$\nu = \frac{\nu_1 + \nu_2}{2}$$

The resulting oscillation is periodic but not harmonic. Its amplitude varies periodically between 0 and $2x_0$ with a frequency $\Delta\nu/2$. See Figure 9-17.

Application: Two tuning forks having almost the same frequency will produce a tune with periodically changing intensity. This is called a *beat.*

3.1(b). RESONANCE

$$\nu_1 = \nu_2 = \nu, \quad \text{phase difference } \theta = 0$$

$$x = x_1 + x_2 = (x_{01} + x_{02}) \cos(2\pi\nu t) \tag{9.10}$$

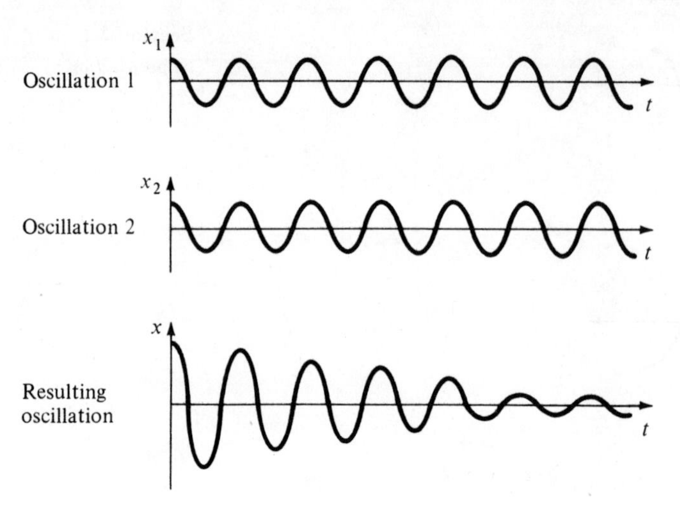

Figure 9-17

The resulting oscillation is periodic and harmonic. Its amplitude is the algebraic sum of the two original amplitudes. See Figure 9-18.

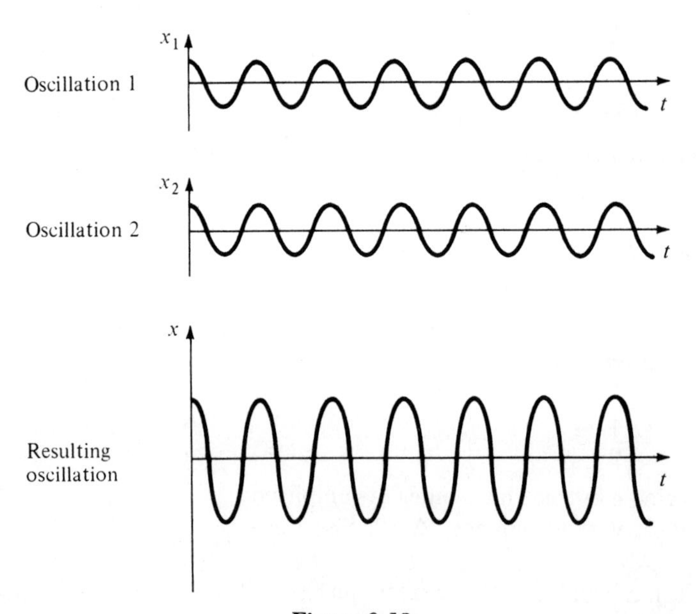

Figure 9-18

3.1(c). CANCELLATION

$$\nu_1 = \nu_2 = \nu$$

$x_{01} = x_{02} = x_0,$ phase difference $\theta = \pi$ (that is, half of a wavelength)

$$x = x_1 + x_2 = x_0 \cos(2\pi\nu t) - x_0 \cos(2\pi\nu t) = 0$$

(Recall that $\cos(x \pm \pi) = -\cos x$.) The amplitude of the resulting wave is zero, that is, both waves cancel each other. See Figure 9-19.

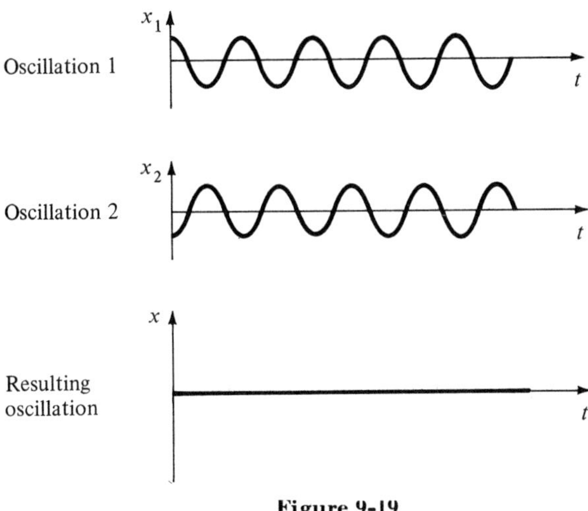

Figure 9-19

3.2 Wave packets

Up to now, we have dealt with continuous waves, having neither a beginning nor an end. This is an idealization of reality. Continuous waves are employed to describe real waves because of mathematical ease of treatment and since the results gained are acceptable approximations of reality.
A finite wave may be described as a *wave packet* or wave group having a finite extension in space. See Figure 9-20. Mathematically, a wave packet is treated as the superposition of an infinite number of harmonic waves, each differing very little in wavelength. The entire packet (enclosed by a dashed line in Figure 9-20)

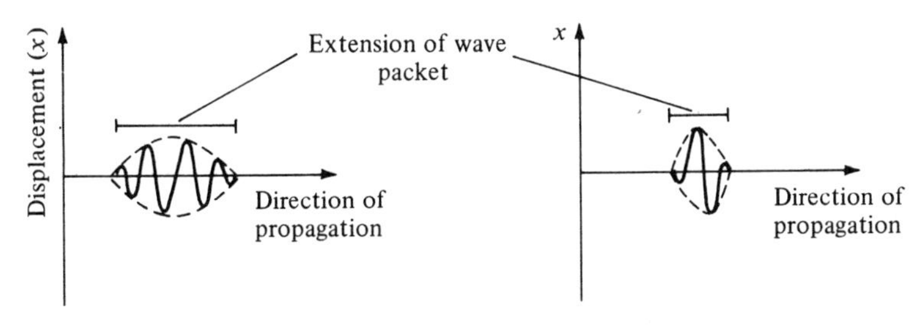

Figure 9-20

propagates with a *group velocity* which is different from the phase velocity of the individual components.

4. WAVES AT BOUNDARIES

A wave encountering a boundary (the division between two media) will be partly reflected (reflection) and partly transmitted into the other medium (refraction). See Figure 9-21.

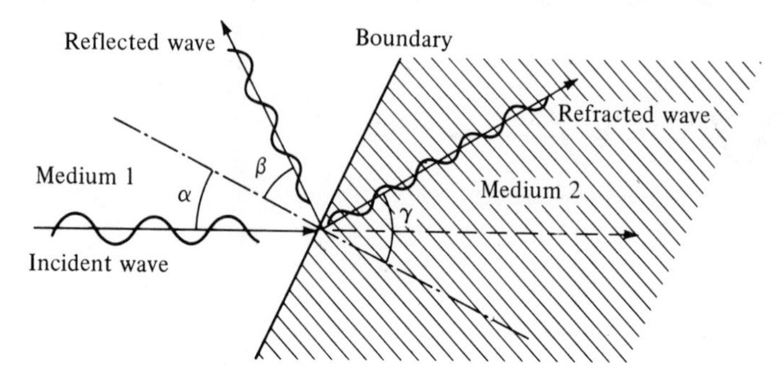

Figure 9-21 α: angle of incidence; β: angle of reflection; γ: angle of refraction.

4.1 Transmission

4.1(a). Refraction. Law of refraction (also named Snell's law):*

$$\frac{\sin \alpha}{\sin \gamma} = \frac{v_1}{v_2} \tag{9.11}$$

where

α: angle of incidence,
γ: angle of refraction,
v_1: phase velocity in medium 1,
v_2: phase velocity in medium 2.

Index of Refraction. Snell's law may also be formulated in the following way,

$$n_1 \sin \alpha = n_2 \sin \gamma \tag{9.12}$$

where

n_1: index of refraction for medium 1,
n_2: index of refraction for medium 2.

From Equations (9.11) and (9.12) we derive the useful relation

$$\frac{v_1}{v_2} = \frac{n_2}{n_1} \tag{9.13}$$

The index of refraction depends on, besides other factors (interesting only to the specialist), the material and wavelength. Its numerical value is always calculated with respect to a standard medium like vacuum,

* W. Snell van Royen (1591–1626), physicist.

air, or water. For numerical values see Tables 1 and 2 in the chapter on Geometrical Optics.

4.1(b). DISPERSION. The angle of refraction depends on wavelength. Consequently, a wave composed of superimposed waves will be separated into its basic harmonic components if refracted. See Figure 9-22. This phenomenon is called *dispersion*.

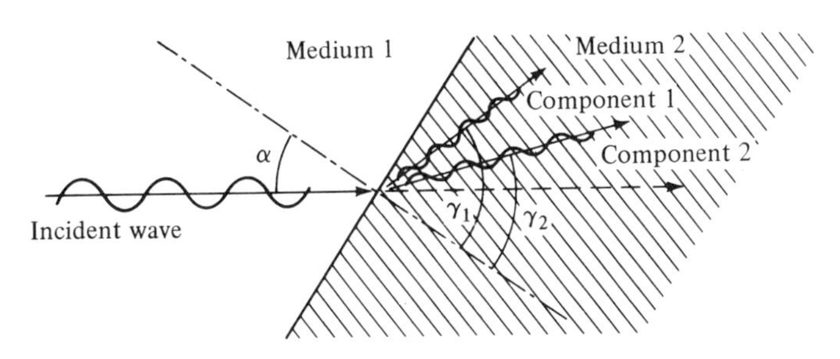

Figure 9-22

EXAMPLE: DISPERSION OF SOUND IN THE INNER EAR

After passing the tympanic membrane, the sound wave is transferred via the ossicular chain to the membrana vestibularis. Due to the mechanical properties of this membrane, the shorter sound waves travel faster than the longer ones. Thus if a complex sound arrives at the ear, its high-frequency components will reach the helicotrema before the low-frequency components. The sound waves are dispersed along the basilar membrane. See Figure 5-11 on page 119.

4.2 Reflection

Law of reflection:

$$\text{angle of incidence} = \text{angle of reflection}$$

4.2(a). REFLECTIVITY. While encountering a boundary, part of the wave's intensity will be reflected. This is described by

$$\text{reflectivity} = \frac{\text{reflected intensity}}{\text{incident intensity}}$$

EXAMPLE:

Eighty-eight percent of the incident light is reflected by a silver mirror, and 71% is reflected by a glass mirror plated with mercury. Consequently the reflectivities are 0.88 and 0.71.

4.2(b). TOTAL REFLECTION. Rewriting Snell's law, Equation 9.11, in the following form

$$\sin \gamma = \frac{v_2}{v_1} \sin \alpha$$

we discover a curious phenomenon. For $v_2 > v_1$, the right side of the equation above may become larger than 1. Clearly, this is impossible since $\sin \gamma$ cannot exceed 1.

Solution: If the angle of incidence reaches or exceeds an angle α_T, then it does not penetrate the boundary at all, but is totally reflected. α_T is called the *angle of total reflection* and may be calculated from the following equation

$$\sin \alpha_T = \frac{v_1}{v_2} = \frac{n_2}{n_1} \tag{9.14}$$

EXAMPLE: TOTAL REFLECTION AT THE HYPOTENUSE OF A RECTANGULAR GLASS PRISM

(Figure 9-23)

$n_2 = 1$ (actually 1.0003)

$n_1 = 1.6$ (crown glass)

$$\sin \alpha_T = \frac{1}{1.6}$$

$$\alpha_T = 38.7°$$

This means that light incident under an angle $\geqslant 38.7°$ will be totally reflected. Since in our example $\alpha = 45°$, the condition for total reflection, Equation (9.14), is met by the prism.

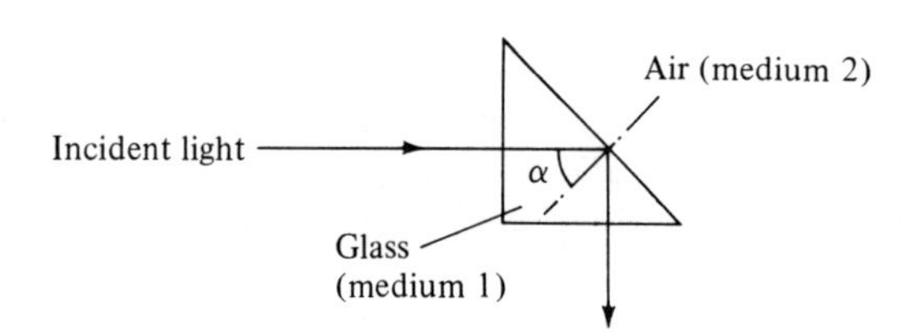

Figure 9-23

The phenomenon just described is widely exploited in optical instruments whenever the direction of light is to be changed by 90°.

5. ATTENUATION

5.1 Spreading

A wave leaving its source spreads out in space. How much of its intensity will decrease due to this purely geometrical effect depends on its spatial appearance. The intensity of plane waves (for example, produced by an oscillating plate; see Figure 9-24) is independent of the source–detector distance.

The intensity of spherical waves (for example, produced by an oscillating point source; see Figure 9-25) is inversely proportional to the square of the source–detector distance.

Figure 9-26 summarizes the spreading of waves.

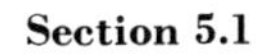

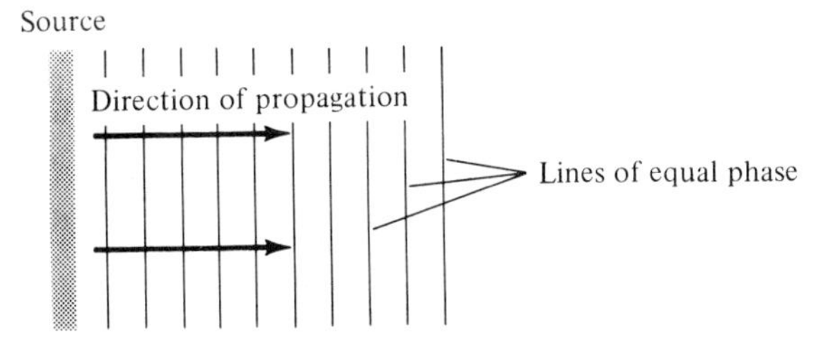

Figure 9-24

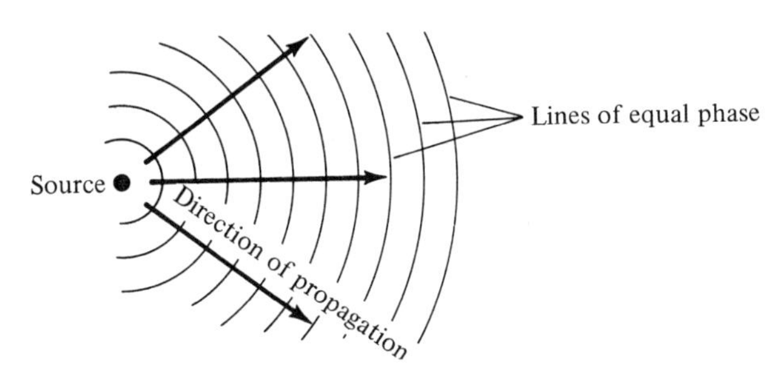

Figure 9-25

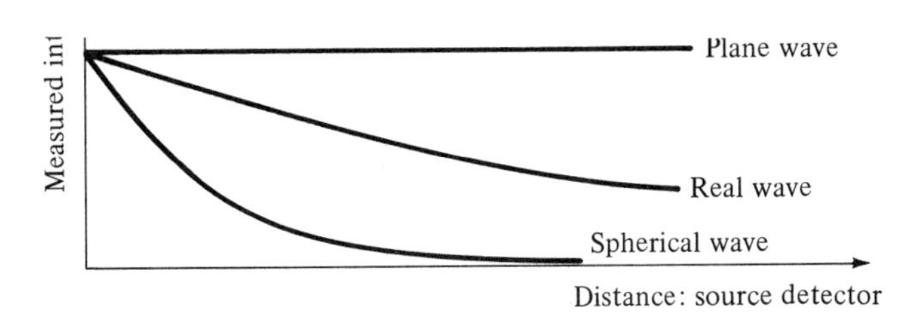

Figure 9-26

5.2 Scattering

Another factor which attenuates a wave is *scattering*. Here the wave is incident on a small obstacle and changes its direction so that it will not reach the detector. See Figure 9-27.

The following treatment is restricted to *elastic* scattering, that is, wavelength and frequency do not change. We may distinguish three cases:

(a) $$\lambda \ll d$$

where

d: extension of obstacle,
λ: wavelength,
φ: scattering angle.

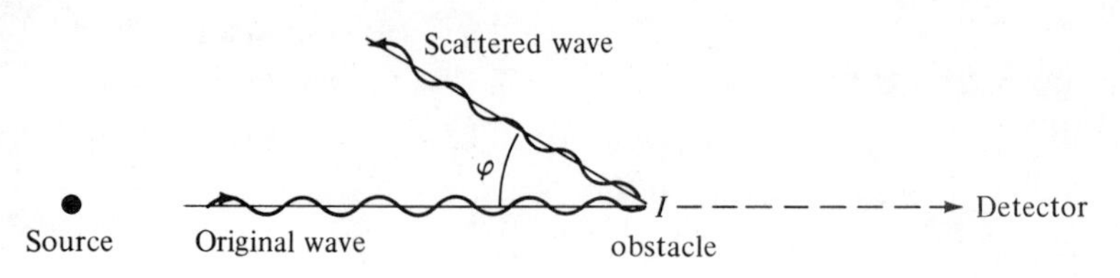

Figure 9-27

The result of this case is *reflection*. The direction changes according to the law of reflection. See Section 4.2.

(b) $$\lambda \approx d$$

The result of this case is *diffraction* or *anisotropic scattering*. The wave is scattered in all directions. The intensity of the scattered wave is proportional to $\sin^2 \varphi$.

(c) $$\lambda \gg d$$

The result of this case is *Rayleigh* scattering* or *isotropic scattering*. The wave is scattered in all directions. The intensity of the scattered wave is independent of the scattering angle. The term *scattering* is often restricted to Rayleigh scattering.

Observing and analyzing the angular distribution of scattered waves leads to information about the size of the scattering obstacles. This information is used to determine the diameter of colloid particles and macromolecules. The physical extension of the atomic nucleus and even its constituting nucleons is also determined by scattering. This method of measurement is limited only by the available sources and detectors of suitable wavelength. For example, to investigate the physical structure of the atomic nucleus, the wavelength must be approximately 10^{-12} cm. Waves having such an extremely short wavelength are only available at the laboratory in the form of matter waves. High-energy ($\approx$ 100 MeV) electrons produced in accelerators are associated with the necessary wavelength.

5.3 Absorption

The intensity of a wave penetrating through matter will decrease due to absorption (Figure 9-28). This is described by

$$I = I_0 e^{-\sigma s} \tag{9.15}$$

where

I: intensity of the wave after penetrating a distance s,
I_0: incident intensity, that is, intensity for $s = 0$,
σ: absorption or extinction coefficient,
s: penetration distance.

* Lord John William Rayleigh (1842–1919), physicist. Nobel Prize winner (1904 for chemistry). Worked mainly in acoustics and optics. More about him in R. J. Strutt, *Fourth Baron Rayleigh*, London, 1924.

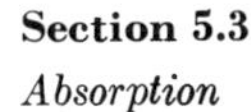

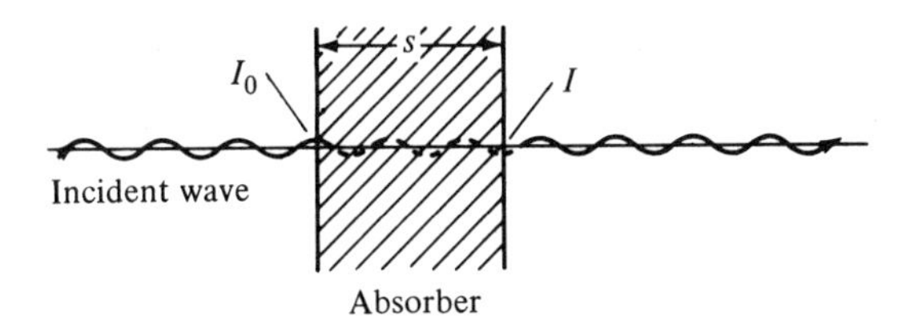

Figure 9-28

The absorption coefficient depends on

1. The type of wave (sound, seismic, light, matter).
2. The wavelength (in general, shorter waves are more strongly absorbed than longer ones).
3. The type and phase of penetrated matter.

The law of absorption as formulated in Equation (9.15) and displayed in Figures 9-29 and 9-30 is truly universal. You will encounter it often.

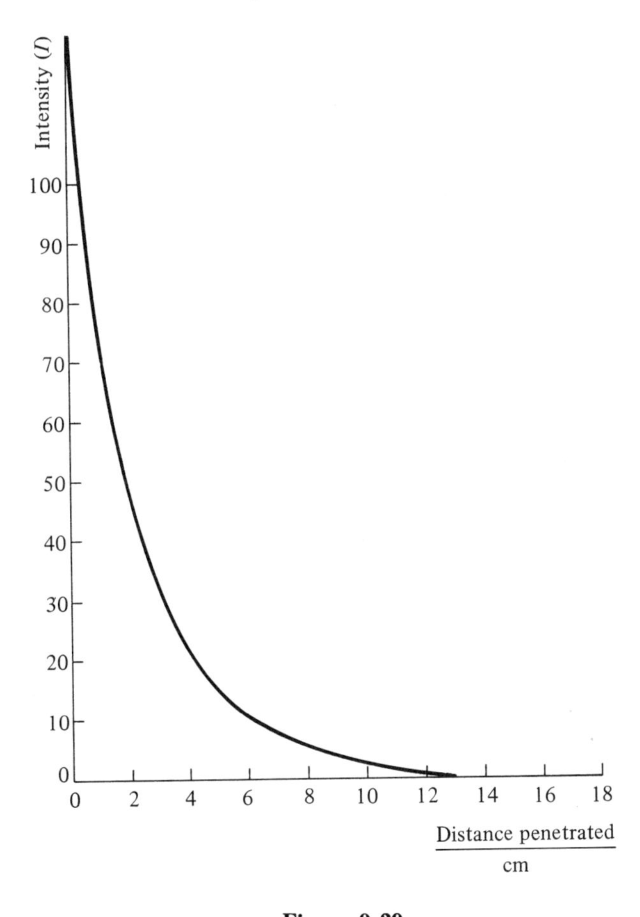

Figure 9-29

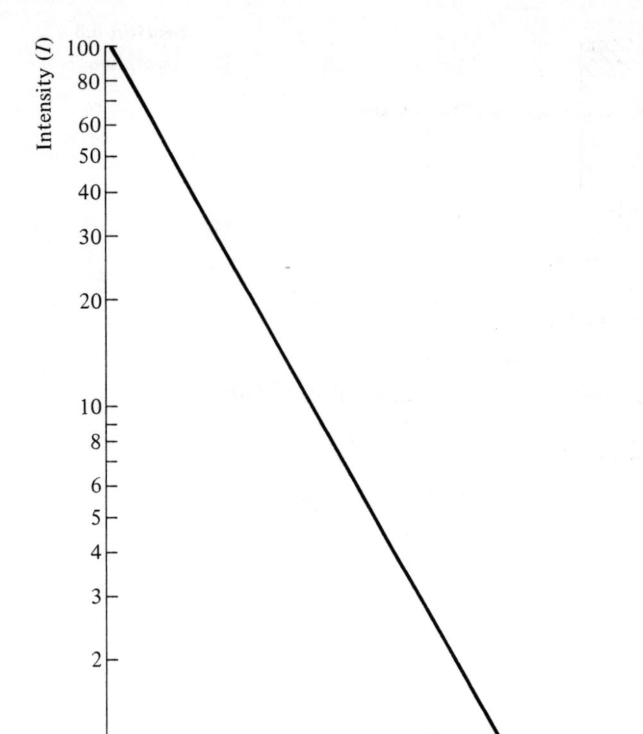

Figure 9-30

Application: Radiation Treatment. Waves will penetrate into the human body. How deep they penetrate depends on the wavelength. This is utilized in medical radiation treatment. Figure 9-31 shows the intensity of various medically employed waves inside the body.

The absorption of sunlight and short waves causes a temperature rise of the absorbing tissue. This is a therapeutic effect. X rays interact

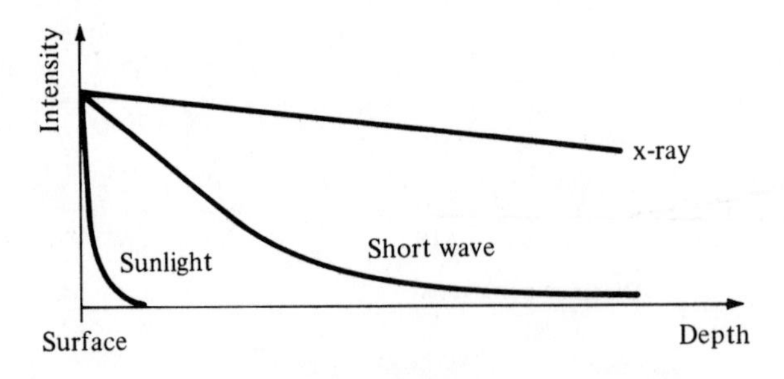

Figure 9-31

with the tissue's molecules and atoms. The resulting absorption is mainly due to ionization. The subsequent effects damage cancer cells more than normal cells.

Another Application: Perceiving the Distance of an Acoustical Source. It was stated earlier that absorption depends on the frequency of the wave. In general (but there are many exceptions), it is proportional to the square of the frequency. This characteristic is used by the human ear.

A thunderclap, like the one produced by lightning, contains many different sound frequencies. Close to the source of the sound, the high-frequency components are clearly audible—a shattering noise. Further away, the high-frequency components are much more damped than the low frequencies. As a result, the bang appears more muffled; the longer wavelengths dominate. From experience, the ear learns to use this effect to perceive distance.

6. SPECTRA

In general, a source emits more than one single frequency simultaneously; it emits a spectrum of frequencies. The extremes are the line spectrum and the continuous spectrum.

6.1 Line spectrum

This spectrum consists of well-separated individual frequencies. Figure 9-32 shows the frequency spectrum of an A-tuning fork. The fundamental tone of 440 Hz dominates; the overtones are less intense.

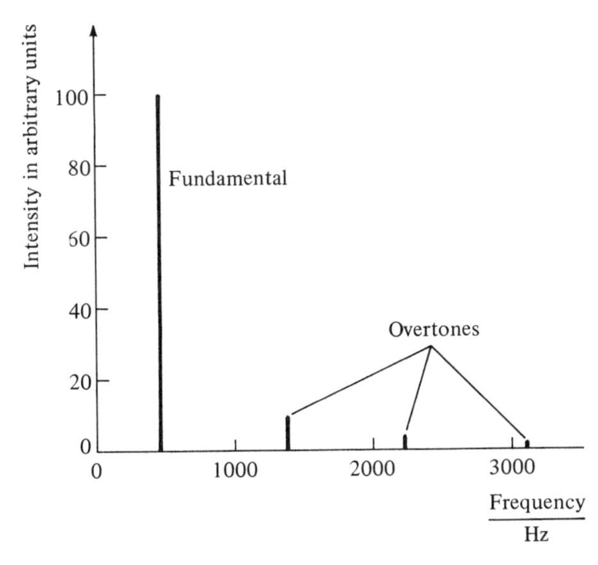

Figure 9-32 Idealized line spectrum emitted by an A tuning fork. The fundamental tone 440 Hz dominates; the overtones are less intense.

6.2 Continuous spectrum

Here the intensity is continuously distributed over a range of frequencies. Figure 9-33 shows an example of a continuous frequency spectrum.

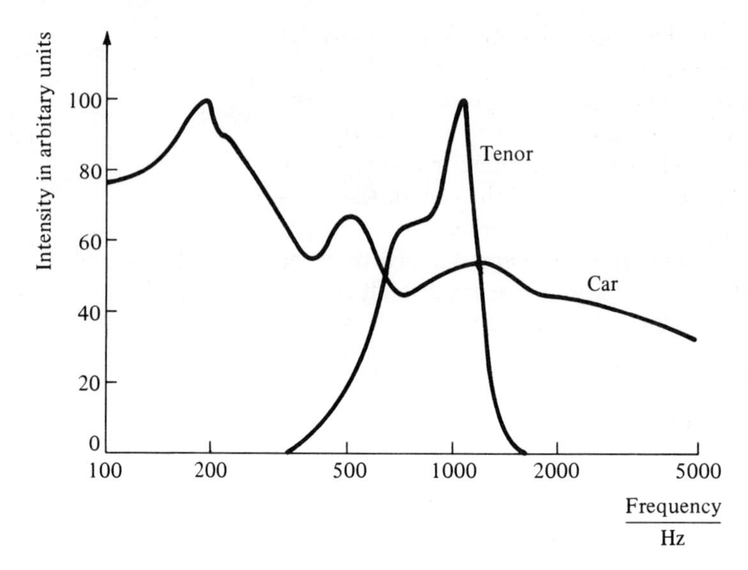

Figure 9-33 Frequency spectra of an opera tenor forming the vowel *a*, and that emitted by a car.

The concept of *spectrum* is not restricted to the analysis of sources emitting waves. Figure 9-34 shows an example of a different kind of spectrum.

6.3 Resolution

An experimentally determined line spectrum never looks like the one shown in Figure 9-32. Even if the source emits a pure line spectrum, the receiver shows a resolution resulting in lines having a finite width. Figure 9-35 shows a line spectrum measured by detectors having various resolutions. The source itself emits a pure frequency (monochromatic source).

The poorer the resolution, the wider the width of the measured spectral line. Detector A shows the best resolution, detector C the worst. If the spectrum of a source is to be measured, we must carefully select a detector having an appropriate resolution. A poor resolution distorts the spectrum and hides information. Figure 9-36 shows an example of that.

By the very nature of a measurement, an ultimate and finite resolution exists. This limitation has little bearing on macroscopic objects,

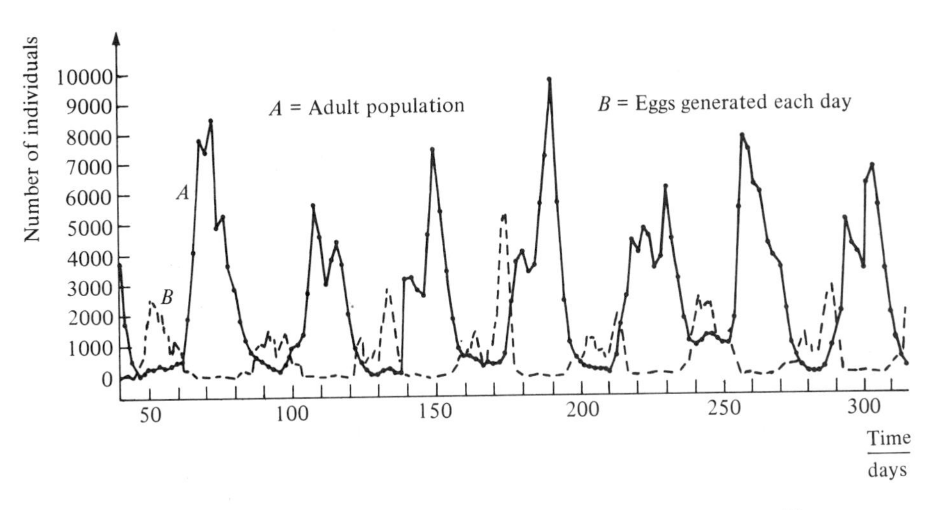

Figure 9-34 Numbers of the blowfly *Lucilia* in a population cage. The spectrum shows a generation cycle of about 40 days.

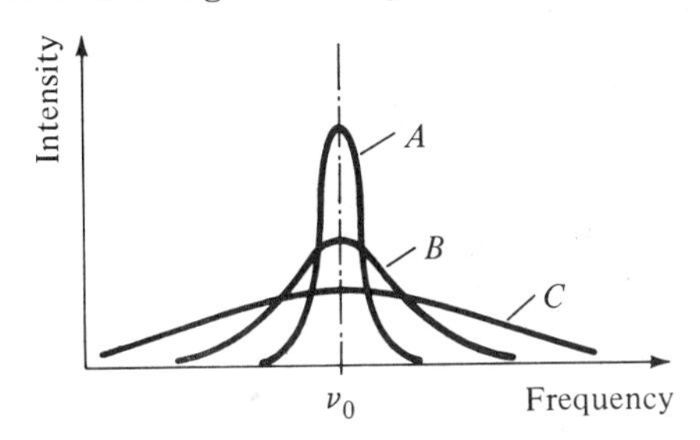

Figure 9-35

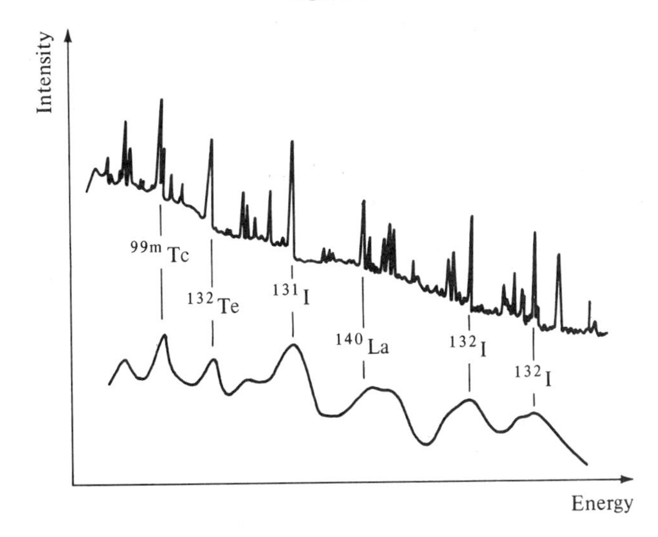

Figure 9-36 Gamma-ray spectrum of gross fission products, measured by a germanium detector (upper curve) and a scintillation detector (lower curve). The better resolution of the germanium detector reveals much more detail.

but in atomic and nuclear physics it is of utmost importance. The uncertainty principle already partly introduced in Chapter 2, Section 1.2(b) deals with this problem.

7. APPLICATIONS

7.1 Echo in nature and technology

7.1(a). AUDIBLE ECHO. A reflected wave is sometimes called an *echo* of the incident wave. The term is used mainly if the reflected wave is a rather short pulse train used to transmit information.

The echo may be employed to measure the distance between receiver and reflecting surface. However, a naturalist will learn more. If a person shouts a one-syllable word toward a perpendicular surface in the mountains, the echo is intelligible only if the listener is at least 17 m away from the surface. The sound train traveled 34 m, and the travel time was 100 millisecond.

For a two-syllable word the distance must be doubled, corresponding to a 200-ms travel time. This experimental result indicates that the human auditory system needs a minimum time span (approximately 50 ms) to perceive a sound. We may call this the *resolution time* of the ear. If more than eighteen short bangs reach the ear within one second, it does not respond fast enough to discriminate between individual bangs. This is in analogy to the human eye: The entire movie industry relies on the fact that still pictures offered in rapid succession will appear to move.

7.1(b). ECHO SOUNDER (SONAR). A short pulse of sound waves directed perpendicularly downward will be reflected by the bottom and be received at a detector (Figure 9-37). If the echo time is t and the speed of sound in water is c_{water}, the boat sails over a depth

$$d = \frac{c_{\text{water}}}{2} t$$

If the echo is continuously produced and recorded by an echo sounder, a profile of the sea bottom emerges. Figure 9-38 shows an example of this.

Highly sensitive echo sounders are used to detect shoals of fish, even

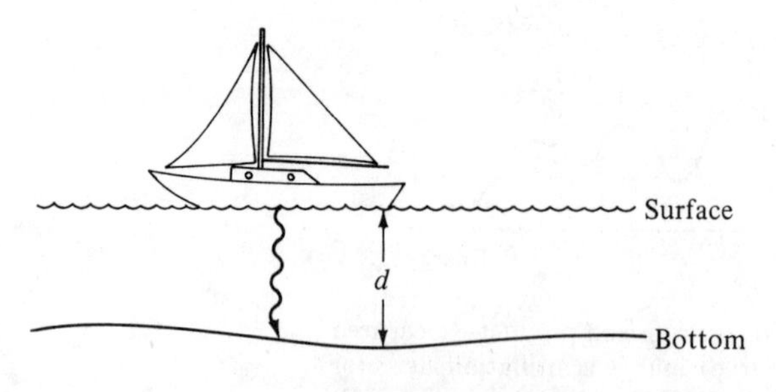

Figure 9-37

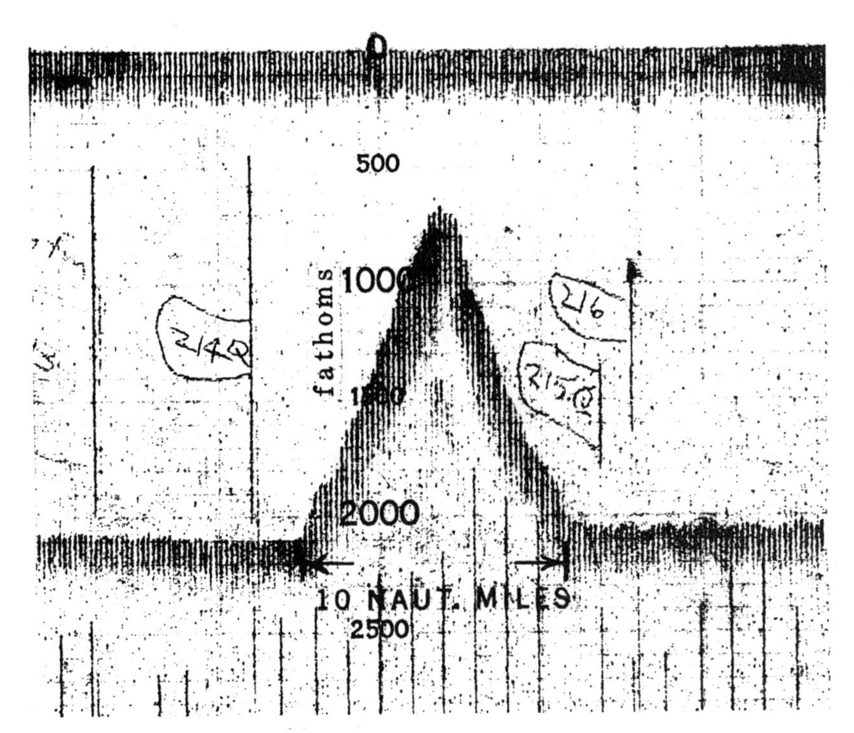

Figure 9-38 Fathogram of a volcanic-cone seamount rising 2500 m from a depth of 4000 m.

individual fishes like tuna are visible. The acoustical fish-finder, named *fishlense,* is by now an indispensable tool in a modern fishing vessel. This sounds like a remarkable achievement of man; however, the dolphin does much better. Its sonar system distinguishes not only small individual fish, but also between different kinds over a distance of 50 m!

Refined forms of the echo sounder are used in medicine to measure the dimensions of the eye bulbus *in vivo.* Ultrasonic waves (frequency > 20 kHz, i.e., the human ear cannot detect these) are allowed to enter the body. They are reflected at each boundary inside. Since the propagation speed of the wave inside the body is known (about 1.5×10^3 m/s), the travel time of the reflected signal allows us to calculate how deep inside the reflecting boundary is. With some technical effort and suitable frequencies (around 10^5 Hz to 10^6 Hz), the reflected waves can be used to form a picture. See Figure 9-39. There seems to be no limit to the applications; even the thickness of the bacon on top of a pig can be determined by the farmer thanks to the echo sounder.

7.1(c). Radar. Short pulses of electromagnetic waves (wavelength in the neighborhood of 10 cm) are emitted and echoes are received by antennas scanning the surroundings. It is the same principle as the acous-

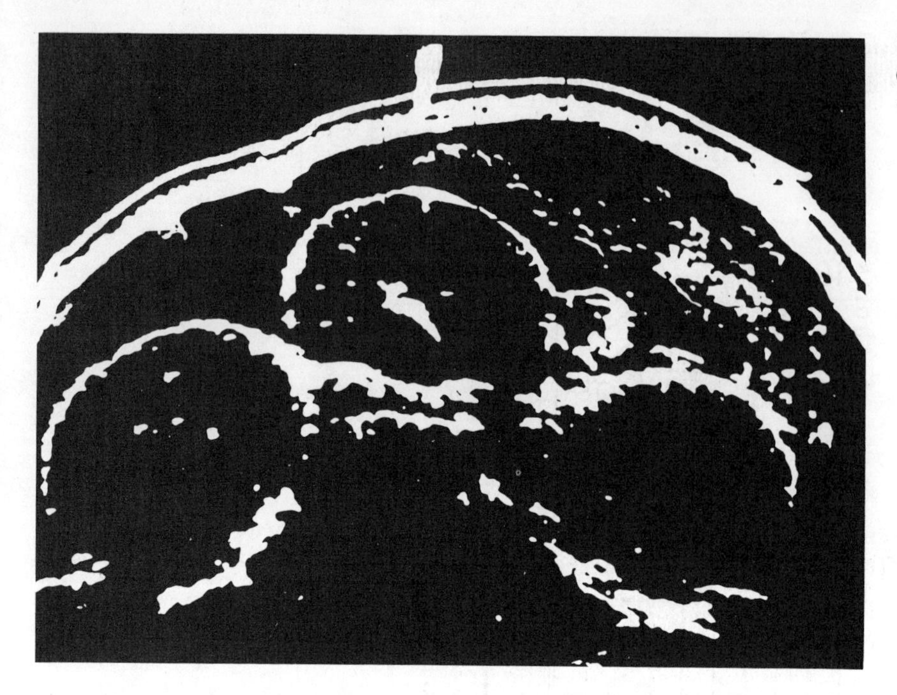

Figure 9-39 Ultrasonic scan of a uterus showing triplets.

tical echo sounder, only the speed of the detecting waves is much higher (3×10^8 m/s compared with 3.3×10^2 m/s).

7.1(d). BATS AND FOG. Bats, dolphins, some species of mice, and certain cave-dwelling birds have developed echo-locating systems (sonar). The frequency of the emitted sounds are mainly in the ultrasonic region ($\nu > 20$ kHz) because short wavelength can be more easily emitted as well-directed beams. The wavelength must also be of the same order of magnitude as the extension of the obstacles or prey to be detected.

Most bats generate a sonic spectrum between 15 kHz and 150 kHz, corresponding to emitted wavelengths of 2.2 cm and 0.22 cm. Just for orientation, a wavelength of 2.2 cm is sufficient. To locate and hunt down insects, the shorter waves are useful.

Bats are rarely observed in fog, although in that kind of weather the number of prey insects is exceptionally large. Why don't the bats hunt in fog? Their sonar systems seem to be all right since the water droplets of fog have diameters around 4×10^{-3} cm. Even the short end of the bat's sonar spectrum contains waves a hundred times larger than the droplet diameter. Fog should be transparent. The answer is that the spherical water droplets have (like every other body) eigenfrequencies. They resonate, as Lord Rayleigh demonstrated at the end of the last century, in a spectral band between 20 kHz and 150 kHz! This means that the eigenfrequencies of the droplets correspond to the frequencies of the bat's sonar system. Consequently, the emitted waves are strongly absorbed, and no echoes will return. Translated into a human picture: the bats fly in a pitch dark space and, although the searchlights are on, nothing can be seen!

7.2 The Doppler effect

A source oscillating with a frequency ν_s emits a wave having a wavelength λ_s. This wave propagates with a speed c. A detector may receive and analyze the approaching waves. The analyzed frequency and wavelength are ν_R and λ_R, respectively.

Until now we have always assumed that source and receiver are at rest, a situation presented in Figure 9-40. Each concentric circle represents the spatial separation between consecutive wave peaks. Since the waves propagate in all directions with uniform speed c, it does not matter where the receiver is placed. It will detect the same frequency and wavelength everywhere, that is:

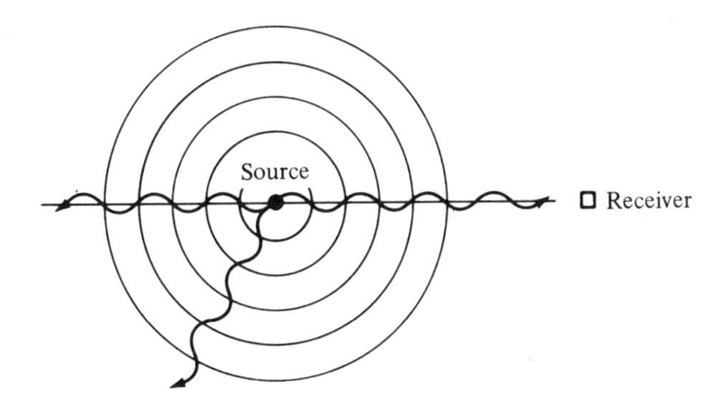

Figure 9-40

$$\nu_s = \nu_R \quad \text{and} \quad \lambda_s = \lambda_R$$

What happens if either source or receiver or both are moving? Consider two basic assumptions:

1. The propagation speed c of the wave remains constant.
2. The direction of motion is restricted to a motion along a straight line connecting source and receiver. (If the actual direction of motion is not along the connecting line, simple geometry will yield appropriate results.)

Figure 9-41 demonstrates what happens if the source moves with a speed v_s to the right. The circles still represent the spatial separation between consecutive waves, but they are not concentric anymore. Now it matters where the receiver happens to be! During the same time span more waves reach the receiver at position R_1 than at R_2; consequently,

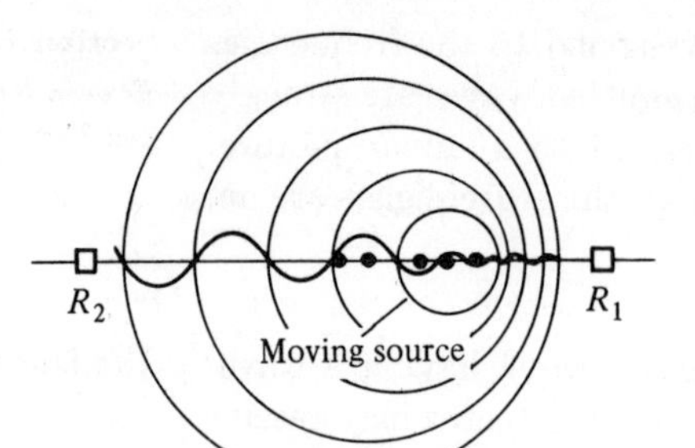

Figure 9-41

the detector will find a higher frequency at R_1 than at R_2. Now if the detector also moves, the situation becomes more complicated. The corresponding frequencies were first calculated in 1850 by Doppler.*

$$\nu_R = \frac{c - v_R}{c - v_s}\,\nu_s \tag{9.16}$$

where

ν_R: frequency as detected by the receiver,
ν_s: frequency emitted by the source,
c: propagation speed of the wave,
v_R: speed of the receiver,
v_s: speed of the source.

(The speed is measured with respect to a reference system in which c is constant. If the source and detector move in the same direction, then v_R and v_s have the same sign. Moving in opposing directions, they have opposite signs.)

Simplification: For $v_R \ll c$ and $v_s \ll c$, Equation 9.16 reduces to

$$\nu_R = \left(1 - \frac{\Delta v}{c}\right)\nu_s \tag{9.17}$$

where

$$\Delta v = v_R - v_s$$

Doppler phenomena are observable practically everywhere in nature; the examples are widespread:

7.2(a). MEASURING THE SPEED OF BLOOD. The speed of blood inside an intact vessel is an important quantity in medicine. By means of Doppler's effect, this quantity is measured from the outside without any disturbance of the organism itself:

Any boundary reflecting part of the wave acts like a source. If this boundary is in motion, a shift in frequency occurs according to Doppler's

* Christian Doppler (1803–1853), physicist and mathematician. He discovered the described frequency shift in acoustics. It is now called *Doppler effect* or *Doppler shift,* and it applies to all wavelengths.

law, Equation (9.16). Although the blood inside a vessel moves slowly (inside the aorta at about 20 cm/s) compared with the speed of the ultrasonic wave, the resulting shift in frequency is measurable. This method is so sensitive that the speed of blood inside a vein of a human fetus may be determined.

Doppler radar, now indispensable in aviation and traffic supervision, works in the same fashion. The employed waves have a frequency of around 3×10^9 Hz and propagate with the speed of light. The frequency shift caused by the intrinsic motion of the observed object is measured, converted into miles/h, and displayed on the radar screen.

7.2(b). HUBBLE EFFECT. Each chemical element features many eigenfrequencies and, if suitably excited, will emit those frequencies. Many of these eigenfrequencies correspond to wavelengths in the visible spectrum and are analyzed in optical spectrometers.

Astronomical objects such as stars, galaxies, clusters, nebulas, and comets emit light which may be analyzed in a spectrometer. Often the observed spectrum contains individual lines, revealing eigenfrequencies of elements and compounds. If those frequencies are compared with eigenfrequencies obtained in the laboratory, the chemical composition of the astronomical object may be determined. But a well-recorded spectrum yields even more information:

Hubble* discovered that, in distant stars and nebulas, the individual spectral lines are systematically shifted toward lower frequencies, that is, toward the red end of the visible spectrum. This shift, called the *Hubble effect*, is interpreted as a Doppler phenomenon, indicating that the observed object is moving.
Figure 9-42 shows the spectrum of a nebula in the Great Dipper. The upper part displays the spectrum of the nebula which is about 10^8 light years away. The

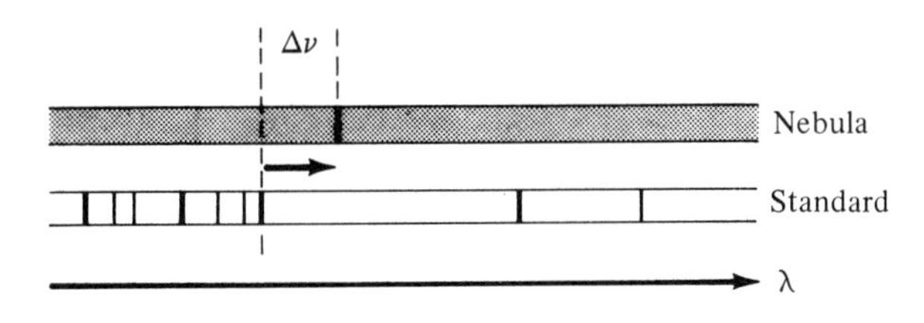

Figure 9-42

isolated spot represents a spectral line (an eigenfunction) of hydrogen. The lower part is occupied, for comparison, by a hydrogen spectrum measured in the laboratory. Clearly, the H-line emitted by the nebula is shifted to the right. Calculating the speed corresponding to the measured frequency shift $\Delta\nu$, we may deduce that the nebula is moving away from us at 1.5×10^7 m/s.
For extremely distant galaxies, the observed Doppler shifts indicate speeds reaching 90% of the speed of light! Of course, these incredible speeds vanish if

* Edwin Powell Hubble (1889–1953), astronomer.

the frequency shifts can be attributed to other phenomena and not to a Doppler effect. But this is open to speculation; at present there are no other convincing explanations.

7.3 Shock waves

Looking at Equation (9.16), you may wonder what happens if the speed of the source is greater than the speed of the propagating waves. Figure 9-43 illustrates this situation. Each ring represents the spatial separa-

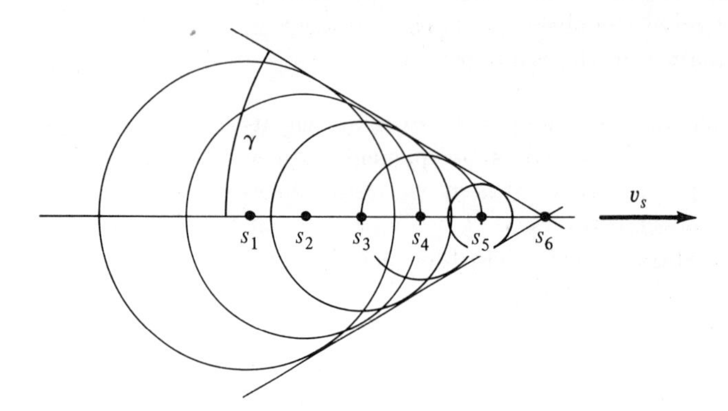

Figure 9-43

tion between consecutive wave peaks. Because the source moves faster than the emitted waves, it is always ahead of them. The surface tangent to all consecutive wave peaks is a cone whose axis is along the direction of motion of the source. The aperture of the cone may be calculated from the following relation:

$$\sin \gamma = \frac{c}{v_s} \tag{9.18}$$

where

γ: aperture of the cone,
c: propagation speed of the wave,
v_s: speed of the source.

The cone is the source of a shock wave moving perpendicular to the surface of the cone. A supersonic airplane or bullet will be trailed by such a shock wave which causes the *sonic boom.*

The typical pattern of a shock wave can also be caused by a high-speed power boat. In this case, the surface water waves propagate slower than the boat.

7.4 Data sheet for the auditory system in man

Earlier we discussed some features of the auditory system of man. This system is summarized below in the form of a technical data sheet. It is most instructive to compare it with technical data of commercial microphones and amplifiers. Keep in mind, though, that most data represent merely orders of magnitude of the indicated quantity.

Section 7.4
Data Sheet for the Auditory System in Man

GENERAL:

Working principle: Frequency dispersion along a membrane (basilar membrane). Two identical detectors (Organs of Corti) separated 22 cm.

Physical dimensions: 10 × 10 × 30 mm.

Power consumption: 10^{-6} watt.

Estimated lifetime: 3.5×10^5 h.

Delay time: 0.18s.

Minimum response energy: 5×10^{-11} erg.

Absolute threshold: 2×10^{-4} dyn·cm^{-2}.

Impedance: 1.5×10^5 g·cm^{-2}·s^{-1}; flexible.

Details:

Frequency response: See Figure 9-44.

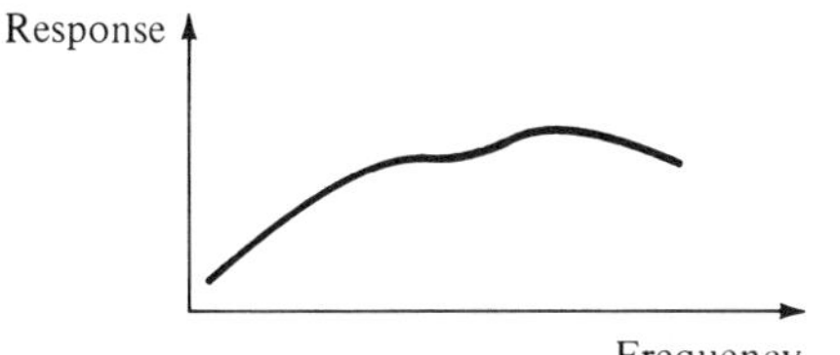

Figure 9-44

Resolution:

(a) Intensity: 8%, depending on frequency and intensity.
(b) Frequency: 0.2%.
(c) Time: 55 ms, depending on intensity.
(d) Direction: 4 degrees.

Load limits:

(a) Minimum: 2×10^{-4} dyn·cm^{-2}.
(b) Normal working range: 10^{-2} to 50 dyn·cm^{-2}.
(c) Temporary permissible overload: 6×10^3 dyn·cm^{-2}.

SPECIAL FEATURES:

Built-in 26-decibel amplifier (middle ear), automatic intensity adaption, built-in on-line computer, high redundance (double components), feedback decoupling (spiral form of receiver and ossicular chain dampens own sound and body vibration sounds).

7.5 Measurements in acoustics

Mankind has become more and more aware of the health hazards connected with noise. Accurate and reproducible methods are sought for determining noise levels. Before that can happen, appropriate units must be defined. For absolute and objective measurements, acoustical pressure (measured in dyn/cm^2) or sound energy flux (measured in watt/cm^2) are obviously sufficient. However, the human ear does not respond proportionally to the sound energy flux. Doubling the flux causes much less than a doubled subjective sensation of noise. Units matching (to a good approximation) the subjective sensation level of sound intensity were consequently introduced.

7.5(a). SOUND INTENSITY LEVEL. In general, the energy flux is measured in watts/cm^2; the same applies to the sound energy flux, which is usually called *sound intensity level.* The numerical value is independent of human response. To account for the (roughly) logarithmic response of the human ear, the sound intensity level is related to the absolute threshold of the human ear and is expressed in the units Bel* and decibel.

$$\text{Bel (symbolized } B) = \log \frac{I}{I_0}$$

where

I: sound energy flux received,
I_0: 10^{-16} W·cm^{-2} (minimum sound energy flux to cause a response in the ear, or absolute threshold).

Conversion:

$$10 \text{ decibels (symbolized dB)} = 1 \text{ Bel}$$

In general, measurements are expressed in decibels not in Bels. A useful relation is:

$$P = \frac{p^2}{2\rho c} \tag{9.19}$$

where

P: sound energy flux in W/cm^2,
p: sound pressure in dyn/cm^2,
$2\rho c$: acoustical impedance,
ρ: density of medium in g/cm^3,
c: speed of sound in cm/s.

Another relation is:

$$P = 2 \log \frac{p}{p_0} \tag{9.20}$$

where

P: intensity level in Bels,
p: sound pressure in dyn·cm^{-2},
p_0: minimum sound pressure causing a response in the human ear (2×10^{-4} dyn/cm^2).

Realize that the Bel and the decibel are absolute quantities because the reference flux is defined as 10^{-16} W/cm^2. Figure 9-45 shows the sound intensity level of various sources.

* The name Bel is coined in honor of Alexander Graham Bell (1847–1922), teacher and physiologist. One of the first to build a telephone.

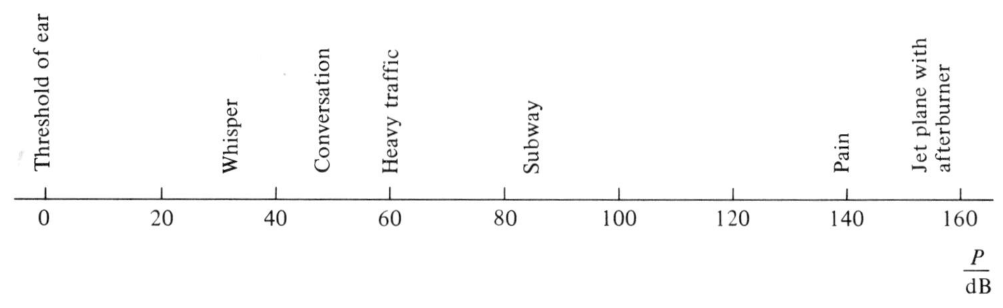

Figure 9-45

7.5(b). Sensation level. This quantity, also called the *loudness level*, is frequency-dependent and takes into account the logarithmic response of the ear. It is measured in the unit *phon*.

phon = number of decibels above the intensity threshold

This means that a sensation level of zero phon corresponds to the threshold of hearing. Figure 9-46 shows the threshold of hearing as a function of frequency.

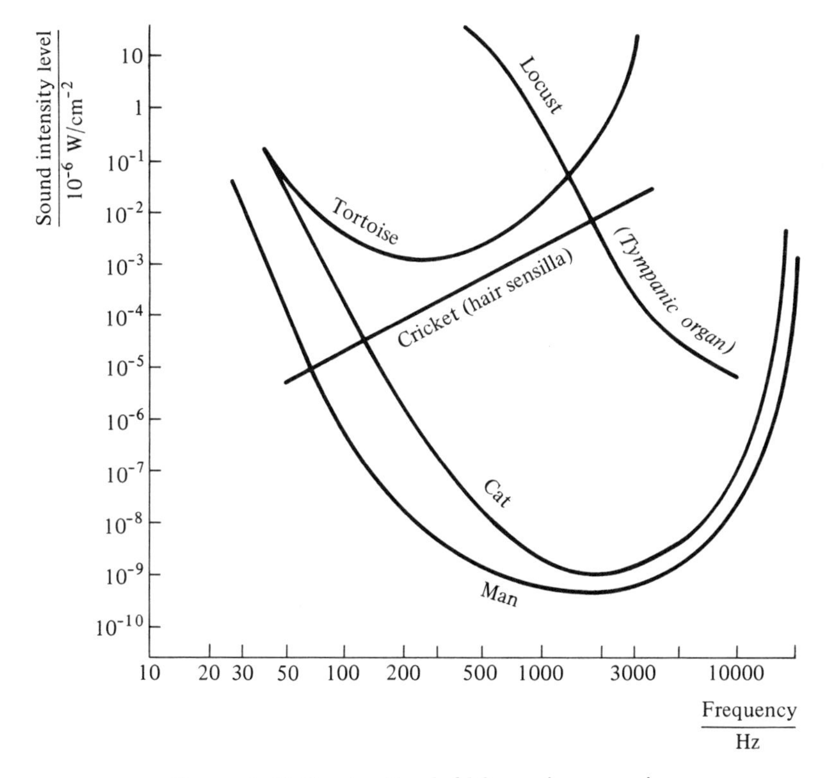

Figure 9-46 Hearing threshold for various organisms.

Conversion:

$$n \text{ decibel} = n \text{ phon at 1 kHz}$$

7.5(c). LOUDNESS. There may be two sources, each individually causing the same sensation level. If both are combined, the subjective sensation does not double. This shortcoming of the phon scale is eliminated using the unit of loudness, sone. n sources each of loudness m (measured in sone) have the combined total loudness $(n + m)$ sone.

Conversion:

$$1 \text{ sone} = 40 \text{ phons at 1 kHz}$$

PROBLEMS

1. Calculate the wavelength, wave number, period, and angular frequency for the concert-pitch A ($\nu = 440$ Hz).
2. Sketch an experimental arrangement for using Doppler's principle to measure the speed of blood. Which factors must be considered to determine the ultrasonic frequency? What limits the employed intensity of the ultrasonic waves?
3. The sonar system of most bats emits ultrasonic signals in a band centered around 70 kHz. Discuss: (a) Why is the center of emission at 70 kHz? (b) How will this sonar system respond to obstacles very small and very large compared with the emitted wavelength?
4. During daytime, bright light comes not only from the sun's direction but also from other directions. Why?
5. Blue light suffers more scattering than the longer wavelengths of the visible spectrum. Explain why the sky appears to be red at dawn.
6. A few drops of milk suspended in a glass of water cause a strange effect. From above, the fluid looks blue; from below it appears to be red. Why?
7. In modern fiber optics the light guides are coated with a medium having a lower refractive index than the guide itself. Why?
8. In Section 4.2(b) a glass prism is used to reflect light. What is its advantage over a conventional mirror? How large will the light intensity be after three reflections in mirrors (reflectivity 88%) and in prisms?
9. Plot the spectrum of Figure 9-32 as a function of the wave number.
10. Extract from Figure 9-34 the generation cycle; estimate the error limits.
11. Absorption is often characterized by the half-value layer HVL, the absorber thickness which will reduce the radiation intensity to one-

half of its incident intensity. Derive a formula connecting HVL and absorption coefficient. Using this formula, find the absorption coefficient for the radiation in Figure 9-29.

12. The phase velocity of a sound in water is 1.5×10^3 m/s. Calculate the index of refraction for water and the angle of total reflection. ($n_{air} = 1.00$).

13. The visible light spectrum extends between 3.85×10^{14} Hz and 8.35×10^{14} Hz. Passing from air into crown glass the corresponding indices of refraction are 1.511 and 1.540. Calculate the spatial separation between the red and ultraviolet ends of the visible spectrum at a penetration depth of 10 cm. Assume an angle of incidence of 30°.

14. What is the frequency of a sound wave sufficient to resolve spatially a distance of 10^{-2} cm inside the eye?

15. The mean velocity of blood in the aorta is 40 cm/s. This velocity is to be measured with the help of the Doppler effect and ultrasonic waves. Calculate the minimum sound frequency to determine the velocity of blood within 10%. (Propagation speed of ultrasound in the body is 1.5×10^5 cm/s).

16. Validity of the simplified Equation (9.17): If the values for v_R and v_S are, respectively, 5 and 10% and 10 and 20% of c, calculate the error in ν_R if Equation (9.17) is used instead of Equation (9.16).

17. Calculate the largest eigenfrequency of the hydrogen atom.

18. Calculate the amplitude ξ_0 of the oscillating molecules for the temporary permissible overload of the ear (6×10^3 dyn·cm^{-2}).

19. Take the technical data sheet for your hifi receiver and compare it with the auditory system in man.

20. The pressure of a 1-kHz sound causing pain is about 1.6×10^3 dyn·cm^{-2}. Calculate the intensity level in decibels and in phons. Calculate the loudness level for the same sound pressure at 50 Hz.

FURTHER READING

G. VON BÉKÉSY, *Experiments in Hearing*. McGraw-Hill, New York, 1960.

J. L. FLANAGAN, *Speech Analysis, Synthesis and Perception*. Springer, Berlin, 1972.

D. R. GRIFFIN, *Listening in the Dark*. Yale University Press, New Haven, 1958.

J. D. PYE, "Bats and Fog," *Nature*, **229** (1971) 572–574.

LEO SZILARD, *The Voice of the Dolphins*. Simon and Schuster, New York, 1961.

J. V. TOBIAS, *Foundations of Modern Auditory Theory*. Academic Press, London, 1970.

T. H. WATERMAN, H. J. MOROWITZ, eds., *Theoretical and Mathematical Biology*, New York, 1965. (Article by G. von Békésy.)

chapter 10

THERMODYNAMICS

1. INTRODUCTION

Before the middle of this century, thermodynamics seemed to be a closed chapter of the sciences, as much as mechanics was a hundred years ago. Apparently all major discoveries were made, and wherever thermodynamics was applied it yielded correct results. That impression has changed very much indeed. Modern thermodynamics is still in full swing. It developed into a field with the most demanding mathematics, especially in the subareas of steady-state and open systems. These are the areas of thermodynamics the student of the life sciences should become familiar with. However, if we apply the cost-benefit judgment, it is fair to say that you only need to work through the very basics of ther-

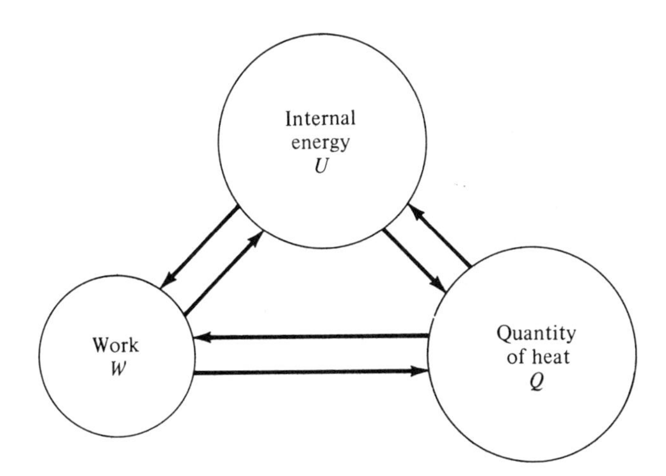

Figure 10-1

modynamics—a rather simple task. But be aware that you are merely looking at the tip of the iceberg. To delve further into this subject, brush up on your math (especially integrals and differential equations) and look into a modern text of applied thermodynamics. Highly recommended is van Holde's book listed at the end of this chapter, but there are others as well.

Thermodynamics investigates the interrelation of work (W), quantity of heat (Q), and internal energy (U) of systems. Work is done by the system or received by it. The quantity of heat is either emitted or absorbed by the system. The internal energy of the system refers only to those forms of energy which may be absorbed or emitted readily as heat or work.

Examples of systems are:

1. A car on the road with a driver: Work is produced by the system; heat is emitted. The system's internal energy comprises its kinetic energy, potential energy, and the chemical energy stored in the fuel.
2. An amoeba: It moves around, thus producing work; it receives heat from its surroundings and emits heat—the unavoidable by-product of its metabolism. The internal energy is its stored food supply and any readily available reserves.

1.1 Isothermal and adiabatic processes

The change of the internal energy of a system is isothermal or adiabatic:

$$\textit{isothermal process:}\quad T_{\text{initial}} - T_{\text{final}} = \Delta T = 0$$

The temperature T of the system remains constant.

$$\textit{adiabatic process:}\quad Q_{\text{initial}} - Q_{\text{final}} = \Delta Q = 0$$

The system neither emits nor receives a quantity of heat Q.

Note: To indicate very small (infinitesimal) changes, the Greek letter Δ is replaced by the letter d.

Thermodynamics (except for some most modern branches) applies to multibody systems. This means that the system under investigation must contain many components. A cell does; so does even a small amount of car fuel. An individual chemical reaction involving only a handful of molecules is outside the scope of thermodynamics.

It does not matter how the system changes in detail. We look at it in the beginning and note a system in its initial condition. It may have temperature, mass, volume, a certain phase, or other characteristic properties. Applying the laws of thermodynamics correctly, we calculate the final condition of the system with new values for its characteristic properties. What happens during the transition does not concern us. This overall approach is not by choice: even a tiny bacterium or a living cell is already such a vastly complex system that modern science cannot explain or predict in detail what goes on during transitions. Experience teaches that this general approach of thermodynamics yields acceptable results.

1.2 Open and closed systems

A *closed system* does not interact with its surroundings. This is true only if the system encompasses the entire universe. However, there are less extended systems which are good approximations to a closed system. Whenever the interaction between system and surroundings is negligible, the system is rated a *closed system*.

The biosphere of the earth is such a closed system. Whatever is produced inside originates from internal components and remains there. Some components will be redistributed, but still not leave the biosphere. This is the heart of the pollution problem.

A space capsule is also a closed system. The approximation is that any heat loss or uptake from the outside is negligible. Indeed, initiating the propulsion system is not an essential characteristic of this closed system—a miniature biosphere.

A virus in its dormant state is also a closed system.

An *open system* interacts with its surroundings. By enclosing its surroundings into the system, every open system becomes a closed system. See Figure 10-2. Classical thermodynamics applies to closed systems

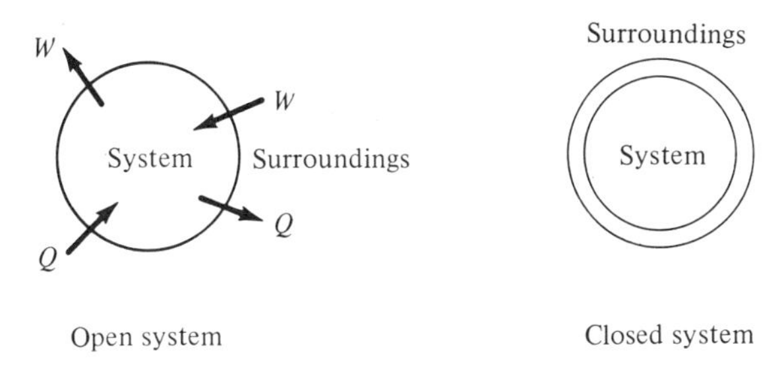

Figure 10-2

only. Whenever it fails to correctly predict a final condition, it was applied to an open system.

Man is an open system. He takes food and oxygen from his surroundings and emits heat and substances. Obviously, the influence of his surroundings cannot be neglected. Man, investigated within the biosphere, is a closed system. Classical thermodynamics may be applied to man if he is placed inside an artificial biosphere—a caloric chamber (see Chapter 6), for example.

1.3 Time

The quantities characterizing a closed system are time-independent; in an open system they change with time. Those are the two extremes. Real systems find their place in between.

Equilibrium	*Quasiequilibrium*	*Nonequilibrium*
Closed system	Steady-state system	Open system
Time-independent Scope of classical thermodynamics.	Time-independent for short periods only	Time-dependent

Life in any of its innumerable forms is a steady-state system; it is in quasiequilibrium. An amoeba may serve as example: Observed over a time span that is short compared with its regeneration time, the amoeba does not change its contents, volume, or any other gross feature. It interchanges substance and heat with its surroundings, but very slowly. Any change is bound to be small and insignificant.

Conclusion: Before applying the laws of classical thermodynamics, it is important to know whether or not the system under investigation can be treated as open or closed. Steady states are covered by classical thermodynamics if the observation time is short compared with the total lifetime of the system.

1.4 Limitations

The following treatment is limited to:

1. Closed systems.
2. Time-independent changes.
3. Adiabatic and isothermal processes.

That is classical equilibrium thermodynamics, mainly concerned with reversible energy changes between initial and final conditions of a system. The change in energy is investigated because, in general, it is too difficult to determine the total energy content (due to heat, volume, pressure, composition, etc.) of a system.

2. FIRST LAW OF THERMODYNAMICS

2.1 Formulation

The first law of thermodynamics is:

$$U_i + Q = U_f + W \tag{10.1}$$

where

U_i: internal energy of the initial system,
U_f: internal energy of the final system,
Q: quantity of heat absorbed or emitted by the system,
W: work done or received.

It extends the law of mechanical energy conservation by including Q and W.

In general, the law is concerned with readily available internal energy, such as kinetic or potential energy. That excludes, for example, the energy of a system stored in the form of its mass.

The first law of thermodynamics expressed in differential form is:

$$dU = dW - dQ \tag{10.2}$$

where

$dU = U_i - U_f$: change of internal energy,
dQ: transferred amount of heat,
dW: work done or received.

The letter d indicates that all quantities are very small (infinitesimal quantities).

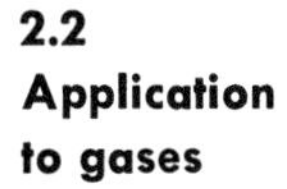

2.2 Application to gases

The system under investigation is a gas. This is of interest to us because a living system—like a cell—may be treated successfully as an aqueous solution of gases. A gas-filled container closed by a movable piston is a suitable illustration for a system in thermodynamics. See Figure 10-3. Only the gas is of importance; the influence of container and piston is neglected.

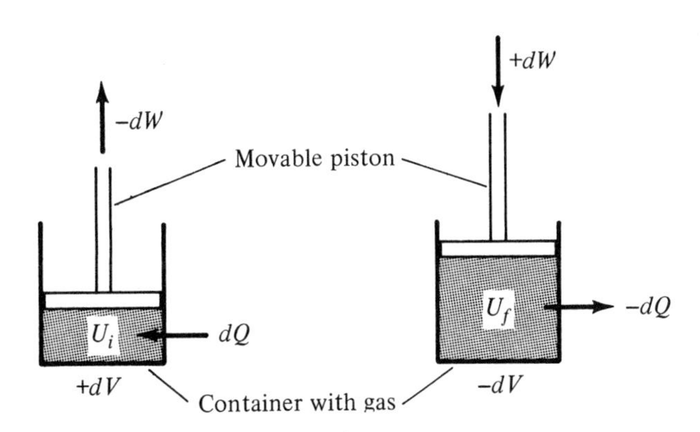

Figure 10-3

If the gas changes its volume, the piston moves up ($+dV$) or down ($-dV$). A quantity of heat emitted ($-dQ$) or received ($+dQ$) by the system is expressed by suitably directed arrows. Work done ($+dW$) on the system pushes the piston in; work exerted ($-dW$) drives the piston outward.

As an illustration of the first law of thermodynamics, we shall investigate what happens during a small change of volume (dV). We shall idealize that the volume changes by an isothermal process ($dT = 0$) or an adiabatic process ($dQ = 0$). In real life both changes will occur simultaneously. However, the real process can be approximated if we first investigate the adiabatic part and then the isothermal part, or vice versa.

As described in Chapter 6, the internal energy U of a gas changes by the amount dU if its temperature changes by dT:

$$dU = m \cdot c \cdot dT$$

where

dU: very small change of the internal energy of gas,
m: mass of the gas,
c: specific heat capacity of the gas,
dT: small temperature change of the gas.

Substituted into the differential form of the first law of thermodynamics, Equation (10.2), this becomes:

$$m \cdot c \cdot dT = dW - dQ \tag{10.3}$$

For a gas, we already know (see Chapter 4, work) that $dW = p \cdot dV$; hence

$$m \cdot c \cdot dT = p \cdot dV - dQ \tag{10.4}$$

where

p: pressure of gas, equal to the outside pressure,
dV: very small change in volume,
dQ: transferred amount of heat (very small).

If the right side of the equation is negative, the internal energy of the gas decreases; if it is positive, then U increases. The change in volume is easily observed. To gain insight, we investigate this change of volume for the two extreme cases: $dT = 0$ and $dQ = 0$.

2.2(a). Isothermal process

$$dT = 0, \qquad \text{no temperature change}$$

The temperature of the system remains constant, and the container walls are transparent to thermal energy (Figure 10-4).

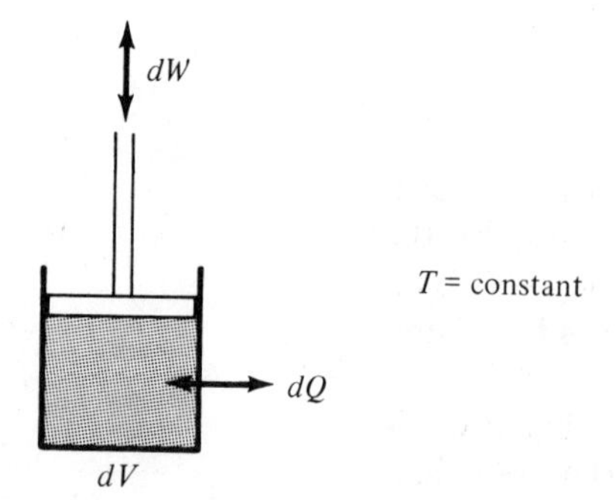

Figure 10-4

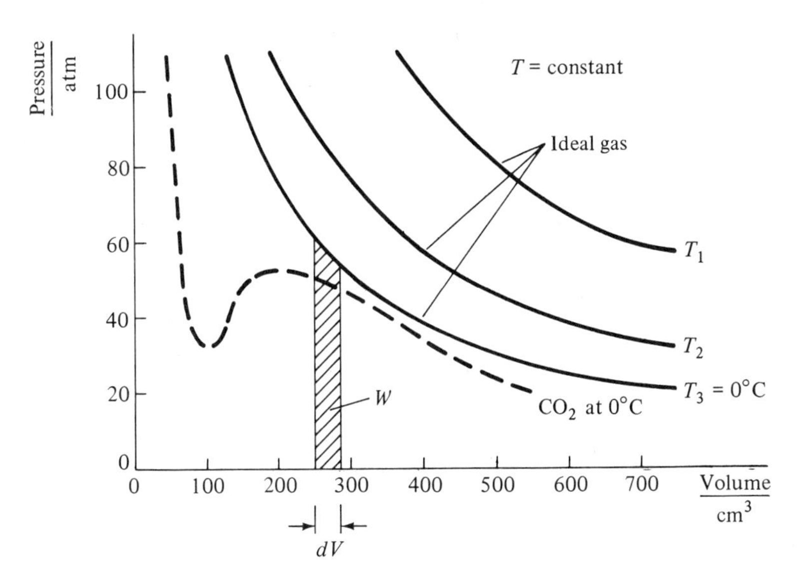

Figure 10-5 Pressure-volume diagram for an ideal gas during isothermal transformation (T = constant). $T_1 > T_2 > T_3$. For comparison, the isothermal of a real gas (CO_2) is included (dashed line).

The volume changes isothermally. Figure 10-5 shows this isothermal expansion or contraction. The isothermal is a hyperbola, p = constant/V. The shaded area under the curve is proportional to the work (W) occurring during the process, or in mathematical terms:

$$W = \int dW = \int p \cdot dV \tag{10.5}$$

The quantitative value for the work is computed by replacing p under the integral with either Boyle-Mariotte's law (ideal gas) or van der Waal's law (real gas). For one mole of ideal gas we can compute:

$$W = RT \ln \frac{V_2}{V_1}$$

where

R: molar gas constant,
T: absolute temperature,
V_1: initial volume,
V_2: final volume.

2.2(b). Adiabatic process

$$dQ = 0$$

No thermal energy is exchanged; the container walls are opaque to heat (Figure 10-6).

The volume changes adiabatically. Figure 10-7 shows this adiabatic expansion (or contraction) of a gas. The adiabatic is described by p =

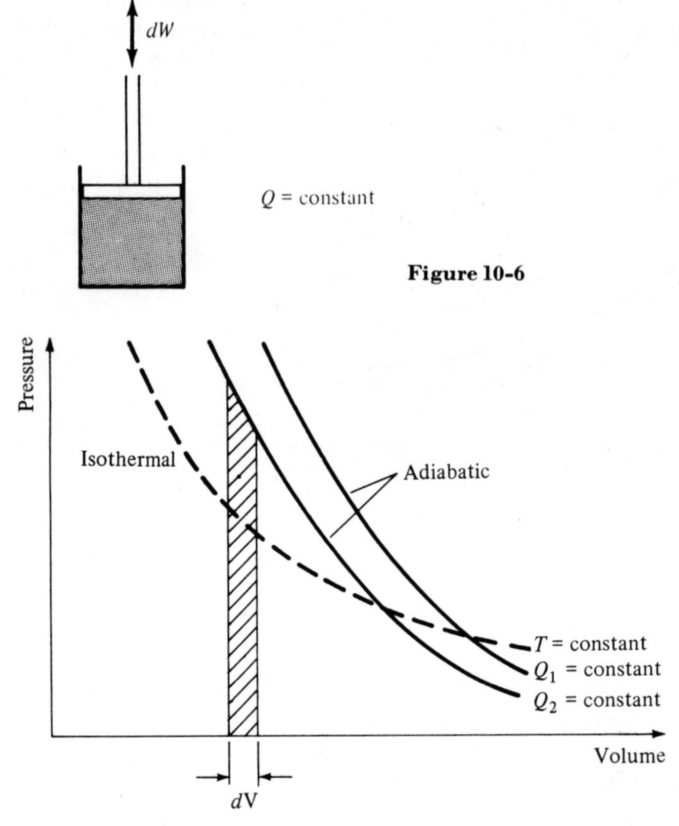

Figure 10-6

Figure 10-7 Pressure-volume diagram for an ideal gas during adiabatic transformation (Q = constant). For comparison, a corresponding isothermal is shown (dashed line).

constant/$V^{\varkappa}$ where $\varkappa$ = material constant ($1.2 < \varkappa < 1.7$). Again, the shaded area under the adiabatic is proportional to the work occurring during the process. For one mole of an ideal gas,

$$W = \frac{R(T_1 - T_2)}{\varkappa - 1}$$

where

W: work during an adiabatic process for an ideal gas,
R: molar gas constant,
T_1: initial temperature,
T_2: final temperature.

Don't lose sight of the first law of thermodynamics as we discuss all these new terms. The flow diagram in Figure 10-8 is a guidepost for the previous pages.

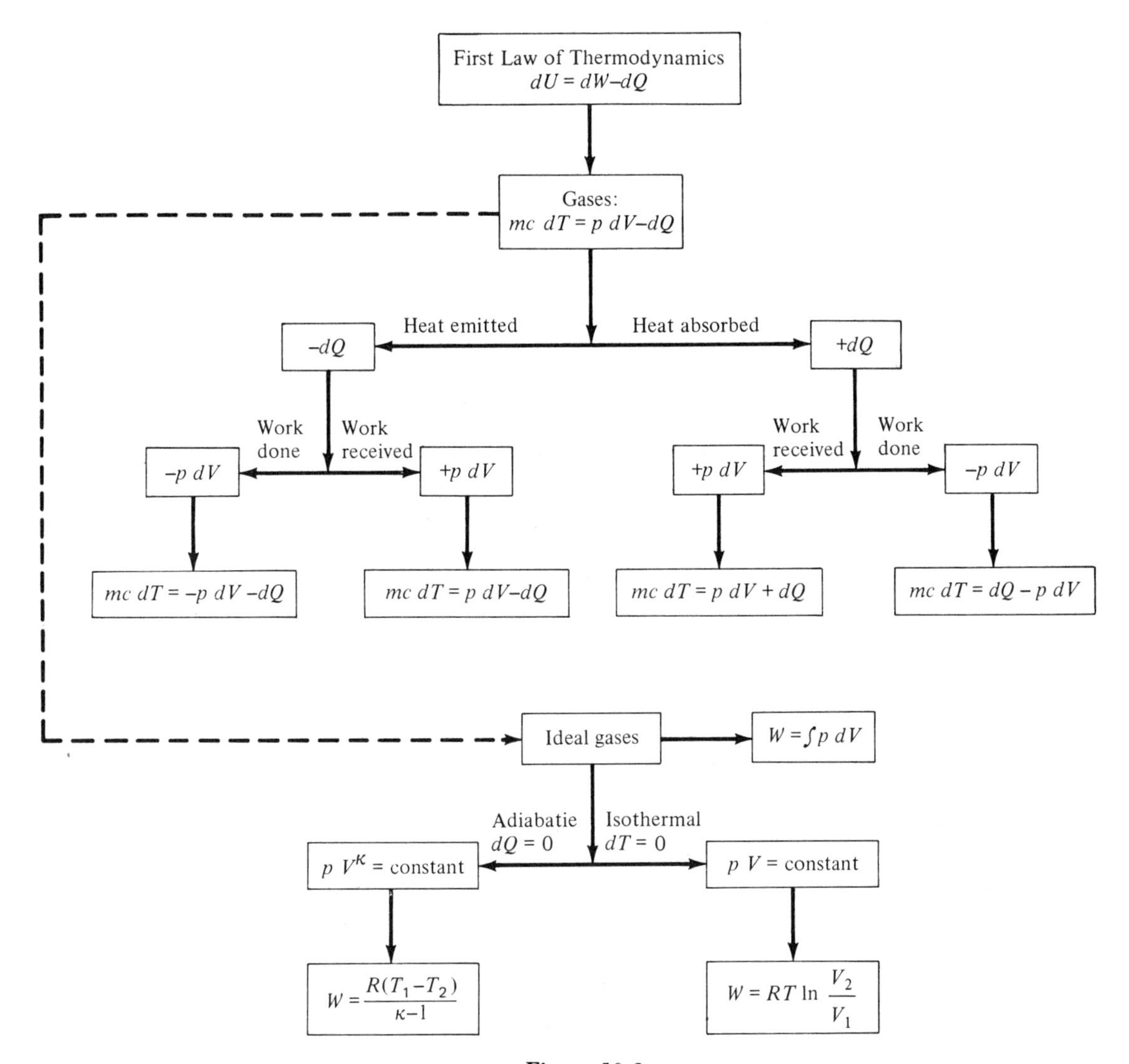

Figure 10-8

2.3 Application to life

The first law of thermodynamics is an application of the more general law of energy conservation: The mechanical and thermal energies emitted by a system are equal to its energy consumption. Lavoisier* demonstrated that this law also applies to processes in the animated world. Food burned inside a calorimetric bomb yields the same amount

* Antoine Laurent Lavoisier (1743–1794), chemist. His analytical methods founded modern organic chemistry. Because he held the office of head tax-collector, he lost his life under the guillotine. More about him in D. McKil, *Antoine Lavoisier*, London, 1952.

of heat as food digested! Moreover, all energy spent by a living system has its ultimate source in its food. He proved this in a direct calorimetric measurement. See Figure 10-9.

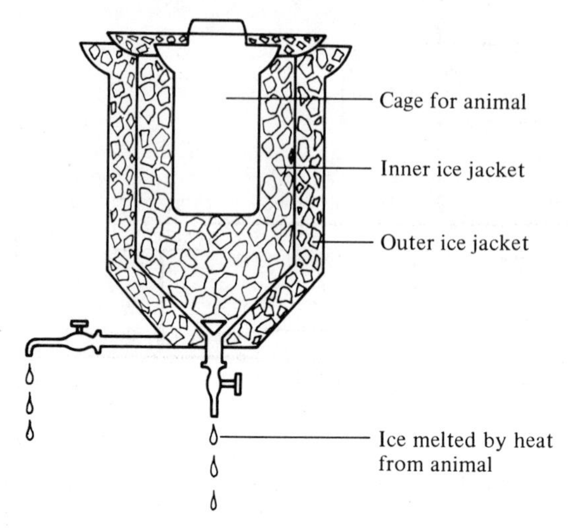

Figure 10-9 Animal calorimeter devised and used by Lavoisier. Its ingenious, double-walled design allows accurate measurements.

In Figure 10-9, the outside ice-container keeps the inner one always at 0°C; no thermal energy can penetrate into the inner chamber. The calorimeter must be adiabatic ($\Delta Q = 0$, no heat exchange with the outside), to compare the gained results with bombcalorimetric measurements. The thermal energy produced by the resting animal melts ice on the inside. From the amount of water, the emitted calories are calculated. Modern calorimeters also have a calibrated treadmill inside which allows a measurement of the work done by a moving animal. It all proves that—at least from the viewpoint of energetics—no difference exists between animated and nonanimated life.

Another blow against the *vis vitalis* (the special spirit supposed to animate nature) was dealt in 1897 by the Nobel-laureate (1907, chemistry) Eduard Buchner (1860–1917) and his brother Hans (1850–1902). They proved that the processes inside a living cell are governed by enzymes. These enzymes could be produced outside the cell and even synthesized by man.

2.4 Circular process

Many processes in thermodynamics are circular. This means that the system starts from an initial condition and reaches (after some steps) its final condition which is identical with the initial condition. Any periodic engine moves through such a cycle.

Let us investigate a circular process with the help of a *gedankenexperiment*. (*Gedankenexperiment*, a term mainly used in quantum theory, is an experiment which may be performed in principle, although most likely not in practice. Nevertheless, it is most enlightening to get at the root of physical processes. We shall encounter gedankenexperiment again.)

Figure 10-10 shows a four-stage circular process. The system contains an ideal gas and is in principle the most simple heat engine, named the *Carnot engine.**

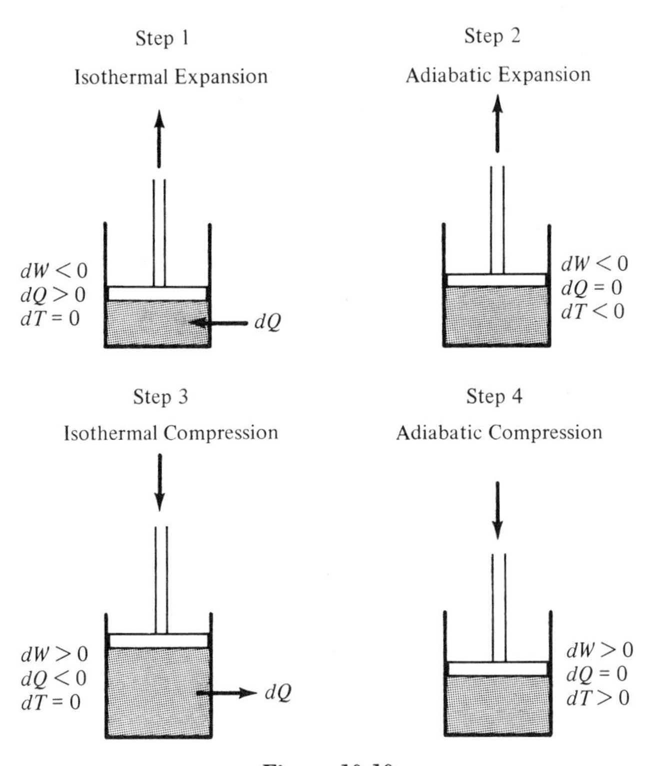

Figure 10-10

Figure 10-11 demonstrates all four steps in a pressure-volume diagram. The process is circular since the initial condition of step 1 is identical with the final condition of step 4. The enclosed area is proportional to the work done during this *Carnot cycle.*

The efficiency (for its definition, see Chapter 4) of the Carnot engine:

$$\eta = \frac{T_1 - T_2}{T_1} \qquad \text{Carnot's theorem} \qquad (10.6)$$

* N. L. Sadi Carnot (1796–1832), physicist. Devised the hypothetical Carnot engine and calculated its efficiency. The first one to calculate the mechanical equivalent of heat.

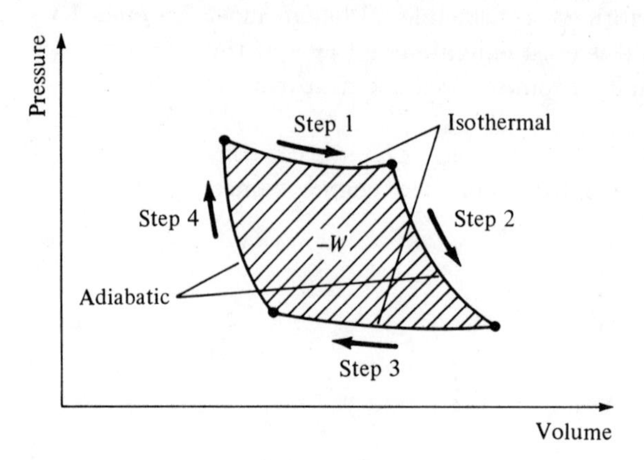

Figure 10-11

where

η: efficiency of the Carnot process,
T_1: initial temperature measured on the absolute scale,
T_2: temperature at the end of step 3.

Obviously, the greater the temperature difference, the higher the efficiency.

The Carnot engine is an ideal heat engine (no friction, perfect isothermal and adiabatic transitions). Its efficiency is the highest possible. Real engines have a lower efficiency. Nevertheless, the point to remember is that the maximum temperature difference during the cyclic process determines the efficiency.

In Figure 10-11, the cycle runs clockwise. It represents a heat engine like a combustion motor (work done: $W < 0$). If the cycle runs counterclockwise, it represents a refrigerator (work received from outside the system: $W > 0$).

3. SECOND LAW OF THERMODYNAMICS

In an isolated system, spontaneous heat exchange occurs from areas with higher temperatures to areas with lower temperatures. A heat transfer toward higher temperature would not invalidate the first law of thermodynamics. But experience teaches that a preferred direction for the transfer of thermal energy exists. The second law of thermodynamics summarizes this experience. To formulate it, we will first introduce a new quantity, *entropy*.

3.1 Entropy

A closed system does work by converting one form of energy into another. Mechanical energy (potential, kinetic) can be fully converted into thermal energy. The reverse is not true: thermal energy cannot be completely converted into mechanical energy.

The available (or free) energy of a closed system is:

$$F = U - E_{\text{waste}} \tag{10.7}$$

where

F: energy fully convertible into mechanical energy,
U: internal energy of the system (mechanical + thermal energy),
E_{waste}: amount of energy not available for conversion into mechanical energy.

E_{waste} is not always wasted energy. Take a warm-blooded animal. The amount of energy from the metabolic processes which is not free energy (about 50%) appears as the heat necessary to keep its body warm—to keep it at a temperature which allows the chemical reactions to proceed with optimum speed. Thus, thermal energy supplies the energy of activation for many life-sustaining reactions.

Experience teaches that E_{waste} is proportional to the absolute temperature of the system,

$$E_{\text{waste}} \sim T$$

Converting this into an equation, we obtain:

$$E_{\text{waste}} = ST$$

where S is a proportionality factor, called *entropy*. Inserting this into Equation (10.7), we obtain:

$$F = U - ST \quad \text{or} \quad F = U - TS \qquad \text{Helmholtz's function*} \tag{10.8}$$

where

S: entropy of system, expressed in cal/°K,
T: temperature of system, expressed in °K.

Entropy is thus a measure for the amount of thermal energy not convertible into mechanical energy. It is a characteristic quantity for a system and can be used to describe it in the same sense as by temperature, volume, pressure, energy, etc. Entropy is additive; thus if S_1, S_2, S_3, . . . , S_n are the entropy values of subsystems within the closed system, then

$$S = \sum_{i=1}^{n} S_i$$

The total entropy of the system is the sum of the entropies of the subsystems.

* Hermann von Helmholtz (1821–1894), physician and physicist. A truly universal scholar. He made essential contributions to anatomy of nerves, physiology of music, optical instruments, thermodynamics of chemistry, free energy, conservation of energy, meteorology.

3.1(a). THE SYSTEM ICE-LIQUID WATER. Entropy is such a basic and celebrated quantity that we shall dwell more upon it. For a closed system, let us take a container filled with H_2O in its solid and liquid states. At zero degree Celsius, both states occur simultaneously (Figure 10-12).

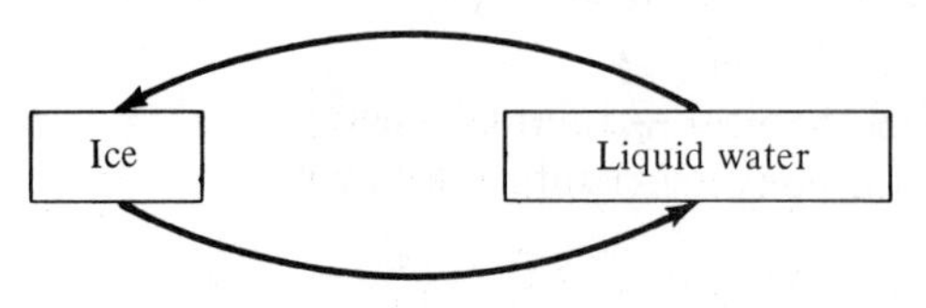

Figure 10-12

To make everything simple, we shall assume that H_2O has only a solid (ice) and a liquid (water) state. Which state has the higher entropy?

Entropy of ice: Rewriting Equation (10.8), we obtain

$$\frac{F - U}{T} = -S$$

or, since $U > F$,

$$S = \frac{U - F}{T}$$

where

$$F = \text{heat of fusion} - \text{heat of formation}$$
$$= (79.71 - 68.36) \text{ cal} = 11.35 \text{ cal}$$

This amount of free energy is utilized by the system to build the lattice structure of ice out of the unordered molecules of liquid water.

The system's temperature is $T = 0°\text{C} = 273.15°\text{K}$.

$$S_{\text{ice}} = +\frac{79.71 - 11.35}{273.15} \text{ cal}\cdot°\text{K}^{-1} = 0.251 \frac{\text{cal}}{°\text{K}}$$

Now the same formal computation for water:

$$F = \text{heat of formation} - \text{heat of fusion}$$
$$= (68.36 - 79.71) \text{ cal} = -11.35 \text{ cal}$$

Formally this is a free energy, but it has a negative sign; thus it is an energy to be supplied.

$$S_{\text{water}} = \frac{68.36 + 11.35}{273.15} \text{ cal}\cdot°\text{K}^{-1} = 0.293 \frac{\text{cal}}{°\text{K}}$$

Result:

$$S_{\text{water}} > S_{\text{ice}}$$

or

$$\Delta S = S_{\text{water}} - S_{\text{ice}} = 0.042 \frac{\text{cal}}{°\text{K}}$$

This result makes sense because potential energy is stored in the lattice structure of ice, and this potential energy is not present in water. This means that ice carries more free (or available) energy which can be converted into work; hence its entropy is smaller than the same amount of liquid water having the same temperature.

3.1(b). IDEAL GAS. With the help of the following formulas, the entropy of a system, or its entropy change while transferring from condition 1 to condition 2 within a system, can be calculated. It does not make much sense to derive these formulas here; you can find the derivations elsewhere.

Ideal gas:

$$S = m\left\{c_v \ln T + \frac{R}{M}\ln\frac{V}{m}\right\} + S_0 \tag{10.9}$$

where

m: mass of gas,
c_v: specific heat capacity at constant volume,
T: temperature measured in °K,
R: molar gas constant $= 8.314 \times 10^7$ erg·°K^{-1}·mol^{-1}
M: molecular weight,
V: volume,
S_0: a constant which depends on the kind of gas. It is not important for calculations because we are usually interested in an entropy change ΔS of the same gas. In that case, S_0 cancels out.

Ideal gas (isothermal change):

$$\Delta S = \frac{m}{M} R \ln\frac{V_2}{V_1} \tag{10.10}$$

where

V_2: final volume of gas,
V_1: initial volume of gas.

3.2 Reversible and irreversible processes

Within a closed system a transformation is called *reversible* if, after describing a full cycle, the system returns into its initial condition. It is an *irreversible* process if a full cycle does not lead back to the initial condition without energy supplied from outside the system. Using the new term entropy, we can express a reversible process as:

$$S_{\text{initial}} = S_{\text{final}} \curvearrowright \Delta S = 0$$

An example is the isothermal change of an ideal gas.

An irreversible process may be expressed:

$$S_{\text{initial}} \neq S_{\text{final}} \curvearrowright \Delta S \neq 0$$

An example is any process observed in nature.

3.3 Formulating the law

The preceding was a lengthy excursion, but it enables us to understand a pregnant formulation of the second law of thermodynamics.

The Second Law of Thermodynamics: within a closed system

$$\Delta S \geqslant 0 \tag{10.11}$$

This means that the entropy of a closed system never decreases. In general, any change will lead to an entropy increase.

The second law of thermodynamics assigns each process a preferred direction: the system will change in such a way that its entropy increases. If entropy has reached a maximum value, then the system does not change spontaneously anymore. It has reached its equilibrium.

EXAMPLE: THE CLOSED SYSTEM ICE-WATER

If there is no heat exchange with the outside, ice will spontaneously convert into water. This ends as soon as all the ice is melted because now the entropy has reached its maximum value.

Another example: The thermal energy stored in the waters of the oceans is formidable (about 2×10^{27} cal). It would not contradict the first law of thermodynamics to build an engine which extracts part of this energy and converts it into mechanical energy. The water would have a lower temperature after this extraction. But the second law of thermodynamics states that during this change the entropy will decrease; thus this process is not possible.

4. ASPECTS OF ENTROPY

4.1 Entropy and probability

In a subfield of thermodynamics—statistical thermodynamics—entropy is interpreted in a different way: An ordered system has a lower entropy than a random system. The larger the degree of order, the smaller its entropy. In other words, entropy is a measure for the randomness of arrangement.

As an example, a crystal of NaCl is a highly ordered system compared with melted NaCl. Thus the entropy of crystallized NaCl is lower than that for the same amount of melted NaCl. A living cell has a higher degree of orderliness than a dead one of the same kind. This means, after death, the entropy increases.

It is difficult to find a suitable measure for the degree of order of a state, and so the entropy is related to the probability of this state:

$$S = k \ln W \qquad (10.12)$$

where

S: entropy of the state of the system,
k: Boltzmann's constant* $= 3.29 \times 10^{-24}$ cal/°K,
W: probability of the state.

Notice that S still has the dimensions cal/°K.

EXAMPLE:

Two gases inside a closed system (Figure 10-13) are separated by a wall. If the wall is removed (Figure 10-14), both gases will mix via diffusion. After awhile, a homogeneous mixture is established—both gases are uniformly distributed over the entire system.

The probability that one gas will concentrate in the left half and the other in the right

* Ludwig Boltzmann (1844–1904), physicist.

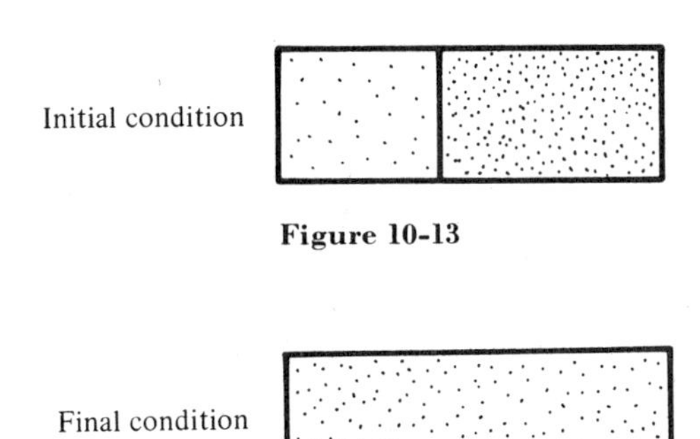

Figure 10-13

Final condition

Figure 10-14

half is extremely unlikely. It would represent a certain degree of order within the system. The most probable state is the homogeneous mixture. This is the state with the highest entropy, the equilibrium state.

Returning to the previously discussed system, ice-water, it is clear that ice has a lower entropy than water because its lattice structure represents a higher degree of order. With no interference from the outside, this system will change spontaneously toward a more probable state, a pure water system. This is the state with the highest entropy.

Wherever in nature a system proceeds toward more than one direction, we can calculate the preferred direction by looking into the entropy of possible future states. If the system is not isolated, we must enlarge it so that it encloses the pertinent outside. Transport phenomena, thermal conduction, and radiation effects are areas where entropy considerations are applied often. We shall not delve into that, but instead present a few paragraphs about the connection of entropy with life, information, and the end of the universe.

4.2 Other aspects

4.2(a). Life. Life is a very complicated and highly ordered system; consequently its entropy is low. However, we should keep in mind that it is not a closed system.

From a thermodynamic point of view, it operates at constant temperature, volume, and pressure. It is not possible to compare it with a heat engine, since the essential feature of a heat engine, a temperature gradient, does not exist. Also, the living cell is by no means in equilibrium with its surroundings. Its steady state with small and slow changes is somewhat similar to an equilibrium. That gives us confidence to apply terms such as entropy.

There are interesting aspects of life connected to entropy. Highly recommended is the short essay by Schrödinger: "What Is Life?" Refer to the end of this chapter. Quoted from page 76 of the Nobel laureate's book: "The essential thing in metabolism is that the organism succeeds in freeing itself from all the entropy it cannot help producing while alive."

4.2(b). INFORMATION. Entropy and information are interrelated. Ordered systems contain plenty of information. This information can be calculated (at least for simple examples) and expressed in the unit of information, in bit. On the other side, entropy is a measure for the randomness of the system. We may say that information is something like negative entropy. These few lines should serve as a stimulus to read more about entropy and information.

4.2(c). DOOM. It is an inescapable consequence of the second law of thermodynamics that eventually all processes will stop. If, and only if, the universe is a closed system, it will reach the ultimate equilibrium with maximum entropy. We don't need to worry though; that lies far ahead in the future.

Figure 10-15

SUMMARY

Thermodynamics covers the interchange of mechanical energy (kinetic and potential), thermal energy, and the work done within a system. The system can be any well-defined arrangement, ranging from the molecules of a gas inside a container to the entire universe. It is a closed system if no energy in any form enters or leaves it. The thermodynamics of open systems—and really all systems within the realms of the life sciences are open systems—are a very modern development and are not covered here.

A system is said to be in equilibrium if its characteristic quantities such as temperature, volume, and pressure are time-independent. Real systems depend on time. Most systems in nature are in steady state, that is, their rate of exchange with the outside is low and their time-dependence is negligible. These systems are described in good approximation by classical, equilibrium, closed-system thermodynamics.

First law of thermodynamics: For a closed system,

$$U_i + W = U_f + Q$$

where

U_i: initial internal energy of the system,
W: work done or received,
U_f: final internal energy,
Q: quantity of heat absorbed or emitted.

This is a general formulation of the law of energy conservation.

While the kinetic and potential energies are fully convertible into thermal energy, the reverse is not true. Only the free energy of a system can be converted into work:

$$F = U - ST$$

where

F: free energy of the system,
U: final internal energy,
S: entropy of the system,
T: absolute temperature.

The term ST describes the amount of the system's energy not available for conversion into work. This *wasted energy* appears as heat.

The quantity S, the entropy, is a characteristic property of thermodynamic systems like temperature, volume, and pressure.

Systems (like an engine) performing work operate in cycles. After intermediate steps, the initial state is reached again. The efficiency of such a circular process is

$$\eta \leq \frac{\Delta T}{T_i}$$

where

ΔT: maximum absolute temperature difference during the cycle,
T_i: initial absolute temperature.

The equal sign applies for the ideal Carnot engine.

During a reversible process, initial and final entropies are equal,

$$\Delta S = 0$$

where ΔS is the entropy difference between the initial and final states.

For an irreversible process,

$$\Delta S \neq 0$$

Second law of thermodynamics: For a closed system,

$$\Delta S \geq 0$$

Entropy never decreases within a closed system. Since every system can be enlarged to be a closed one (ultimate limits: the universe), entropy increases all the time. This determines a unique direction to energetically possible processes: the process yielding the highest amount of entropy is preferred.

A system is in equilibrium if its entropy has reached a maximum value.

PROBLEMS

1. Plot a pressure-volume diagram of the changes of an ideal gas for (a) an isothermal process, (b) an adiabatic process, and (c) an isobaric process.

2. Calculate the work done by a gas expanding isobarically.
3. One mole of an ideal gas at 0°C and 760 mm pressure expands to four times its initial volume. How much work is done by the gas (a) for isobaric expansion (b) for isothermal expansion, and (c) for adiabatic expansion? (Assume $\varkappa = 1.5$)
4. Calculate the work done during the isothermal process shown in Figure 10-5 for (a) an ideal gas at 0°C, (b) CO_2 at 0°C.
5. Why is it not possible to compare thermodynamically a living system with a Carnot engine?
6. How can the efficiency of a Carnot engine be maximized? State the condition for which η approaches one.
7. Treat the engine of your car as a Carnot engine and calculate its efficiency. Compare with actual efficiency. Explain the difference.
8. Calculate the entropy change for an ideal gas after it is expanded isothermally to four times its original volume.
9. What is the entropy change of an ideal gas during an adiabatic process?
10. Experiments show that if a gas expands into an evacuated container its temperature remains constant. What happens to the entropy of the gas?
11. Calculate the entropy difference between water in its liquid state (heat of formation: 58 cal/g) and in its vapor state (heat of vaporization: 540 cal/g) at 100°C. If this water-vapor system is a closed system, in which direction will it change spontaneously?

FURTHER READING

L. VON BERTALANFFY, *Problems of Life.* Wiley and Sons, New York, 1952.

K. E. VAN HOLDE, *Physical Biochemistry.* Prentice-Hall, Englewood Cliffs, N.J., 1971.

I. KLOTZ, *Energetics in Biochemical Reactions.* Academic Press, New York, 1957.

ERWIN SCHRÖDINGER, *What Is Life?* Cambridge University Press, Cambridge, Mass., 1969.

chapter 11

GEOMETRICAL OPTICS

1. INTRODUCTION

The old dispute whether light is a wave or a particle seems to be decided by now: it is neither. It depends on the experimental arrangement how we best describe its result. Sometimes the wave properties are dominant; sometimes the outcome of an optical experiment is most easily visualized by assuming light particles. The emphasis of the previous sentence rests on the word *visualized.* Our language is not flexible enough, and our everyday experience is not pertinent enough to visualize all optical phenomena without employing contradicting and mutually exclusive words. Using the language of mathematics, there are no ambiguities.

In this chapter we are concerned with rays. These are infinitesimal thin bundles of light, propagating along a straight line. A change in direction occurs abruptly at boundaries. We are not concerned about what happens in detail at the boundaries but are satisfied with a gross description.

The treatment is not limited to classical, that is, geometrical, optics. The sonar of the dolphin, the radar of the bat, and the beam of a particle accelerator may be described by rays and are thus covered under the heading *geometrical optics.* As long as the wavelength of the radiation is short compared to obstacles and distances traveled, we achieve useful results applying the laws of geometrical optics.

Waves and their properties were already described in Chapter 9 (Oscillations, Waves, and Spectra); some overlapping will be unavoidable. However, the phenomena (like reflection and refraction) appear under a different light.

As you can imagine, much emphasis is given to optical instruments. In practice it is difficult to use these as black boxes (a very successful approach in electronics) since you would be unaware of their merits and limitations. Also, there is no need for that. The construction of the basic optical instruments is simple. The details that the engineer has to contend with do not concern us. Before we delve into the physics, let us look at some interesting historical facts.

Geometrical optics is one of the oldest sciences. Looking at a water surface, man was naturally curious about his mirror image. In ancient Greece, Pythagoras (about 580 to 496 B.C.) and his followers even founded a cult around geometrical optics. They assumed that light rays leave the eyes and are returned from the object back into the eyes. As we shall see, the direction of a ray does not matter in geometrical optics, and consequently the optical laws derived in antiquity are correct.

The first extensive treatment involving 58 theorems was done around 300 B.C. by Euclid. It certainly was in the possession of Archimedes (about 287 to 212 B.C.) when he defended his hometown Syrakus against the invading fleet of Rome. It is not a historic fable that he burned enemy ships using a big spherical mirror. The naturalist Georges von Buffon (1707–1788) reconstructed this far-reaching weapon in 1747 and proved its effectiveness. Diocles published a book in the first century B.C. (two thousand years ago) on focusing mirrors. Parts of Claudius Ptolemy's (2nd century A.D.) *Optics* could be incorporated in this text. He even presented tables with experimental values for the refractive index.

While in Europe the dark medieval times descended on the sciences, the Arab Ibn al-Haitham (965–1039) wrote a comprehensive, very mod-

Figure 11-1 Meditating man with concave and convex mirrors. After a fresco on the walls of a cave temple in the Chinese province of Tunhuang. Thang period (618–906 A.C.).

ern volume on optics. He was the first to show that the objects are lit by sources other than the human eye, that light enters the eye and produces the sensation of seeing.

2. RAYS AT BOUNDARIES

2.1 Notation

A ray traces the path of emitted energy. It may be the energy of light quanta, the energy transmitted by sound, or energy in form of moving particles. A ray is graphically represented by a pointed arrow and has the following properties.

1. Propagation along a straight line. A change in direction occurs only at boundaries.
2. Boundaries and obstacles are very large compared with any wavelength associated with the physical appearance of the emitted energy.
3. Along the same trace, the direction of the ray is reversible. For ease of description, a direction is usually introduced and indicated by a pointed arrow.
4. The absorption of rays is neglected.

A ray coming from medium 1 encounters a boundary at the angle of incidence α. See Figure 11-2. At the boundary the ray is split into:

1. Reflected ray, with angle of reflection β. It remains in medium 1.
2. Refracted ray, with angle of refraction γ. It passes the boundary into medium 2.

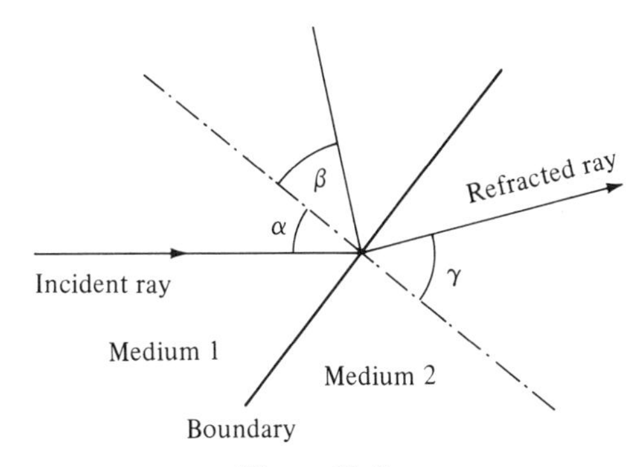

Figure 11-2

Notice that all angles are determined with respect to the vertical to the boundary. In general, angles in optics are identified by letters of the Greek alphabet.

Application: The Beam Splitter. It is tiring to look through a microscope or any other optical instrument having only one eyepiece. A beam splitter divides the upcoming light into identical rays, each entering another eyepiece. See Figure 11-3. No additional information is gained by this arrangement, but observing through a binocular microscope is much easier. Beam splitters are also widely used in photographic cameras. There, part of the ray traveling toward the film is diverted into a light sensitive cell to set the aperture of the lens.

Double Refraction. Actually, this notation is applicable only if medium 2 is an isotropic medium. This means that medium 2 is structually uni-

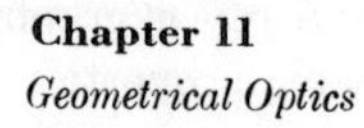

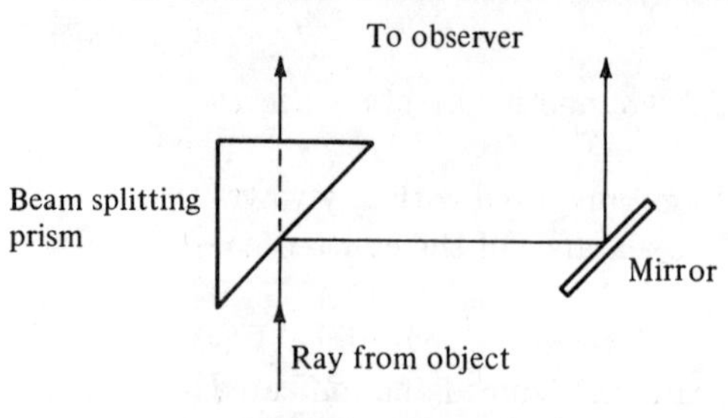

Figure 11-3

form in all directions. A crystal, for example, is not uniform in space; there are always preferred directions due to the regular arrangement of the molecules and atoms. In an anisotropic medium 2 (in nature practically all media are anisotropic to some degree), there are two refracted rays. The ordinary ray and the extraordinary ray each display different angles of refraction. If not stated otherwise, this treatment is restricted to the ordinary refracted ray.

2.2 Reflection

2.2(a). LAW OF REFLECTION. In Figure 11-4,

$$\alpha = \beta \tag{11.1}$$

The angle of incidence α is equal to the angle of reflection β. The rays are in one plane.

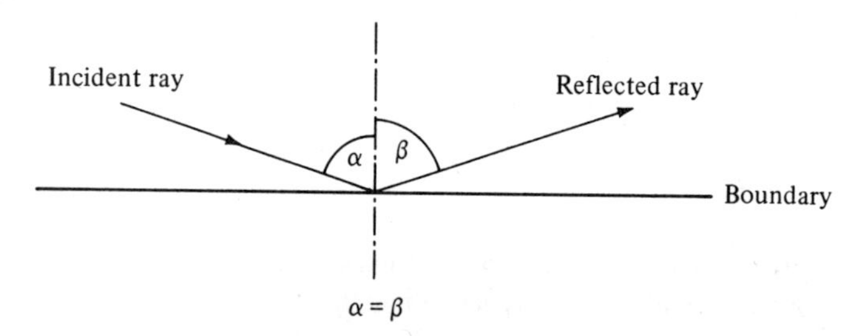

Figure 11-4

Diffuse Reflection:

Because ideally a ray has no physical extension, the law of reflection is also valid for each point on an irregular boundary. In this case, parallel incident rays are reflected in various directions. Figure 11-5 demonstrates this diffuse reflection.

Mirrors and Camouflage:

A mirror is a specially prepared plane or curved surface with a high reflectivity, that is, the overwhelming part of the incident energy is reflected. In open nature, a highly reflective boundary looks silvery due

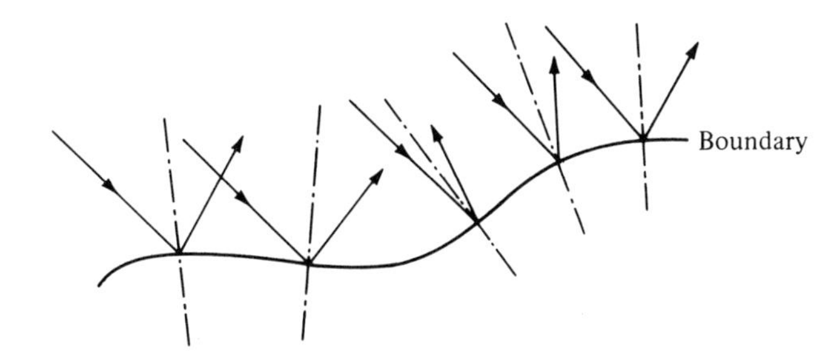

Figure 11-5

to its uniform illumination by the sky. This is utilized for camouflage by fish. Looking sideways or from below—as a predatory fish would—the water surface appears silvery. Reflective platelets at the sides and belly of herring return 75% of the incident light. They are difficult to discover against the silvery background. At the back those platelets are so arranged to reflect only 20% of the incident light. Seen from above, the herring looks dark and blends into the bottom.

In ancient China, the Taoists used mirrors as demifuge armor whenever they ventured into the mountains, the seat of evil demons.

Insects escaping into banks of fog are not pursued anymore by hunting bats. The water droplets cause a diffuse reflection of the acoustical rays of the bat, and the ray reflected from the prey disappears in the background.

2.2(b). Curved surfaces. There are many ways to analyze what happens if rays are incident on a curved reflective surface. The simplest one is to follow a number of rays incident parallel to the axis of symmetry.

Concave Spherical Surface (Figure 11-6):

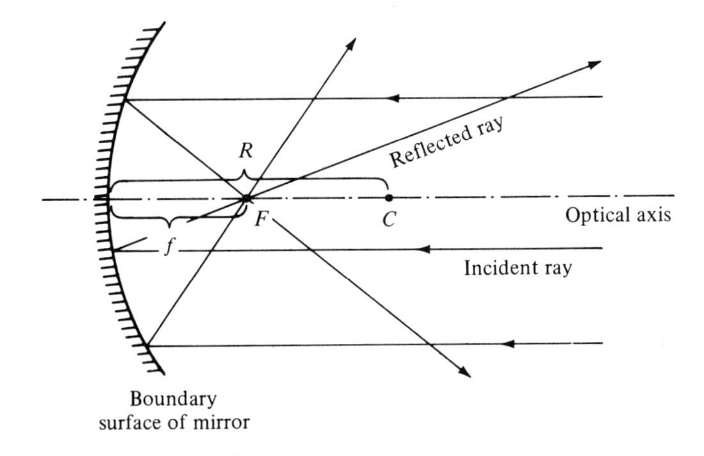

Figure 11-6 Rays incident on a concave mirror.

C: center of curvature,
R: radius of curvature,
F: focal point,
f: focal distance.

Rays incident parallel to the axis are reflected through the focal point F. Reversing the direction, rays originating at the focal point and propagating toward the surface are reflected parallel to the axis.

Relation:

$$f = \frac{R}{2} \tag{11.2}$$

where

R: radius of curvature of the surface,
f: focal distance or focal length.

Restriction: Only rays close to the axis will meet at the focal point. Rays farther away converge between F and the surface. If the surface is a paraboloid, all parallel rays will meet at F.

Convex Spherical Surface (Figure 11-7):

C': center of curvature,
R': radius of curvature,
F': virtual focal point,
f': virtual focal distance.

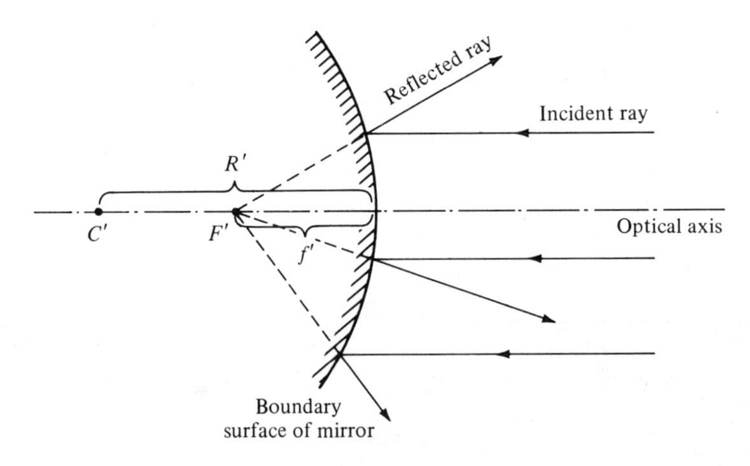

Figure 11-7 Rays incident on a convex mirror.

Rays incident parallel to the axis are reflected as if they came from a point F' (virtual focal point) behind the reflecting surface.

Relation:

$$-f' = \frac{R'}{2}$$

where

R': radius of curvature of surface,
f': virtual focal distance.

2.2(c). Applications

Headlights. The bulb is placed at the focal point of the concave mirror inside the lamp. (See Figure 11-8.) Most of the light leaves parallel to the axis of symmetry. With suitably arranged absorbers, nonparallel rays are eliminated.

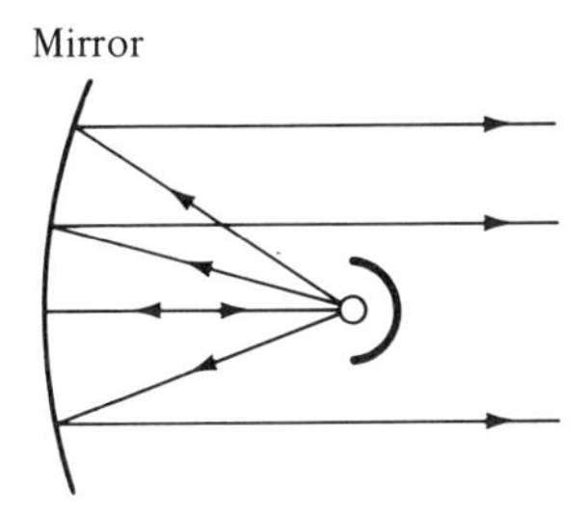

Figure 11-8

Sonar Transmitter of Bats. Bats use a sonar system for orientation. To save energy, some species focus the emitted sound on a cylindrically shaped skin organ and scan their surroundings with those parallel rays.

A cylindrically concave surface is suitable if the rays are emitted from a line in space instead of from a point source. Conversely, rays incident on a cylindrically concave surface focus on a line parallel to the cylinder's axis and halfway between this axis and the surface. If you ever go to Casablanca, Morocco, do not miss the camping place. There, water pipes are placed at the focal line of a concave cylindrical mirror. As long as the sun is shining, you will have hot water. That is not new: Plutarch (about 50 to 125 A.D.) reports about fire-making in Rome's temples with concave mirrors.

2.3 Refraction

2.3(a). Law of refraction (Snell's law)

$$\frac{\sin \alpha}{\sin \gamma} = \frac{n_2}{n_1} = \frac{c_1}{c_2} \tag{11.3}$$

where

n_1: index of refraction of medium 1,
n_2: index of refraction of medium 2,
c_1: propagation speed of ray in medium 1,
c_2: propagation speed of ray in medium 2.

Note that the index of refraction is related to the propagation speed of the ray inside the medium:

$$\frac{n_2}{n_1} = \frac{c_1}{c_2} \tag{11.4}$$

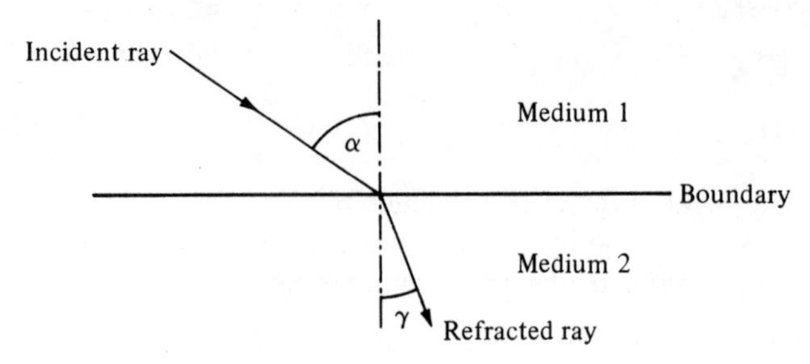

Figure 11-9

The numerical value for the index of refraction depends on the medium, the kind of ray (light, sound), and on the wavelength. Tables 1 and 2 display some values of n.

In general, the index of refraction is expressed in comparison to a standard medium, for example, vacuum or air.

$$n_{\text{medium}} = n_{\text{standard}} \cdot \frac{c_{\text{standard}}}{c_{\text{medium}}} \tag{11.5}$$

For $n > 1$, it follows that the propagation speed of the ray is larger in the medium than in the standard medium.

Table 1 INDEX OF REFRACTION (n) FOR YELLOW SODIUM LIGHT (wavelength $\lambda = 5893$ A). Standard medium: vacuum

Medium	n
vacuum	1.000 (definition)
air	1.000 3
water	1.33
vitreous body of eye	1.33
ethyl alcohol	1.36
eye lens of man	1.41
olive oil	1.47
crown glass	1.52
flint glass	1.61
diamond	2.42

Table 2 INDEX OF REFRACTION (n) FOR AUDIBLE SOUND WAVES. Standard medium: air.

Medium	n
air	1.000 (definition)
glass	0.06
copper	0.092
water	0.23
alcohol	0.27
cork	0.65
methane	0.76
ether vapor	1.85
rubber	6.1

EXAMPLE:

Table 1 shows for yellow light in flint glass that $n = 1.61$. This refers to the following situation (see Figure 11-10). The speed of light in vacuum is 3×10^{10} cm/s; consequently the speed of light in flint glass is given by Equation (11-4):

$$c_{\text{flint glass}} = 1.9 \times 10^{10} \text{ cm/s}$$
$$= \frac{n_{\text{vac}}}{n_{\text{flint glass}}} \times c_{\text{vac}}$$
$$= \frac{1}{1.6} \times 3 \times 10^{10} \frac{\text{cm}}{\text{s}}$$

If the boundary is not between vacuum and another medium but between medium 1 and medium 2, for example olive oil and water (see Figure 11-11), then the speed of light in water is:

$$c_{\text{water}} = 1.1 c_{\text{oil}}$$

The speed of light in water is 10% higher than in olive oil. We obtain the actual value for c_{water} as in the previous example:

$$c_{\text{water}} = 2.3 \times 10^{10} \text{ cm/s}$$

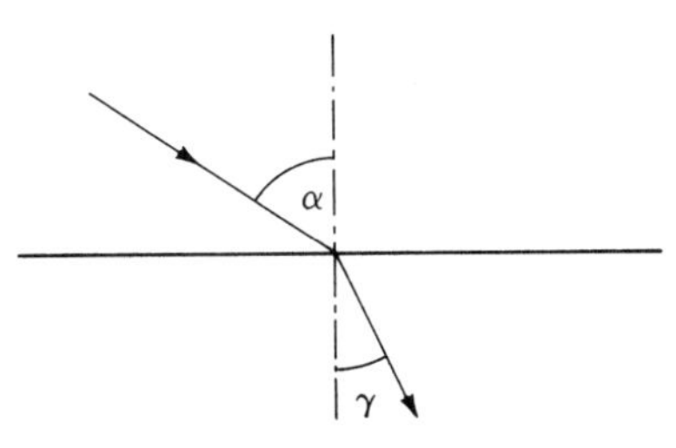

Figure 11-10

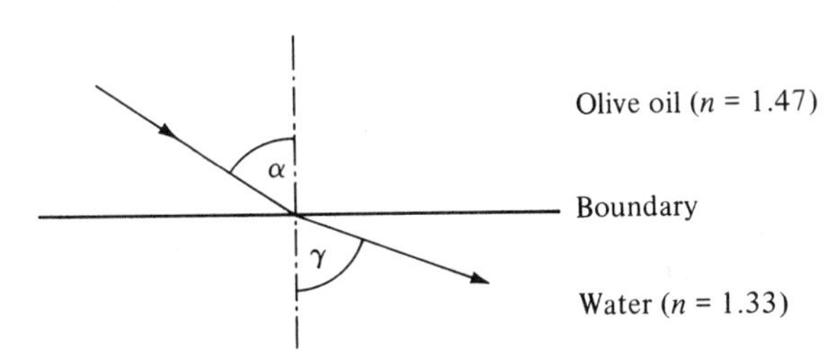

Figure 11-11

2.3(b). TOTAL REFLECTION. Equation (11.3) indicates that if $n_1 > n_2$, it follows that $\gamma > \alpha$. There is a critical angle of incidence α_c for which γ becomes 90° (that is, $\sin \gamma = 1$).

$$\sin \alpha_c = \frac{n_2}{n_1} \tag{11.6}$$

α_c γ

Medium 1

Boundary

Medium 2

If the angle of incidence exceeds α_c, the ray does not enter medium 2 but is reflected back (because the maximum possible value for the sine is 1). This phenomenon is named *total reflection* and is exploited very much in nature and technology.

EXAMPLE· LIGHT GUIDE IN THE APPOSITION EYE

The visual apparatus of some insects and crustacea consists of many bundled individual eyes. Light incident on the surface of the individual cornea lens is guided toward the sense cell inside a crystalline cone. See Figure 11-13. The medium containing the distal pigment envelopes each crystalline cone. The index of refraction for this medium is smaller than n for the crystalline cone itself, thus causing total internal reflection for most penetrating light rays. In this manner, the loss of light between lens and sense cell is minimized.

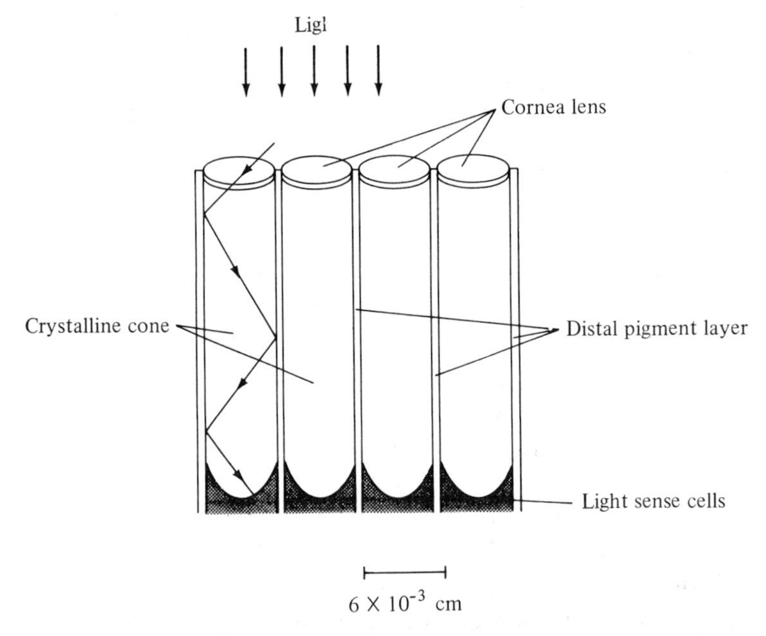

Figure 11-13 Schematic drawing of the structure of the apposition eye. Notice the scale.

Application: Glass Fiber Optics. Glass fibers having a diameter of a few hundredths of a millimeter are flexible. If bundled together, a flexible light guide is formed. Light rays inside an individual fiber are totally reflected at the boundary glass-air (or plastic cover). If the flexible

Figure 11-14

glass fiber bundle is bent, light is directed around corners. In medicine this is used, for example, to observe the inside of the stomach. A flexible light pipe is swallowed, and light is sent downward along the circumferential fibers, while the center fibers are used for observation. This is the first time that man can look inside a body without performing surgery.

A glass fiber cable may also be used to produce an unbreakable code: The light pipe (containing millions of individual fibers) is twisted before the fibers are fixed with respect to each other. A document photographed through a normal light pipe appears intelligible; through a twisted pipe it looks quite different. See Figure 11-15. Only if you look at the photograph through the same light pipe can the picture be deciphered.

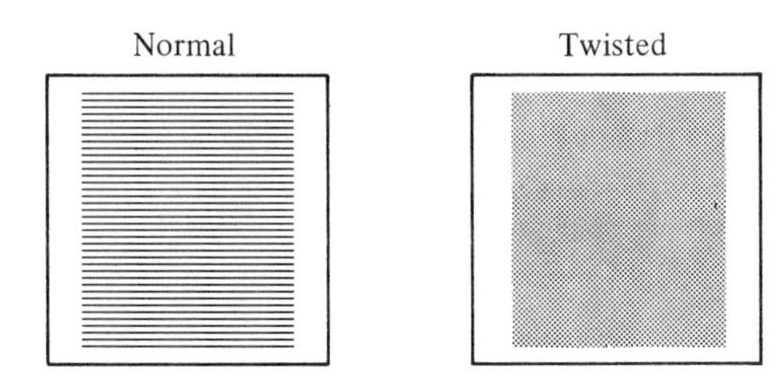

Figure 11-15

A perfect coding device: a bundle of parallel glass fibers are fixed at both ends. Then the middle is twisted and fixed in position. Looking through this light pipe you will not discover that the fibers run parallel only at both ends.

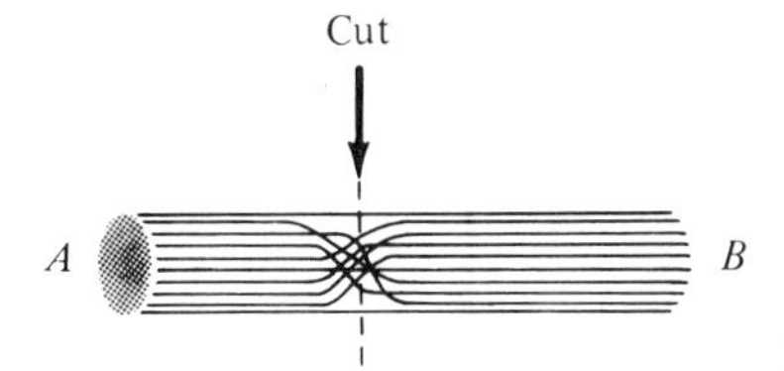

Figure 11-16

Now the device is cut in the middle. A document placed at the cut surface and photographed through A or B (see Figure 11-17) will appear to be just an irregular gray sheet. The picture placed at B or A can be de-

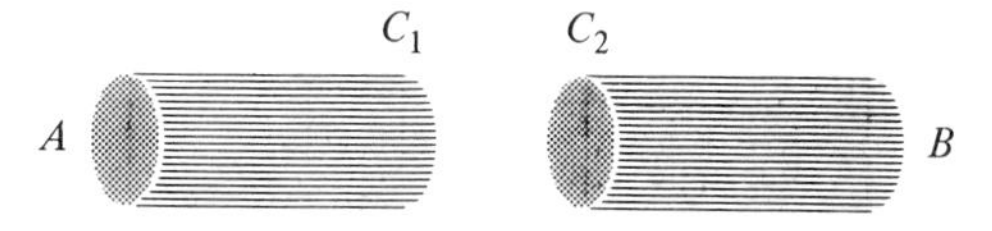

Figure 11-17

ciphered by looking through C_2 or C_1, respectively. The twisting may be done in such a way that the fibers pass through the cut area at random. Each of those coding devices is truly unique; its code is unbreakable.

2.3(c). Plane surfaces. The path of a ray through media bound by plane surfaces is just a multiple application of Snell's law. *Parallel boundaries* are shown in Figure 11-18. The result of this arrangement is

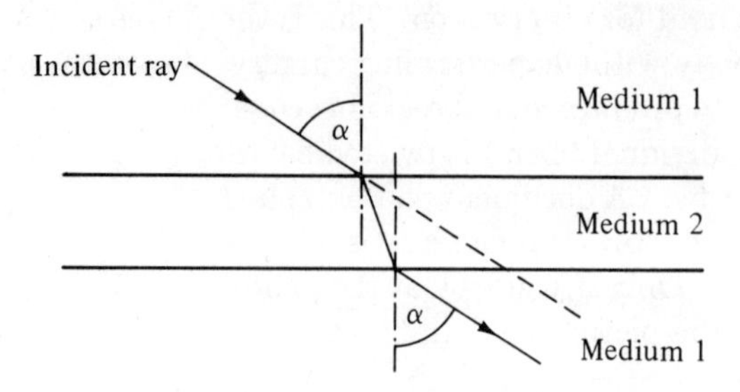

Figure 11-18

that the ray is shifted parallel to its original direction. *Oblique boundaries* are shown in Figure 11-19. The result of this arrangement is that the twice-refracted ray changes its direction with respect to the incident ray.

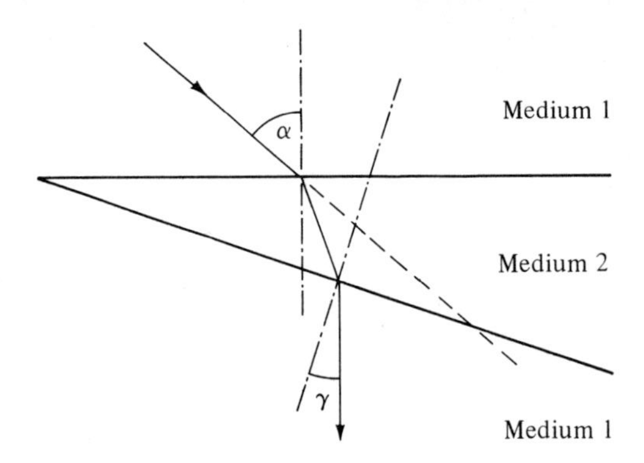

Figure 11-19

2.3(d). LENSES. Various combinations of plane and curved surfaces form lenses. Figure 11-20 shows the notation used. Notice that convex lenses are thickest at the center and concave lenses are thinnest at the center.

Rays at Convex Lens (Figure 11-21). Rays incident parallel to the optical axis are refracted and converge at the focal point F. Reversing the direction: Rays originating at the focal point and propagating toward the lens leave parallel to the optical axis. A lens has two focal points, one on each side. The following relation is known as the lens-maker's equation:

$$\frac{1}{f} = (n - 1)\left(\frac{1}{R_1} + \frac{1}{R_2}\right) \tag{11.7}$$

where

f: focal distance (or focal length) of lens,
n: index of refraction of lens material,
R_1: radius of curvature of front surface,
R_2: radius of curvature of back surface.

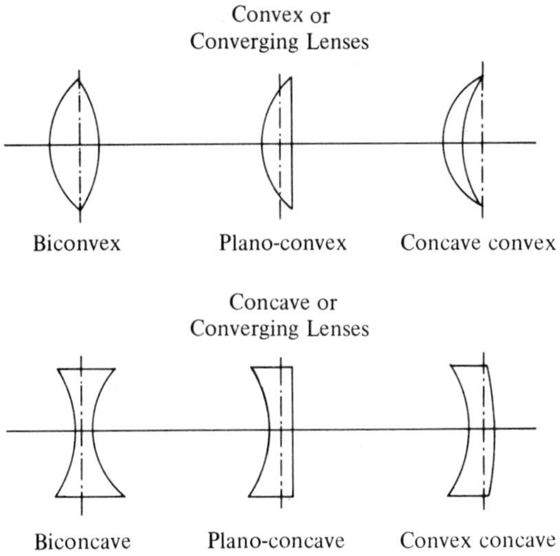

Convex or
Converging Lenses

Biconvex   Plano-convex   Concave convex

Concave or
Converging Lenses

Biconcave   Plano-concave   Convex concave

Figure 11-20

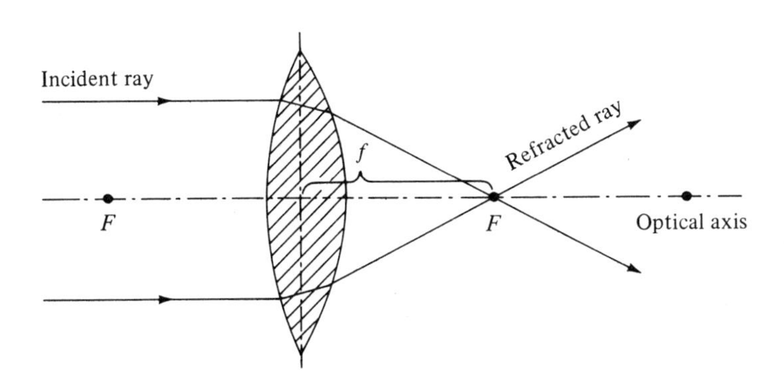

Figure 11-21 A convex lens. f is focal distance; F are focal points.

Rays at Concave Lens (Figure 11-22). Rays incident parallel to the optical axis leave the concave lens as if they came from the virtual focal point F'.

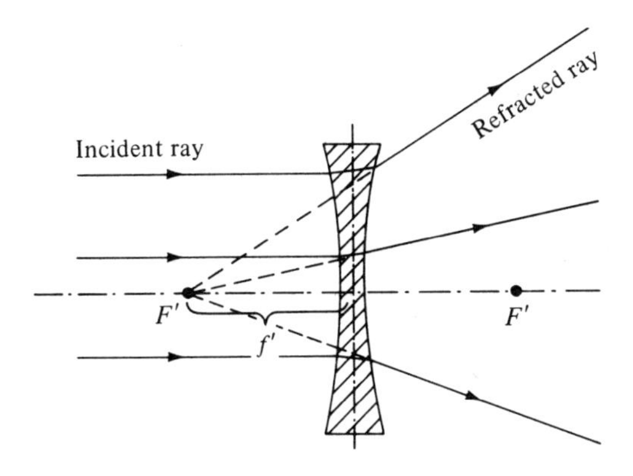

Figure 11-22 A concave lens. F' are virtual focal points; f' is virtual focal length.

Power of a Lens. The (refractive) power of a lens is $1/f$. There is a special unit for it: *diopter.*

$$\frac{1}{\text{focal distance in } m} = \text{number of diopters}$$

The virtual focal distance of a concave lens is expressed in negative diopters.

EXAMPLE:

The relaxed human eye has a focal distance of 17 mm; hence its power is $1/0.017 = 59$ diopters. If the lens is taken out, the remaining system has a power of 43 diopters. This is not sufficient to produce a focused image at the retina of the eye.

Combination of Lenses. For thin lenses, the total power of an optical system is equal to the sum of the powers of its elements. Remember in this connection that concave lenses have negative power, most important in prescribing eyeglasses.

2.3(e). LENS ABERRATIONS. Real lenses may display the following aberrations:

1. *Spherical aberration.* Rays farther away from the optical axis converge between focal point and lens. Correction: Very thin lenses or nonspherical surfaces.
2. *Chromatic aberration.* The index of refraction depends on the wavelength. The focal distance of blue light is shorter than the focal distance for red light passing through the same lens. Correction: Combination of concave and convex lenses, made of materials having different indexes of refraction.
3. *Astigmatism.* One or both surfaces are cylindrical. Parallel rays converge at one or more lines in space. Correction: Additional compensating lenses.

2.4 Applications

2.4(a). LENSES WITH VARIABLE FOCAL DISTANCE

The Eye of Man. The lens of the eye forms an image of outside objects. This image is detected at the retina and analyzed in the brain. The spatial position of the image depends to some degree on the distance from eye to object. There are two ways to achieve a well-focused image of objects having variable distances from the eye. Either the lens changes its focal distance or the lens-retina distance is altered. The lens-retina distance for the human eye is fixed at 24 mm; accommodation of the lens leads to sharp images. Figure 11-23 shows a cross section of the focusing apparatus of the human eye.

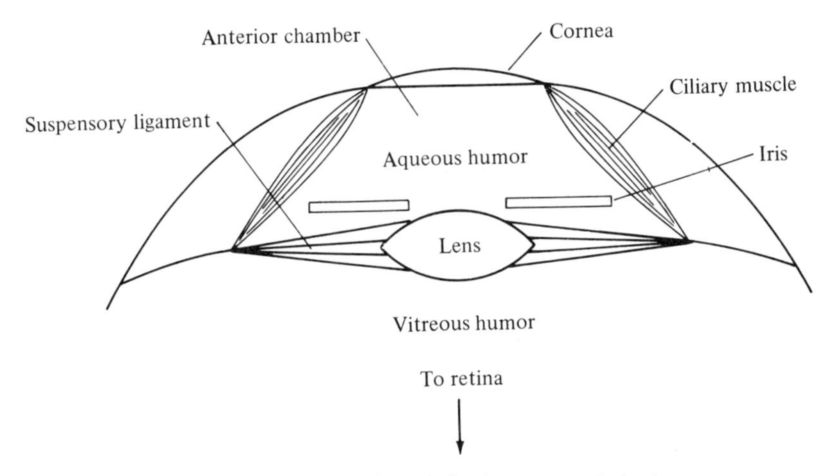

Figure 11-23 Schematic drawing of the front part of the human eye and its auxillary apparatus.

The lens is elastic. If the ciliary muscle contracts, the tension of the suspensory ligaments decreases, the radius of curvature increases, and a shorter focal distance results. The completely relaxed lens has a focal distance of about 17 mm; maximum accommodation reduces this to about 10 mm.

The refractory system of the eye consists of more than one element: the cornea, the aqueous humor inside the anterior chamber, the lens itself, and the vitreous humor. In fact, the lens has only half of the refractive power of the cornea. Details about this most complicated system are referred to special books.

Zoom Lenses. The focal distance of a zoom lens can be continuously changed within a certain range, approximately by a factor of 6. This is not necessary for focusing (a camera focuses like a frog's eye by changing the position of the lens), but it is used to achieve images of different sizes. The larger the focal distance, the larger the formed image of the same object.

Compared with the human eye, a modern zoom lens is a complicated and awkward arrangement of optical elements, as you can see in Figure 11-24. Table 3 shows an interesting comparison in this connection.

Table 3 Some technical data for the eye and a zoom camera lens

	Eye	Zoomar
type	anastigmat achromat	anastigmat achromat
No. of elements	4	14
focal distance range	17 to 10 mm	82 to 36 mm
f-stop range	1:1 to 1:8.5	1:2.8 to 1:16
power	59 diopters	28 diopters

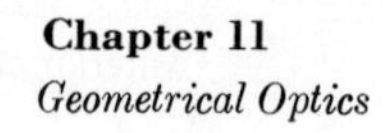

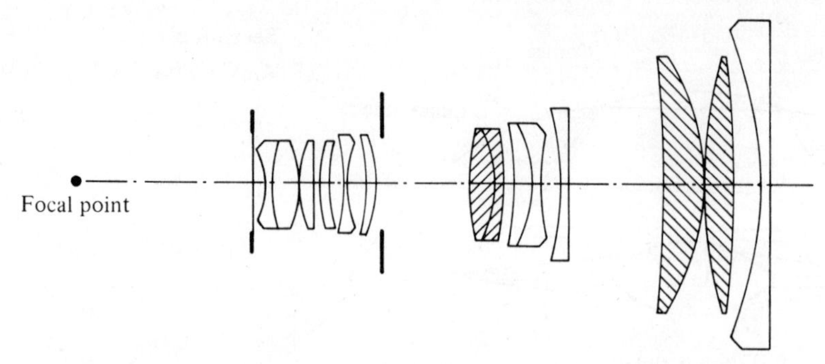

Figure 11-24 A modern zoom lens: Zoomar (Voigtländer) 1:2.8, shown in position $f = 82$ mm. The focal distance reduces continuously to 36 mm if the hatched elements are moved to the left.

2.4(b). EYEGLASSES. If the eye lens and length of the bulbus do not match, glasses are placed in front. The physical principle is the same for both spectacles and contact lenses.

Nearsightedness (Myopia). The focal distance is too short; the refractive power of the optical system is reduced by an additional concave glass. See Figure 11-25.

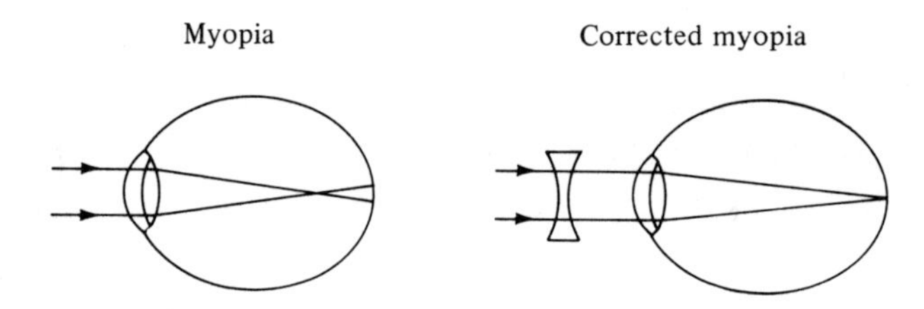

Figure 11-25

Farsightedness (Hyperopia). The refractive power of the eye is too small; an additional convex lens reduces the focal distance. See Figure 11-26.

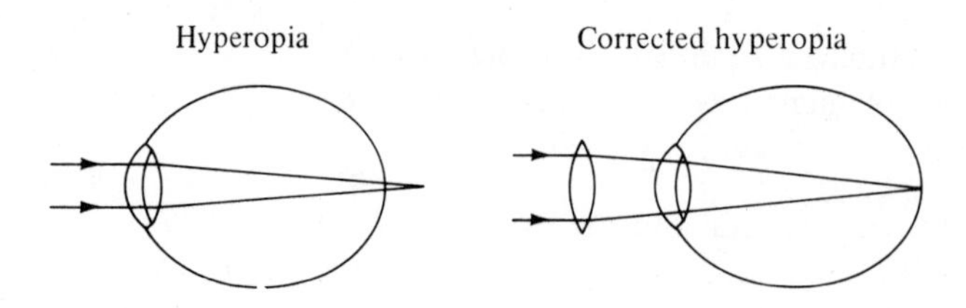

Figure 11-26

Astigmatism. Here the curvature of the cornea is not identical in all directions, and distorted images result. A corrective glass can be designed to compensate for the abnormal curvature.

3. IMAGES

3.1 Notation

Mirrors and lenses are used to form images of objects. To construct an image on paper, rays are drawn from the object toward the optical system. The corresponding image point of an object point is where the rays coming from the same point cross over. There are real and virtual images.

Real Image. Rays from the same object point meet at a corresponding image point. A real image can be fixed (on a film for example) and again be imaged by another optical system.

Virtual Image. Rays from the same object point diverge; they do not meet. The backward extension of these rays meet at a corresponding virtual image point. Virtual images cannot be fixed or imaged again. They exist only in the autonomous image-analyzing system of the brain. The image from a plane mirror is a virtual image. See Figure 11-27.

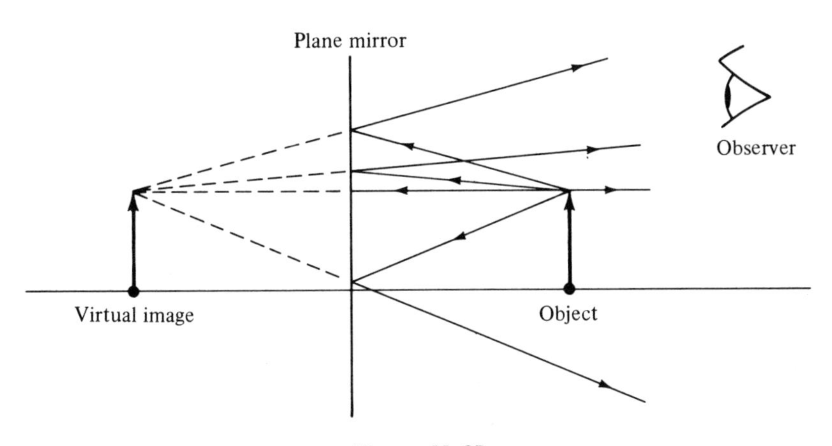

Figure 11-27

The analysis of the incident rays takes place in an autonomous, i.e., consciousness-independent, area of the brain. For the analyzing system, there is no way to know whether diverging rays are reflected from a mirror surface or originated directly from a point source. That is the reason why diverging rays appear to be extended backwards. The resulting virtual image exists only in the brain. If our consciousness controlled this image-analyzing system, only truly converging rays would form images. That would be a different looking world!

3.2 Images formed by curved optical elements

To construct an image on paper, it is convenient to employ three principal rays.

1. *Parallel ray,* incident parallel to the axis of the optical element.

2. *Focal ray,* passes through the focal point before it reaches the element.
3. *Center ray,* passes through the optical center of the element and emerges without change of direction.

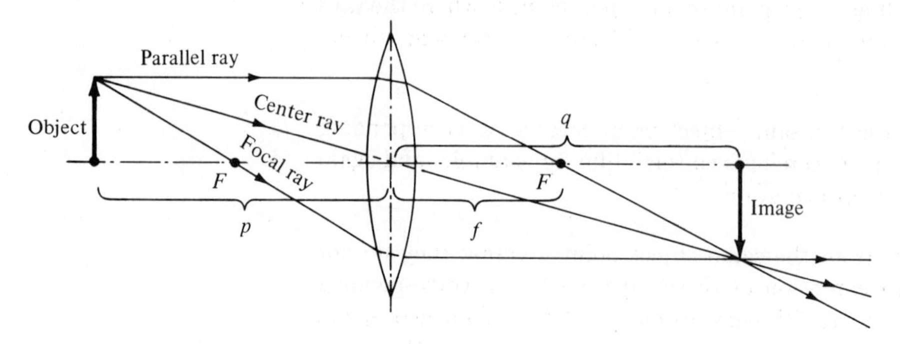

Figure 11-28

Figure 11-28 shows these three principal rays incident on a lens.

The type, size, and position of the image depends on the focal distance and the position of the object. For all possible combinations, the following relation holds:

$$\frac{1}{p} + \frac{1}{q} = \frac{1}{f} \tag{11.8}$$

or

$$f = \frac{pq}{p + q}$$

where

p: optical element–object distance,
q: optical element–image distance,
f: focal distance of element.

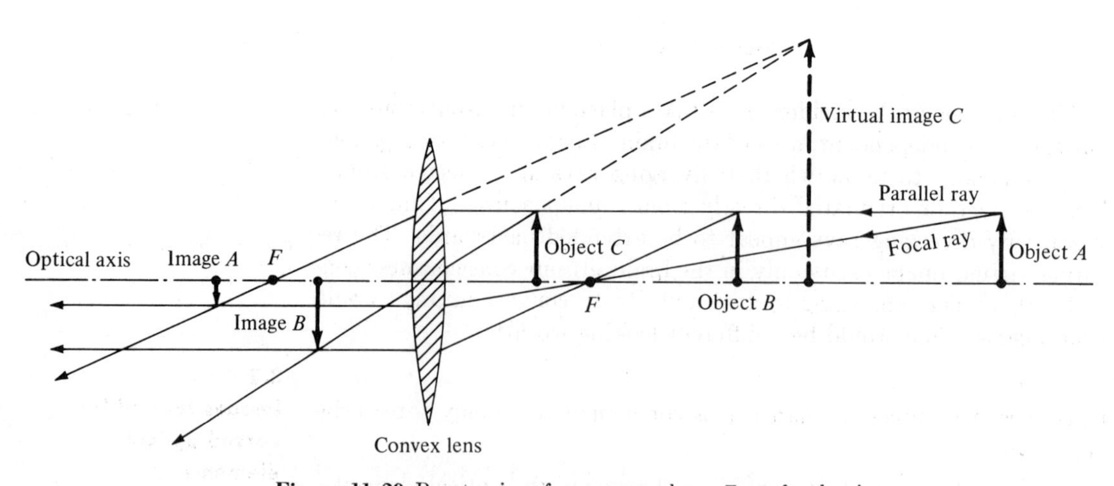

Figure 11-29 Ray tracings for a convex lens. F are focal points.

Restriction: Equation (11.8) is valid only for thin lenses and parabolic mirrors.

The following two figures show ray tracings for a convex lens (Figure 11-29) and a concave mirror (Figure 11-30). The object is an upright

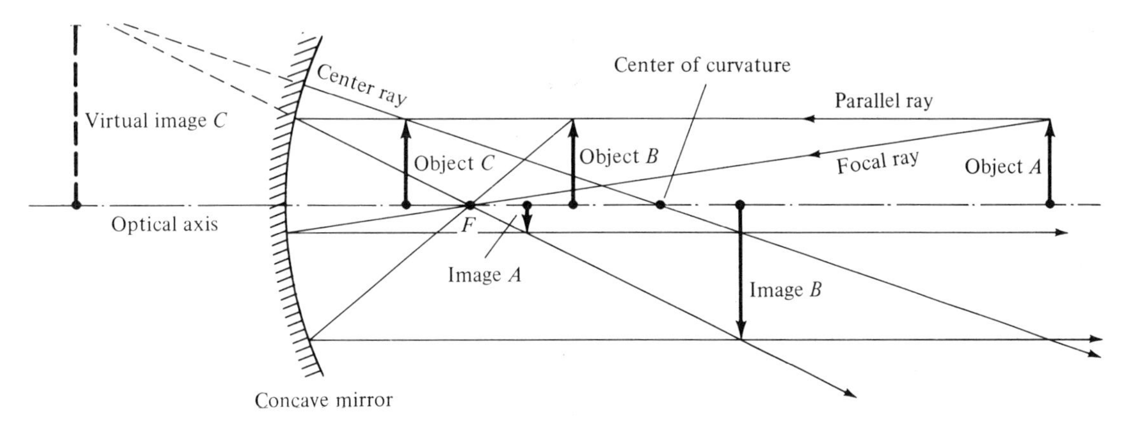

Figure 11-30 Ray tracings for a concave mirror. F is focal point.

radiating arrow. Only the ray tracings for one object point—the tip of the arrow—are drawn. For any other object point, the tracing is similar. Notice the various positions of the object and its corresponding images. To avoid confusion, just two of the three principal rays are drawn for each image. To complete the treatment, ray tracings for a convex mirror and a concave lens are shown in Figures 11-31 and 11-32.

Table 4 summarizes the ray tracings.

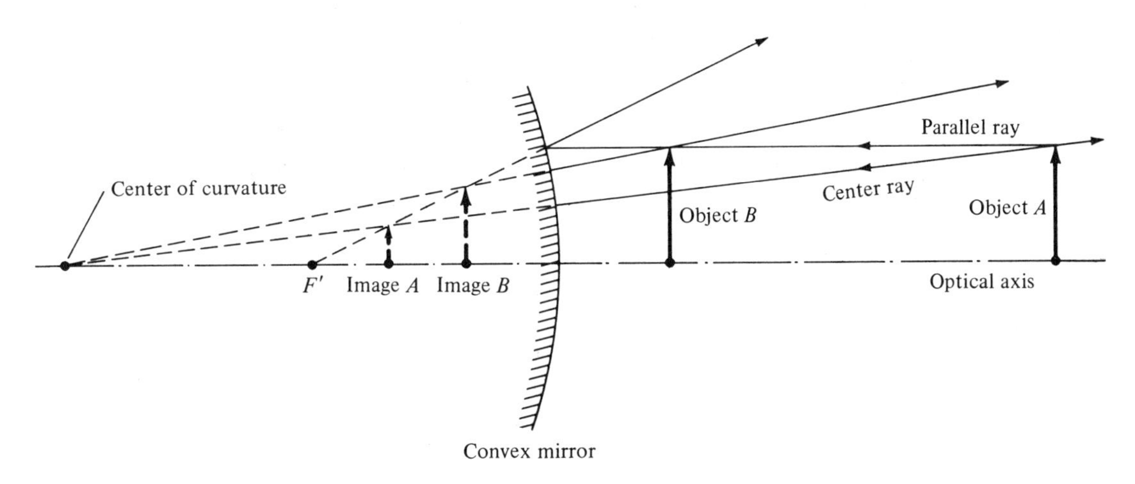

Figure 11-31 Ray tracings for a convex mirror. F' is virtual focal point. All images are virtual.

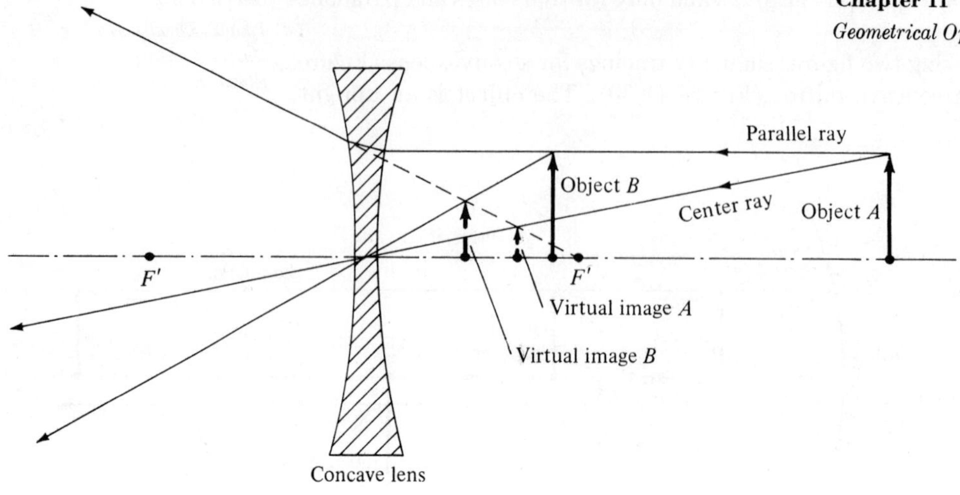

Figure 11-32 Ray tracings for a concave lens. *F'* are virtual focal points. All images are virtual.

Table 4 TYPE, SIZE, AND POSITION OF IMAGES PRODUCED BY VARIOUS OPTICAL ELEMENTS AND VARIOUS OBJECT DISTANCES

(p = object distance, f = focal distance)

Optical Element	Distance Object to Element				
	$p > 2f$	$p = 2f$	$2f > p > f$	$p = f$	$p < f$
convex lens concave mirror	real reduced inverted	real equal inverted	real magnified inverted	no image	virtual magnified upright
concave lens convex mirror	virtual reduced upright	virtual reduced upright	virtual reduced upright	virtual reduced upright	virtual reduced upright

3.3 The pinhole

Because rays propagate along a straight line, a tiny hole in an otherwise opaque sheet will produce an image on a screen. See Figure 11-33. The

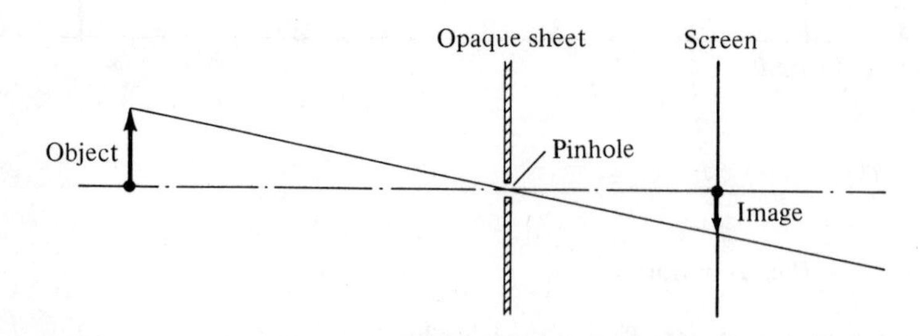

Figure 11-33

resulting real image is reduced in size compared with the object, and it is of very low intensity. Decreasing the size of the hole increases the sharpness of the image but also reduces its intensity. For visual light, the optimum hole diameter is 0.4 mm. The image from a smaller hole will be blurred due to diffraction effects resulting from the wavelike character of light.

During a solar eclipse, you may pierce a piece of cardboard and use the formed pinhole to project an image of the spectacle upon a screen. The brightness of the sun is such that the image is sufficient to observe all stages of the eclipse. Also look at the shadow of a tree; you will find very many pinholes and consequently many images of the sun.

You can observe a similar application in x-ray diagnostics. Rays leaving the anode of an x-ray tube pass through a tiny opening in a lead shield. The pinhole image of the radiating anode is detected by a photographic plate. This is the only way to find out whether or not the x rays emerge from a point source. And only a point source will produce x-ray photographs of adequate resolution.

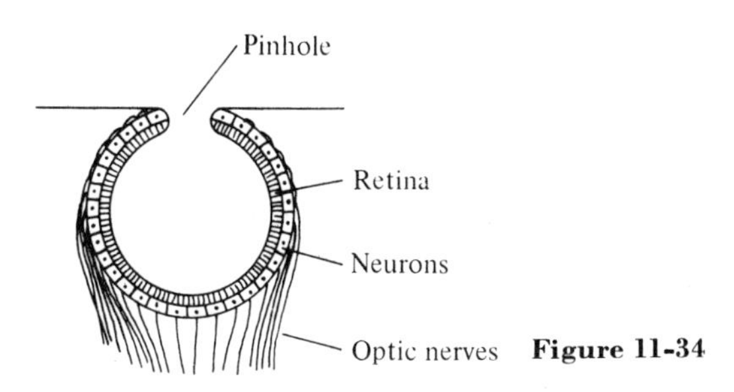

Figure 11-34

The chambered nautilus uses the pinhole principle as a visual instrument. See Figure 11-34.

3.4 Image converter

Optics in the broad sense of the word is not restricted to image formation by visible light. If a formed image cannot be detected directly by the eye, its interpretation could be difficult. The eye and its almost perfect interface with the brain is still the most versatile analyzing device of man. To facilitate it, various image converters exist. A well-known image converter is the photographic plate. Its disadvantage is the unavoidable time delay between formation of the latent image and its fixed and observable counterpart.

In general, image converters use either of two processes:

1. *Fluorescence.* The emitted radiation has a longer wavelength than the absorbed radiation. The time delay between the absorbed and re-emitted radiation is shorter than 10^{-8} s. Uranium- and platinum-barium compounds dissolved in glass are efficient converter materials.
2. *Phosphorescence.* The emitted radiation has a longer wavelength

than the absorbed radiation. The time delay between absorbed and re-emitted radiation ranges from 10^{-3} s to days or weeks. This delay depends not only on the compounds employed (e.g., zinc sulfite), but also on its temperature. Low temperature corresponds with long time delays.

X-Ray Image Converter. In the life sciences it is important to know how much energy is needed for the conversion. The normal x-ray screen is rather inefficient, and consequently an intense radiation source is necessary. The resulting radiation hazards to object and observer led to the development of the *image-amplifier tube.* See Figure 11-35.

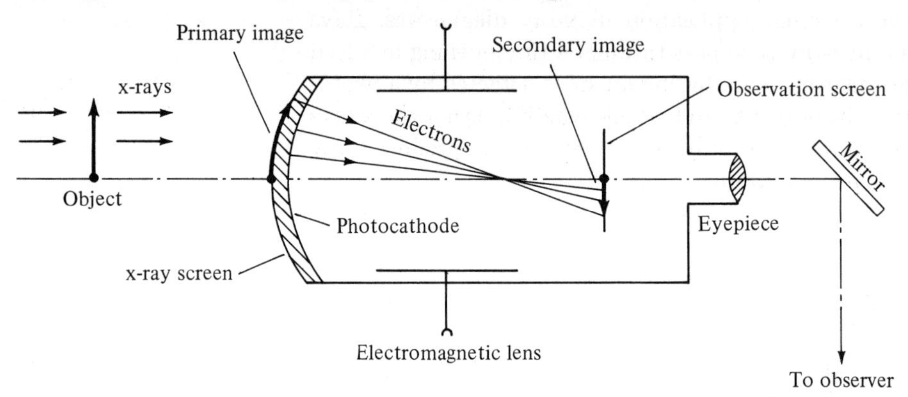

Figure 11-35 Image converter and amplifier for x rays.

The x rays produced in the usual way form a primary image on the fluorescent screen. The radiation from each primary image point knocks out electrons from the attached photocathode. An electromagnetic lens focuses the electrons on the observation screen. It functions in the same way as an optical lens: each object point corresponds uniquely with an image point. The electrons forming the secondary image cause another fluorescence process in the visible range. This image is then magnified by a glass lens and observed.

An advantage of the system is that only a small percentage of the intensity of a normal x-ray observation is sufficient to achieve a satisfying image. This is of utmost importance in medicine. The unavoidable radiation dose received by patient and doctor is acceptably low.

Ultrasonic Image Converter. Ultrasound (frequency $> 2 \times 10^4$ Hz) is a valuable diagnostic tool in medicine. We have already presented a method (time-of-flight) for using ultrasonic echoes to produce an image. A less complex way to achieve and analyze ultrasonic images is offered by a direct converter. However, its simplicity is achieved at the price of low resolution.

Ultrasonic rays illuminate and pass through the object; see Figure 11-36. The ultrasonic lens (made of plexiglass, for example) forms an

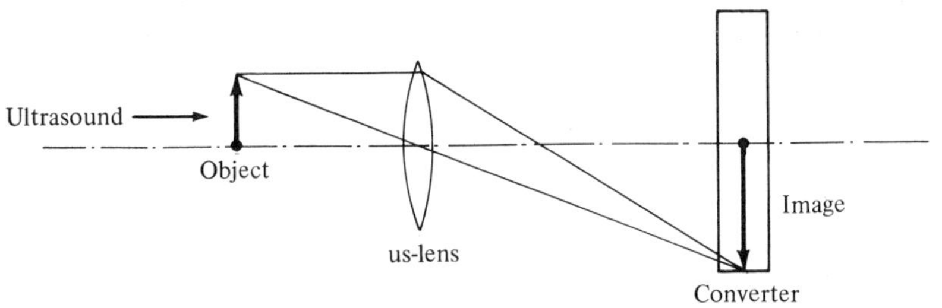

Figure 11-36

image situated inside the suitably positioned converter. There, tiny flakes suspended in a liquid align perpendicularly to the direction of the impinging ultrasonic rays and in this way form an observable image. The ultrasonic intensity needed to form an acceptable image is only about 10^{-7} watt/cm^2. For this reason, child-bearing women are not x-rayed but examined instead by ultrasound.

4. MAGNIFICATION AND RESOLUTION

4.1 Magnification

An optical instrument is an aid for the eyes. It either allows them to penetrate into the interior of an object (x rays, us-picture) or to observe more detail. In the latter case, everyday language states that the image shows a magnification of the object. The term *magnification* should be avoided in conjunction with optical instruments because it can be misleading. If you are planning to buy an optical instrument, ask for its resolution, not for its magnifying power.

4.2 Spatial resolution

The spatial resolution is the smallest distance between two adjacent object points that can be distinguished. Assuming a perfect optical system, the resolution is limited by the wave properties of the radiation employed.

$$d = \frac{\lambda}{n \sin \alpha} \qquad \text{(Abbé's formula)*} \qquad (11.9)$$

where

d: smallest distinguishable distance between two object points (resolution),
λ: wavelength of employed radiation,
n: relative refractive index $= n_1/n_2$,
n_1: refractive index of medium around object,

* Ernst Abbé (1840–1905), physicist and social reformer. The first to apply wave optics to the calculation of microscope lenses.

n_2: refractive index of medium around image,
α: angular aperture of lens. See Figure 11-37.

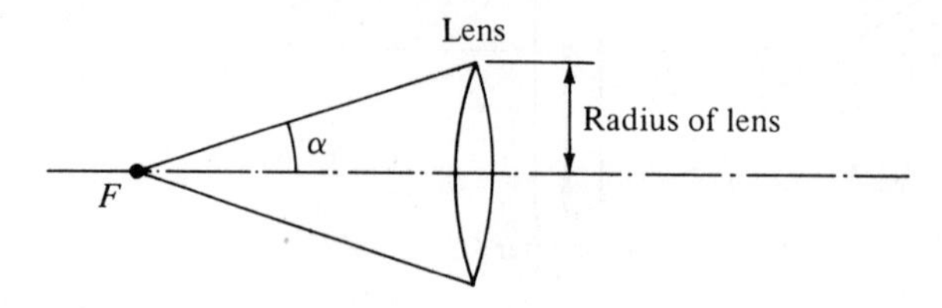

Figure 11-37

The product $n \sin \alpha$ is called *numerical aperture*.

Spatial resolution of the eye: Under optimal viewing conditions, we can employ Equation (11.9). Refer to Figure 11-38:

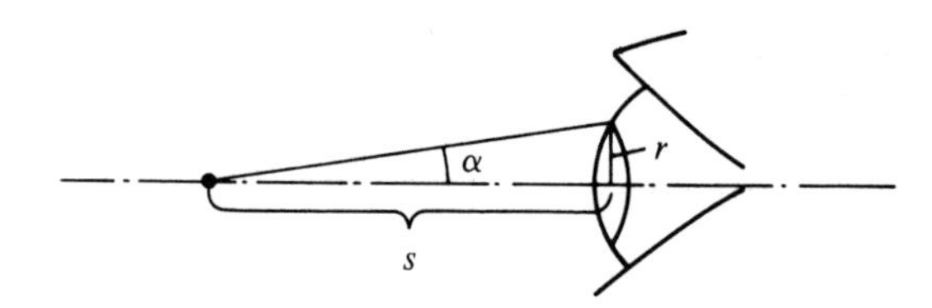

Figure 11-38

$$\tan \alpha = \frac{\text{pupil's radius } (r)}{\text{best viewing distance } (s)} = \frac{0.25 \text{ cm}}{25 \text{ cm}} = 0.01$$

hence $\alpha = 0.6$ degree. $n_1 = 1.00$ (air) and $n_2 = 1.33$; consequently $n = 0.75$. For an average wavelength of the visible spectrum $\lambda = 6 \times 10^{-5}$ cm, we get for the resolution of the unaided eye:

$$d_{\text{eye}} = \frac{6 \times 10^{-5} \text{ cm}}{0.75 \times 0.01} = 8 \times 10^{-3} \text{ cm}$$

Another calculation shows that two object points spaced about 0.1 mm apart have an image separation at the retina of about 5×10^{-3} mm. This is also the distance between two elementary light detectors (cones). The optical and nervous systems of the eye are well matched.

Abbé's formula is the simplest expression for the resolving power. It does not take into account the minimum difference in contrast that is detectable. Look at Figure 11-39. It shows two neighboring opaque objects and the intensity distribution across the corresponding image. The larger the contrast, the better the chance to discern between both images. The minimum contrast necessary depends on the recording system and on the observer. However, Equation 11.9 is still approximately correct.

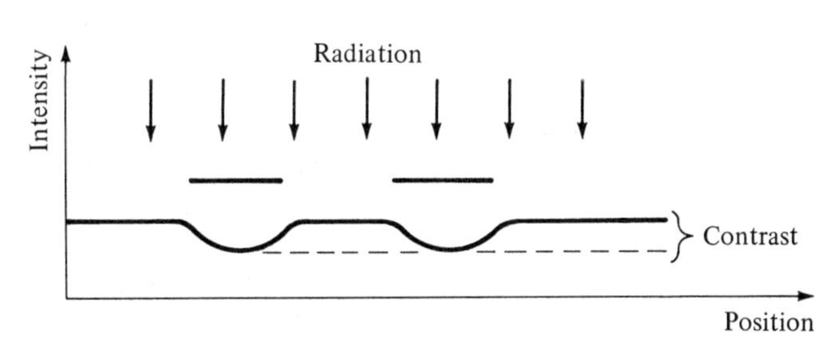

Figure 11-39

The spatial resolution of an optical instrument can be improved, i.e., the resolving power increased, by:

1. Illumination with radiation of short wavelength (ultraviolet-, electron-, or proton-microscope).
2. Increasing the index of refraction (immersion lenses in microscopy).
3. Enlarging the angular aperture (short focal distance of the object lens of a microscope).

Table 5 SPATIAL RESOLUTION, EXPRESSED IN cm, OF A FEW OPTICAL INSTRUMENTS

	Resolution
human eye	8×10^{-3}
light microscope	2×10^{-5}
uv-microscope	1×10^{-5}
electron microscope	2×10^{-8}
linear accelerator	10^{-14}

Table 5 shows the spatial resolution of a few instruments. Electron microscopes and accelerators do not achieve the high resolution which is possible theoretically. The reason is that the aperture is kept very small to avoid excessive aberrations. Another reason is the occurrence of radiation damage to microstructures if the employed intensity is high.

4.3 Time resolution

If objects appear intermittently, the image-analyzing device needs a time resolution that is short compared with the time spacing of the object's occurrence. The time resolution is the shortest time span between the appearance of two objects which can still be distinguished. For example, two light flashes separated by 0.01 s cannot be resolved by the human eye. They are perceived as one flash. But a fast photomultiplier tube can perceive such flashes. There are special types having a resolution of 10^{-8} s. This means that two light flashes spaced in time by a hundred-millionth of a second are resolved! This is not the place to investigate the time resolution of optical instruments in detail. We will restrict it to the human eye.

EXAMPLE: TIME RESOLUTION OF THE EYE

A simple experiment demonstrates that the time resolution of the eye is limited: Look at a neon light and move your eyes sideways for a split second. You will notice a bandlike after-image with dark and light fields. A neon lamp radiates intermittent light. Its frequency is mostly equal to the power line frequency. The time resolution of the eye is described as

$$\text{fusion frequency} = \frac{1}{\text{time resolution}}$$

If the time resolution is expressed in seconds, the fusion frequency is measured in hertz.

The fusion frequency is low and depends on the light intensity reaching the retina. Figure 11-40 shows the measured fusion frequency as a function of light intensity. Observe that the time resolution is different for the periphery (black and white vision) and the area around the fovea centralis (the area of color vision).

The entire movie industry depends on the finite and low time resolution of the eye. Two slightly different pictures presented within a time span shorter than the time resolution will fuse, and a continuous motion results. Eighteen pictures per second intermittently projected are sufficient.

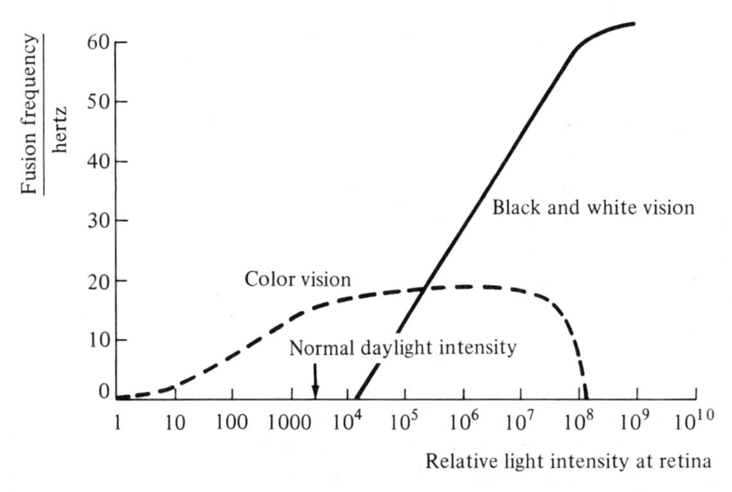

Figure 11-40 The fusion frequency of the human eye varies with the intensity of the illuminating light.

The Stroboscope. Looking at a marked, rotating white disc, it will appear uniformly gray if its rotation frequency is larger than the fusion frequency. Now, if the disc is illuminated by an intermittent light source, stroboscopic effects are observed.

For

$$\nu_s = n \cdot \nu_{\text{rot}} \qquad \text{(synchronizing condition)}$$

where

ν_s: flash frequency of the light source (flash duration is very short compared with $1/\nu_s$),

ν_{rot}: rotation frequency of the disc,

n: 1, 2, 3, . . . , succession of integral numbers,

the mark appears fixed in position. Both frequencies are said to be *synchronized*. A slight deviation from the synchronizing condition lets the image of the mark rotate slowly. This effect is familiar to the movie watcher: the spokes of turning spoked wheels sometimes appear to stand still or even to rotate backwards.

A stroboscope is an instrument which produces light flashes of variable, preset frequency. It is used, for example, to measure the rotation frequency of fast-turning objects or to produce slow-moving images of fast-rotating objects.

5. OPTICAL INSTRUMENTS

The following is a selection of optical instruments presented in a schematic fashion to underline their main features. For more information

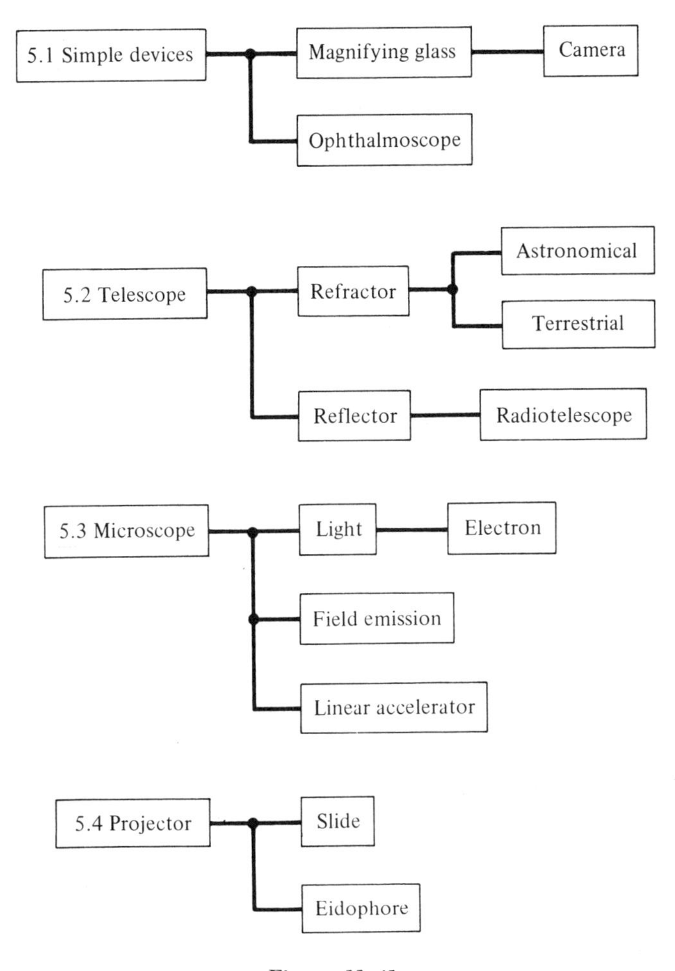

Figure 11-41

about any of the instruments presented, see the references at the end of this chapter. About a dozen schematic drawings are presented here. Pay particular attention to the last one, the eidophor. If you understand its performance, you are familiar with geometrical optics. The diagram (Figure 11-41) will guide you through the next pages.

5.1 Simple devices

MAGNIFYING GLASS. It increases the angle of view, that is, the numerical aperture.

Convention: Its magnification is the size of the image divided by the size of the object as seen at a 25-cm distance by the unaided eye.

Image: virtual, upright, magnified. Object between lens and focal point. Useful magnification up to 20 times.

PHOTOGRAPHIC CAMERA. Object outside the focal distance. The image is fixed on the photographic film. The product of shutter speed and lens

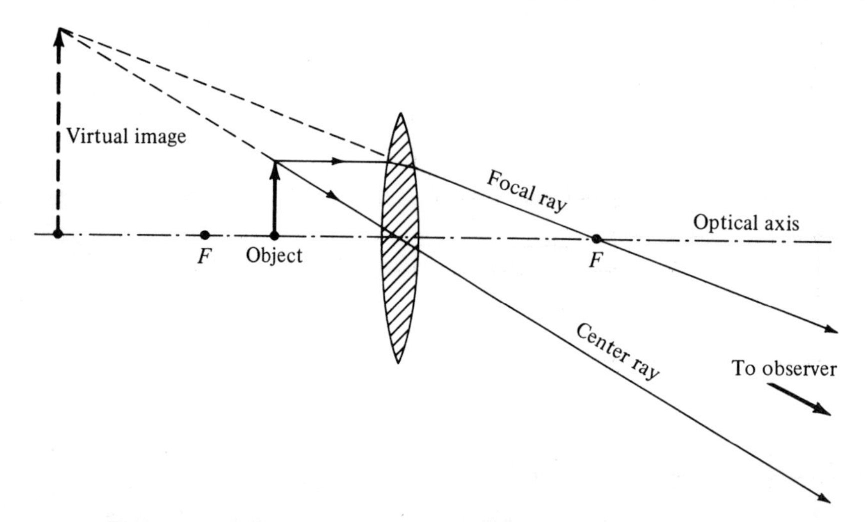

Figure 11-42 Ray tracings for a magnifying glass. F are focal points.

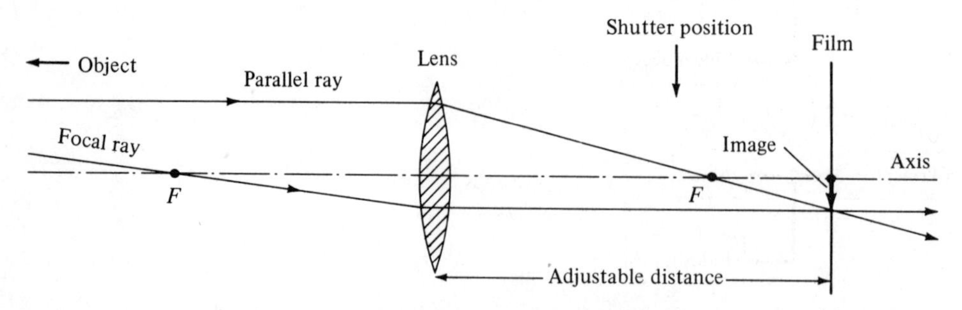

Figure 11-43 Ray tracings for a camera. F are focal points.

opening determines the brightness of the image. The resolution is, in general, determined by the grain of the film.

First mentioned: D. Barbaro, 1568.

Literature: C. E. Engels: *Photography for the Scientist.* New York, 1968.

Examples for widely used camera lenses are shown in Figure 11-44.

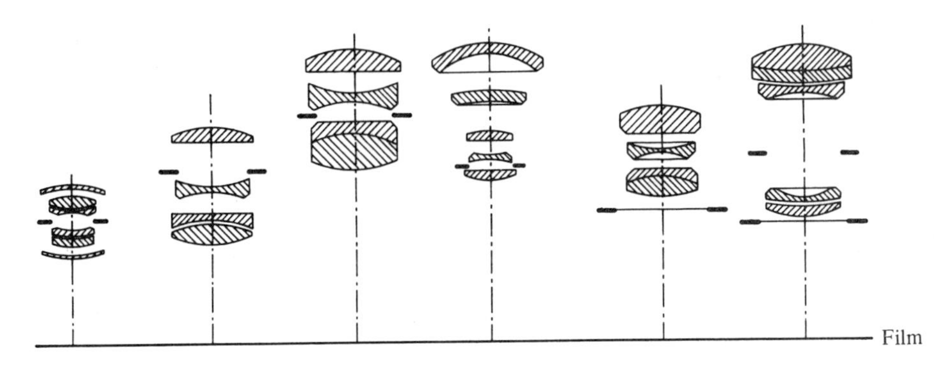

Figure 11-44 Cross sections of six modern camera lenses. All elements are drawn to the same scale.

Ophthalmoscope. Used to observe the interior of the eye, especially the retina. A semitransparent mirror (or a mirror with a center opening) illuminates the eye's background and allows the observer to see through. The lens of the patient's eye is used as a magnifying glass. For that reason, the optical apparatus is relaxed by means of drugs and consequently its focal point is beyond the retina.

System A is shown in Figure 11-45.

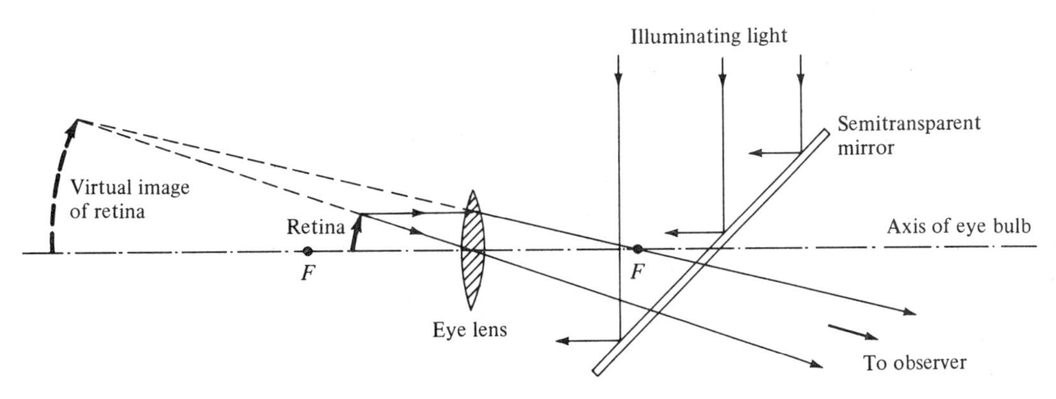

Figure 11-45 Ray tracings for the ophthalmoscope (system A). The semitransparent mirror may be replaced by a mirror with a center opening.

Image: virtual, upright, magnified (12 to 15 times).

System B is shown in Figure 11-46.

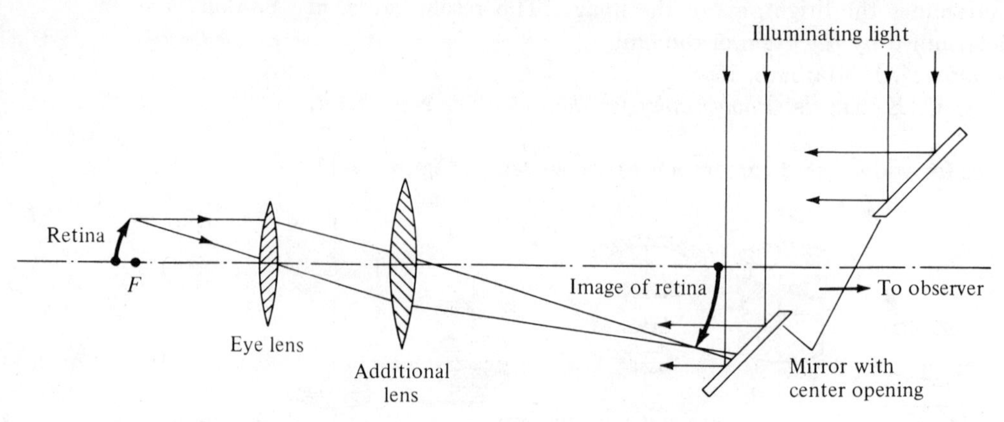

Figure 11-46 Ray tracings of system B for the ophthalmoscope.

The patient's lens and an additional lens act together like a projector.

Image: real, inverted, magnified (5 to 6 times).

Advantage: larger field of view than system A. The patient's eye does not need to be relaxed. The image can be photographed.

First mentioned: H. von Helmholtz, 1850.

5.2 Telescope

Used to observe distant objects.

Magnification: f_1/f_2, where

f_1: focal distance of front lens,
f_2: focal distance of eyepiece.

The magnification is limited by the thermal motion of the air between object and front (objective) lens. For this reason, astronomical telescopes are placed at high altitudes, mounted on balloons or built into orbiting satellites.

First mentioned: In Holland at the beginning of the 17th century.

Astronomical refractor. Maximum front lens diameter is about one meter, determined by the structural stability of glass. The front lens produces a real intermediate image at F_1. A scaled reticule and an aperture may be placed there. The eyepiece acts as a magnifying glass. See Figure 11-47.

Final image: virtual, inverted, magnified.

Galilei refractor. Oldest type of refractor. The field of view is very small. Still in use as an opera glass (magnification 2 to 3 times). See Figure 11-48.

Final image: virtual, upright, magnified.

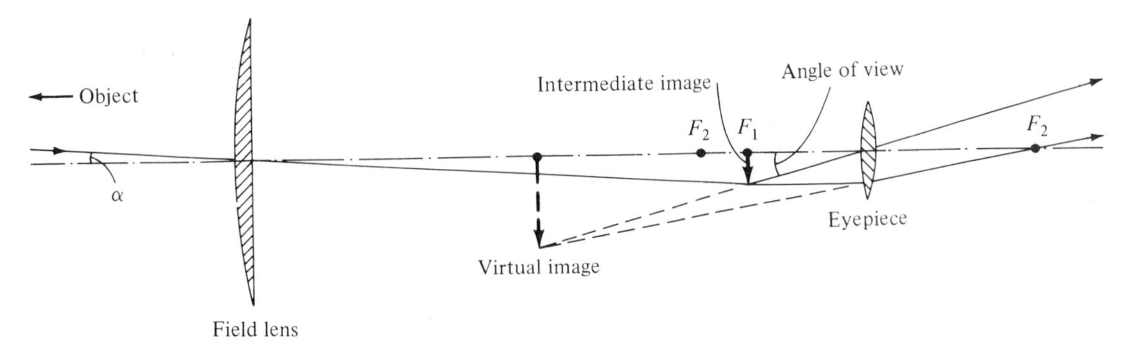

Figure 11-47 Ray tracings for an astronomical telescope. F_1 is focal point for front lens; F_2 is focal point for eyepiece. This is a very simplified figure: in general, both optical lenses consist of a series of elements.

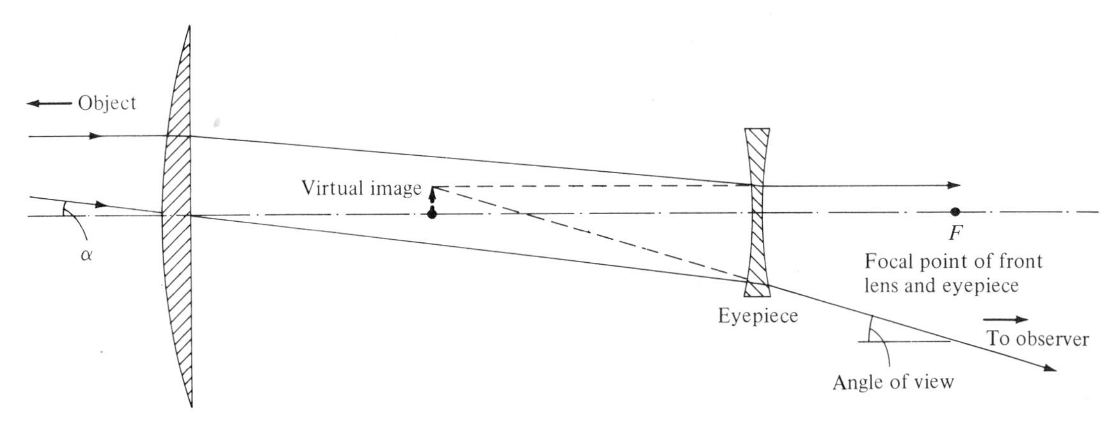

Figure 11-48 Ray tracings for the terrestrial (or Galilei) telescope. The first telescope to be constructed.

Reflector. A parabolic mirror replaces the front lens. The mirror produces an image at F_1. The eyepiece acts as a magnifying glass. See Figure 11-49.

Final image: virtual, inverted, magnified.

Advantage: large possible diameter (example: Hale reflector at Mount Paolmar with 5-m diameter), lower cost, relatively compact size, no absorption in front lens.

Radiotelescope. Parabolic antenna to observe extraterrestrial electromagnetic waves between 1 cm and 20 m wavelength. As long as the width of mesh is much smaller than the observed wavelength, the antenna may be a parabolically shaped wire netting. The radiation is focused upon a dipole antenna situated at the focal point. The received electric signals are amplified and analyzed.

Angular resolution:

$$\alpha = \frac{\lambda}{d} \times 70$$

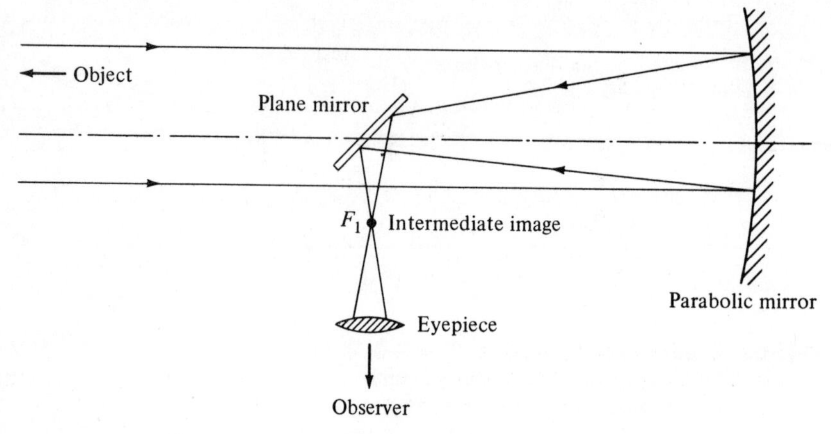

Figure 11-49 Ray tracings for an astronomical reflector. Shape and position of the auxillary (plane) mirror varies. This is Newton's construction.

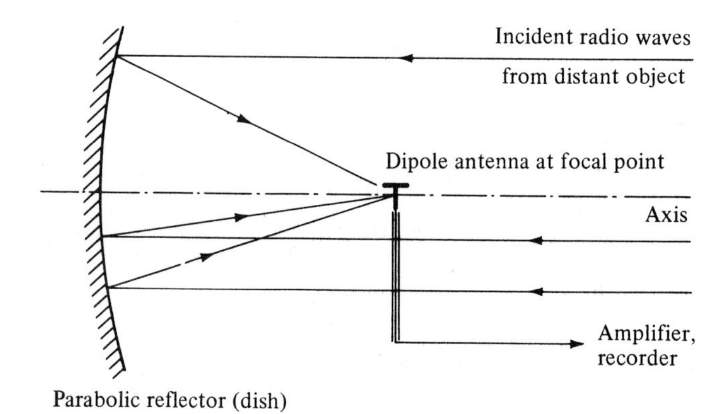

Figure 11-50 Schematic for a radiotelescope. The eyepiece of the light telescope is replaced by the antenna of the analyzing electronics.

where

α: resolution measured in degrees,
λ: observed wavelength, expressed in meters,
d: diameter of dish, expressed in meters.

The resolution is very poor compared with light telescopes.

First mentioned: K. Jansky, 1931.
Literature: J. D. Kraus: *Radio Astronomy.* New York, 1966.

5.3 Microscope

Used to observe very small objects.

Magnification: magnification of front lens times magnification of eyepiece. It is limited by the wavelength of the illuminating radiation.

First mentioned: Z. Jansen, 1590.

Visible light microscope. The object lens (focal distance a few millimeter) forms a real, inverted, magnified (up to 100 times), intermediate image. This is viewed by the eyepiece acting as a magnifying glass (magnification up to 25 times). Useful total magnification is about 1000 times.

Final image: virtual, inverted, magnified. See Figure 11-51.

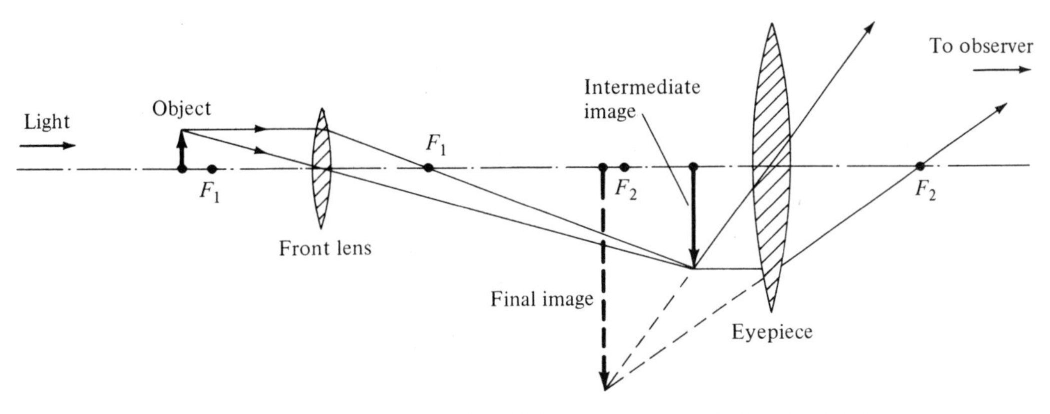

Figure 11-51 Ray tracings for a light microscope. This is a simplified drawing; actually, complicated lens systems are employed to correct for all aberrations.

Electron microscope. The ray tracing is analogous to the visible light microscope. Electrons replace the light, electrostatic or magnetic lenses replace the glass lenses, and an image converter replaces the observer's eye.

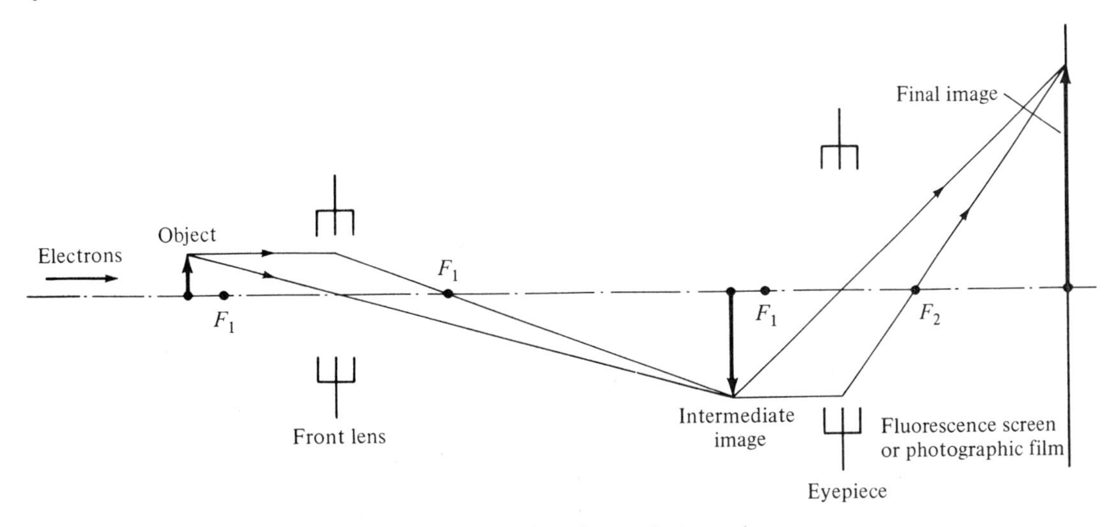

Figure 11-52 Ray tracings for an electron microscope.

Electrons display a wavelike behavior with a wavelength of

$$\lambda = \sqrt{\frac{150}{E}} \times 10^{-8} \text{ cm} \qquad (11.10)$$

where E is the energy of electrons, expressed in electron volts. For a modern electron microscope, $E = 10^5$ eV and thus $\lambda = 3.9 \times 10^{-10}$ cm.

The resolution is limited by the aperture of the lenses. Magnification up to 5×10^5, corresponding to a resolution of up to 5×10^{-8} cm. See Figure 11-52.

The front lens forms a real, inverted, magnified intermediate image. This image is picked up by the eyepiece and projected onto a screen or photographic plate.

Final image: real, upright, magnified.

Disadvantage: Very difficult technique is required to prepare objects against the destructive heat produced by the illuminating electrons.

First mentioned: E. Ruska, 1933.

Literature: P. Grivet: *Electron Optics.* Oxford, 1965.

FIELD EMISSION MICROSCOPE. In principle, a simple technique to image the tip of an electrode. Ions or electrons leaving the tip are accelerated along a straight line toward the fluorescent screen. Magnification up to 5×10^6 times, corresponding to a resolution of around 5×10^{-9} cm. This allows detection of the position of individual atoms and molecules! Main application: structural analysis of crystals and large molecules. See Figure 11-53.

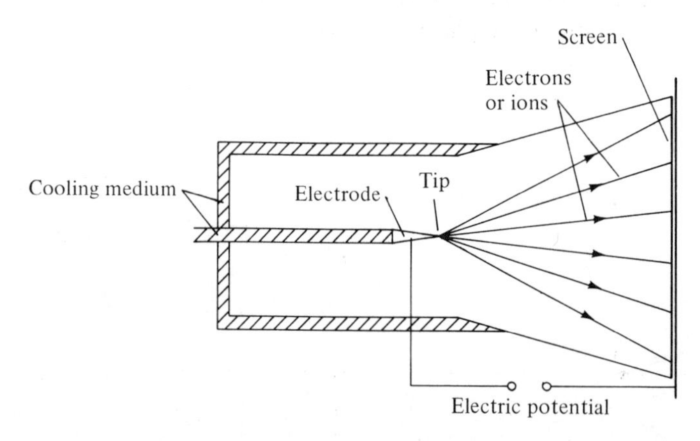

Figure 11-53 Ray tracings for a field emission microscope. The substance to be analyzed is placed at the tip of the electrode.

LINEAR ACCELERATOR. Electrically charged particles (electrons, protons, ions) are accelerated in successive stages. (See Figure 11-54.) Their scattering at individual atoms and nuclei is analyzed. Except for financial considerations, there are no limits to the possible resolution. Modern linear accelerators (up to 2×10^{10} eV energy and a few thousand meters long) analyze the fine structure of the proton; the corresponding resolution is about 10^{-14} cm!

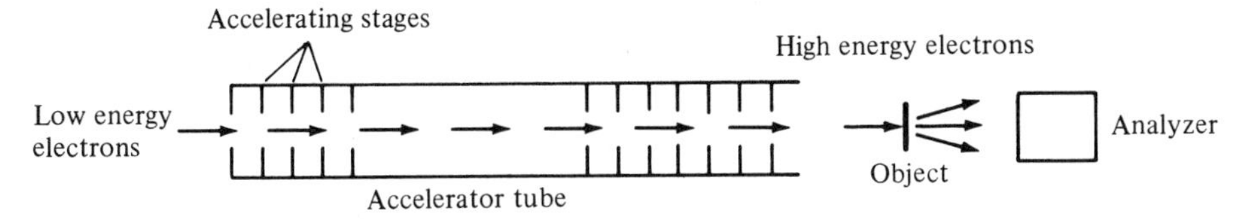

Figure 11-54

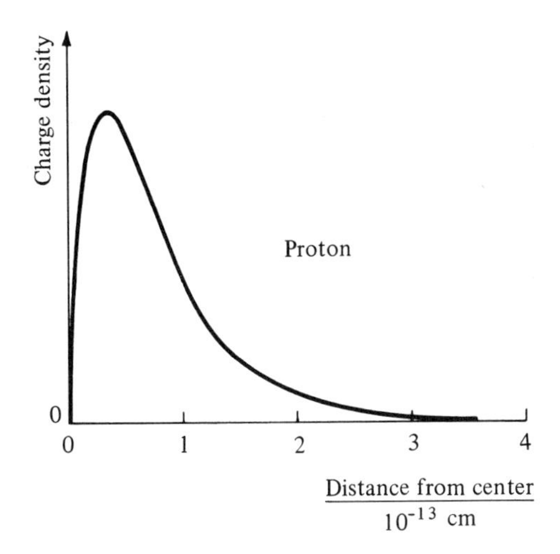

Figure 11-55 The charge distribution of a proton as measured by a high-energy linear accelerator. The proton is not uniform in charge density; it is difficult to determine its radius.

Figure 11-55 shows the image of the charge distribution of the proton as observed by scattered electrons from a high-energy linear accelerator.

5.4 Projector

SLIDE PROJECTOR. An instrument to magnify a transparent photograph and project it on a screen to make it visible to a large audience. See Figure 11-56.

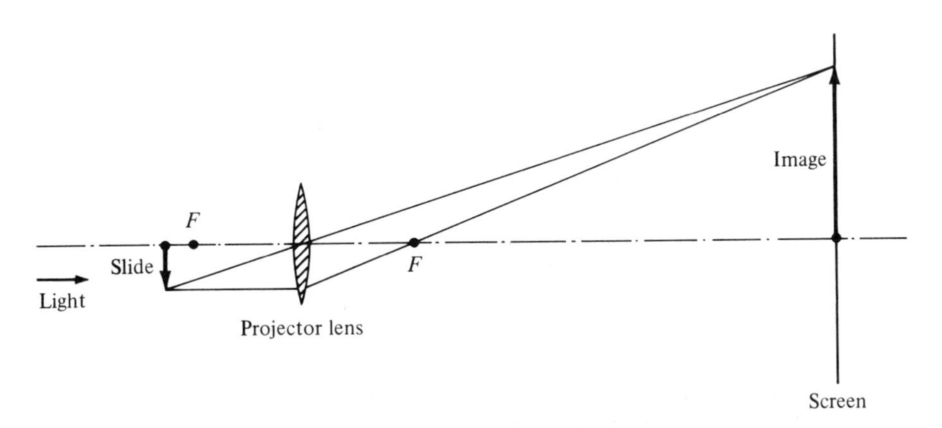

Figure 11-56 Ray tracings for a slide projector.

Image: real, inverted, magnified.

Limits to magnification: Temperature sensitivity of object. To get a sufficient light intensity on the screen, the object must be illuminated by a powerful light source.

EIDOPHOR. A special instrument to magnify and project pictures directly from a television transmission. See Figure 11-57.

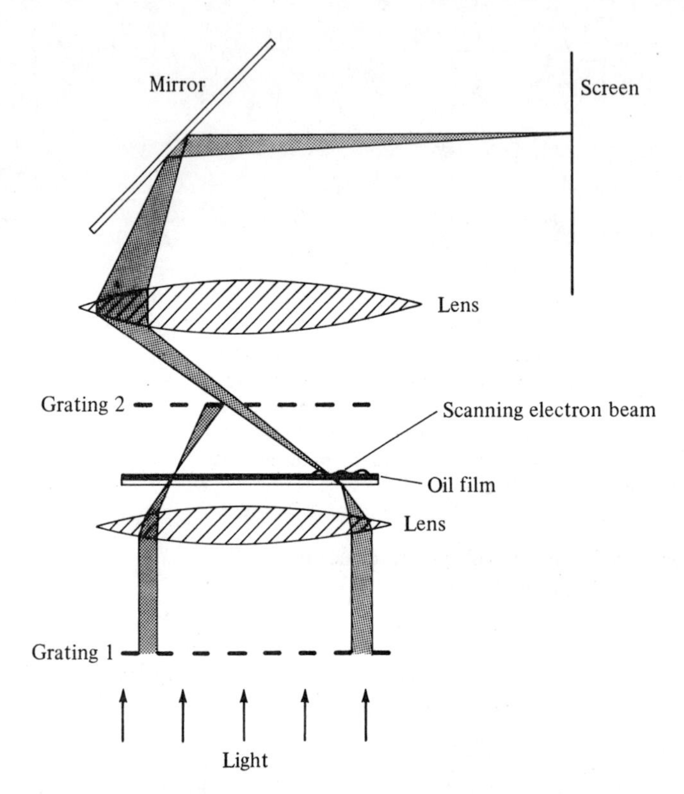

Figure 11-57 Principle ray tracings for an eidophor. The rays on the left side pass the film which is not activated at this position; consequently, they are absorbed by grating 2. The rays on the right side are bent from their rest position by the activated spot at the oil film to reach the screen.

Light from a very intense source passes through a parallel grating 1. This grating is imaged onto another grating 2 in such a way that no light passes through directly. Between both gratings is placed a high-viscosity oil film (the eidophor) on a rotating glass support. As long as the eidophor has a plane surface, it has no effect on the passing light. If a fine beam of electrons hits the oil film, the film becomes electrically charged. The electrical force will bend the surface of the eidophor; at this spot it will act like a lens. The light coming from grating 1 and passing through the activated spot of the eidophor will consequently be bent, miss its corresponding opaque spot on grating 2, and reach the projection screen.

The electron beam scans the eidophor in the same way as on a television screen. The system is so designed that the amount of light passing the eidophor is proportional to the intensity of the electron beam. The magnified image on the projection screen corresponds to the television image.

Literature: Journal of the Institution of Radio Engineers, Vol. 12, 1952, p. 2.

SUMMARY

In the field of geometrical optics, the working principles of optical instruments are demonstrated by ray tracings: infinitesimally thin bundles of light, propagating along a straight line. Its direction changes abruptly at boundaries. The intrinsic wavelike nature of light is neglected because the physical extensions of encountered boundaries are assumed to be very large compared with the wavelength of light.

A ray meeting a boundary is reflected or refracted or both.

Law of reflection:

$$\text{angle of incidence } (\alpha) = \text{angle of reflection } (\gamma)$$

The angles are bounded by the ray and the perpendicular to the boundary.

Law of refraction:

$$\sin \alpha = n \sin \gamma$$

where the index of refraction (n) is a number characterizing the media before and behind the boundary.

Optical elements such as lenses and mirrors are employed to form images from objects. These elements encompass a wide variety of shapes and materials, depending on desired features and types of rays. The radius of curvature and the focal distance are important, characterizing numbers.

Images may be real or virtual, and its size may be larger or smaller than the observed object. Optical instruments are combinations of optical elements. They are generally employed to enlarge the angle of view under which an object appears to the eye of the observer. The magnifying glass, microscope, and telescope are well-known examples.

The resolution of an optical instrument, that is, the smallest still-separable distance between two object points, is in principle limited by the wavelength of the radiation employed. For visible light, the spatial resolution is about 10^{-4} cm; employing elementary particles (like electrons or protons), the resolution reaches the order of 10^{-14} cm.

The time resolution of an optical instrument is important if the investigated object emits pulsed radiation, like those from quasars and pulsars. The time resolution of the human eye is of the order of 10^{-1} s. Modern devices such as photomultiplier tubes reach time resolutions of about 10^{-8} s.

Section 5 is devoted to brief descriptions of optical instruments such as the microscope, telescope, and projector. Examples of optical instruments discovered in the realm of nature are spread throughout the entire chapter.

PROBLEMS

1. Even if illuminated by a lamp, the pupil of man appears to be black. Why? This is not the case with cats and dogs.
2. Why is vacuum not suitable as the standard medium for the index of refraction of sound?
3. The boundary between two media having the same index of refraction is invisible. Why?
4. See Figure 11-3: The ray tracing through the beam splitting prism is idealized. Draw a real beam splitter where both rays emerge parallel.
5. A water pipe is placed at the focal line of a perfectly reflecting cylindrical mirror. Sunlight is focused on the pipe and heats the water flowing inside. By how many degrees will the water temperature rise for: cylinder radius = 40 cm, cylinder length 200 cm, and water flow rate = 50 $\text{cm}^3 \cdot \text{s}^{-1}$? The sun supplies 1.9 $\text{cal} \cdot \text{cm}^{-2} \cdot \text{min}^{-1}$.
6. In Section 4.3, a simple experiment to observe the finite resolution time of the eye is described. If the light source is a normal light bulb, the experiment will not work. Why?
7. The bad guys in the movie are chasing a car, being driven by the good guys. Suddenly the spokes appear to stand still. How fast are the wheels turning at this instant? Assume eight spokes and twenty frames per second.
8. Construct a two-element eyepiece which achieves upright images if used in the astronomical telescope.
9. A ray passing through oblique plane surfaces, as shown in Section 2.3(c), changes its direction. Use plane geometry to derive a formula for a symmetric passage.
10. What is the screen-hole distance for an image formed by a pinhole?
11. A concave mirror has a focal length of 20 cm. An object is placed at the following distances from the surface: 15 cm, 20 cm, 25 cm, 40

cm, and 100 cm. Calculate and determine graphically the position and type of each of these images.

12. A light ray strikes a 5-mm thick glass sheet (index of refraction = 1.45) at an angle of 50°. By how much is the transmitted ray shifted in parallel?

13. An object is placed 50 cm away from a converging lens. Its image is found to be 120 cm away. Calculate the focal length of the lens.

14. A bird having a wing span of 1.5 m is observed at a distance of 25 m. What focal length is required for a 35 mm camera to get a frame-filling image of the bird?

15. The crystalline cones of the apposition eye act as light pipes having an index of refraction of 1.55. The enveloping distal pigment cells have an index of refraction of 1.33. What is the maximum angle between a ray inside a cone and the side of the cone that will keep the light inside by total reflection?

16. A child can vary the refractive power of its eyes between 60 and 74 diopters; an adult between 60 and 68 diopters. Calculate the variation in focal length for both cases.

17. An object is held 10 cm away from the relaxed eye (refractive power = 60 diopters). Calculate the power of the additional lens necessary to achieve a sharp image at the retina.

18. The distance lens-retina inside the eyeball is about 17 mm. If the refractive power of the lens is fixed to 60 diopters, what is the reading distance? What type and power of eyeglasses must be used to bring the reading distance to 30 cm?

19. What is the time resolution of the human eye for color vision and black-and-white vision for a relative light intensity ten times the normal daylight intensity? See Figure 11-40.

20. Calculate the spatial resolution of the human eye for red light ($\lambda = 7.8 \times 10^{-5}$ cm) and blue light ($\lambda = 3.6 \times 10^{-5}$ cm).

21. The front lens of an astronomical refractor has a focal length of 9 m. A magnification of 85 is desired. Calculate the focal length of the eyepiece and the total length of the telescope.

22. In some light microscopes the space between front lens and object can be filled with oil (immersion system). What is the advantage? Why can the human eye not utilize this method? Opening the eyes under water yields blurred images.

23. Two sufficiently bright stars can be distinguished in the sky if they are separated by x degrees. Calculate x. Calculate the diameter of the dish of a radiotelescope operating at a wavelength of 10 cm and having the same resolution.

FURTHER READING

H. AUTRUM ET AL., "Handbook of Sensory Physiology," *Vision*, **VII/1–4**, Springer, Berlin, 1972.

J. R. BREEDLOVE, "Molecular Microscopy: Fundamental Limitations," *Science*, **170** (1970) 1310.

E. DENTON, "Reflectors in Fishes", *Scientific American* (Jan., 1971) 65.

D. R. GRIFFIN, *Listening in the Dark*. Yale University Press, New Haven, 1958.

K. N. OGLE, *Optics: An Introduction for Ophthalmologists*, 2nd ed., C. C. Thomas, Springfield Ill., 1968.

chapter 12

ELECTROSTATICS AND MAGNETOSTATICS

1. INTRODUCTION

Physics grows irregularly. Some fields like mechanics and time measurements had an early start; others like electricity and magnetism reached the quantitative stage not before the 18th or even 19th century. The resulting confusion of units is still with us despite numerous conferences. The confusion is especially great in electromagnetism, although all measuring systems are metric!

Up to now we have used the *cgs system* (centimeter-gram-second) and the *mks system* (meter-kilogram-second): 1 m = 100 cm and 1 kg = 1000 g.

Now we are reaching the crossroads; the signpost shows the following directions:

1. Continue the cgs system and introduce new units like the statcoulomb, abcoulomb, statfarad, oersted, and gauss. This is the *symmetric* or *Gaussian* system. A number of equations will look more simple than in other systems, but outside of the National Bureau of Standards and textbooks, you will never encounter those units. (For example, you may sometimes read that the magnetic field strength is expressed in gauss—a rather common mistake in the specification sheets of industrial equipment. The magnetic field strength is measured in oersted in the cgs system.)
2. Use the *Giorgi system* which extends the mks system with units like the coulomb, farad, volt and tesla. Although tesla may not be as familiar as gauss, we shall use it in the following chapters. It is internationally the most widely accepted system and recommended by the American Institute of Physics.

There are more (also metric) systems, but we shall use the extended mksA system (meter-kilogram-second-ampere) and once in a while convert it into the units of the extended cgs system.

Electricity and magnetism are conveniently treated together. Both are described in equivalent terms (e.g., field, potential, inductance) and are intimately connected at the subatomic level. However, there is a basic difference: electrical properties are intrinsic to matter while magnetism is caused by electric charges.

Throughout the chapter, electromagnetic phenomena are time-independent; we deal with electro- and magnetostatics. Time-dependent phenomena such as electric current and the action potential of a cell membrane are reserved for the next chapter. The resulting ease of representation is the reason for this division. In nature, all phenomena are time-dependent.

The flow diagram on page 297 shows the organization of the chapter.

2. ELECTROSTATICS

2.1 Electric charge

Material bodies appear to the observer to be very different. Their size, color, density, shape, consistence, etc., varies within wide limits. But there is one property they all have in common: gravity. It is observed as an attractive force between matter and inseparably connected to it.

There is another basic property of matter: it may be electrically charged. This electric charge is also inseparably connected to it. However, the analogy stops right there. Only one kind of gravity is known, but matter may be charged in two basically different ways. It carries either a positive or a negative charge. The existence of these charges is proven by its interactions with the surroundings, for example, by the occurrence of attractive or repellent electric forces. Matter which does not show this electric property is named—also by tradition—electrically neutral. It appears electrically neutral only because the observable effects of the positive and negative charges cancel each other.

Looking at the atomic and subatomic level, we notice that matter is in general electrically positive or negative. The positive atomic nucleus is surrounded by a cloud of negative electrons. As observed from a distance, the entire atom is electrically neutral. Even the neutron—together with the proton making up the atomic nucleus—is electrically neutral seen from outside. Modern scattering experiments show that the positive inside of the neutron is electrically neutralized by an outer negative layer. The great variety of electric effects is caused by this trifold occurrence: positive, negative, and neutral. Although the details will be presented later, it is worthwhile to note that all electric charges are composed of a number of elementary charges. An elementary charge is the smallest amount of electricity possible. Positive and negative elementary charges are identical in magnitude.

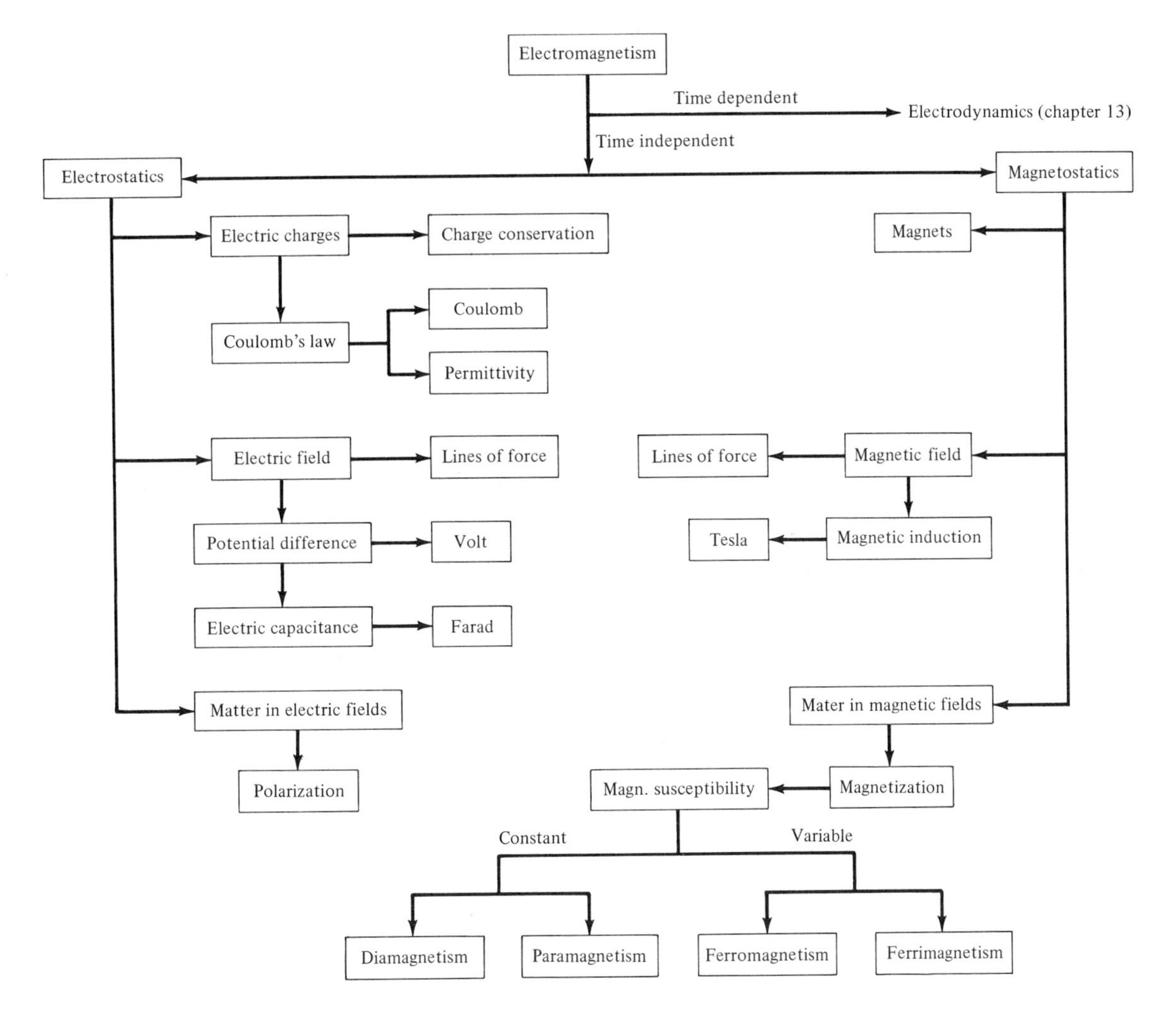

2.2 Conservation of charge

Experiments prove that, within a closed system, no one elementary charge can appear or disappear alone. There are two famous phenomena in nuclear physics demonstrating the conservation of charge: annihilation and pair production.

Annihilation. If two elementary charges of opposite sign and equal mass interact, they will vanish. However, energy conservation (see Chapter 4) still holds.

Figure 12-1 demonstrates the annihilation process for two nuclear reactions: (e^-, e^+) and $(p, \overline{p})$. The mass of the elementary particles electron (e^-), positron (e^+), proton (p), and antiproton $(\overline{p})$ are converted into radiation energy, called *annihilation radiation.* Realize that the electrical charge of the charged particles has vanished together with its mass.

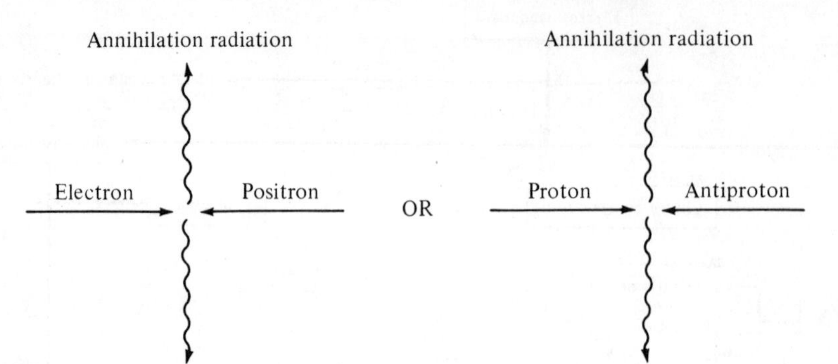

Figure 12-1

Pair production: Radiation of sufficiently high energy may convert into electrically charged particles. This conversion takes place in the vicinity of an atomic nucleus. The radiation must carry a minimum energy, called *threshold energy*, sufficient to satisfy the law of energy conservation. The rest mass of an electron or positron is 9.1×10^{-31} kg, corresponding to an energy of 0.511 MeV. The rest mass of a proton or antiproton is 1.67×10^{-27} kg, corresponding to an energy of 938 MeV. Note that two charges of identical magnitude but of opposite sign are always created (Figure 12-2).

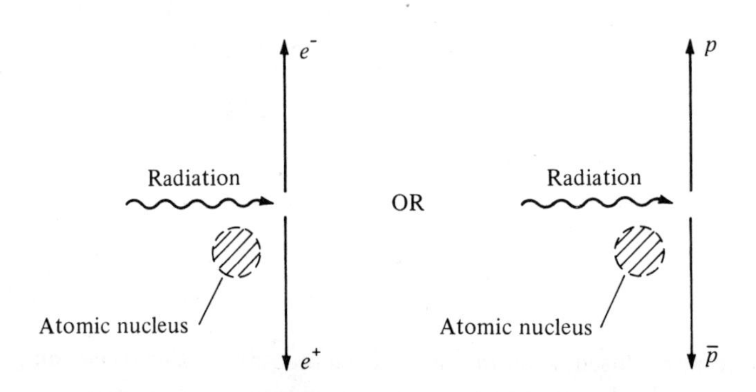

Figure 12-2

2.3 Coulomb's law

2.3(a). FORMULATION. Electric charges exert a force on each other. This force is called *Coulomb force* and it acts along the straight line connecting those charges. Its magnitude is

$$F_c = \gamma \frac{Q_1 Q_2}{r^2} \qquad \text{Coulomb's law*} \qquad (12.1)$$

* Charles Augustin Coulomb (1736–1806), physicist and engineer-officer. Developed the first high-sensitivity torsion balance and used it to make the first quantitative determination of electrostatic and magnetostatic interactions.

where

F_c: magnitude of force acting between the charges 1 and 2,
Q_1, Q_2: amount of electricity carried by charge 1 and charge 2, respectively,
r: separation between both charges,
γ: proportionality factor.

The direction of the Coulomb force depends on the signs of the charges Q_1 and Q_2. If Q_1 and Q_2 have the same sign, i.e., both are negative or both positive, then we have a *repellent* force. If Q_1 and Q_2 are of different signs, that is, one charge is positive and the other one is negative, then the Coulomb force is an *attractive* force.

Note again the difference between electric force and gravitational force. Although Newton's law (Eq. 3.15) looks formally like Coulomb's law, Equation (12.1), there are only attractive gravitational forces! This still holds true for the gravitational interaction between matter and antimatter; the gravitational force is always attractive.

2.3(b). UNIT OF CHARGE. The amount of electricity carried by a charge is measured in coulombs (symbolized C).

DEFINITION: *Two charges situated in a vacuum and separated by a distance of one meter carry the charge of one coulomb each if they exert a force of magnitude* 8.99×10^9 *newtons on each other.*

This is a rather abstract definition. You will get a better picture if you realize that the elementary charge (symbolized e), the smallest charge possible, is

$$e = 1.602 \times 10^{-19} \text{ C} \qquad (12.2)$$

This means that 6.25×10^{18} elementary charges together carry just one coulomb of charge.

2.3(c). ELECTRIC PERMITTIVITY. The definition for the unit of charge is derived from Coulomb's law, Equation (12.1), with the proportionality factor expressed in the following form

$$\gamma = \frac{1}{4\pi\epsilon} \qquad (12.3)$$

where ϵ is the *electric permittivity* of the medium separating the electric charges. This expression for the proportionality factor is used because it leads to simple mathematical expressions in electromagnetism.

The permittivity ϵ is a numerical quantity characterizing a medium. To get values easy to remember, the following relation is convenient:

$$\text{relative permittivity } (\epsilon_r) = \frac{\text{permittivity } (\epsilon)}{\text{permittivity of vacuum } (\epsilon_0)} \qquad (12.4)$$

ϵ_r, sometimes called the *dielectric constant*, has numerical values between 1 and 100. Table 1 presents ϵ for various materials.

Table 1 NUMERICAL VALUES FOR THE PERMITTIVITY OF SOME MATERIALS

(ϵ is expressed in $C^2 \cdot m^{-2} \cdot N^{-1}$)

Material	Permittivity
vacuum	8.86×10^{-12}, symbolized ϵ_0
air	8.94×10^{-12}
transformer oil	1.98×10^{-11}
benzene	2.02×10^{-11}
asphalt	2.38×10^{-11}
sucrose	2.94×10^{-11}
urea	3.10×10^{-11}
sulfur	3.55×10^{-11}
acetone	1.90×10^{-10}
diamond	1.45×10^{-10}
water (100°C)	4.25×10^{-10}
water (0°C)	7.80×10^{-10}

2.4 Electric field

An electric charge is always surrounded by an electric field. The presence of this electric field is detected by the force acting on another electrically charged body (a test charge) entering this field. It is characterized by its electric field strength, symbolized **E**.

DEFINITION:

$$\mathbf{E} \equiv \frac{\mathbf{F}}{Q} \tag{12.5}$$

where

E: *electric field strength*, a vector quantity pointing in the same direction as **F**,
F: electric force at the position of **E**,
Q: quantity of electrical charge experiencing the electric field.

The electric field strength is a vector and is measured in newtons/coulomb. There is no special unit for it.

2.4(a). LINES OF FORCE. An electric field can be visualized if we draw its lines of force. A line of force is traced out by a particle moving under the influence of a force—here, under the influence of an electric force.

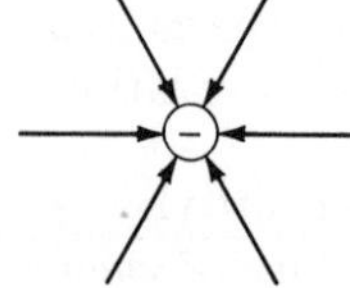

Figure 12-3 Electric field lines for isolated positive and isolated negative charges. The field lines are perpendicular to the charge.

By convention, an electric line of force (also named electric field line) begins at the positive charge (source) and always ends at a negative charge (sink). The denser the lines of force, the larger the electric field strength. Parallel lines of force indicate a homogeneous field, that is, the electric field strength is constant at any position. Otherwise, the field is called nonhomogeneous; its electric field strength is different for each position. Examples of lines of force follow.

Single point charges are shown in Figure 12-3.

Two separated point charges of the same magnitude are shown in Figures 12-4 and 12-5.

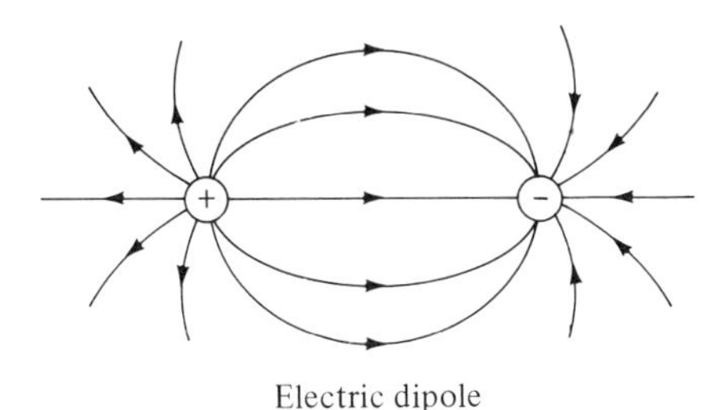

Figure 12-4 Electric field lines for neighboring charges of equal magnitude and opposite sign.

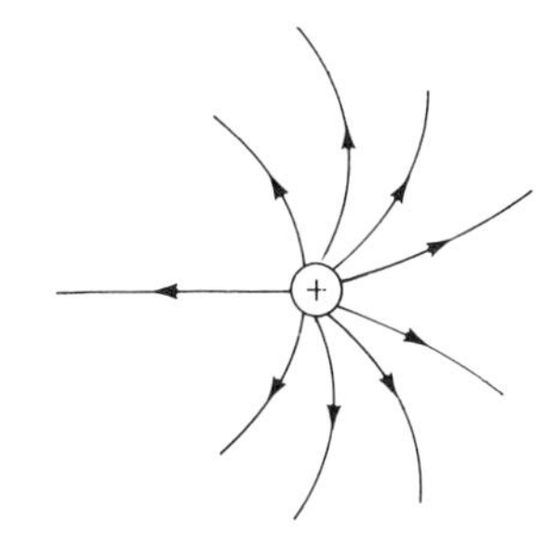

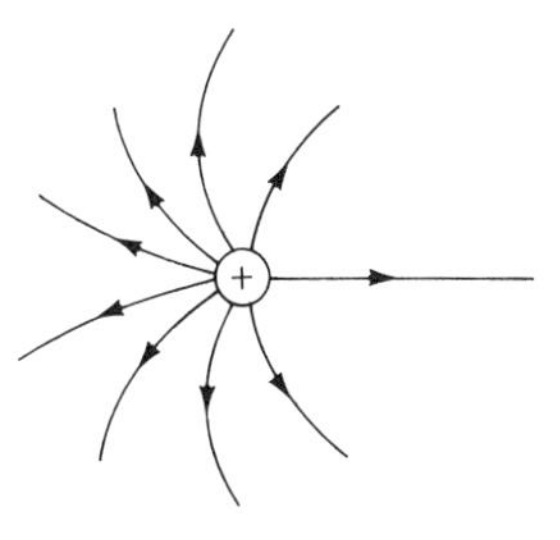

Figure 12-5 Electric field lines for neighboring identical charges.

Charged parallel plates are shown in Figure 12-6.

An electric fish is shown in Figure 12-7.

An electric field is influenced by objects within it. This fact is used by the Nile pike, an electric fish which produces an electric field around itself and has tiny organs which signal any change in the pattern of the

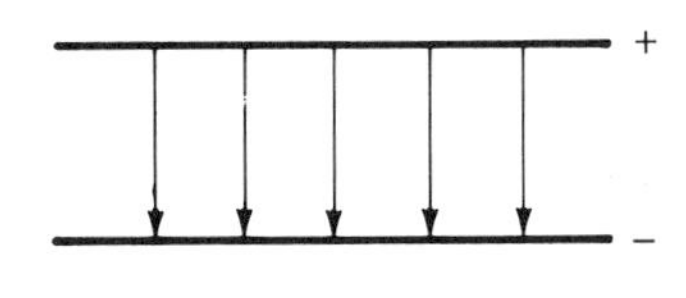

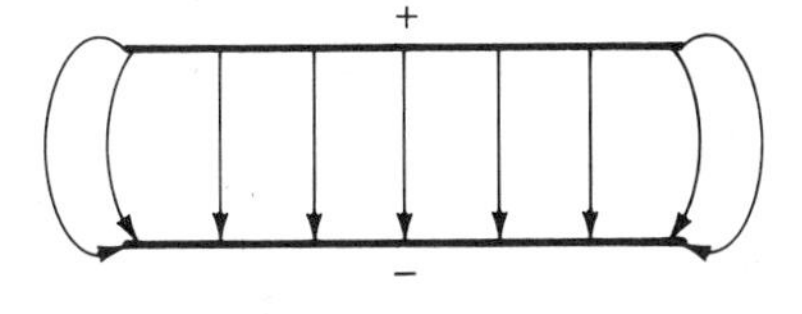

Figure 12-6 Electric field lines between two electrically charged plates: left, infinitely long; right, finite length. Notice the curved field lines at the boundaries of the latter.

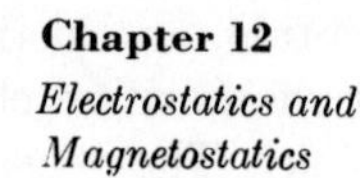

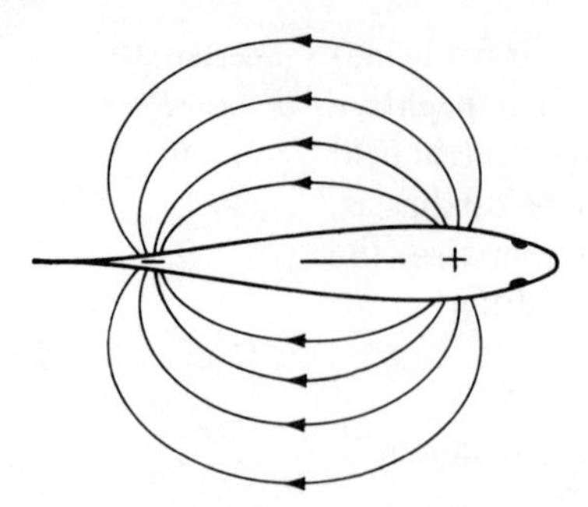

Figure 12-7

electric field. In this way the fish orients itself in the murky waters and may even locate its prey. The electric organs are too weak to be used as a weapon.

Figure 12-8 shows a hollow metallic body. Note that there is no electric field inside the hollow body. This arrangement is called a *Faraday*

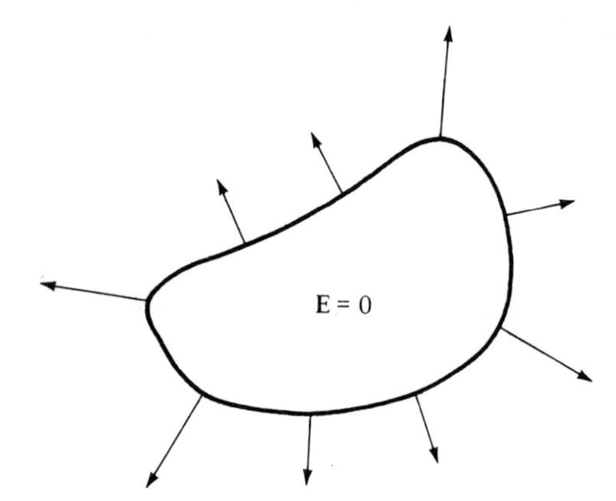

Figure 12-8 Electric field lines of a positively charged hollow object. Notice that no field lines extend into the interior.

cage and is widely used to screen-off unwanted electric fields. A radio will not receive signals inside a car unless an antenna penetrates to the outside because the car's body shields the electric waves produced by the transmitting radio station. For the same reason, nothing happens to you at the moment your car gets hit by lightning.

2.4(b). Potential difference. Electrostatic forces are present between charged bodies. It is more difficult to investigate the actions of electric forces than the actions created by gravitational force. Gravitational forces are always attractive and may be assumed to originate at a center of gravity. Electric forces are attractive or repellent and may originate at a point, a plane, or a spatially curved surface. The electric field therefore varies greatly from situation to situation. Merely for the special case of an isolated electric point charge, the formalisms for treating both forces are similar. Another basic difference is that, in addition, electric forces may be time-dependent (this is the case in electrodynamics) and are influenced to a large degree by objects situated inside the electric field.

Keep these very basic and substantial differences in mind if you try to compare the actions of gravitational and electric forces. To investigate electrostatic forces it is convenient to introduce the *electric potential difference* or *electric tension* between two positions inside an electric field.

The *potential difference* (symbolized U) is defined as

$$U \equiv \frac{W}{Q} \tag{12.6}$$

where W is the work gained or supplied to move the quantity of charge Q from position 1 to another position 2 in the electric field.

The potential difference may be positive, negative, or zero, depending on the direction of the lines of force and the sign of the charge. There is a derived unit for the electric potential difference, the volt* (symbolized V):

$$1 \text{ volt} = \frac{1 \text{ joule}}{1 \text{ coulomb}} \tag{12.7}$$

Thus if we observe an object carrying a charge of one coulomb and changing its energy content by one joule, we conclude that the potential difference between its initial and final positions in space is one volt.

Instruments which measure the potential difference are called *voltmeters*.

In atomic and nuclear physics it is convenient to connect the definition of the potential difference with the elementary charge in order to gain an appropriate new unit for energy, the electron volt (symbolized eV):

$$1 \text{ electron volt} = 1.602 \times 10^{-19} \text{ coulomb} \times 1 \text{ volt} \tag{12.8}$$

This means that an electron or proton will change its energy content by one electron volt if it moves through a potential difference of one volt.
Conversion:

$$1 \text{ eV} = 1.602 \times 10^{-19} \text{ J}$$
$$1 \text{ J} = 6.242 \times 10^{18} \text{ eV}$$
$$1 \text{ eV} = 10^{-6} \text{ MeV}$$

EXAMPLES: MEMBRANE POTENTIAL

A very thin membrane separates the inside and outside of a cell. As long as the cell is alive, it exhibits a selective permeability for certain ions. This means that some charged molecules or atoms penetrate more easily into the cell than out of the cell. Others show a preferred outward movement. As a result, the ion concentration on both sides will be different. Experiments show that the inside of a cell is electrically negative and the outside is positive. Figure 12-9 shows this situation in an octopus nerve.

The charge distribution on both sides of the nerve is determined by K^+-ions. Outside,

* Alessandro Volta (1745–1827), physicist. Built in 1796 the first electric battery and invented the electrometer to measure its electric tension.

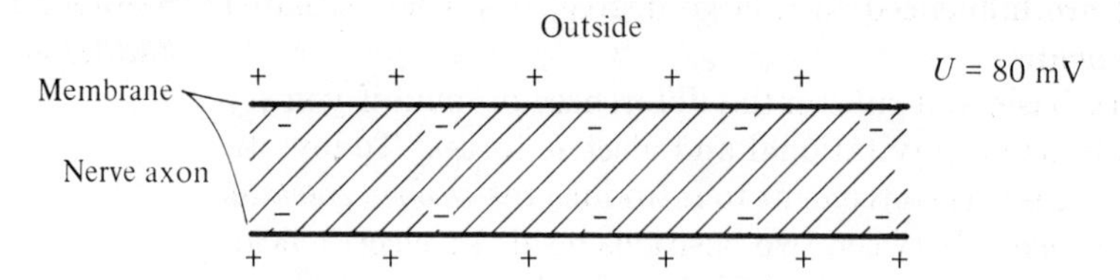

Figure 12-9 Double layer of charge on a biological membrane that separates the exterior and interior of a nerve cell.

the concentration of K^+-ions is about forty times as high as inside. The resulting potential difference across the cell membrane is about 8×10^{-2} volt = 80 mV. It is called the *rest potential* of the cell membrane and can be measured directly by means of microelectrodes and a sensitive voltmeter. This rest potential is of utmost importance to life. If excited, the cell membrane suddenly changes its permeability for K^+-ions (and others besides), the potential difference drops to practically zero, and a nerve impulse results. After that, the permeability returns to its initial value, and the potential difference of 80 mV is reestablished. Within a few milliseconds the membrane is ready for the next excitation. This electric discharge moves along the nerve axon and is one of the means of transferring information within an organism.

ANOTHER EXAMPLE: ELECTRIC BATTERIES

Chemical processes between two metals or their compounds are used in technology to produce a potential difference. The Fe-Ni battery cell shown in Figure 12-10 yields a potential difference of 1.36 volts.

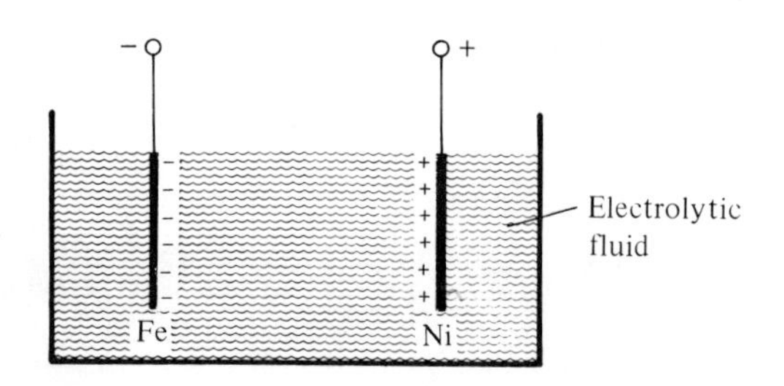

Figure 12-10

In electrical circuitry, a battery is symbolized by —|⊢—. The numerical value for the potential difference of a battery cell depends on the material of the two electrodes, here Fe and Ni. It may be increased by connecting the poles of equal cells in series. Figure 12-11 shows such a series connection. The resulting potential difference U is the sum of the potential differences of the individual elements:

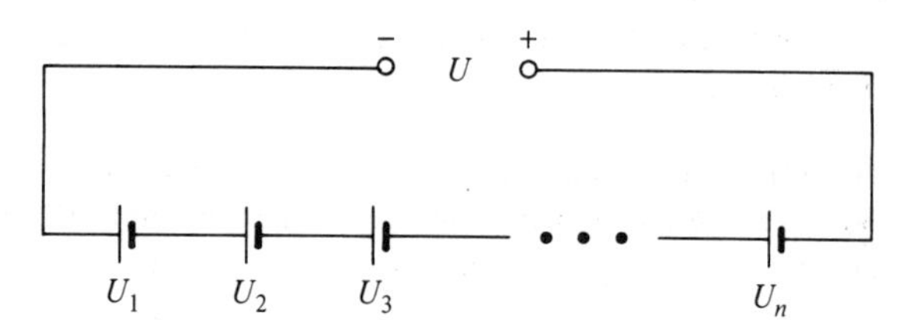

Figure 12-11 Individual battery cells connected in series.

$$U = \sum_{i=1}^{n} U_i$$

$$= U_1 + U_2 + U_3 + \cdots + U_n \quad (12.9)$$

where n is the number of elements.

The other possible way to interconnect

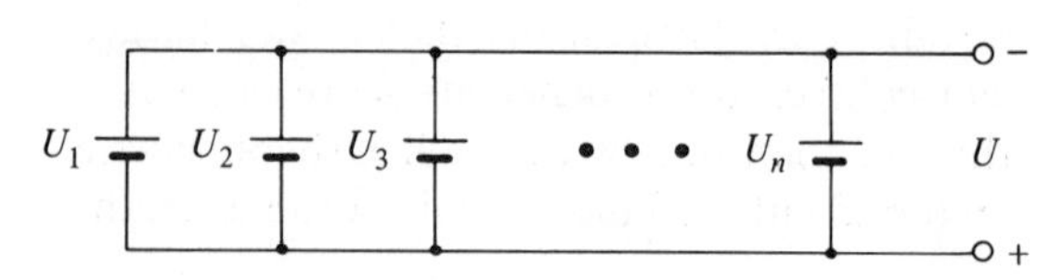

Figure 12-12 Individual battery cells connected in parallel.

battery cells does not yield a higher potential difference. Figure 12-12 shows parallel connection:

$$U = U_1 = U_2 = U_3 = \cdots = U_n \quad (12.10)$$

The resulting potential difference equals the potential difference of the individual element.

Long before man built batteries for his own use, nature had developed its own. A cell and its surroundings are a battery having an electric tension of about 80 mV. In general, this battery is utilized as an energy source for the transmission of information. Electric fishes have banks of specialized cells which are interconnected. Although each individual cell yields only 70–80 mV, the total battery contains around a thousand volts. This indicates that more than ten thousand cells are connected in series. An electric discharge of such a battery will stun small fishes. This method is also copied by man and named *electrofishing*.

2.4(c). ELECTRIC CAPACITANCE. If we place electric charges on concentric spheres, a set of parallel plates, on straight wires, or any other physical arrangement, an electric field is created. The measureable potential difference will be proportional to the charge. It is convenient to express this in the following way:

$$U = \frac{1}{C} Q \quad (12.11)$$

where

U: potential difference between the positions of the electric charges, measured in volts,
Q: quantity of electricity of these charges, measured in coulombs,
C: proportionality factor called the *electric capacitance* of the arrangement.

The electric capacitance is measured in farads* (symbolized F):

$$1 \text{ farad} = \frac{1 \text{ coulomb}}{1 \text{ volt}} \quad (12.12)$$

In general, the capacitance of an arrangement is measured indirectly, for example, by measuring U and Q and using Equation (12.11).

The farad is a very large unit, as you will observe in the following example.

EXAMPLE: CAPACITANCE OF PARALLEL PLATES

The electric capacitance of an arrangement depends on its configuration and dimensions. Two equal and parallel plates (see Figure 12-13) have a capacitance of

$$C = \epsilon \frac{A}{d} \quad (12.13)$$

where

A: surface area of one plate,

* In honor of Michael Faraday (1791–1867), physicist and chemist. He discovered electromagnetic induction and introduced the concept of electromagnetic lines of force. Founder of electrochemistry and discoverer of diamagnetism. More about him in L. Pearce Williams, *Michael Faraday*, London, 1965.

d: separation between plates,

ϵ: permittivity of the space between both plates.

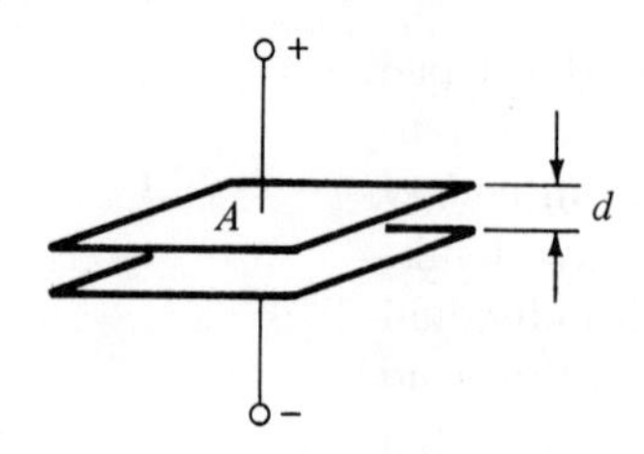

Figure 12-13

Observe that C depends only on physical arrangement and permittivity. If the capacitance of the above arrangement, in this case called a *parallel-plate capacitor*, is to be 1 farad, we can easily calculate the necessary surface area. Let us assume

$$d = 1 \text{ cm} \quad \text{and}$$

$$\epsilon = 8.94 \times 10^{-12}\, C^2 \cdot \text{m}^{-2} \cdot \text{N}^{-1} \quad \text{(air)}$$

We get

$$A = \frac{Cd}{\epsilon} = \frac{1 \times 0.01}{8.94 \times 10^{-12}} \cdot \frac{F \cdot \text{m} \cdot \text{m}^{-2}}{C^2 \cdot \text{N}}$$

$$= 1.12 \times 10^9 \text{ m}^2$$

This capacitor has a surface equal to one third of the Great Salt Lake in Utah!

In electrical circuitry a capacitor is symbolized by —||— and expressed in

$$1 \text{ nanofarad} = 10^{-9} \text{ F}$$

$$1 \text{ picofarad} = 10^{-12} \text{ F}$$

Any interconnection of capacitors can be reduced to two basic arrangements: series and parallel connection.

Series connection is shown in Figure 12-14. The resulting total capacitance C is calculated from the following formula:

$$\frac{1}{C} = \sum_{i=1}^{n} \frac{1}{C_i} = \frac{1}{C_1} + \frac{1}{C_2} + \frac{1}{C_3} + \cdots + \frac{1}{C_n} \tag{12.14}$$

where n is the number of elements. Note that the total capacitance is smaller than any individual capacitance.

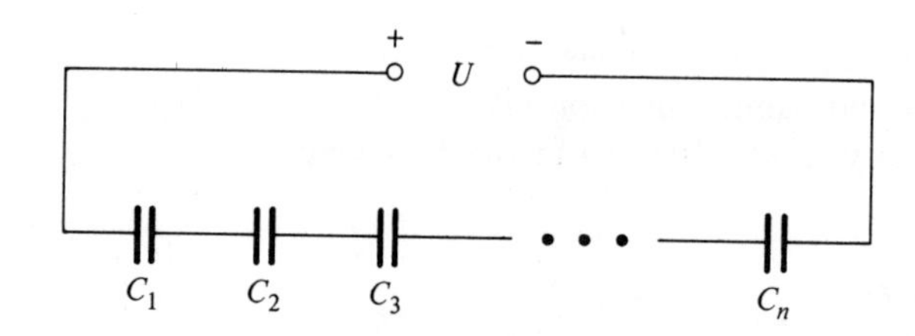

Figure 12-14

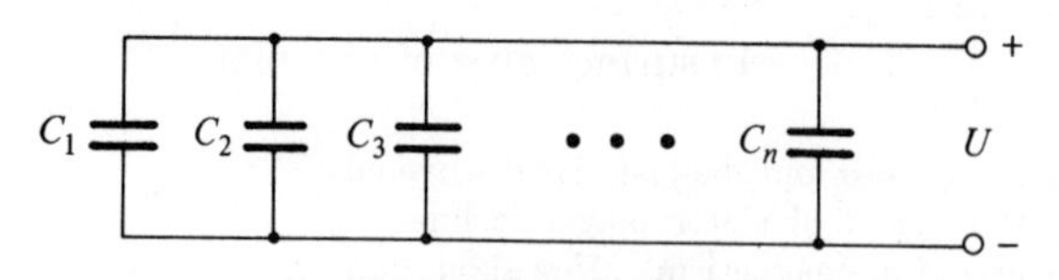

Figure 12-15

Parallel connection is shown in Figure 12-15. The resulting total capacitance C is the sum of all individual capacitors:

$$C = \sum_{i=1}^{n} C_i = C_1 + C_2 + C_3 + \cdots + C_n \qquad (12.15)$$

The capacitance of biological membranes is important because it influences the time needed for the discharge of a nerve cell. A low capacitance leads to a fast nerve impulse. This in turn allows more nerve impulses to be emitted during a given time span, thus increasing the information flow.

2.5 Matter in electric fields

2.5(a). Polarization. If we place a nonmetallic substance into an electric field, for example, between the plates of a capacitor, we observe the appearance of electric charges at the surface of the material. See Figure 12-16. These charges are induced by the electric field and disappear as

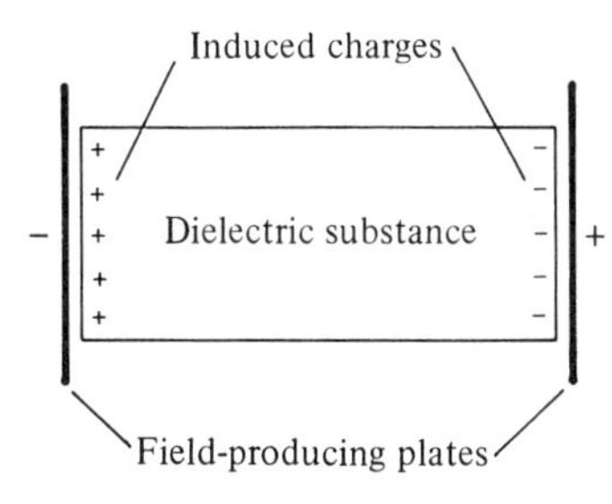

Figure 12-16

soon as the inducing electric field vanishes. The phenomenon described is called *polarization*. The material is called a *dielectric*. Polarization is explained as follows.

Due to slight structural asymmetries, many molecules are *electric dipoles*. This means that each molecule can be represented by two opposite and equal charges separated by approximately the molecular diameter. In general, the molecules inside a substance are arranged at random. The electric field produced by each of these tiny dipoles compensates—no electric charge is observed from outside.

Placed in an electric field, the dipoles tend to align parallel to the lines of force, that is, parallel to **E**. See Figure 12-17. This alignment

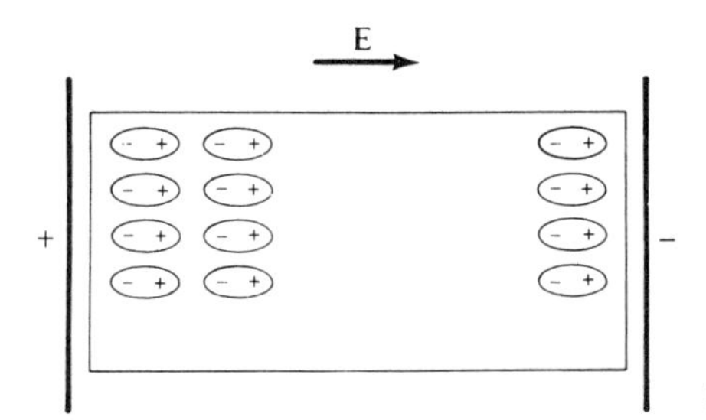

Figure 12-17

is not perfect or complete because it is counteracted by mutual interaction and thermal motion. Inside the substance, the aligned opposite charges compensate; we observe electric charge only at the surface. In addition, the electric field distorts the charge distribution of otherwise nonpolar molecules, thus inducing more electric surface charges. These polarized charges are fixed in position and disappear as soon as the electric field is turned off.

2.5(b). ELECTRIC DISPLACEMENT. To describe polarization, a new physical quantity is introduced, the *electric displacement* (symbolized **D**). It is related to the inducing electric field by

$$\mathbf{D} = \epsilon \mathbf{E} \tag{12.16}$$

where

D: electric displacement,
E: electric field strength,
ϵ: electric permittivity, introduced in Section 2.3(c).

The electric displacement is a vector quantity and is expressed in coulombs/m^2. It has the dimensions of charge density (charge per area). Its numerical value allows the calculation of how large a quantity of electricity is induced at the surface of the substance.

2.5(c). APPLICATIONS

Cleaning of Air. Tiny dust particles are easily polarized in an electric field. See Figure 12-18. This technique is used to clean air. The pol-

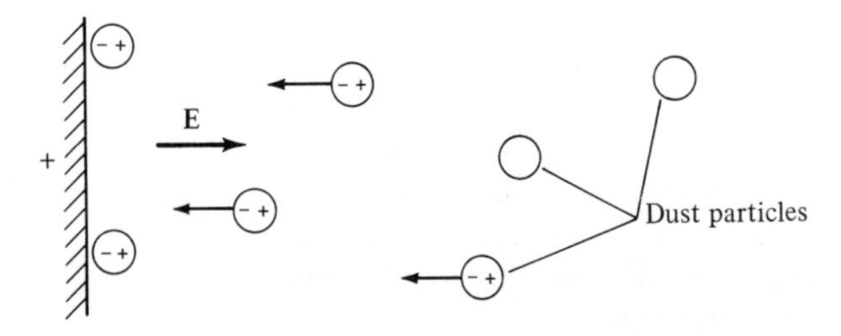

Figure 12-18

luted air is guided toward electrically charged surfaces. Within its electric field, a polarization is induced. A polarized particle will move under the resulting electrostatic force toward the surface and stick to it. Once in a while the electric field is turned off, and the accumulated dust collected.

Electrostriction. Some crystals like quartz and tourmaline show a polarization if mechanically deformed. This *piezo effect* can be reversed. When its surface is charged, the crystal will contract or expand, depending on the sign of the charge. This electrostriction is used to produce ultrasonic waves. A piece is cut from a quartz crystal as shown in Figure 12-19. The slab is placed between two parallel plates which are alter-

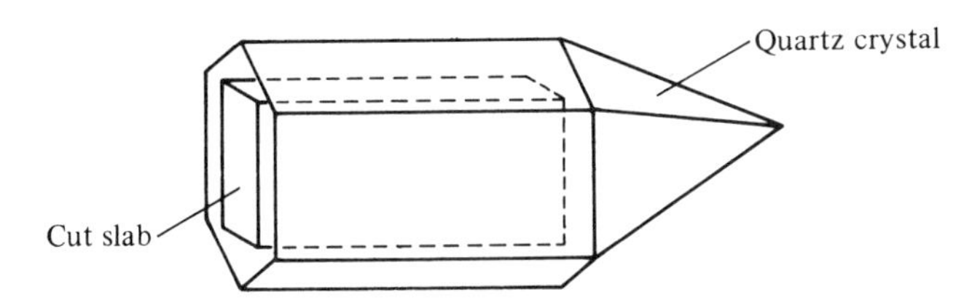

Figure 12-19 Piezocrystal cut from a naturally grown quartz crystal.

nately charged. Whenever the sign of the plates is switched, the piece of quartz changes its dimension perpendicular to the direction of the electric field (Figure 12-20).

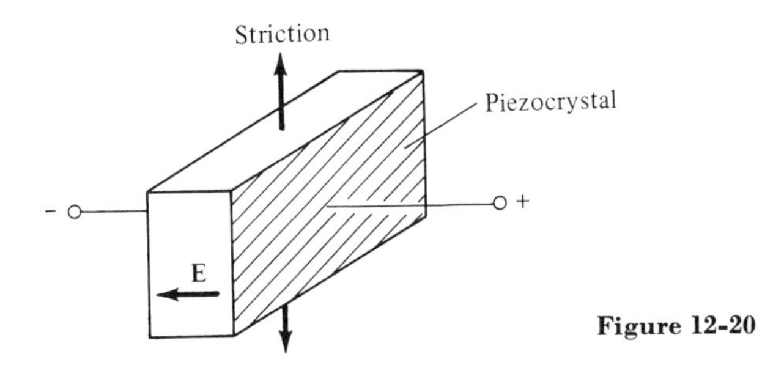

Figure 12-20

If placed in a medium (like water or air), this *piezocrystal* will emit sound waves having a frequency equal to the switching frequency. Ultrasonic waves (frequency > 20 000 Hz) which are widely used in medical diagnostics and therapy are produced in this fashion.

The piezo effect was discovered in 1880 by Jacques and Pierre Curie. (One of the brothers is the famous physicist who, together with his wife, Marie, discovered the radioactive element radium.)

3. MAGNETOSTATICS

3.1 Magnets

As far back as written tradition goes, man knew that certain metallic ores such as magnetite (Fe_3O_4) exert forces on iron. These substances are called *magnets*. A closer inspection shows that each magnet can attract or repel a piece of iron, depending on which end of the magnet points toward the piece. Traditionally the ends of a magnet are named *north magnetic* (N) and *south magnetic* (S) *poles*. Like poles repel each other; unlike attract. It would be most convenient if magnetism could be described in close analogy to electricity. Unfortunately this is only partly possible because we cannot separate north magnetic and south magnetic poles as we did for positive and negative electric charge.

A magnet broken between its poles becomes a pair of individual magnets, each exhibiting its own poles. See Figure 12-21.

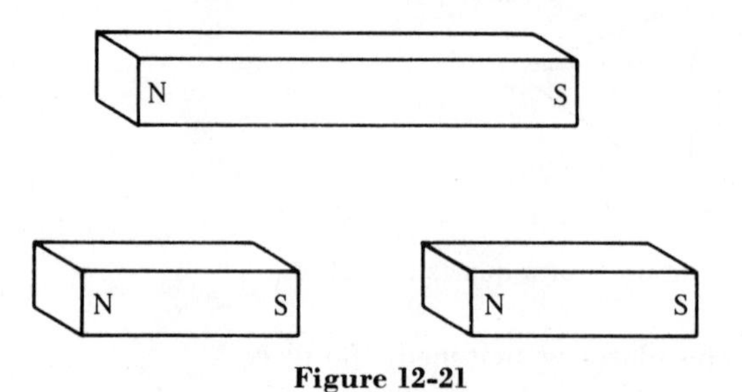

Figure 12-21

Magnetism is caused by the motion of an electric charge. Because moving electric charges produce a magnetic field, we are not restricted to substances which are magnetic by nature. Any electric coil with an electric current (moving electrons) passing through it acts as a magnet. Figure 12-22 shows such a solenoid with its north and south magnetic poles.

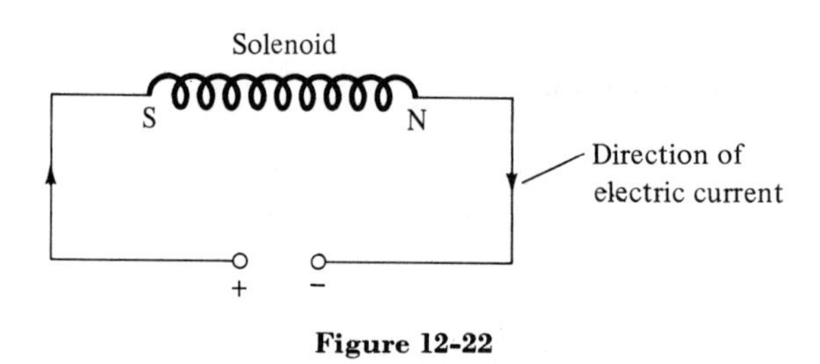

Figure 12-22

3.2 Magnetic field

A magnet is always surrounded by a *magnetic field*, as an electric charge is surrounded by its electric field. The presence of this field may be detected by the force acting on a suitable object entering this field. This test body could be another magnet. But more convenient is a moving, electrically charged particle such as an electron, which is itself a tiny magnet.

3.2(a). MAGNETIC INDUCTION. The magnetic field is characterized by the value of its *magnetic induction* at various positions in space. The magnetic induction (symbolized **B**) is measured in teslas* (symbolized

* Nicholas Tesla (1856–1943). Investigated very-high-frequency currents and fields.

T). Still in use in the cgs system is the gauss (symbolized G) as a measure for the magnetic induction.

Conversion:

$$1 \text{ tesla} = 10^4 \text{ gauss}$$

The magnetic induction is a vector quantity. The instrument used to measure it is called a *magnetometer*. Figure 12-23 summarizes some meth-

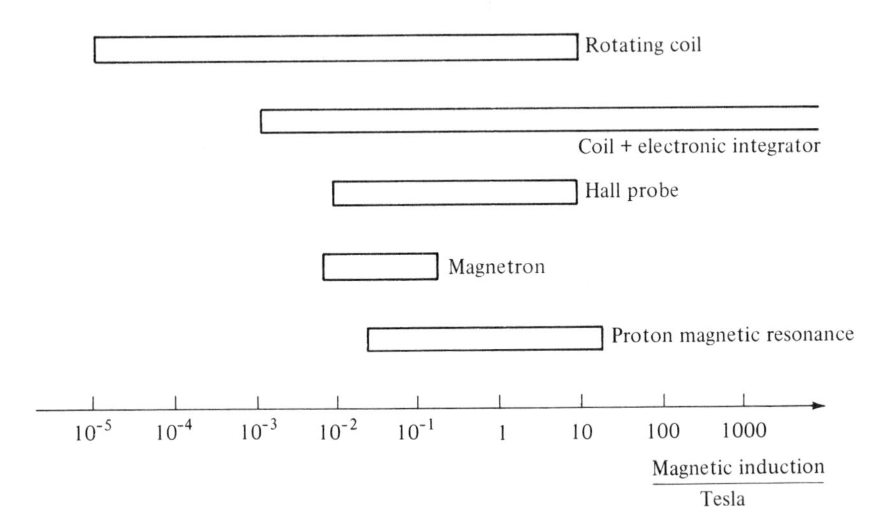

Figure 12-23 Range of some methods used to measure magnetic induction. The accuracy of the proton magnetic resonance method is one part in 10^4; for all other methods it is about 0.1%.

ods employed in magnetometers together with their ranges of measurement. Table 2 presents the numerical values of the magnetic induction of various fields.

Table 2 NUMERICAL VALUES OF THE MAGNETIC INDUCTION **B** FOR SOME NATURAL AND MAN-MADE MAGNETIC FIELDS

Field	Magnetic Induction in teslas
interstellar space	10^{-8} to 10^{-10}
earth's field around 50° N	2×10^{-5}
solenoid (40 turns, 5 amperes)	2×10^{-3}
average electromagnet	0.1
surface of sunspot	1 to 10
pulsed electromagnet	10^3
neutron star	10^8 to 10^9

Man has no organ for detecting magnetic fields. There are some experiments which indicate that plant sprouts have a preferred direction of growth if raised inside a magnetic field. Migrating birds seem to orient themselves with respect to the magnetic field of the earth.

3.2(b). LINES OF FORCE. The magnetic field may be visualized if we draw its lines of force (magnetic field lines). A magnetic field line is traced out by another magnet under the influence of the magnetic field. Each magnetic field line is closed in itself; it has neither source nor sink. The direction of the magnetic field lines of force is from south magnetic to north magnetic. The denser the field lines, the higher the magnetic induction at this position. Within a homogeneous magnetic field, the magnetic induction is constant.

EXAMPLE: MAGNETIZED BAR

As shown in Figure 12-24, the magnetic field lines are not parallel. The magnetic field is nonhomogeneous.

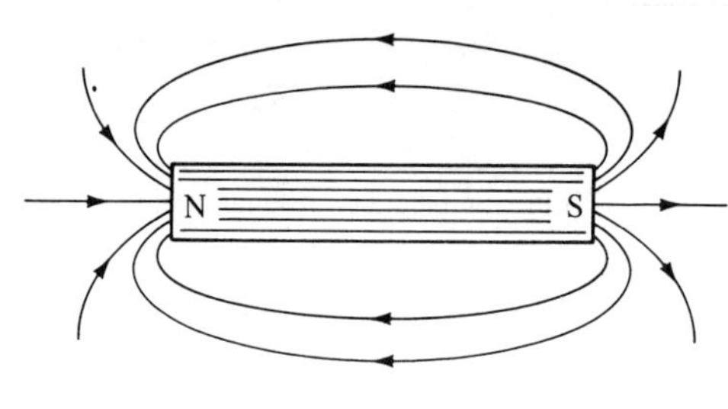

Figure 12-24

EXAMPLE: HORSESHOE MAGNET

As shown in Figure 12-25, between the pole faces, the lines of force are parallel. This indicates a homogeneous magnetic field.

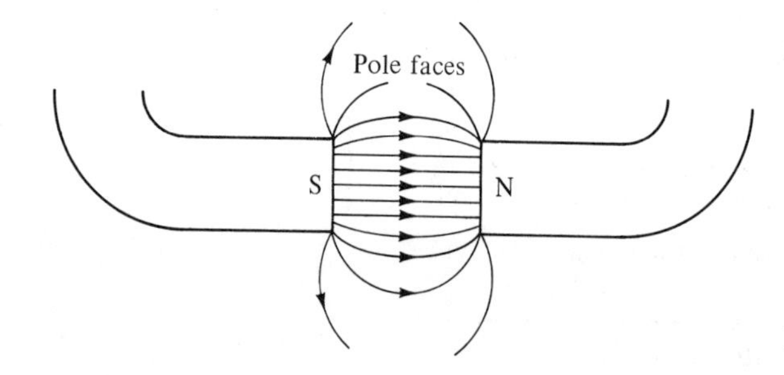

Figure 12-25

3.3 Matter in magnetic fields

The source of magnetism is moving electric charges. Orbiting electrons are part of each atom and molecule; they act as tiny elementary magnets. In most materials these electron orbits are oriented at random. The individual elementary magnets cancel each other, and no magnetism is observed. If matter is placed inside a magnetic field, the elementary magnets tend to align parallel to the magnetic field lines. The material becomes magnetized. Observe that this *magnetization* is similar to the *polarization* of matter placed inside an electric field.

3.3(a). MAGNETIZATION. If we place a substance into a magnetic field, the magnetic field lines will penetrate it and cause its magnetization. The following relation holds:

$$\mathbf{M} = \frac{\chi_m}{\mu_0(1 + \chi_m)} \mathbf{B} \tag{12.17}$$

where

M: magnetization of substance, a vector quantity,
$\mu_0 = 1.26 \times 10^{-6}$ m·kg·C^{-2}: a constant physical quantity,
χ_m: magnetic susceptibility, a numerical value characterizing the substance,
B: magnetic induction within the substance.

Some remarks are called for: The magnetization has been formulated according to Equation (12.17). The term on the right-hand side preceding **B** looks complicated. Conventionally, the magnetization is introduced in a different fashion which appears at first sight to be more simple. At a second glance, however, you would discover that additional quantities were introduced too. We can spare those additional quantities; there are already enough in magnetism. The physics is not affected.

We distinguish between *dia-*, *para-*, *ferro-*, and *ferrimagnetism*, depending on the value of χ_m.

Diamagnetism. The value for the magnetic susceptibility is negative and very small. Consequently, the magnetization is weak and opposes the magnetic induction.

Range:

$$10^{-9} < -\chi_m < 10^{-5}$$

Diamagnetism is common to all substances and independent of temperature.

Paramagnetism. The values for χ are positive and small. The magnetization points in the same direction as the magnetic induction.

Range:

$$10^{-6} < +\chi_m < 10^{-3}$$

Paramagnetism is restricted to certain substances, for example, oxygen, aluminum, magnesium, and platinum. It is a much stronger effect than diamagnetism and often masks it. Paramagnetism depends on temperature: its magnetic susceptibility is inversely proportional to the temperature of the substance.

Note: Equation (12.17) gets very simple if you consider that for dia- and paramagnetism $\chi_m \ll 1$ and thus $(1 + \chi_m) \approx 1$. Consequently

$$\mathbf{M} = \frac{\chi_m}{\mu_0} \mathbf{B} \tag{12.18}$$

Ferromagnetism. The values for χ_m are positive. They depend on the magnetic induction and whether or not the substance was magnetized before. The magnetization points in the same direction as the magnetic induction.

Range:

$$+\chi_m \quad \text{up to } 10^3 \text{ and more.}$$

Note: Again, Equation (12.17) becomes simple if you consider that for $\chi_m \gg 1$ follows $(1 + \chi_m) \approx \chi_m$. Consequently,

$$\mathbf{M} = \frac{\mathbf{B}}{\mu_0} \tag{12.19}$$

Ferromagnetism is restricted to a few substances: iron, cobalt, nickel, gadolinium, and some alloys. Ferromagnetic substances retain a *residual magnetization* if the magnetizing field is switched off. Permanent magnets—such as the horseshoe magnet made out of iron—must be magnetized, otherwise they will *not* display a residual magnetic field.

Ferromagnetism depends on temperature. For each ferromagnetic material a temperature exists—the *Curie temperature*—above which the substance is no longer ferromagnetic. Table 3 gives the Curie temperature (symbolized T_c) for ferromagnets.

Table 3 CURIE TEMPERATURE T_c OF FERROMAGNETIC SUBSTANCES

Substance	T_c in °K
cobalt	1400
iron	1043
nickel	631
ferrites	about 500
gadolinium	289

Ferrimagnetism. The first magnet known to man was a ferrimagnet, the lodestone (Fe_3O_4). Ferrimagnetic substances, usually called *ferrites,* exhibit a permanent magnetic field without prior magnetization. This is a consequence of the crystalline structure of ferrites. Their behavior inside a magnetic field is similar to that of a ferromagnetic substance having a magnetic susceptibility of the order one.

3.3(b). APPLICATIONS

Magnetic Memory. Ferromagnetic substances exhibit a property which we can compare with memory. Once magnetized, they retain a residual magnetization. They remember the fact that they were in a magnetic field. This remembrance can be erased by placing the same ferromagnetic substance into a magnetic field pointing in the opposite direction from the previous magnetizing field. It is a reliable memory because the only other way to make it forget is to raise its temperature above the Curie temperature.

A single ferro- or ferrimagnet can be used to store the information "1" (magnetized) or "0" (demagnetized). It will be demonstrated in a later chapter (Information Handling) that any information can be expressed in a binary code. This code uses only the two symbols 1 and 0. Consequently, a string of magnets can store information.

The core memory of a digital computer is made up of many individual magnets. For practical reasons (speed, reliability, low energy requirement), tiny ferrites are preferred to store information. Each of the ringlike ferrite cores is smaller than a millimeter, spatially ordered in a matrix, and pierced by a network of electric wires. A short electric pulse passing through a ferrite core will magnetize it, and thus the symbol "1" is stored. The core will be demagnetized if another short electric pulse of opposite sign passes through it. Then the symbol "0" is stored.

Core memories of computers are made up of thousands of ferrite cores, each one enlarging the memory capacity by one bit. Actually, the operation is more complicated. Figure 12-26 demonstrates the structure and functioning of a core memory.

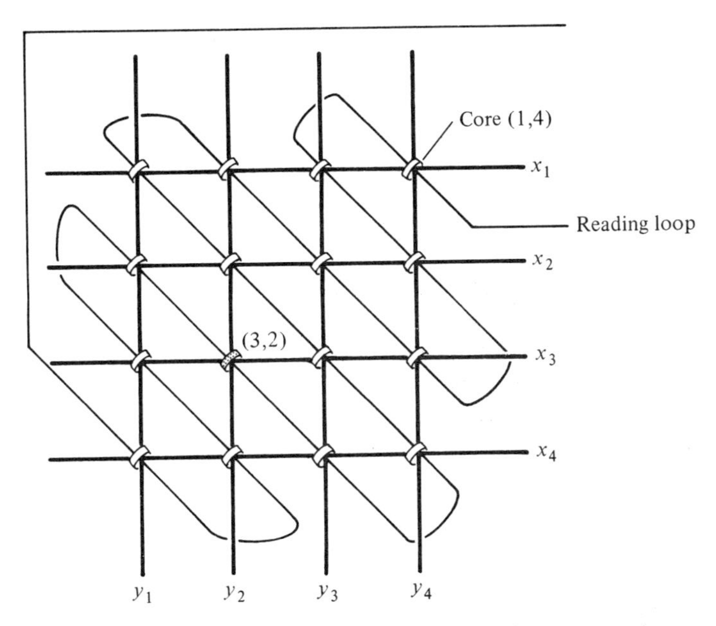

Figure 12-26 Part of a ferrite core memory; 16 elements are shown. Each element has x-y coordinates, determined by a horizontal (x) and a vertical (y) wire. A reading loop passes each core in succession. Information is stored in the shaded core (coordinates 3, 2) if an electric pulse passes along wires x_3 and y_2. The strength of each individual pulse is such that one alone cannot change the magnetization of a core. At core (3, 2) the strength of both pulses adds and causes magnetization. All other cores are unaffected. To read out information, pulses of opposite sign scan the x and y wires of the entire memory. Again, core (3, 2) becomes demagnetized only if two pulses read it at the same time.

Compass. Experiments show that the earth acts as a huge magnet. Its magnetic field is roughly similar to a magnetic dipole. At the magnetic poles the lines of force are vertical.

The magnetic poles of the earth change their position and polarity; the reasons for that are still unknown. The poles are not well-defined positions on the globe but are spread over an area. For the north magnetic pole, this area is about 75° north latitude and 100° west longitude.

The detecting element of a compass is a magnetized needle which aligns parallel to the direction of the lines of force produced by the earth's magnetic field.

Declination. The site of the north magnetic pole is separated from the geographic north pole. See Figure 12-27. The declination of the compass

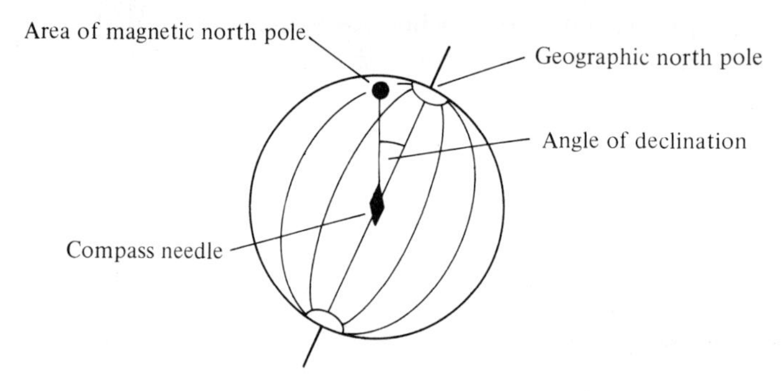

Figure 12-27

is the angle between geographic north and the direction in which the compass needle points. Its value differs at various positions of the globe. An unpredictable secular change of the declination is observed; its explanation still awaits discovery. In the neighborhood of the magnetic poles, compass readings are unreliable because the pole is not a definite determinable point. In addition, the needle will point toward the vertical (*inclination* of the compass) since it aligns parallel to the magnetic lines of force.

SUMMARY

The phenomena described in this chapter are independent of time; thus we use the terms *electrostatics* and *magnetostatics*. Electric charge is a property intrinsic to matter. Matter is electrically charged and carries a quantity of electricity Q. There are two complementary types of charge, conventionally distinguished as *positive* and *negative*. Some matter may appear electrically *neutral*, i.e., neither negative nor positive, only because it contains the same quantity of negative and positive electricity.

The quantity of electricity is measured in coulombs (C). In a closed system, charge is conserved.

Coulomb's law describes the force $\mathbf{F}_c$ acting between two charges Q_1 and Q_2, separated by a distance r. The magnitude F_c of $\mathbf{F}_c$ is

$$F_c = \frac{1}{4\pi\epsilon}\frac{Q_1 Q_2}{r^2} \qquad (12.1)$$

Each electric charge is surrounded by an electric field which may be visualized by its electric field lines.

Magnetism is created by moving electric charges. Its formal treatment is often analogous to that for electricity, but there are basic differences: Positive and negative charges can be separated spatially, the *north magnetic* and *south magnetic poles* of a magnet cannot. In a magnetic field, the magnetic field lines are closed in themselves while the electric field lines originate at the positive charge (source) and end at the negative charge (sink).

For historical and practical reasons, the analogous quantities of many electrostatic quantities either were not developed for magnetostatics or are not used anymore. The *potential difference* U between two positions within the same electric field is one of those.

U (measured in volts) and the charge Q creating the electric field are related by

$$Q = CU \qquad (12.11)$$

The *electric capacitance* C has no analogy in magnetism.

Most substances brought into a field become polarized. Even if it was neutral before, it now displays magnetic or electric or both properties. Within an electric field we observe induced surface charges, measured as *electric displacement*, which weaken the original inducing electric field.

The *magnetization* $\mathbf{M}$ of matter brought into a magnetic field may weaken or strengthen the field. The change depends on the sign and numerical value of the *magnetic susceptibility*, χ_m. It is characteristic for each substance and serves to distinguish between various types of magnetization.

Diamagnetism: $10^{-9} < -\chi_m < 10^{-5}$
Paramagnetism: $10^{-6} < \chi_m < 10^{-3}$
Ferro- and *ferrimagnetism:* $10^{-1} < +\chi_m < 10^{3}$ (variable for the same substance)

Only ferrimagnets such as the lodestone exhibit a natural magnetization. Ferromagnets such as iron or nickel exhibit a *residual magnetization* if removed from the magnetizing field.

Table 4 presents new quantities introduced in this chapter.

Table 4 NEW QUANTITIES

Quantity	Symbol	Units
quantity of charge	Q	coulomb (C)
electric permittivity	ϵ	$C^2 \cdot m^{-2} \cdot N^{-1}$
dielectric constant	ϵ_r	pure number
electric field strength	$\mathbf{E}$	$N \cdot C^{-1}$
potential difference	U	volt (V)
electric capacitance	C	farad (F)
electric displacement	$\mathbf{D}$	$C \cdot m^{-2}$
magnetic induction	$\mathbf{B}$	tesla (T)
magnetization	$\mathbf{M}$	$C \cdot m^{-1} \cdot s^{-1}$
magnetic susceptibility	χ_m	pure number

PROBLEMS

1. The quantity of electricity Q stored in a capacitor can be measured directly by a special balance, as shown in Figure 12-28. Derive a relation for Q.

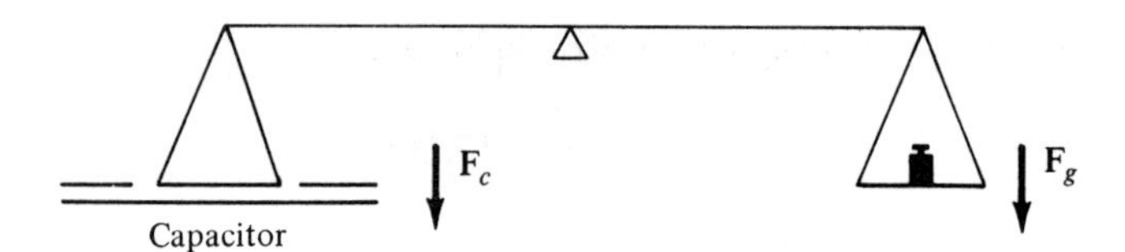

Figure 12-28

2. Pair production occurs only in the neighborhood of an atomic nucleus. The reason is a basic conservation law. Which law is that and is there an experiment to prove it?
3. Compare the magnitudes of the Coulomb force and the gravitational force between electron-positron and proton-antiproton pairs.
4. In Section 2.5(c), a method for the cleaning of air was described. It works for small dust particles but why not for water droplets?
5. Three capacitors are connected as shown in Figure 12-29. Calculate the total capacitance C of the arrangement. To check the derived formula, note that $C = 6.67 \times 10^{-13}$ F if $C_1 = C_2 = C_3 = 1$ picofarad.
6. Sketch the electric field lines if a spherical dielectric is brought into a parallel-plate capacitor.
7. If a dielectric is placed in a capacitor, its capacitance changes. Why and by how much?

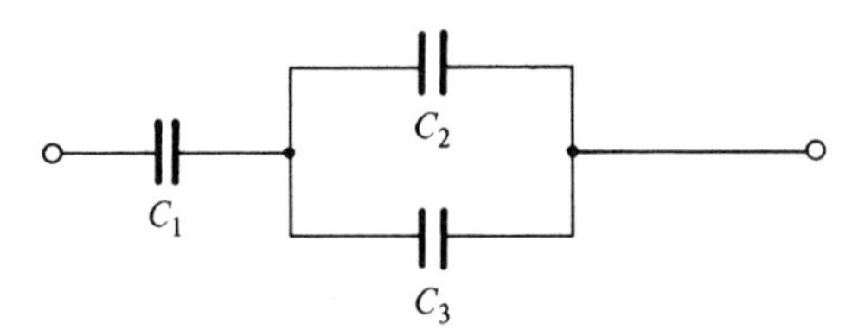

Figure 12-29

8. The Coulomb force between charges can partly or totally be screened. Why is this so and which material will be the most effective?

9. A spherical capacitor is an arrangement of two concentric spheres. Its capacitance is:

$$C = \epsilon \frac{R_1 R_2}{R_2 - R_1}$$

where R_2 and R_1 are the radii of the spheres. Derive a simplified formula for C valid for a very thin spherical capacitor.

10. Calculate the charge density at the plates of a parallel-plate capacitor of 5 cm^2 area. The dielectric between the plates is transformer oil; the potential difference is 1000 V. Compare this with the resulting charge densities if the oil is replaced by water.

11. The permittivity of water depends on temperature. Calculate the sensitivity of a thermometer working on this principle. Use interpolated values of Table 1.

12. Take the first five values for permittivity from Table 1, page 300, and calculate the appropriate dielectric constants.

13. A magnetic needle is placed in a magnetic field. Its axis forms an angle α with the magnetic lines of force. For which angle is the mutual interaction a minimum? (See Figure 12-30.)

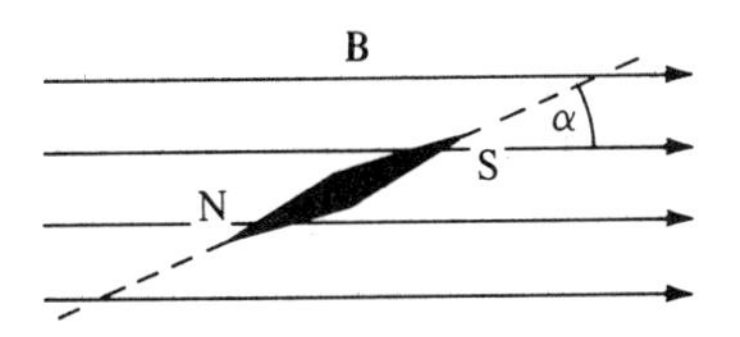

Figure 12-30

14. The density of magnetic lines of force is a measure of the magnetic induction. How will this density change if a paramagnetic or a diamagnetic substance is brought into the magnetic field? Sketch the situation. Assume that the original magnetic field was homogeneous.

15. The atomic nuclei of hydrogen molecules carry one elementary charge each. Calculate the repelling Coulomb force and the attractive gravitational force. The distance between nuclei is 8×10^{-13} cm.

16. Three point charges are arranged in vacuum to form an equilateral triangle. The individual charges are 3 C, 7 C and -2 C. The mutual distance is 10 m. What is the magnitude of the Coulomb force at the negative charge? What is the electric field strength at the position of the negative charge? Calculate the same quantities for charges in water of 0°C.

17. Positronium is formed by a positron and an electron orbiting their common center of mass. If their mutual distance is 10^{-8} cm, what is their speed?

18. A 15 eV electron beam enters a parallel-plate capacitor through a hole perpendicular to the plates. The plates are 3 cm apart. Calculate the potential difference necessary to completely stop the electrons. What is the electric field strength?

19. A biological membrane (thickness 2×10^{-6} cm, diameter 7.5×10^{-4} cm) exhibits a rest potential of 80 mV. Calculate the quantity of electricity causing this membrane potential. How many ions are involved?

FURTHER READING

M. F. BARNOTHY, ed., *Biological Effects of Magnetic Fields.* Plenum Press, London, 1969.

C. GERMAIN, "Biographical Review of the Methods of Measuring Magnetic Fields," *Nucl. Instr. Meth.*, **21** (1963) 17–46.

A. S. PRESMAN, *Electromagnetic Fields and Life.* Plenum Press, New York, 1970.

Magnetic Poles and the Compass. U. S. Department of Commerce, Washington. Serial 726.

M. H. SHAMOS, *Great Experiments in Physics.* Holt, Rinehart and Winston, New York, 1959.

chapter 13

ELECTRODYNAMICS AND CIRCUITRY

1. INTRODUCTION

In the previous chapter, time was not considered. But look around; everything under the sun is time-dependent. This means that all quantities describing aspects of nature change their values as time progresses.

Electro- and magnetostatics already display a host of phenomena. You can imagine that by introducing a new variable, time, into it, this multitude will rapidly increase. It is difficult to strip the vast field of electrodynamics down to a skeleton relevant to the student of the life sciences. You will find in this chapter only the very basics. However, working through this chapter you will not only learn about currents, resistors, and inductors, but also gain some insight into one of the most important phenomenon of animated matter: the action potential of a cell. This electric discharge across a polarized membrane is known as a *nerve pulse*, and it allows communication between the components of a multicellular organism. There are other means of information exchange within a body, but this one is fast and reliable in the gathering, storage, and interpretation of information. The propagation of these pulses along specialized cells can by no means be compared with currents flowing inside an electric cable. It is more sophisticated than that.

The structure of this chapter is dictated by convenience. First we shall look into the phenomena caused by a constant flow of electric charges, called a *direct current* (dc). Time does not enter explicitly. Oscillating charges introduce an *alternating current* (ac); here the time dependence is very important. The last section deals with a few aspects of electric circuitry, presented mainly as a source of reference.

2. DC PHENOMENA

A constant flow of electrically charged particles is called *direct current*, abbreviated dc. Which particles sustain this current and how it may be understood at the submicroscopic level does not concern us in general. A macroscopic description of phenomena connected with direct currents is sufficient. Figure 13-1 shows what you will encounter in this subsection.

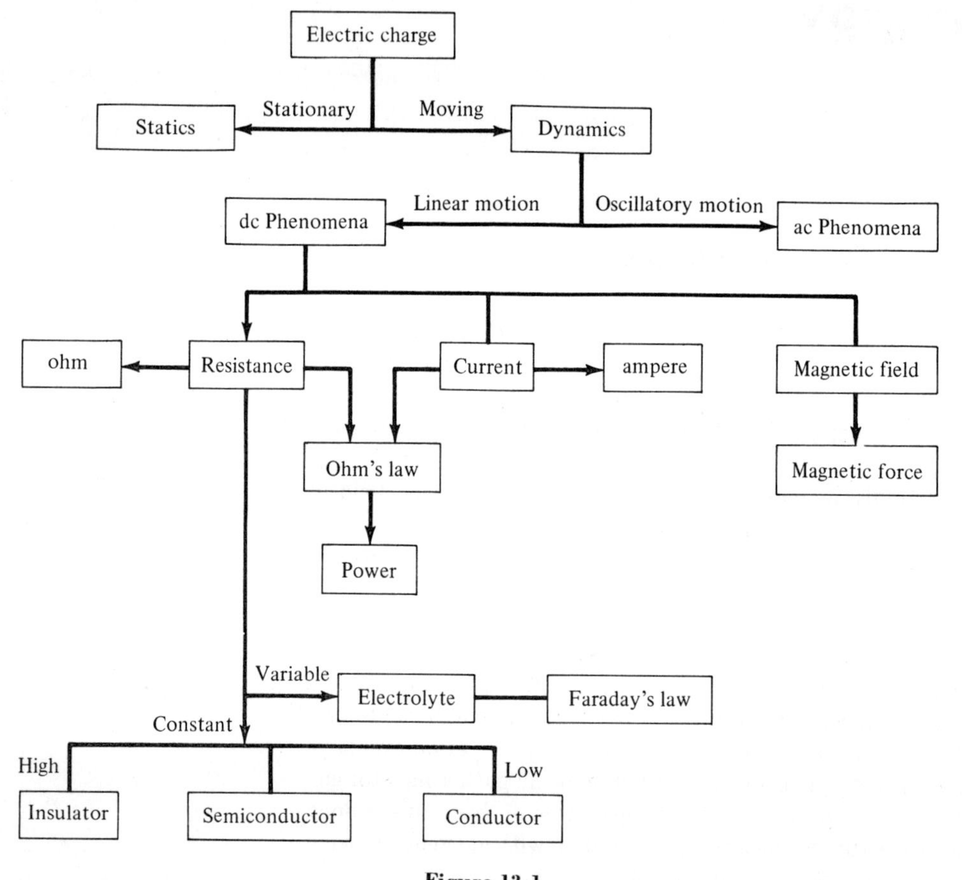

Figure 13-1

2.1 Basics

2.1(a). CURRENT. Moving electric particles constitute an electric current (symbolized I). The electric current is measured in amperes* (abbreviated A). The ampere is one of the four basic units of the International System of Units or mksA system.

* André Marie Ampère (1775–1836), mathematician and physicist. Discovered the interaction between electric currents. Founder of electrodynamics.

$$1 \text{ ampere} = \frac{1 \text{ coulomb}}{1 \text{ second}} \tag{13.1}$$

This means that, if the amount of charge of one coulomb moves along a given path within one second, an electric current of one ampere flows. In general a current is caused by a potential difference (see Chapter 12, Section 2.4). The direction of a current in electric circuits is, by definition, from plus to minus.

Instruments used to measure the electric current are called ammeters. Very sensitive ammeters are named *galvanometers.* Table 1 lists some currents encountered in nature and technology.

Table 1 SOME TYPICAL ELECTRIC CURRENTS

(Sometimes the current lasts much less than one second; this is indicated by the remark *peak*.)

	Current in amperes
one elementary charge per second	1.6×10^{-19}
across a synapse	10^{-11} (peak)
giant squid axon	10^{-5} (peak)
electron accelerator	$10^{-3} - 10^{-6}$
x-ray tube	10^{-3}
flashlight	0.2
electric light bulb	0.1 to 1
oven	10
electric fish	up to 100 (peak)
electric motor of street car	200
welding	500
lightning	10^4 (peak)
electric furnace	10^5
plasma generator	10^6

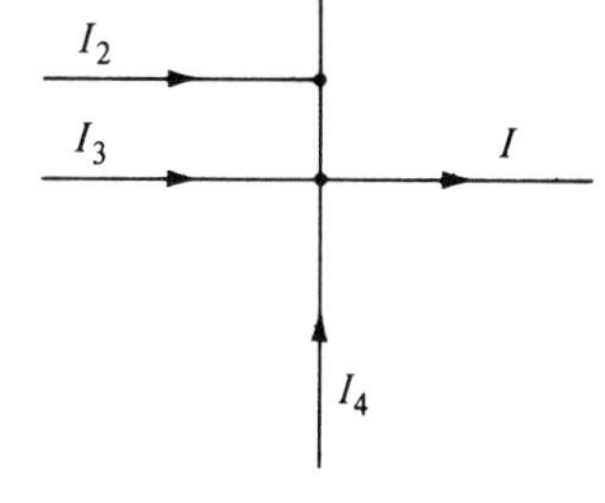

Figure 13-2

Network of Currents. If a current branches out or if current branches converge, the total current is conserved. In circuitry an overlayed dot marks a branching point. In Figure 13-2, $I = I_1 + I_2 + I_3 + I_4$. This means

$$I = \sum_{i=1}^{n} I_i \tag{13.2}$$

This is a consequence of charge conservation (Chapter 12, Section 2.2).

2.1(b). RESISTANCE. An electric current flows from its source (+) to its sink (−) under the influence of a potential difference. The material

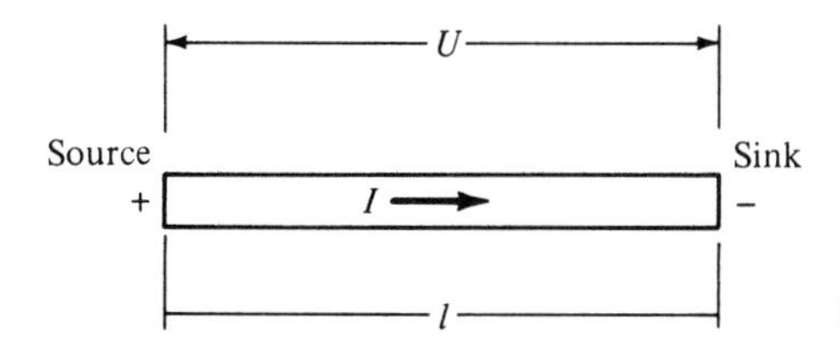

Figure 13-3

connecting source and sink offers a resistance to this flow. Experiments show that the current is proportional to the potential difference and to the cross section of the connecting material. It decreases with the length of the connection. Hence we may write

$$I = \gamma \frac{A}{l} U \tag{13.3}$$

where

I: electric current, measured in amperes,
A: cross section of connecting material,
l: length of connection,
U: potential difference, measured in volts,
γ: proportionality factor, named *conductivity*.

Equation (13.3) may be reformulated as

$$I = \frac{1}{R} U \tag{13.4}$$

This is known as *Ohm's law*.*

The quantity

$$R = \frac{1}{\gamma} \frac{l}{A} \tag{13.5}$$

is called *electrical resistance* and is measured in ohms (symbolized Ω).

$$1 \text{ ohm} = \frac{1 \text{ volt}}{1 \text{ ampere}} \tag{13.6}$$

Thus if a current of one ampere is observed in an electric circuit having a constant potential difference of one volt, we conclude that the resistance between source and sink is one ohm.

Instruments used to measure the electrical resistance are called *ohmmeters*.

The electrical resistance of a material depends on its dimensions [see Equation (13.5)] and on its conductivity, a characteristic numerical value. It is usual to list not the conductivity but its inverse, the *resistivity* (symbolized ρ). The two quantities are related by

$$\rho = \frac{1}{\gamma} \tag{13.7}$$

Resistivity is measured in ohm·meter. Table 2 shows some typical values for ρ.

According to its resistivity, each material belongs to one of three categories: *conductor* (low resistivity), *insulator* (high resistivity), and *semiconductor*. It is worthwhile to note that substances having a low electrical resistivity are also good heat conductors.

* Georg Simon Ohm (1787–1854), physicist. Discovered the linear relationship between electric current and potential difference. He also contributed to acoustics and wave optics.

Table 2 RESISTIVITY ρ OF SOME MATERIALS

Material	ρ in $\Omega\cdot$m
Conductor:	
silver	1.6×10^{-8}
copper	1.7×10^{-8}
aluminum	2.8×10^{-8}
tungsten	5.5×10^{-8}
constantan	4.4×10^{-7}
carbon	3.6×10^{-5}
Semiconductor:	
squid giant axon	0.3
germanium	45
silicon	6.2×10^{4}
Insulator:	
maple wood	3×10^{8}
porcelain	3×10^{12}
teflon	10^{13}
glass	10^{10}–10^{14}
mica	10^{11}–10^{15}
hard rubber	2×10^{16}
paraffin	5×10^{16}
quartz	7.5×10^{17}

Temperature Dependence. The resistance of a conductor increases with its temperature. This phenomenon is used to build sensitive and rapidly responding thermometers. If the temperature of a conductor drops below a characteristic threshold value (in general, below 10°K), its resistivity vanishes suddenly. This most interesting effect is called *superconductivity* and was discovered in 1911 by K. Onnes.* This phe-

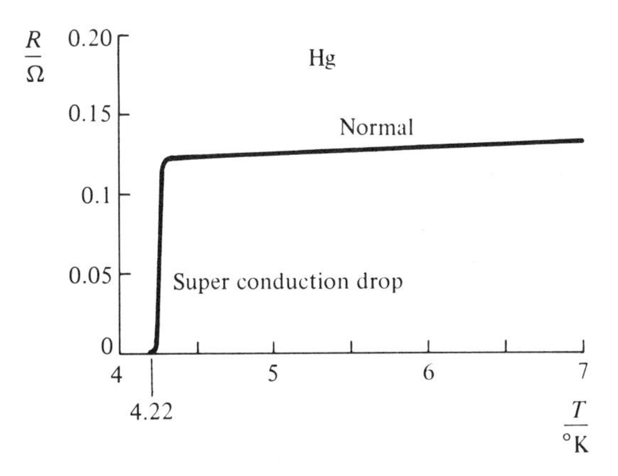

Figure 13-4 The resistance R of a mercury column in the neighborhood of its threshold temperature.

* Heike Kamerlingh Onnes (1853–1926), physicist. Nobel laureate 1913. The first to liquify helium.

nomenon has intrigued physicists and engineers ever since. It took more than forty years to develop a quantitative explanation. The quantum-mechanical BCS theory won the Nobel prize for John Bardeen, Leon N. Cooper, and John R. Schrieffer.

The applications of superconductivity are still limited to the laboratory because it is very expensive to keep materials at the required low temperatures. But there is hope for the future: organic solids like (TTF) (TCNQ) seem to superconduct at 58°K, and there are theoretical considerations that one day some synthetic organic macromolecules will show superconductivity at room temperature!

Network of Resistors:

A *resistor* is an element of electric circuitry. It is symbolized by —▭— or —⋀⋀⋀—. Resistors may be connected in series or parallel.

Series connection is shown in Figure 13-5.

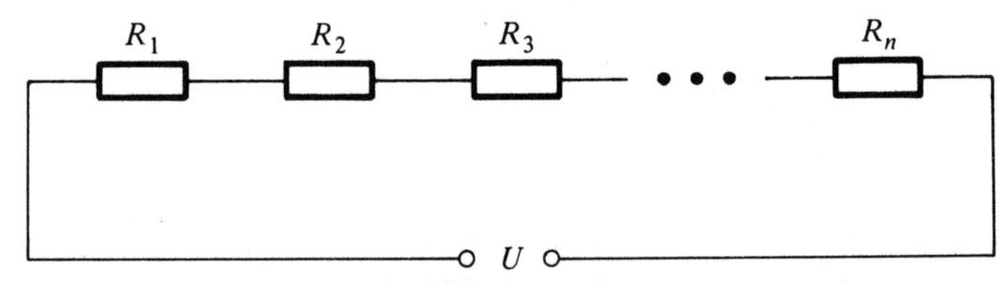

Figure 13-5

$$R = \sum_{i=1}^{n} R_i = R_1 + R_2 + R_3 + \cdots + R_n \tag{13.8}$$

where

R: total resistance of circuit,
R_i: resistance of the ith resistor.

Parallel connection is shown in Figure 13-6.

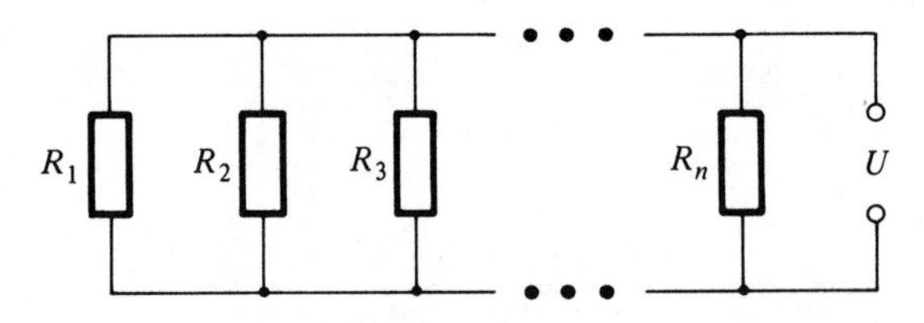

Figure 13-6

$$\frac{1}{R} = \sum_{i=1}^{n} \frac{1}{R_i} = \frac{1}{R_1} + \frac{1}{R_2} + \frac{1}{R_3} + \cdots + \frac{1}{R_n} \tag{13.9}$$

2.1(c). Electric power. A current flowing through matter dissipates energy; consequently *electric power* is required to maintain the current. The power dissipated is

$$P = UI \quad (13.10)$$

where

P: electric power dissipated,
U: potential difference,
I: electric current.

The electric power is measured in watts; see Chapter 4, Section 5.1.

As long as Ohm's law is valid, we may rewrite Equation (13.10) in the alternative form

$$P = RI^2 \quad (13.11)$$

where R is the electrical resistance expressed in ohms.

Heat Produced by Dissipated Electric Power. An electric current will heat a resistor. The dissipated power appears as heat:

$$Q = 0.239Pt \quad (13.12)$$

where

Q: quantity of heat produced, measured in calories,
P: electric power,
t: time, expressed in seconds.

EXAMPLE:

An average electric heater has a power consumption of $2\ \text{kW} = 2 \times 10^3\ \text{W}$. The amount dissipated per hour is

$$Q = 0.239 \times 2 \times 10^3 \times 3.6 \times 10^3\ \text{W}\cdot\text{s}$$
$$= 1.76 \times 10^6\ \text{cal} = 1760\ \text{kcal}$$

This is equivalent to burning 250 g of coal.

2.2 Currents and magnetic fields

2.2(a). Magnetic field produced by a current. A moving electric charge has a magnetic field around it. Consequently an electric current is enclosed by magnetic field lines. See Figure 13-7. The direction of the

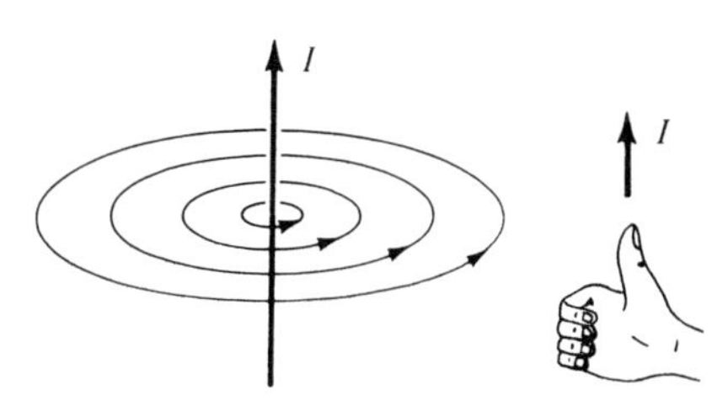

Figure 13-7

magnetic induction **B** is indicated by the right-hand rule: If the erect thumb of your right hand points in the direction of the current, your closed fingers indicate the direction of **B**. The magnitude of **B** is proportional to the current I and decreases linearly with the distance l to the current.

$$B \sim \frac{I}{l}$$

2.2(b). Magnetic force. A charged particle at rest inside a magnetic field experiences no force. Since it carries an intrinsic magnetic field, a current (moving charges) will interact with another magnetic field. See Figure 13-8. The resulting force, called *magnetic force*, acting on the current in a homogeneous magnetic field is

$$F_m = IlB \sin \theta \tag{13.13}$$

where

$F_m = |\mathbf{F}_m|$: magnitude of magnetic force,
l: length of path of the current inside the magnetic field,
B: magnitude of magnetic induction,
θ: angle between direction of current and magnetic induction.

Note: If I is parallel to $\mathbf{B}$, that is, $\theta = 0°$ or $180°$, then the magnetic force is zero.

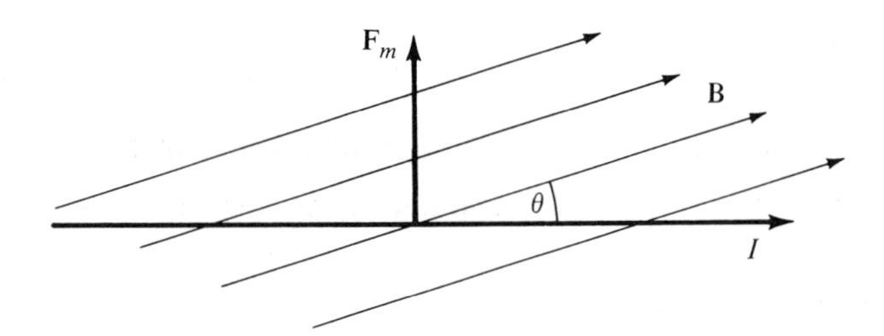

Figure 13-8

The direction of the magnetic force is perpendicular to $\mathbf{B}$ and I. It is indicated by the three-finger rule of the left hand. See Figure 13-9.

Application: The galvanometer is a sensitive current-measuring device arranged as shown in Figure 13-10. A very thin wire is suspended between the pole faces of a magnet. The position of the wire can be

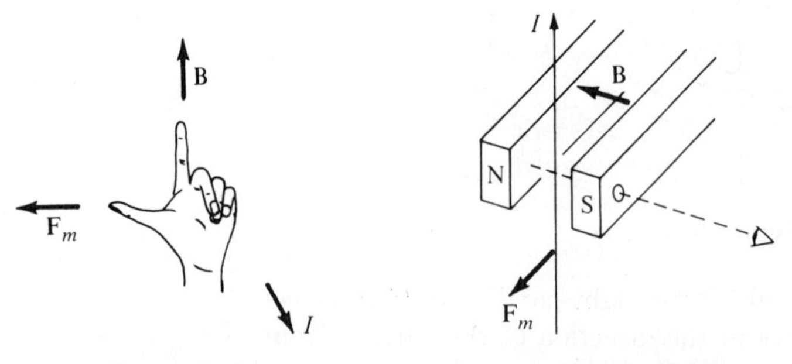

Figure 13-9 Figure 13-10

observed through an opening. If a current flows through the wire, the occurring magnetic force will bend it. The resulting shift in position of the wire is proportional to the passing current and is observed on a micrometer scale.

Northern Lights. The sun emits a stream of electrically charged particles, the solar wind. Its electrons and protons move with a speed of about 10^5 m/s. In the neighborhood of the earth, they encounter the earth's magnetic field and consequently experience a magnetic force. Their direction of flight will change as described by the three-finger rule of the left hand. In the plane of the earth's magnetic equator, the charges move perpendicularly to the magnetic lines of force and hence they will follow a circular path. (See Figure 13-11.) Because the field is non-

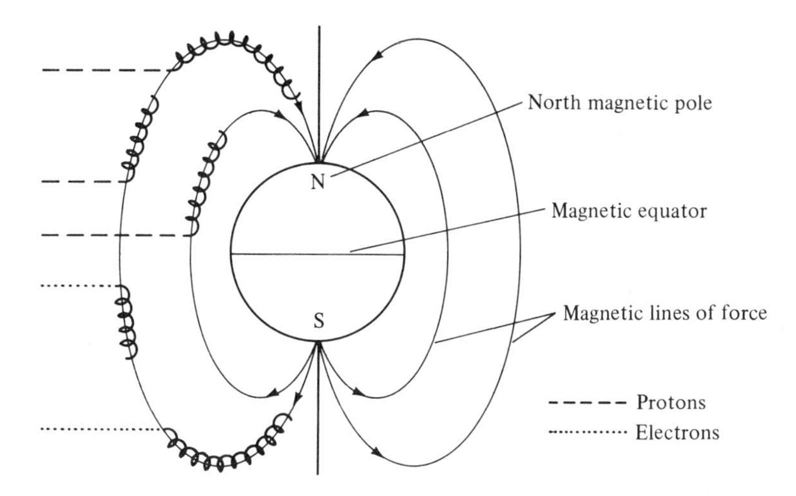

Figure 13-11

homogeneous, the actual orbit is a spiral. The charged particles wrap themselves around the magnetic lines of force, become trapped, and contribute to the so-called *Van Allen belt.*

Electrons or protons encountering the earth's magnetic field high above or below the plane of the magnetic equator also spiral, but may reach the earth's atmosphere. The interaction of these charged particles with the earth's high atmosphere (at about 100 km altitude) causes the northern lights (or southern lights).

During periods of strong solar activity, many charged particles reach the earth at irregular intervals. The magnetic fields carried by these currents interact with the earth's magnetic field and give rise to magnetic storms.

2.3 Electrolysis

Many substances (mainly acids, bases, and salts) dissolve into ions if brought into solution. This process is called *electrolytic dissociation,* and the solution is an *electrolyte*. Ions are electrically charged particles and will constitute an electric current if moving. Placed between a potential difference, the ions will propagate parallel to the electric field lines.

In Figure 13-12, the potential difference U is set between two plates (electrodes) submerged into the electrolyte. The negative electrode is

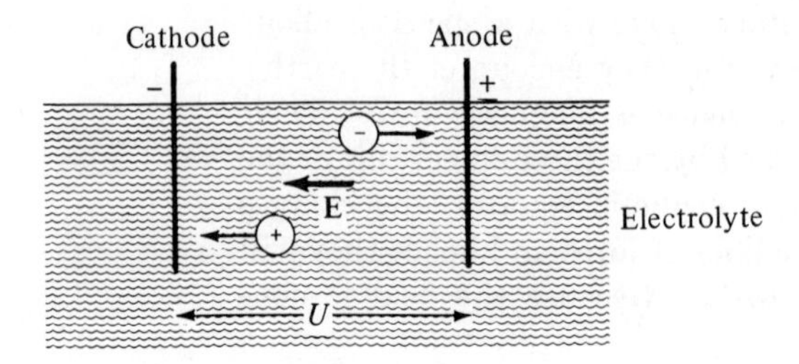

Figure 13-12

called the *cathode*, and the positive electrode is the *anode*. Negative ions travel toward the anode; positive ions travel toward the cathode.

2.3(a). Resistivity. Due to the strong mutual interaction between dissolved ions and the solvent, the resistivity of an electrolyte (ionic solution) is high compared to metals. The resistivity is inversely proportional to the concentration of the ions. This is obvious if you realize that many inorganic chemical compounds dissolve completely, with each molecule being split into a negative and a positive ion. The higher the concentration, the larger the number of current-carrying charges. The values for the resistivity of electrolytes are between those of metallic conductors (about 10^{-6} to $10^{-8}\,\Omega\cdot\text{m}$) and isolators (about 10^{16} to $10^{12}\,\Omega\cdot\text{m}$). For example, the resistivity of an aqueous NaCl solution having a concentration of 10^{-4} mol/cm^3 is about $1\,\Omega\cdot\text{m}$.

2.3(b). Faraday's law. The ions of an electrolyte move toward their respective electrodes and form deposits there. This process is called *electrolysis*. The mass of the substance deposited at each electrode is calculated according to Faraday's law:

$$M = zIt \tag{13.14}$$

where

M: amount of substance deposited, measured in kg,
z: electrochemical equivalent weight, expressed in kg/C,
I: electric current passing the electrolyte,
t: time.

The electrochemical equivalent weight z is the mass of the ion divided by its charge. Let us calculate z for the Mg^{++} ion.

The mass of Mg^{++} ion is equal to the atomic weight of Mg times 1/12 of the mass of carbon-12:

$$= 24.3 \times 1.67 \times 10^{-27}\ \text{kg}$$
$$= 4.07 \times 10^{-26}\ \text{kg}$$

Charge of Mg^{++} ion $= 2 \times$ elementary charge $= 3.2 \times 10^{-19}$ C. Hence

$$z = \frac{4.07 \times 10^{-26}}{3.2 \times 10^{-19}}\ \text{kg}\cdot\text{C}^{-1} = 1.27 \times 10^{-7}\ \frac{\text{kg}}{\text{C}}$$

Observe that, since $It = Q$, Faraday's law can be utilized to measure the amount of charge crossing the electrolyte. Thus the mass deposited allows an absolute current measurement.

2.4 Application

2.4(a). ELECTROMAGNETIC PUMPING OF BLOOD. Heart-lung machines, artificial kidneys, and other apparatus are indispensable tools in modern medicine. To drive the blood through its circuits, pumps are needed. Conventional pumps have moving parts and consequently are difficult to seal. There is also the danger of damaging the blood particles while they are passing through. Using the magnetic force solves the problem; the blood is pumped electromagnetically. See Figure 13-13. A magnetic

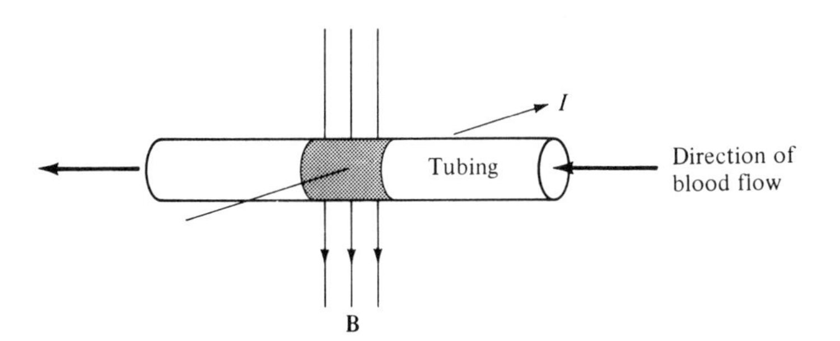

Figure 13-13

field passes through the shaded part of the tube. At right angles to the field, a potential difference is applied. The ions normally present in blood are sufficient to sustain an electric current. The resulting magnetic force (see Section 2.2) drives the blood to the left. In this fashion a pump with no moving parts and no leakage problems is achieved.

3. AC PHENOMENA

In physics most phenomena have their symmetric counterparts. Thus if the observed effect is reversed, a symmetric phenomenon appears. Take, for example, the piezo effect described in the preceding chapter, Section 2.5. If a suitably cut quartz crystal is compressed, a potential difference is observed at its surface. The symmetric effect: If a potential difference is applied from outside between two opposing faces of the crystal, a contraction or expansion is observed.

Another example of symmetric phenomena is the *thermoelectric* effect. Two different conductors (1 and 2) are soldered together as shown in Figure 13-14. If the two connecting areas are at different temperatures T_1 and T_2, then a thermoelectric current I_{th} will flow, causing a potential

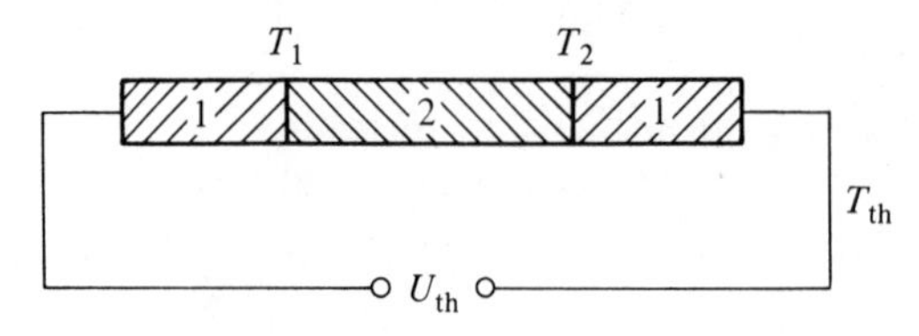

Figure 13-14

difference U_{th}. This arrangement is used as a thermometer. The symmetric effect was discovered in 1834 by Peltier: An electric current passing through the described arrangement causes a temperature difference between both connecting areas. With modern semiconducting materials, the Peltier effect* is applied to build compact coolers having no moving parts.

This general symmetric behavior of physical phenomena is well developed in electromagnetism. Figure 13-15 shows the structure of this section.

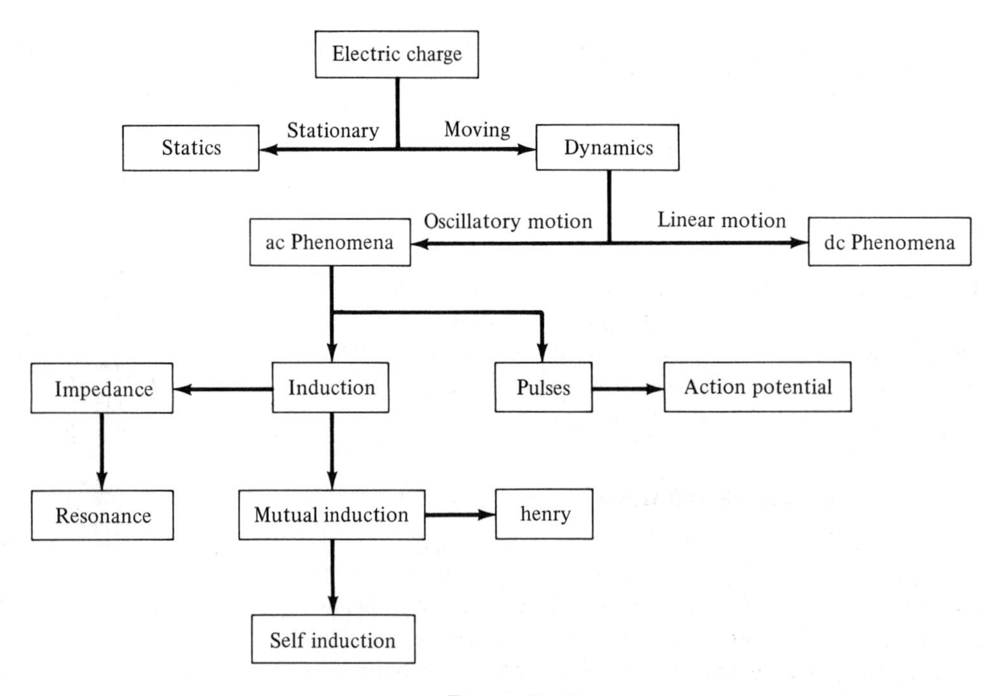

Figure 13-15

* Jean Charles Athanase Peltier (1785–1845), watchmaker and scholar.

Moving electric charges produce a magnetic field. Thus a current-carrying wire has a magnetic field around it. If the current disappears, the field will vanish.

The symmetric effect is known as *induction*. Electric charges under the influence of a *changing* magnetic field will move. We can observe an electric current and consequently a potential difference along a conducting wire placed into a *varying* magnetic field. There are two ways to achieve this:

1. The distance between the wire and a constant magnetic field varies.
2. The position of the wire is fixed, and the magnetic field strength varies with time.

The potential difference induced across a conductor moving through a magnetic field is

$$U = Blv \tag{13.15}$$

where

U: induced potential difference, measured in volts,
$B = |\mathbf{B}|$: magnitude of magnetic induction at the position of the conductor, expressed in teslas,
l: length of the conductor perpendicular to $\mathbf{B}$,
v: speed of the conductor moving perpendicularly to the magnetic lines of force.

The direction of the induced current is indicated by the three-finger rule of the right hand. See Figure 13-16. If the conductor moves back

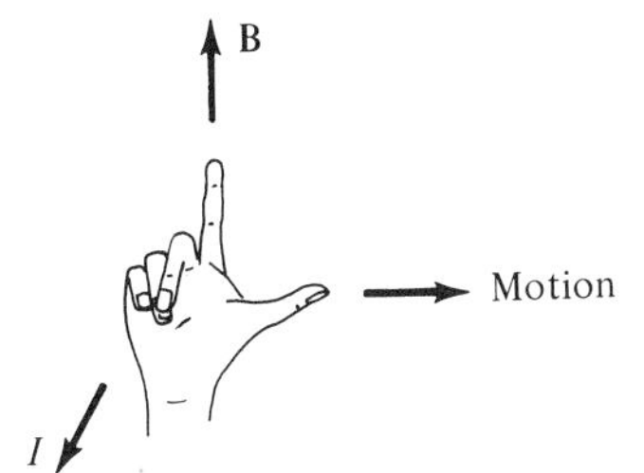

Figure 13-16

and forth or rotates as a loop inside the magnetic field, the induced potential difference will vary in magnitude and sign. An alternating current results. Ac voltages are measured with ac voltmeters. It is necessary to distinguish between ac and dc meters because, in general, an ac quantity varies its polarity so rapidly that a dc meter will just show the time-averaged values, which are zero. The time dependence of an alternating potential difference is easily displayed on an oscilloscope.

Curiosity: If you visit an electric power plant and walk by one of the huge transformers, you will be told not to fold your arms. The folded arms form a conducting loop and you are inside a varying magnetic field. In this way you could induce an electric current across your chest which might influence your heart action.

Application: Sinusoidal Potential Difference. If a conductor moves with uniform speed through a homogeneous magnetic field, the induced potential difference is independent of time. The electric charges inside the conductor move with uniform speed, and a *direct current* (dc) results.

Reversing the direction of motion of the conductor inside the magnetic field reverses the direction of the induced potential difference as well. For a wire fixed in position, a change in polarization of the inducing magnetic field causes the same effect. A periodic change of either motion or field produces a periodically changing potential difference U. The charges inside the conductor will oscillate back and forth; this is an *alternating current* (ac).

How U and I depend on time is determined by the technical arrangement for producing them. The utility network and most other sources supply a sinusoidal potential difference. Its time dependence can be described by:

$$U = U_m \sin (2\pi\nu t) \tag{13.16}$$

where

U: instantaneous potential difference, measured in volts,
U_m: maximum potential difference,
ν: frequency, expressed in hertz,
t: time.

For the current drawn from the utility net, an equivalent formula holds:

$$I = I_m \sin (2\pi\nu t) \tag{13.17}$$

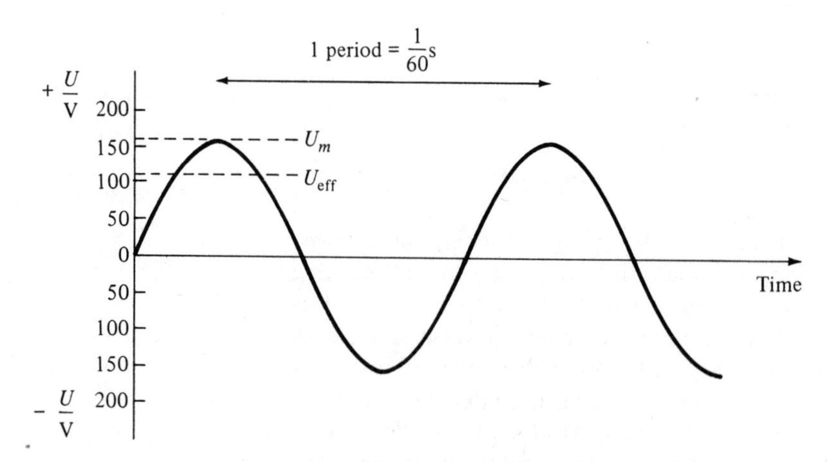

Figure 13-17 The variation of the potential difference U with time t for the 110 V utility net: U, instantaneous value; U_m, maximum value; U_{eff}, root-mean-square value. The frequency is 60 Hz.

Figure 13-17 displays this time dependence of U.

The observed potential difference varies between $+U_m$ and $-U_m$. It passes through zero 120 times per second (twice the line frequency). It is common practice to quote not the maximum potential difference, but its root-mean-square value (U_{eff}).

Conversion:

$$U_{\text{eff}} = \frac{U_m}{\sqrt{2}} \tag{13.18}$$

and

$$I_{\text{eff}} = \frac{I_m}{\sqrt{2}}$$

Thus the potential difference of the electric utility network varies be-between $+156$ volts and -156 volts.

The line frequency is much larger than the resolution frequency (approximately 18 Hz) of the human eye. Therefore we are unable to detect directly any optical effects caused by the line frequency. For example, modern fluorescent light bulbs of the discharge type do not emit a continuous light. They flash with the line frequency or multiples of it. Because the time response of our eye is too slow, we perceive it as a continuous light source.

3.2 Mutual induction

A current produces a magnetic field around itself. An alternating current carries a field of alternating magnetic field strength. The frequencies of the current and the field are identical. This changing magnetic field induces a current in a neighboring conductor, and the induced current will also alternate. Figure 13-18 demonstrates the situation for two parallel wires.

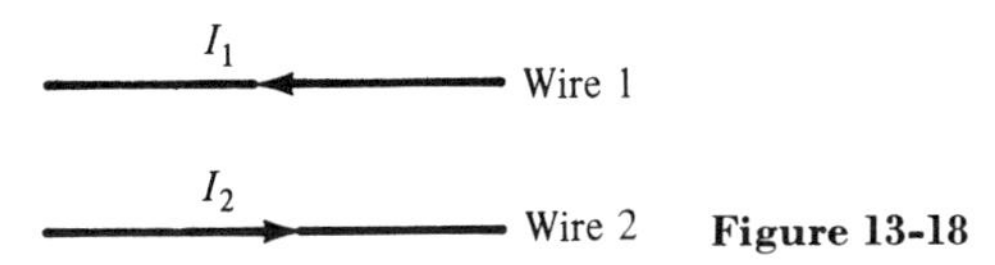

Figure 13-18

The ac I_1 induces an ac I_2 in wire 2. This induced current in turn feeds back and induces a current in wire 1, and so on. This interaction between two alternating currents is known as *mutual induction.*

The magnitude of I_2 depends on how fast the inducing current I_1 varies with time. The physical parameters such as distance, size, and shape of the two wires also influence the magnitude of the induced current. This influence is described by the inductance of the arrangement, a quantity of importance in electric circuitry. Note that both currents have opposing directions.

3.2(a). Unit of inductance. The *inductance* (symbolized L) is measured in henries (abbreviated H).*

* Joseph Henry (1797–1878), physicist. Discovered induction and self-induction independently of Faraday. Devised the first transformer. Father of the American

$$1 \text{ henry} = \frac{1 \text{ volt} \cdot 1 \text{ second}}{1 \text{ ampere}} \tag{13.19}$$

A current change of one ampere per second will induce a potential difference of one volt if the arrangement has the inductance of one henry. In circuitry, an inductor is symbolized by

3.2(b). TRANSFORMER. In a transformer mutual induction is used to change the voltage in ac circuits. Coils are wound around a closed iron core as indicated in Figure 13-19. The alternating current through coil 1 caused by the potential difference U_1 induces a potential difference U_2 across coil 2.

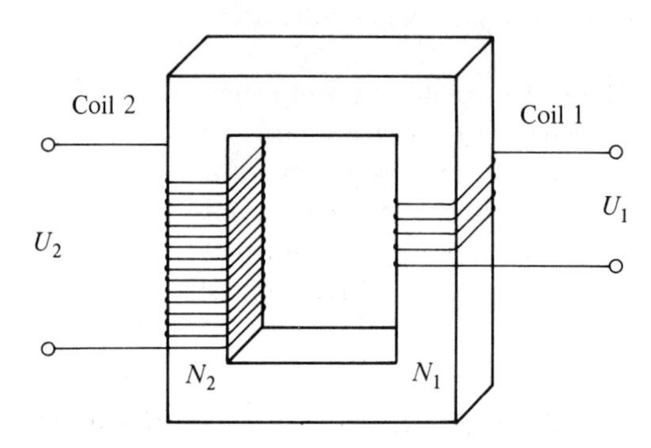

Figure 13-19

Relation:

$$U_2 = \frac{N_2}{N_1} U_1 \tag{13.20}$$

where

U_1, U_2: potential differences at coils 1 and 2,
N_1, N_2: number of turns in coils 1 and 2.

The transformer offers a simple method to alter the voltage in ac circuits. Setting the relation N_2/N_1, we can change the voltage by the desired factor. Note that transformers can increase or decrease a potential difference.

3.2(c). SELF-INDUCTION. The changing magnetic field of an alternating current commands an influence on its own moving charges. The charges do not "know" whether the changing magnetic field is produced externally or by themselves. This feedback into itself is named *self-induction* and is described by the self-inductance of the conductor. It is measured in henries.

---

weather-forecast system. More about him in T. Coulson, *Joseph Henry,* Princeton, 1950.

3.2(d). Time constants. As long as an electric circuit is open, no current flows. If the switch of a dc circuit is closed, a current appears with a value determined by Ohm's law [Equation (13.4)]. See Figure 13-20.

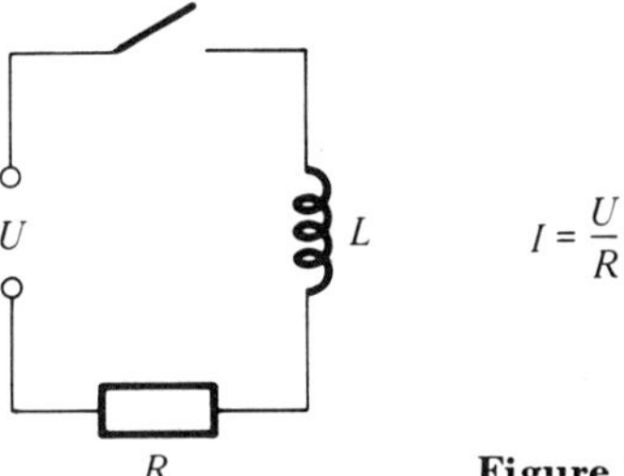

Figure 13-20

The current does not appear instantaneously with its full value I. Instead, the current reaches its value I only after a time delay. Induction explains why: Every electric circuit has a self-inductance because, even without explicitly containing a coil, it is by its very nature a coil with one turn. Its self-inductance produces a countercurrent as soon as the switch closes. The current rises as follows,

$$I = \frac{U}{R} - \frac{U}{R} e^{-(R/L)t} \tag{13.21}$$

where

I: current passing through the circuit at any moment, measured in amperes,
U: potential difference (assumed to be constant),
R: resistance of the circuit,
L: self-inductance of the circuit, expressed in henries,
t: time, expressed in seconds.

Figure 13-21 demonstrates the rise of the current. The current increases exponentially until it reaches its final value U/R. The rise is characterized by the

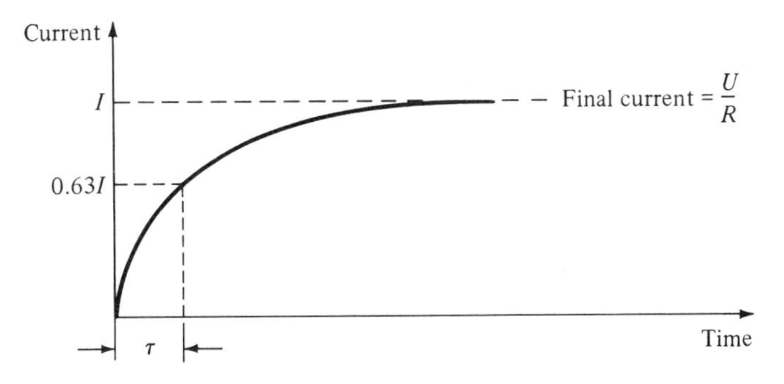

Figure 13-21

relation R/L in Equation (13.21). The inverse of this relation is the time constant of the circuit (symbolized τ_L):

$$\tau_L = \frac{L}{R} \tag{13.22}$$

The time constant is measured in seconds. Its numerical value is the time needed for the current to reach 63% of its final value. If a closed circuit is interrupted, τ_L is the time the current needs to drop to 37% of its initial value.
The time constant is a convenient measure for characterizing a circuit. For example, the self-inductance of a circular wire loop of 10 cm radius is about 10^{-6} henries. If the wire has a resistance of 2 ohms, the time constant is $\tau_L = 5 \times 10^{-7}$ s. Thus the current reaches 63% of its full value half a microsecond after the circuit is closed.
In a circuit having a capacitance C, we observe a similar phenomenon. Refer to Figure 13-22. If the capacitor C is charged and then the switch is closed, the capacitor will discharge across the resistor R. The resulting current does not reach its final value instantaneously but rises exponentially with a time constant, where

$$\tau_c = RC \tag{13.23}$$

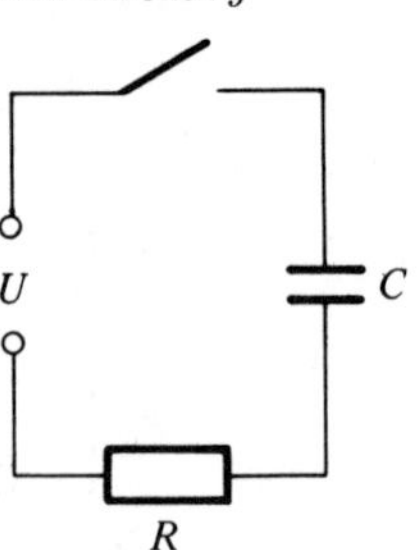

Figure 13-22

It is worthwhile to keep the time delays caused by inductance and capacitance of an electric circuit in mind if you investigate ac phenomena. The rise of the action potential of an excited cell membrane can be compared with the switching-on of a dc current. A coiled lead or the capacitance of an electrode may distort the shape of the action potential considerably.

3.3 Pulses

3.3(a). PULSE SHAPES. The sinusoidal shape of the potential difference as displayed in Section 3.1 is only one possible shape of an alternating potential difference. However, it is the most widely used.

Any rise and subsequent fall in potential difference is named a *voltage pulse*. Successive pulses form a *pulse train*. Figure 13-23 displays voltage pulse trains of various shapes. Each pulse is characterized by its *pulse height* (the maximum potential difference) and by its *pulse width.*

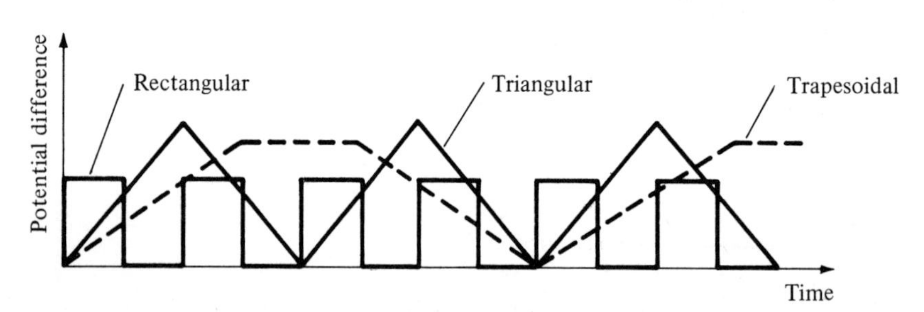

Figure 13-23

In nature, trains of electric pulses are very important. Voltage pulses drive the action of the heart. They are observed indicating the activity of the brain or as triggers to shorten muscular fibers. These pulses propagate along the nerves, but they also reach through the surrounding tissue and can be detected by placing suitable electrodes at the surface of the skin. Height, shape, and spacing of the observed pulses are analyzed for diagnostic purposes. Figure 13-24 shows voltage pulses measured between electrodes on the chest of a man. This is called an

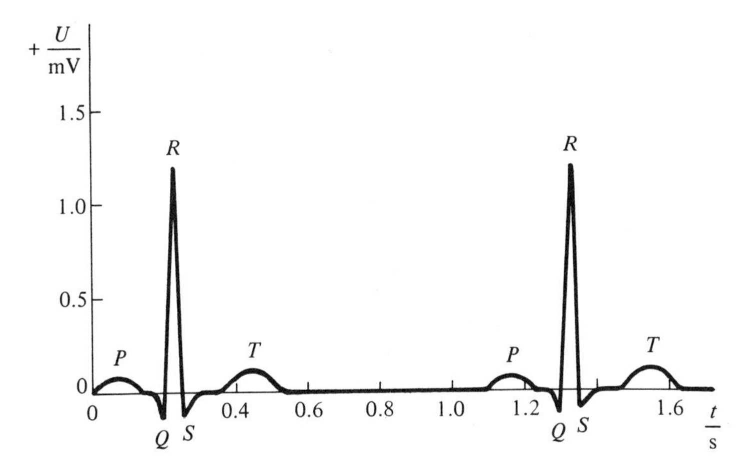

Figure 13-24

electrocardiogram. Note that the measured pulse height is of the order of a few millivolts. In the electroencephalogram (pulses taken at the head) the pulse height is only a few microvolts.

3.3(b). ACTION POTENTIAL OF A CELL MEMBRANE. In the life sciences the most important electric phenomenon is the *action potential* of an excited cell membrane. It causes a voltage pulse traveling along the nerves and enables fast information transfer. This information is coded in pulse trains of variable spacing in time. Figure 13-25 displays an average action potential.

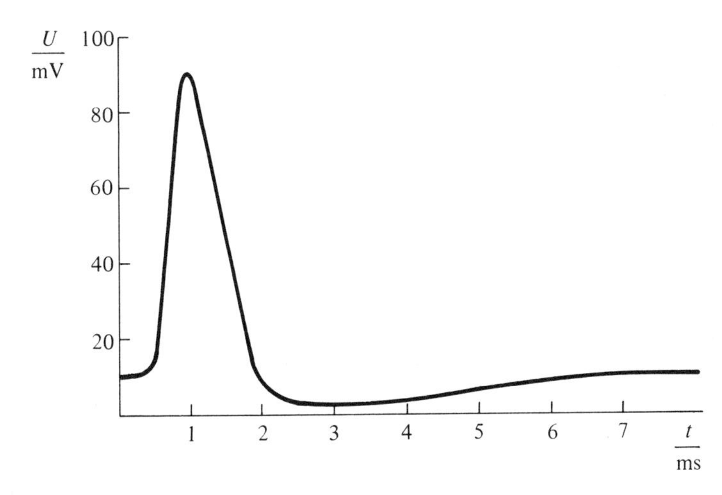

Figure 13-25 Time dependence of a discharge across a cell membrane. The entire pulse is called *action potential.* Its gross features are the same for all types of cell membranes. Notice that the pulse rises within a fraction of a millisecond while its decay lasts for a few milliseconds.

Note: It is difficult to express the shape of the action potential in a closed mathematical expression. Here, Fourier analysis is profitably employed. The pulse shape is approximated by a sum of trigonometric functions, as described in Chapter 9, Section 3.2, for a wave packet. This method is widely used in theoretical research.

3.3(c). NERVOUS CONDUCTION. The transmission of a voltage pulse through a physical conductor, e.g., along a metallic wire, can be described in the same fashion as already done for ac phenomena. It is more complicated because the actual pulse shape must be synthesized as a sum of sinusoidal potential differences, but there are no big difficulties. The conduction of the action potential or a pulse along a nerve fiber is entirely different. An electric pulse propagates along a wire with the speed of light, whereas a nerve pulse is conducted with speeds of the order of meters per second. This is a difference of seven to eight orders of magnitude!

The details are interesting: see Figure 13-26. Nerves are bundles of neurons. Each neuron is the long extension of an individual nerve cell.

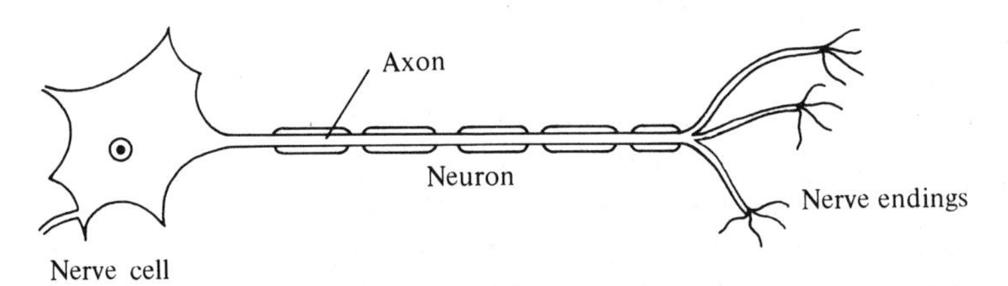

Figure 13-26

The length of a single neuron may reach about one meter. Anatomical studies reveal that some kinds of nerves, for example the motor nerves, are wrapped in a myelin sheath. Sympathetic nerves do not have this sheath. Table 3 shows the measured speeds of conduction of the action potential in various nerves.

Table 3 TYPICAL VALUES FOR THE SPEED OF CONDUCTION IN A NERVE

Nerve	Speed of Conduction in m/s
motor of mammals	60–120
motor of fish	15–30
sensory	5–25
vegetative	2–10
sympatic	0.2–2

Figure 13-27 shows a longitudinal section of a neuron. Normally a potential difference exists across the membrane (rest potential). If excited at position A, the membrane discharges (depolarizes) and a nerve

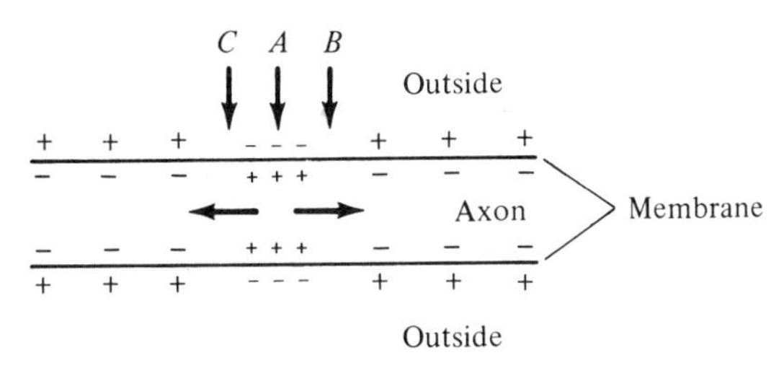

Figure 13-27

pulse (action potential) appears. The discharge occurs over a spot of a few millimeters in length around position A. The depolarization in turn excites a short section (again of the order of a few millimeters) to the left (C) and to the right (B). While the membrane at A quickly reestablishes its rest potential, an action potential is now triggered at positions B and C. The diffusion of ions across the membrane determines the speed of propagation of the action potential.

The speed is rather low, but realize that the shape and magnitude of the nerve pulse are preserved while spreading along the axon! This is so because at each position a new pulse is created. Figure 13-28 demonstrates the advantage of this kind of conduction.

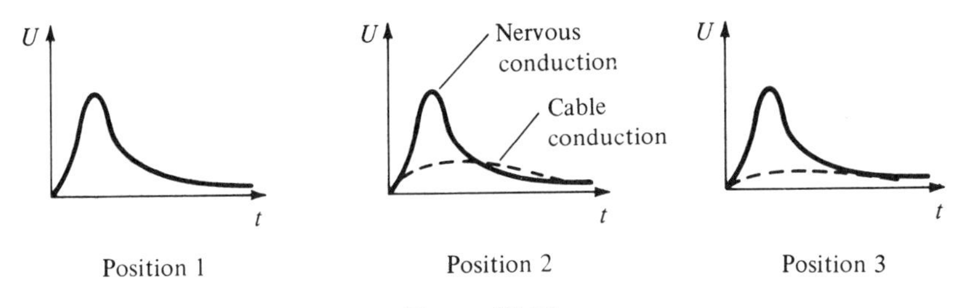

Figure 13-28

At position 1 the nerve pulse displays the indicated shape. After propagating some distance to positions 2 and 3, the nerve pulse still has the identical shape, the same height and width. A pulse of the same shape but conducted along an electric cable arrives at positions 2 and 3 much earlier; but its shape is very much changed, with reduced height and enlarged width.

This type of nervous conduction is observed in axons having no myelin sheath. The transmission speed is about one meter per second. Can this speed be increased? Muscular action of mammals calls for faster information transfer. One way is to increase the cross section of the nerve. You can observe this in the giant axon of the squid. Its diameter of almost one millimeter allows a propagation speed of about 25 m/s. Economy limits further development in this direction.

Nature invented an ingenious way to increase the speed and still keep the axon's diameter as low as a thousandth of a centimeter. See Figure 13-29. The axon is wrapped in a myelin sheath. If excited, the membrane covered with the myelin sheath will not depolarize. To facilitate conduction, the myelin sheath has gaps at regular intervals

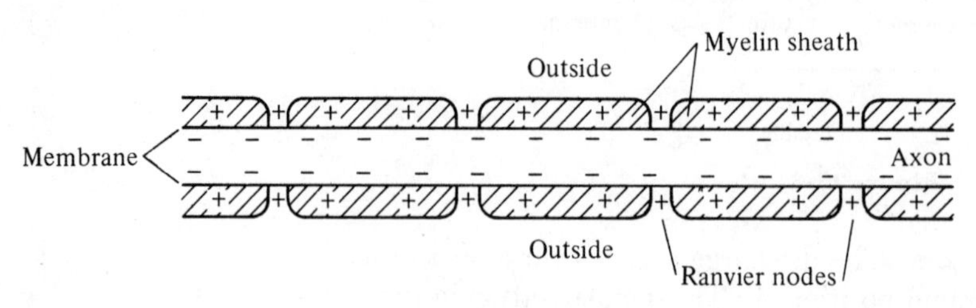

Figure 13-29

(Ranvier nodes). The membrane will discharge only at these nodes. The excitation travels in both directions, and the action potential jumps from node to node. In this fashion, the motor nerves of mammals achieve a high speed of conduction.

EXAMPLE:

The diameter of the nervous saphenous of the dog (it activates the muscles from the lower limbs of the leg) is 0.02 mm and displays a nervous conduction speed of 80 m/s.

This was a lengthy excursion but one of real importance. We will now examine some very simple circuits.

4. CIRCUITRY

Electric currents are driven by a potential difference between the source and sink of an electric circuit. This circuit is composed of various elements as follows: source of potential difference (for example, a battery), resistors, capacitors, inductors, and connecting leads. Figure 13-30 shows a series circuit with one element of each kind.

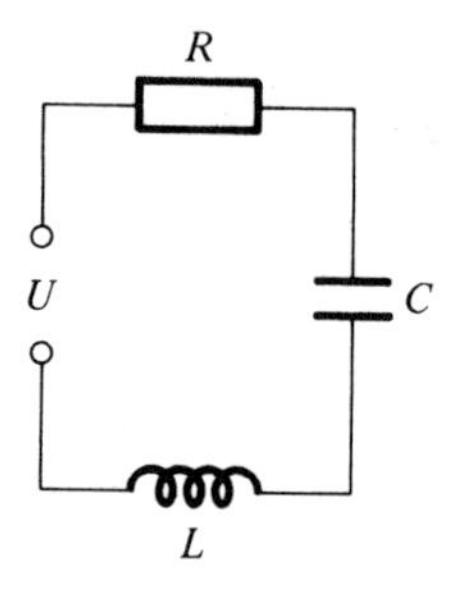

Figure 13-30

Note: The leads connecting the elements also have resistance, capacitance, and inductance. It is customary to lump these values together with the values of R, C, and L of the actual elements. In most instances the influence of the connecting leads is neglected.

4.1 DC circuits

A direct current cannot pass through a capacitor. The inductor merely acts as a resistor because the current is time-independent. Hence, the essential dc circuit is as shown in Figure 13-31. The current I flowing is governed by Ohm's law, Equation (13.4).

$$I = \frac{U}{R}$$

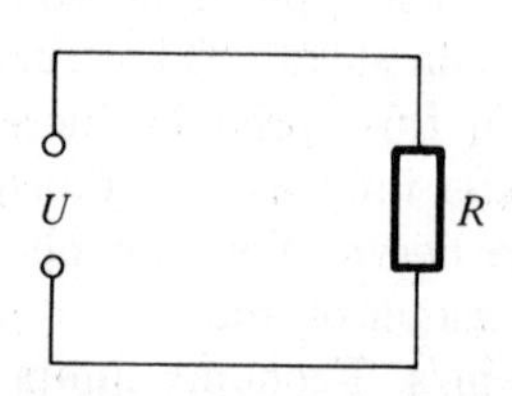

Figure 13-31

4.2(a). IMPEDANCE. The charges sustaining the electric current oscillate back and forth, in effect crossing the capacitor. They also induce a countercurrent in the inductor. As a consequence, the instantaneous current flowing in the same ac circuit depends on time, frequency, potential difference, and physical arrangement. Nevertheless, an equation similar to Ohm's law can be formulated:

$$I = \frac{U}{Z} \tag{13.24}$$

where

I: instantaneous electric current, measured in amperes,
U: instantaneous potential difference, measured in volts,
Z: impedance of the circuit.

The *impedance* is the instantaneous total resistance of the ac circuit; its numerical value is

$$Z = \sqrt{R^2 + \left(\omega L - \frac{1}{\omega C}\right)^2} \tag{13.25}$$

where

Z: impedance of the circuit, measured in ohms,
R: resistance (also called ohmic resistance), expressed in ohms,
L: inductance, expressed in henries,
C: capacitance, expressed in farads,
ω: $= 2\pi\nu$,
ν: frequency of potential difference, expressed in hertz.

Note that $Z \geq R$ always.
Replacing Z in Equation (13.24) we get

$$I = \frac{U}{\sqrt{R^2 + \left(\omega L - \frac{1}{\omega C}\right)^2}} \tag{13.26}$$

and assuming a sinusoidal shape of the potential difference,

$$I = \frac{U_m}{\sqrt{R^2 + \left(\omega L - \frac{1}{\omega C}\right)^2}} \sin(\omega t) \tag{13.27}$$

where U_m is the maximum value of the potential difference.

This current periodically changes its sign and magnitude. If the frequency is ν, it drops 2ν times per second to zero.

The expression for a current in an ac circuit simplifies if either C or L or both are zero. The flow diagram in Figure 13-32 shows the various possibilities.

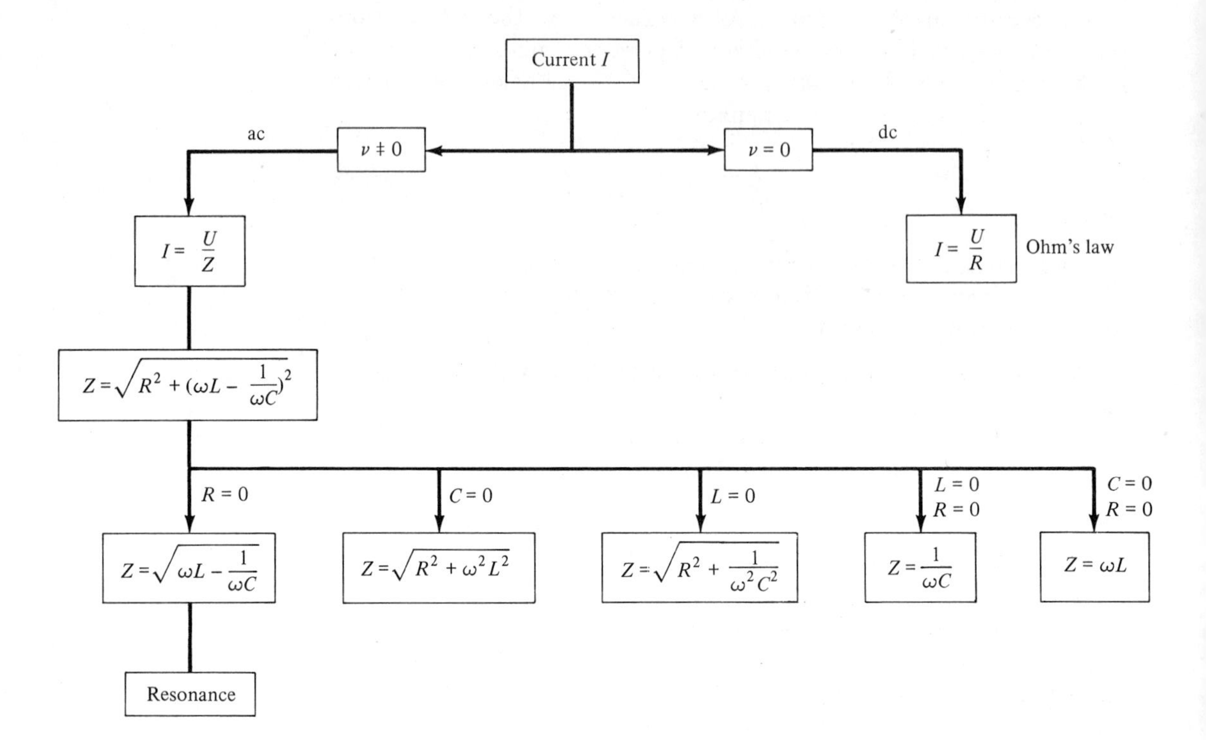

Figure 13-32

4.2(b). RESONANCE. The current flowing through an ac circuit oscillates with a frequency equal to the frequency of the driving potential difference. The maximum value of the current is determined by the impedance Z of that circuit. In Equation (13.25), the term in parentheses is interesting because, if it becomes zero, the impedance reduces to R and the current is a maximum. Obviously, this term is zero for a frequency ν_0, called the *resonant frequency* (the eigenfrequency) of the circuit.

$$\nu_0 = \frac{1}{2\pi\sqrt{LC}} \tag{13.28}$$

Note that the eigenfrequency ν_0 depends only on the values of L and C in the circuit.

The circuit is said to *resonate* if the frequency of the driving potential difference is equal to its eigenfrequency.

Resonance condition:

$$\nu = \nu_0 \tag{13.29}$$

where

ν: frequency of the driving potential difference,
ν_0: eigenfrequency of the circuit.

EXAMPLE: RADIO RECEPTION

The radio waves of a transmitting station induce an alternating potential difference in the antenna of a radio. The reception is best if the eigenfrequency of the antenna circuit is at resonance with the frequency of the station. The station is tuned in. Usually the inductance of the antenna circuit is fixed; we vary its capacitance. For example, suppose we want to receive the Voice of America at 9.5 megahertz. The induction of the antenna circuit of the radio may be 2.8×10^{-6} H. See Figure 13-33. We must tune the variable capacitor until $C = 1 \times 10^{-10}$ F to adjust the circuit to an eigenfrequency of 9.5×10^6 Hz. Now the circuit resonates with the emitter, and the induced antenna circuit current I reaches a maximum. Assuming an induced voltage of 5 μV, the current at resonance is calculated from Equation (13.26) to be 10^{-6} A. The antenna current is then amplified in the subsequent stages of the radio and fed into the loudspeaker.

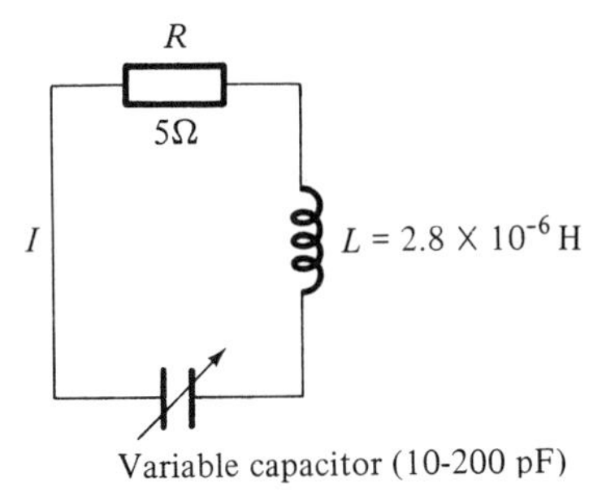

Figure 13-33

Realize that a slight change in the value of C will detune the circuit. The resonance condition, Equation (13.29), is not fulfilled anymore, the impedance rises sharply, and the antenna current drops drastically.

SUMMARY

A constant potential difference will drive a direct current (dc) I between its source and sink. This current is sustained by moving electric charges and measured in amperes (A). The connecting material offers a resistance R, measured in ohms (Ω), to this current. Ohm's law connects all three quantities.

$$I = \frac{U}{R} \tag{13.4}$$

An electric current produces a magnetic field around itself. The magnetic induction decreases proportionally to the distance. The symmetric phenomenon is also observed: If electric charges experience a varying magnetic field, the magnetic force induces a potential difference. The time dependence of this potential difference may be sinusoidal

$$U = U_m \sin(2\pi\nu t) \tag{13.16}$$

or may assume any other shape.

Neighboring alternating currents (ac) influence each other by mutual induction or, if only one conductor is present, by self induction. This feedback is characterized by the inductance L of the physical arrangement and is measured in henries (H).

Ohm's law which governs dc circuits is extended to ac circuits if R is replaced by the impedance Z of the electric circuit.

$$Z = \sqrt{R^2 + \left(\omega L - \frac{1}{\omega C}\right)^2} \tag{13.25}$$

Any ac circuit has a characteristic frequency (its eigenfrequency):

$$\nu_0 = \frac{1}{2\pi\sqrt{LC}} \tag{13.28}$$

New quantities introduced in this chapter are given in Table 4.

Table 4 NEW QUANTITIES

Quantity	Symbol	Units
electric current	I	ampere (A)
electric resistance	R	ohm (Ω)
electric resistivity	ρ	$\Omega \cdot m$
electric conductivity	γ	$\Omega^{-1} \cdot m^{-1}$
electric power	P	W
magnetic force	$\mathbf{F}_m$	N
induction	L	henry (H)
time constant	τ	s
impedance	Z	Ω
resonant frequency	ν_0	Hz

PROBLEMS

1. How many calories are dissipated in the form of heat by an electric motor of 1 kW power consumption? The efficiency is 76%.
2. A resting man needs about 2000 kcal to maintain his body functions. Most of this is dissipated as heat. Calculate the dissipated power in watts.
3. A ship's compass is illuminated by a light bulb. The current flowing may disturb the compass reading. How can this be avoided with a minimum of effort?
4. The earth's magnetic field shields us to some degree against particles coming from outer space. This shield is best at the magnetic equator. There are two holes in this shield. Why and where?
5. How much magnesium is deposited at the cathode if a current of 5 amperes flows for 10 seconds through an electrolyte with Mg^{++} ions?

6. Aluminum is produced by electrolysis of bauxite. The current flowing through one production unit is 9×10^4 A; the potential difference is 4 V. Calculate the hourly production of Al per unit and per kWh.

7. Calculate the number of Na^+ ions transported across a synapse per nerve pulse. The action potential is 80 mV; the current is 2×10^{-11} A. How much thermal energy is dissipated during the transmission?

8. An average lightning bolt is characterized by $U = 10^8$ V and $I = 10^3$ A; it lasts for about 10^{-4} s. Calculate the dissipated heat and the quantity of electricity.

9. A friction-free bearing is achieved if the shaft is suspended inside a magnetic field. Calculate the current passing through the shaft ($m = 3$ g, $l = 1$ cm) necessary to keep it suspended in a field of 10^{-2} tesla.

10. The direction of an induced current always opposes the direction of the inducing motion. Why?

11. The transformer presented in Section 3.2 has an iron core to link the inducing magnetic fields. Would the arrangement still work with the core removed? What would be the difference?

12. In general, the leads at the exit of a high-voltage transformer are thinner than the ones at the entrance side. Why?

13. Calculate the total resistance and the currents I and I_2 of the circuit in Figure 13-34.

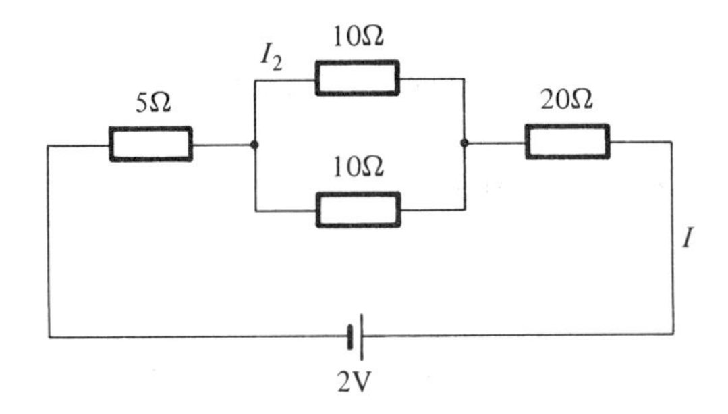

Figure 13-34

14. Draw the circuit diagram of a flashlight. If the resistance of the bulb is 10 Ω, how large a current is drawn from the 1.8-V battery?

15. The specification sheet of an electromotor lists its power consumption as 12 W. If it runs at 110 V, how large a current will flow? If its efficiency is 80%, how large is the power output of the motor? What happens to the difference?

16. To get an impression of how precisely you tune in to a radio station, recalculate the example of Section 4.2, but assume that $C = 1.01 \times 10^{-10}$ F. That means the capacitance is 1% larger than at resonance. Calculate the maximum current under this condition.

17. Two battery cells of 1.5 volts each are connected in series and joined by a resistor with 8-ohms resistance. Each cell has an internal resistance of 0.2 ohm. What is the current in this circuit?

18. A power station delivers 200 megawatts through a transmission line of 4 ohms. How much additional power is lost if the power is delivered at 100 kV instead of 400 kV?

19. A capacitor and an inductor (ohmic resistance 0.1 ohm, inductance 10^{-3} henry) are connected in series. This circuit is powered by a 10-volt, 60-Hz ac source. Calculate the circuit current if the capacitor is tuned to resonant frequency.

20. A circuit having a resistance of 670 ohms and a capacitance of 3×10^{-6} farad is connected to a 35-volt source. How long will it take to have a potential difference of 30 volts across the capacitor?

FURTHER READING

E. J. W. BARRINGTON, *Invertebrate Structure and Function*, Chapter 14. Houghton Mifflin, Boston, 1967.

ARTHUR C. GIESE, *Cell Physiology*, Chapters 22 and 23. Saunders, Philadelphia, 1968.

R. K. HOBBIE, "Nerve Conduction in the Pre-Medical Physics Course," *Am. J. Phys.* **41** (1973) 1176–1183.

A. L. HODGKIN, *The Conduction of the Nervous Impulse.* Charles C. Thomas, Springfield, Ill., 1964.

R. W. STACY, *Biological and Medical Electronics.* McGraw-Hill, New York, 1960.

chapter 14

INFORMATION HANDLING AND PROCESSING

1. INTRODUCTION

A demon guards the passage connecting the two containers A and B, both at thermal equilibrium. Molecules of various speeds bombard the entrance to A all the time. The demon watches closely. As soon as an approaching molecule shows a particularly high speed, he lifts the barrier and allows this molecule to enter container A. All others are rejected. It takes a while—but time is of no consideration for a demon—until the average speed in A is markedly higher than in B. The demon smiles: he has beaten the second law of thermodynamics! Obviously he did not supply energy to either container, which by now exhibit an energy difference. Just by separating the incessantly moving molecules in a suitable fashion, he created an energy difference which could be used to perform work.

James Clerk Maxwell, the eminent physicist, fathered this special demon; it has carried his name ever since. Until the middle of this century, this *gedankenexperiment* puzzled the physicists. Now we know the solution: the demon supplies information, that is, information about the speed of the molecules. To gather this information the demon must spend energy—unfortunately, as much as it can gain from the separation process. The demon will have no energy source at its disposition, and the second law of thermodynamics remains valid.

Energy is apparently not the only aspect under which we might investigate nature. Information is equally important. We have already touched briefly on information at the end of Chapter 10 in Section 4.2. It is a quantity of outstanding importance in the life sciences since any organism is processing information all the time. It picks up information through its sensors and does an immediate evaluation of it. Most of the information is rejected as irrelevant; a fraction of it is transferred and consciously or subconsciously processed. Together with previously obtained and stored information, the outcome of the evaluation determines its reactions.

Before we look into the quantitative aspects of information handling, let's first look at an interesting carrier of information. Figure 14-1

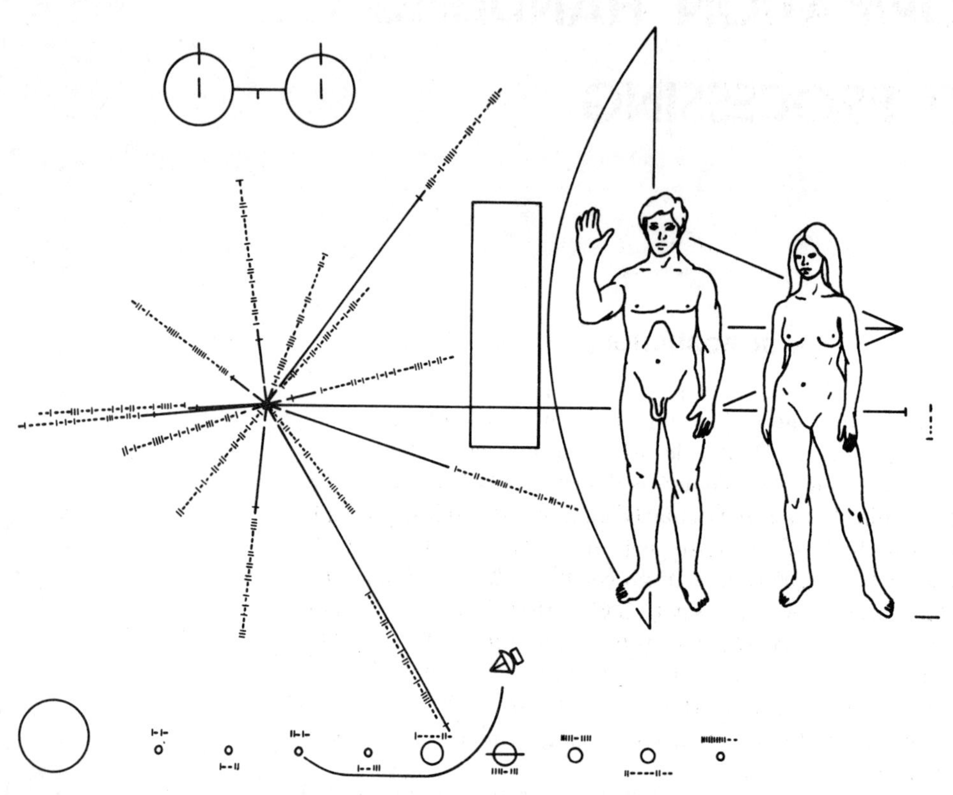

Figure 14-1 Attempt by NASA to transfer information about our earth to intelligent beings who might intercept the Pioneer spacecraft equipped with this plaque.

shows a pictorial plaque attached to a Pioneer spacecraft by NASA engineers to tell scientifically educated inhabitants of other star systems where the deep-space probe was launched from and who did it. Are you able to decipher the message carried?

2. INFORMATION

2.1 Unit of information

Information is a physical quantity; its unit of measurement is the *bit* (contracted from *bi*nary digi*t*). One bit is the minimum information needed to distinguish between two possibilities. This needs further explanation. Imagine a telephone system with a two-digit dial, 0 and 1. If only two stations are connected, then just one bit of information is sufficient to reach a particular station. With four stations connected, we need two bits of information to identify a station, since the numbers are either 11, 10, 01, or 00. In short, the number of bits states the minimum amount of information necessary to uniquely identify an event, a situation, a structure, or a sign.

EXAMPLE: INFORMATION IN A LETTER

How much information is contained in the letter R? It is part of an alphabet having 26 letters. The optimum way to identify R by yes-no reactions is to split the alphabet into halves. The letter is either in the left or right half. The comprising half is then split again. We continue in this fashion until R is uniquely described by a succession of yes-no answers. The number of steps necessary is equal to the number of bits. (See Figure 14-2.)

Identifying sequence: right-left-right-left-right. Thus we need five bits of information to identify the letter R. On the average, only 4.5 bits are needed to identify a letter from our alphabet. If we had chosen P as an example, four steps would have led to a unique sequence of right-left's.

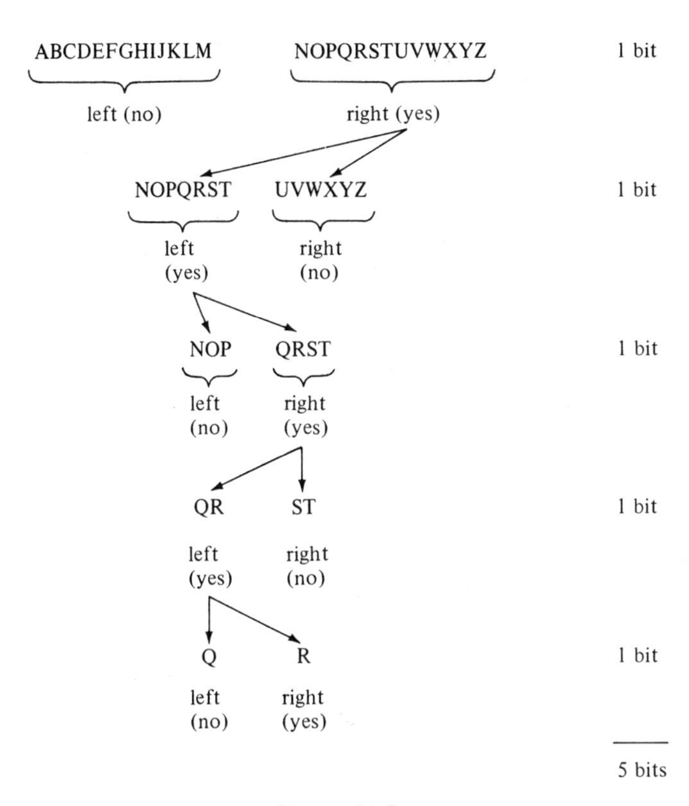

Figure 14-2

In Morse code the letter R is uniquely identified by a sequence of three symbols from a dual system, namely, · — ·. According to the foregoing explanation, its information content is only three bits! But there is no contradiction because the Morse code really encompasses three different signs: dot, dash, and space. With merely · and —, we do not know whether the sequence · — · means · —   · (AE) or ·   — · (EN) or · — · (R).

With that settled, let us return to measuring information in bits. To identify the four-letter word WALL,

$$4.5 \times 4.5 \times 4.5 \times 4.5 = (4.5)^4$$

$$= 410.1 \text{ bits are needed}$$

Actually it is less, because the general structure of our language excludes a number of possible letter combinations such as four vowels. However, the number 410.1 shows the correct order of magnitude.

There is one thing you should keep firmly in mind if you encounter the word *information;* it is a physical quantity measured in bits. The *semantic information* enclosed, in the four-letter sequence WALL, for example, has no bearing in this context.

There are no instruments with which to measure information. It has to be calculated. Table 1 shows the amount of information contained in various structures. An information content of 10^{12} bits for a single bacterium means that this amount of information is necessary to describe its total structure, including the detailed arrangement of each atom and molecule. Naturally, for such a highly complicated structure it is a rough estimate only.

Table 1 INFORMATION CONTENT OF VARIOUS PHYSICAL STRUCTURES

Structure	Information Content Expressed in Bits
electronic flip-flop	1
one-digit number	3.5
letter of the alphabet	4.5
page of a book	1000–5000
core memory of a computer	10^6
disc memory of a computer	10^9
optical memory of a future computer	10^{12}
average bacterium	10^{12}

2.2 Rate of information

The transfer of information is characterized by its rate, that is, transferred information per unit time. The rate of information is expressed in bits/second. Table 2 gives some numerical values.

Table 2 Rate of information handled by various devices

	Information Rate, Expressed in Bits/s
long-term memory storage	1
short-term memory storage	10
piano playing	20
reading	50
conscious information handling	10^2
along a nerve fiber (max.)	800
radio transmission	10^3
telex	10^3
television transmission	10^6
read-in unit of modern digital computer	10^7
output by man to his surroundings	10^7
microwave relay system capacity	10^8
input to man by his surroundings	10^9

2.3 Speed of information

Information is always attached to a carrier. This can be a sound wave, light, a chemical structure, a train of electric pulses, etc. Consequently information travels from one position in space to another with a transfer speed determined by its carrier. The maximum speed is the speed of light, that is, 3×10^8 m/s.

The existence of the upper limit has an interesting application in modern astrophysics: In 1967 a group of astrophysicists from the University of Cambridge discovered a stellar radio source (radio star) emitting short pulses of radio waves in regular intervals. Later, associated light pulses were observed too. A pulsed extraplanetary emitter was not considered at all surprising, but its very short duration and period was. The shortest optical flashes recorded are about 30 microseconds! Since the information necessary to synchronize the molecules of the object to flash simultaneously spreads at most with the speed of light, the flashing object must have a diameter of about 10 km: light travels 10 km in 30 μs. This is an extremely small diameter for a stellar object. Apparently those *pulsars* are neutron stars having a density of about 10^{14} g/cm^3.

3. INFORMATION HANDLING

3.1 General

Any system handling information has three basic components: *input*, *processor*, and *output*. See Figure 14-3.

EXAMPLE: LIGHTMETER

Input: light is converted according to its intensity and wavelength into an electric current. Processor: the photoelectric current activates an electromagnet which in turn

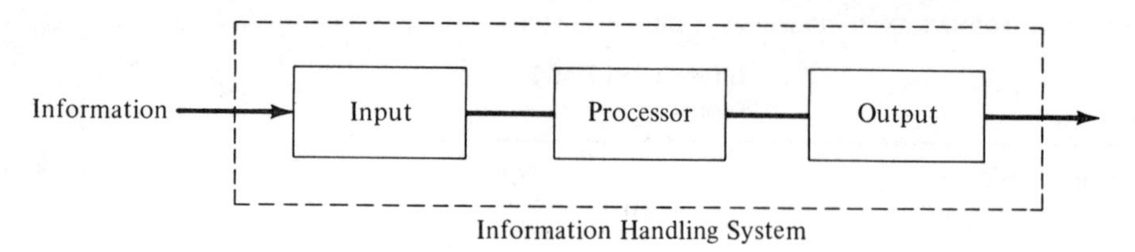

Figure 14-3

exerts a force on a pointer. The electromagnetic force is counteracted by a calibrated spring force. At equilibrium the photoelectric current is compared with the information contained in the calibrated scale. Output: position of the pointer.

EXAMPLE: READING

Input: the pattern of a letter is imaged by the optical apparatus of the eye onto the retina. Each affected retina cell converts its share of incoming light into a train of electric pulses. Processor: the coded information reaches the visual cortex via nerves. The pattern is reconstructed from the incoming coded information, compared with previously stored patterns, and recognized. Output: an order to the eye's motor to shift to the next letter.

Obviously, these examples are oversimplifications. However, they serve as a starting point for further elaboration. The basic components are not independent of each other; they are linked by various feedbacks. The output influences the input, the input processes much of the information arriving, the processor influences the input—to name a few types of feedback.

Each component of the entire information-handling system exhibits limitations and special features which result in loss of information. This is a necessary process because most of the impinging information is not pertinent; the handling system must reduce it. The input especially serves this purpose so that the coded information reaching the processor is reduced by orders of magnitude. Information reduction in man is an example: About 10^9 bits of information reach man's input via eye, ear, and skin per second. But a mere 10^2 bits are accepted consciously in the processor (cortex). However, about 10^7 bits/s are used by automatic processors in man. They mainly control the output via locomotion, mimic, language. In other words, the input of man's information-handling device rejects 99% of the impinging information.

In the next few sections, we will examine some aspects of inputs and processors. We study input because it is essentially a measuring process, and we must know its limitations to draw correct conclusions about the investigated phenomena. We study processors because technical processors, i.e., computers of any kind, are now indispensable not only in physics but in the life sciences as well.

3.2 Input

The input of an information-handling system is conveniently subdivided into *transducer*, *preprocessor*, and *transfer*. See Figure 14-4.

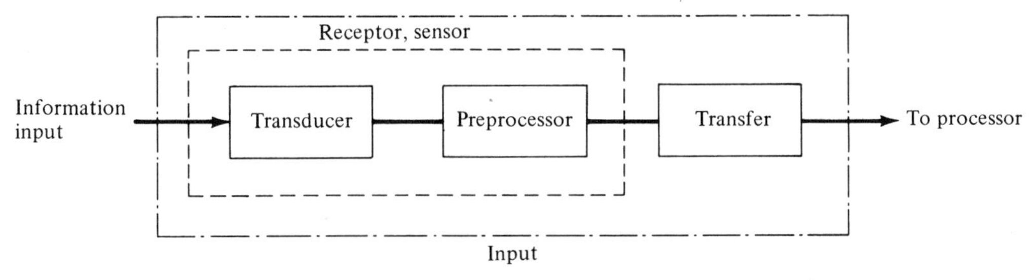

Figure 14-4

3.2(a). TRANSDUCER. Information is always linked to a carrier such as waves, molecules, electric pulses, and magnetic orientation. The transducer transfers the information from one carrier to another one more suitable for handling.

EXAMPLE:

Information carried by sound waves may be transduced in technology by a piezo crystal into electric signals which can easily be transferred and processed. Or: In nature genetic information is transduced by polynucleotides, a very compact carrier and most insensitive to changes.

Table 3 lists some technical transducers.

Table 3 TECHNICAL TRANSDUCERS

Carrier of Incoming Information	Transducer	Carrier of Outgoing Information
magnetic field	rotating coil	electric current
	nuclear resonance	electric pulses
ultrasonic waves	piezo crystal	potential difference
temperature	thermoelectric junction	electric current
time	clock	various types of pulses
light	photographic plate	localized darkening
	photocell	electric current

3.2(b). RECEPTOR. If you compare technical transducers with biological ones such as the ear, eye, or taste buds, you will find that most biological transducers also reduce the incoming information. This means that you cannot expect objective information from a biological transducer, something you rely on when employing technical transducers. Because this is an important question—and not only for philosophers, elaboration is necessary: The eye is a radiation receptor. Its range and sensitivity are very closely matched to the emission of the most important natural radiation source, the sun. Figure 14-5 shows the energy spectrum as emitted by the sun and modified by the earth's atmosphere. The response curve of the human eye is included. Note that both curves are well matched, the result of millions of years of evolution.

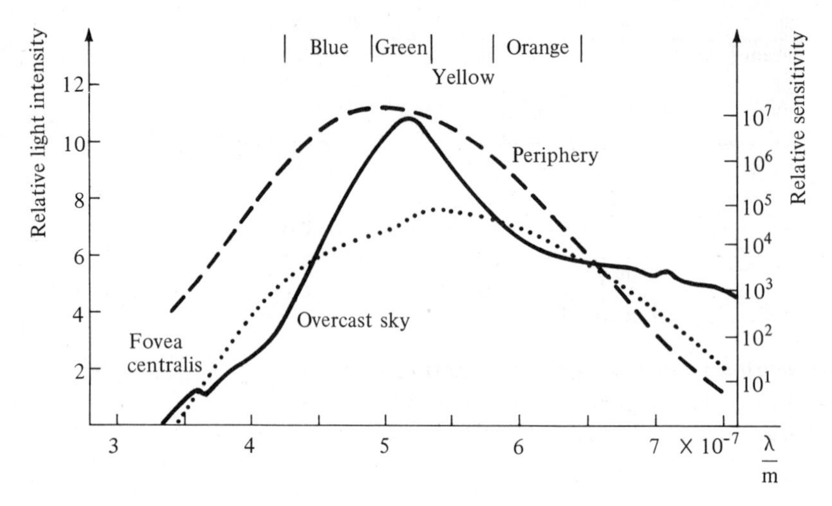

Figure 14-5 The sun's spectrum through an overcast sky (solid curve and left ordinate). Relative sensitivity of periphery and fovea centralis of the human eye (dashed and dotted curves, ordinate at right side). Note that the ordinate scales are very different.

In effect, this very matching distorts incoming information. Formally we can attribute this distortion to a pre-evaluation process. This takes the eye beyond the stage of technical transducer. It is already an information-handling device and should be named *receptor* or *sensor* instead of transducer. We shall not stress the language, but it should be explained that you must be very careful while evaluating received information. This information may already be pre-processed.

Summary: Information needs a carrier. A transducer transfers information from one carrier to another one more suitable for handling. In general, the transducer will distort (or pre-evaluate) this information. If this pre-evaluation is of considerable degree, we call the transducing device a *receptor* or *sensor*.

3.3 Transfer of information

Essential to information handling is its means of transfer. The information emitted by a receptor must reach the information processor. The transfer channel may be a cable, a nerve, a laser beam, a radio wave, etc. The channel capacity is limited; it transfers a limited rate of information. A single myelin-sheathed axon, for example, can transfer up to 800 bits per second, whereas an electric cable can transfer millions of bits per second without mutual interference. As a consequence, thousands of telephone calls (each transferring about 25 bits/s) may be made simultaneously over the same cable. To transfer large rates of information, nature uses parallel-running nerve fibers.

Even without approaching the channel capacity or passing any information-reducing device, information is lost during transfer. This is due to the noise or background always present in the transfer channel.

The background is caused by spontaneous discharges of cell membranes, interference from other transferring channels, remainders of previous information, thermal motion, etc.

3.3(a). REDUNDANCE. Background distorts information. In this connection, the term *redundance* is important: Any message carries a certain amount of information, measured in bits. The message is coded by a transducer, for example, into a succession of electrical pulses. If the information carried by this pulse train has exactly the same numerical value as the message itself, any background or noise in the transfer channel will distort this message. To avoid this, most codes are redundant. That means the transducer adds information in order to secure the content of the original message.

EXAMPLE: REDUNDANT CODE

The International Telegraph Alphabet No. 5 is coded in holes and blanks on an 8-track paper tape. Only seven tracks are used to code numbers, letters, and signs. The eighth track is redundant but is used for checking. It contains a hole if the total number of holes at a position on the tape is odd. Thus the number of holes at any position is either 0, 2, 4, 6, or 8. If a tape is received with an odd number of holes at one position, it is recognized as faulty.

Written language contains much redundance. Its structure determines to some degree the succession of individual letters. It can be calculated that on the average 66% of the letters employed in written language are not essential for the message carried. In other words, 66% of the information carried is redundant. (Remember, this has nothing to do with the semantic information carried by the language.)

3.3(b). AMOUNT TRANSFERRED IN MAN. Man is a most complex information-handling and processing system—more complex than others not only because of physical structure but because man uses his consciousness to evaluate part of the input information subjectively. This is not an easy task if you look at Figure 14-6.

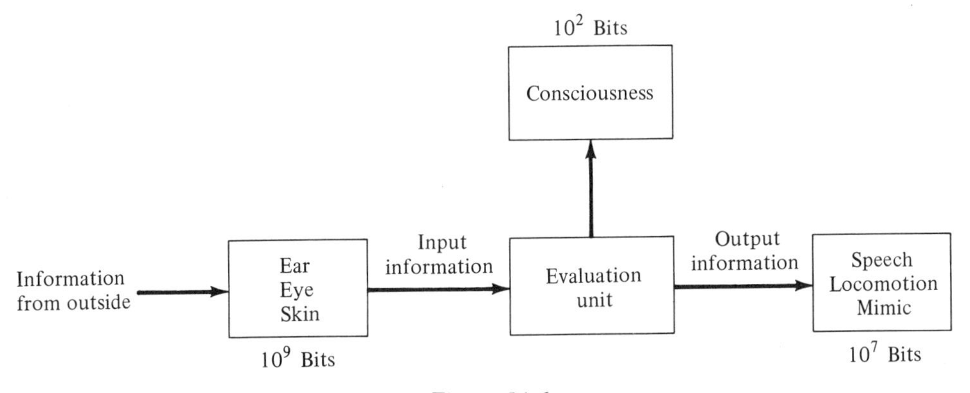

Figure 14-6

Each second, 10^9 bits reach man's sensors. Only about 100 bits of the incoming 10^9 bits is filtered out to his conscious attention every second.

4. INFORMATION PROCESSORS

This is a vast topic and this is not the place to treat it extensively. Of the multitude of information processors—they range from slide rules to hybrid computers in technology and from synaptic transmission to consciousness in the life sciences—only three will be presented: feedback circuits, analog computers, and digital computers.

4.1 Feedback circuit

Feedback means that a fraction of the information leaving an information processor is fed back into its entrance. A voltage amplifier processing a pulse will serve as example to present a few aspects of feedback.

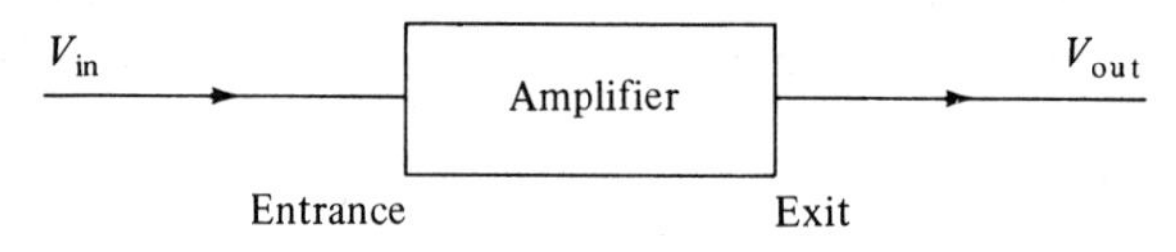

Figure 14-7

The gain of the amplifier characterizes the system in Figure 14-7.

$$G = \frac{V_{out}}{V_{in}} \tag{14.1}$$

where

V_{in}: potential difference of pulse entering,
V_{out}: potential difference of pulse leaving,
G: gain of amplifier.

There is no feedback, and consequently information leaving does not interfere with information entering.

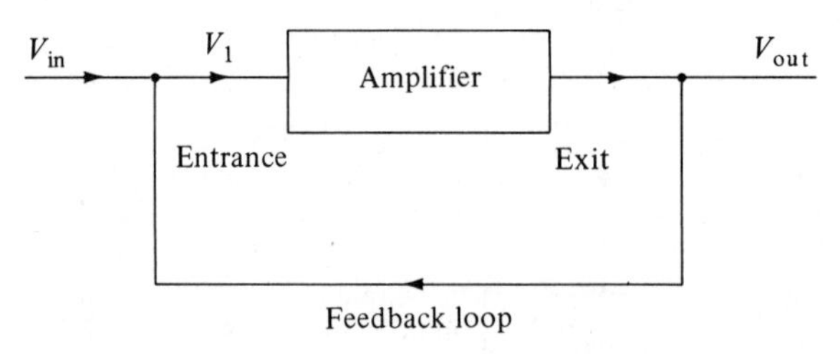

Figure 14-8

Now the system is extended by a return loop through which a fraction of V_{out} is fed back to the entrance. See Figure 14-8. Obviously

$$V_1 = V_{in} + \alpha V_{out} \tag{14.2}$$

and

$$V_{out} = GV_1 \tag{14.3}$$

where

V_1: potential difference entering the amplifier,
αV_{out}: potential difference fed back,
α: feedback coefficient (very small compared with 1),
G: gain of the amplifier measured without feedback.

Solving both equations for V_1, we obtain:

$$\frac{V_{out}}{V_{in}} = G' = \frac{G}{1 - \alpha G} \tag{14.4}$$

G' is the gain of the system including the feedback loop.

Remark: Before we proceed, note that the feedback loop has a stabilizing influence on the entire system. A numerical example will demonstrate that. For $G = 5 \times 10^3$ and $\alpha = 0.01$, we get

$$G' = -102.0$$

The amplification of the entire system is lower than G.
Now, if the amplifier gain varies by 10%, then

$$4500 \leq G \leq 5500$$

but

$$-102.3 \leq G' \leq -101.8$$

Therefore, although G varies by 10%, the gain G' of the entire system changes by only 0.3%. This is important in technique. Because of the feedback loop, the output of the system is less susceptible to outside influences (like a change in the amplifier gain due to temperature variations). The price you pay for this stability is the lower amplification, but that can easily be compensated for.

There are two types of feedback: *negative* and *positive* feedback. Negative feedback opposes any change between input and output of the processing system, while positive feedback enhances it.

4.1(a). NEGATIVE FEEDBACK. If the feedback coefficient of Equation (14.2) is negative, the potential difference fed back opposes V_{in}. It will oppose any change at the input. If V_{in} is rising with time, the negative feedback causes a slowed-down rise in V_{out}. The reverse is also true. The overall effect of the negative feedback is a smoothing action.

In circuitry, negative feedback is achieved by a capacitor C in the feedback loop. See Figure 14-9. In nature, an example of negative feedback action is the control of the eye's iris. See Figure 14-10.

Light passes the central opening of the iris and is transformed into electrical pulses at the retina. The coded information is transferred to the visual cortex. A fraction of this information—information about the intensity of the incident light—feeds back to the motor center which changes the diameter of the iris opening. If the light intensity rises, the

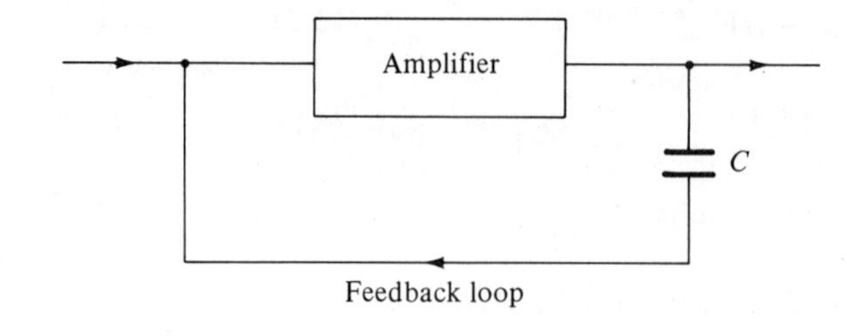

Figure 14-9

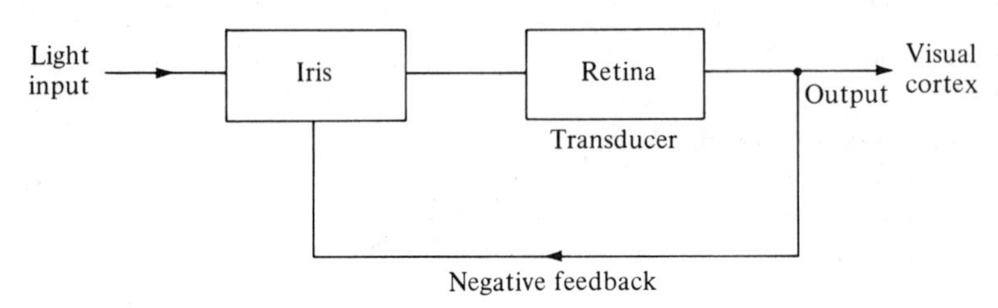

Figure 14-10

aperture is reduced; if it decreases, the motor center enlarges the aperture. As a consequence, the aperture is controlled in such a way that the light's intensity at the retina is optimized for optimum information transfer to the visual cortex.

Realize that this information-processing system is a closed one; the conscious mind has no control over it. It is a *direct reflex*. The reflex which makes you withdraw the probing finger from a hot object is similar, only it is not a totally closed feedback circuit. The conscious mind can interfere and prevent the withdrawal.

4.1(b). Positive feedback. When the feedback coefficient of Equation (14.2) is positive, the potential difference fed back has the same sign as V_{in}. Positive feedback enhances any change at the input. If V_{in} is rising with time, the positive feedback further increases this rise and leads to an even faster rise of V_{out}. The reverse is also true. The overall effect of positive feedback is an unbalancing action. It does not lead to equilibrium but to an avalanche-like increase or decrease of the quantity involved.

In circuitry, positive feedback is achieved by a resistor R in the feedback loop. See Figure 14-11. Positive feedback is rare in nature and is mostly connected with pathological changes.

4.1(c). Application: control of blood pressure. Muscular action of heart and aorta sustains the blood pressure in the veins. Stretch receptors in the walls of the aorta act as transducers. It is a simple information-processing system having negative feedback. Its action is demonstrated in Figures 14-12 and 14-13.

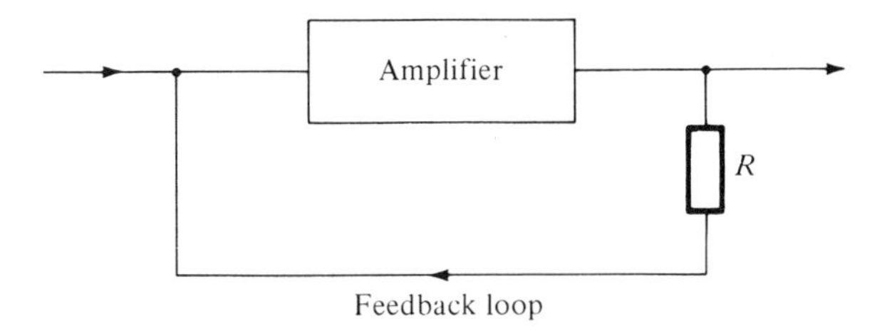

Figure 14-11

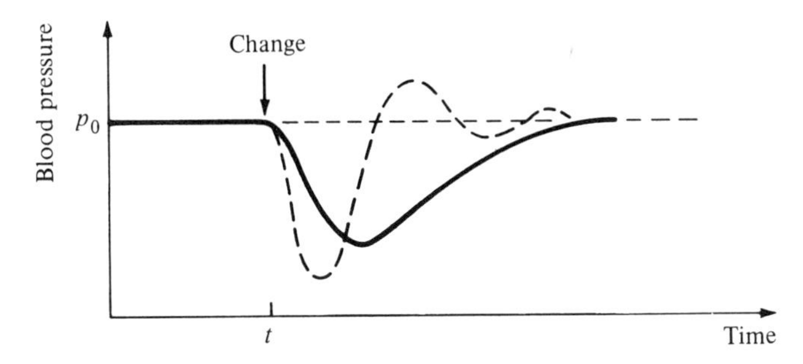

Figure 14-12

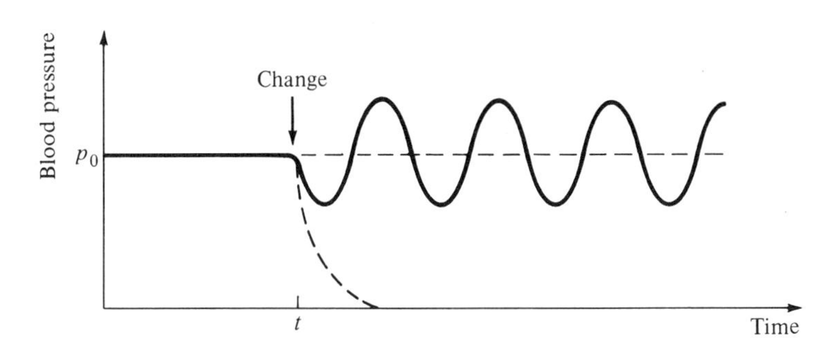

Figure 14-13

The initial blood pressure is p_0. At time t this pressure may change suddenly, for example if a person lying down stands up. The pressure receptors activate the heart's motor center via negative feedback. The falling blood pressure leads to increased activity of the heart. There is some delay (about 10 s), but then the drop in pressure is checked and brought up to p_0. This aperiodic behavior is demonstrated by the solid line in Figure 14-12. In general, however, the counteraction taken by the heart overshoots, so that the pressure-time relation is described by a damped oscillation (dashed line).

If the negative feedback does not function properly (this could be caused by certain drugs), the blood pressure will start to oscillate once it deviates from p_0. Figure 14-13 illustrates this situation (solid line). If the feedback is changed to positive feedback, the blood pressure will drop to zero (dashed line).

4.2 Analog computer

An *analogy* is a model based on similar behavior of two different phenomena. It is used to analyze an unknown system in terms of a known one. This is most fruitful in the life sciences. Such a model is set up by an analog computer. The parameters of the system under investigation are simulated by various electrical components. Electrical components are most convenient, but other modeling like hydraulic simulation is also feasible. Some examples are presented in the following sections.

4.2(a). CAPILLARY FLOW. Obviously, the blood flow through a capillary can be simulated by a current passing through a wire. See Figure 14-14.

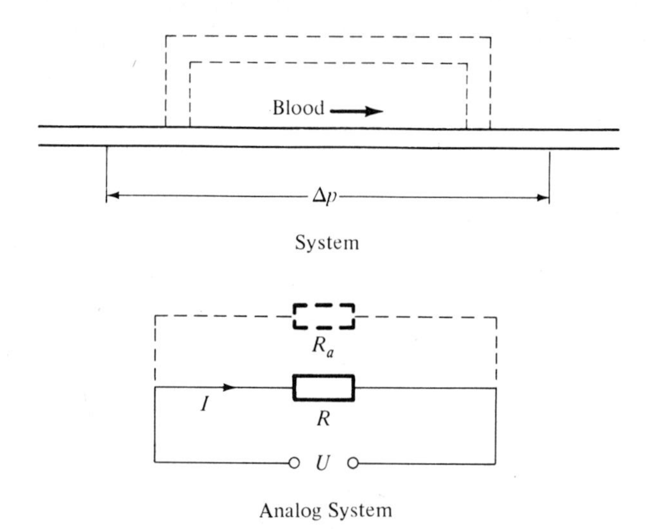

Figure 14-14

The pressure difference Δp driving the blood through the capillaries is simulated by a potential difference U. The counteracting friction is represented by a resistor R. Both circuits are described by Ohm's law (Chapter 13, Section 2.1):

$$I = \frac{U}{R}$$

If a parallel capillary opens, the flow rate through both vessels is simulated by the addition of a resistor R_a in parallel with R. Measuring the current in both loops, we determine the branching ratio of the blood circuit.

The electrical circuit in Figure 14-14 is a simple analog computer. Its advantages are obvious: Once the analog is properly established, many situations can be simulated and the outcome easily measured. This is much more convenient compared with performing the actual experiment on a living system.

4.2(b). DEEP FREEZING. To retain the texture, taste, and other important features of food stored in frozen state, it is important how fast the freezing of the material takes place. There is an optimum time span to reach the desired low temperature. The freezing process is modeled by the analog electrical circuit in Figure 14-15.

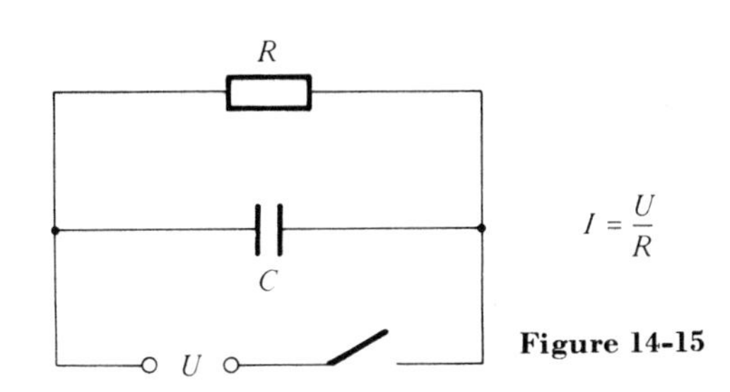

$$I = \frac{U}{R}$$

Figure 14-15

Quantity	Analog Quantity	Symbol
thermal capacity	electrical capacitance	C
thermal conductivity	electrical conductivity	$1/R$
temperature difference	potential difference	U
heat flow rate	electric current	I

The electrical circuit simulates the temperature drop during the freezing process:

$$\Delta T = \Delta T_i e^{-t/\tau} \tag{14.5}$$

where

ΔT: temperature difference between the inside of the freezer and the material,
ΔT_i: initial temperature difference,
t: time,
$\tau = RC$: time constant of the freezing process.

Figure 14-16 shows Equation (14.5) in graphical form. Note that the temperature difference drops to 37% of its initial value after the time RC has passed.

Naturally, the freezing process is more complicated, but that can be accounted for by additional circuit elements. Once the analog model of the freezing process is established and checked against experiments, it is used as an analog computer. By changing the values for C and R and measuring I, we can investigate the freezing for any type and amount of foodstuff.

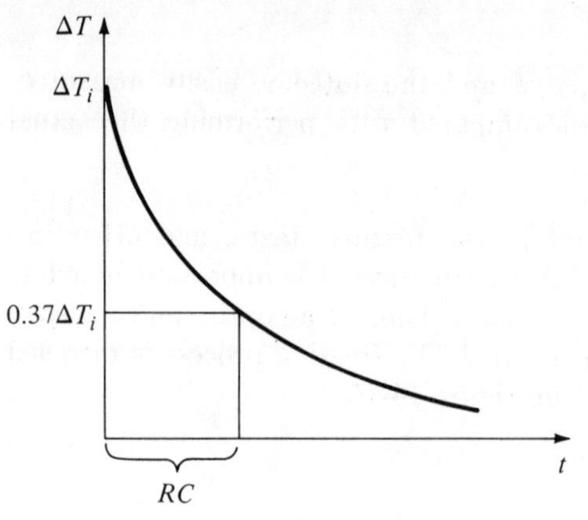

Figure 14-16

4.2(c). RELATIVE MERITS

Advantages of analog computers:

1. Immediate results.
2. Continuous variation of parameters.
3. Simple problems call for simple analog computers.
4. In general, inexpensive.

Disadvantages:

1. Accuracy seldom better than a few percent.
2. Low flexibility. In general, built only for special problems.

4.3 Digital computer

4.3(a). GENERAL CHARACTERISTICS. The information processed by a digital computer is coded into discrete numbers, into *digits*—hence its name. It encompasses the following components (Figure 14-17).

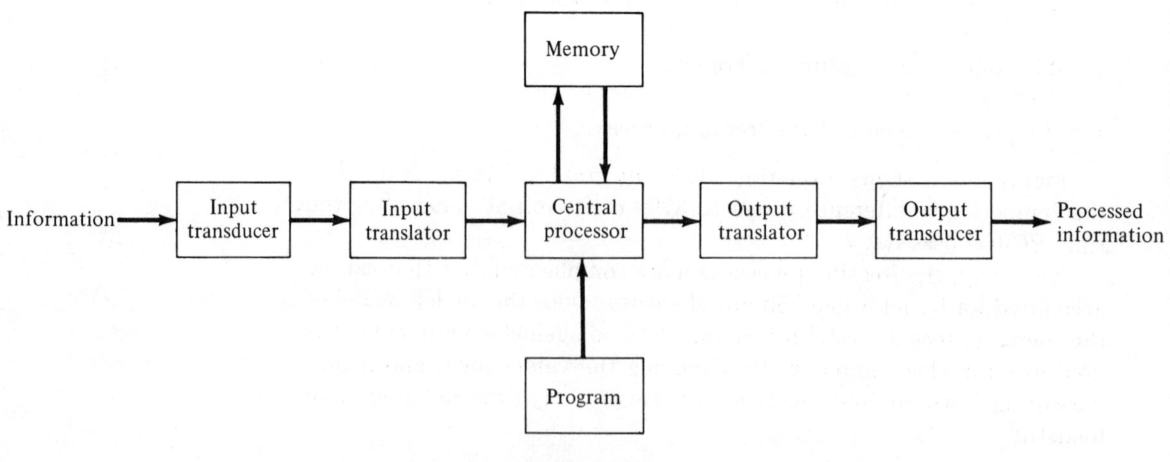

Figure 14-17

Input transducer: Information is fed to the input by means of paper tape, magnetic tape, optical scanner, or typewriter. The information is coded into a train of electric pulses.

Input translator: The coded information is translated into a binary code. For example, decimal numbers are automatically converted into binary numbers according to the scheme in Table 4.

Table 4

Decimal Number	Binary Representation
0	0
1	1
2	10
3	11
4	100
5	101
6	110
7	111
8	1000
9	1001
10	1010
.	.
.	.
.	.
16	10000
32	100000
64	1000000

The binary code is convenient because it can be represented by simple electrical components having two states. For example: capacitor (charged-discharged), switch (on-off), ferrite core (magnetized-demagnetized).

Central processor: It coordinates the action of the entire system. It carries through the actual computation according to the program.

Memory: Arrays of binary electrical components (e.g., ferrite cores) each in one of its two stable states 1 or 0. The arrays are subdivided into individual cells each labeled by an identification number (its address) and containing information. The information-holding capacity of a cell ranges from 2 to a few hundred bits.

EXAMPLE: CORE MEMORY

A 16-bit cell is a group of 16 ferrite cores, each either magnetized (1) or not (0). The decimal number 135 is stored as shown in Figure 14-18. Each square symbolizes one ferrite core of the same memory cell. The memory stores all numerical data, holds intermediate results, and keeps the operating instructions (program) necessary to carry out the computation. As opposed to man's memory, a computer memory will not fail to recall its contents completely. For all practical purposes, it is indestructible by accidental outside influences.

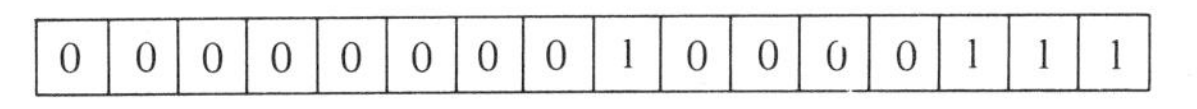

Figure 14-18

Program: Succession of instructions (statements) to control the information processing. It is written by man and stored in binary code inside the memory. The program explains in minute detail all computational operations in proper succession to the central processor. For example: "Put the contents of address 1279 into register 11, put the contents of address 2793 into register 12, add registers 11 and 12, and store the result in address 2797."

Writing such a detailed program takes much time. Modern computer languages like FORTRAN, ALGOL, and PL/1 reduce the effort considerably. The program is written in terms closely related to the usual mathematical terminology and translated by the computer itself into detailed instructions.

Formulated in ALGOL, the above operation reads:

```
C: = A + B;
```

where A and B are data previously fed into the computer. The result C is then called to the output printer by the ALGOL statement:

```
PRINT C;
```

Remark: What you just read about programming a computer is a gross simplification. If you want to know more about it, consult Further Reading or one of the numerous books on programming. Modern computer languages are easy to learn. It is not only profitable but also very enjoyable to communicate with a computer!

Output translator: Information leaving the central processor is binary coded. The translator converts it into a code acceptable to the output transducer.

Output transducer: Device to put out the processed information on a line printer, typewriter, paper tape, magnetic tape, or as a graph.

4.3(b). Relative merits

Advantages of digital computers:

1. Great flexibility in programming.
2. No limit on accuracy (except cost).
3. Very fast. The multiplication of two 8-digit numbers takes only a few microseconds.
4. Fast access to stored information.
5. Reliability.

Disadvantages:

1. In general, much more complex than analog computers even if employed to solve very simple problems.
2. Expensive.

Figures 14-19 and 14-20 are interesting in this context.

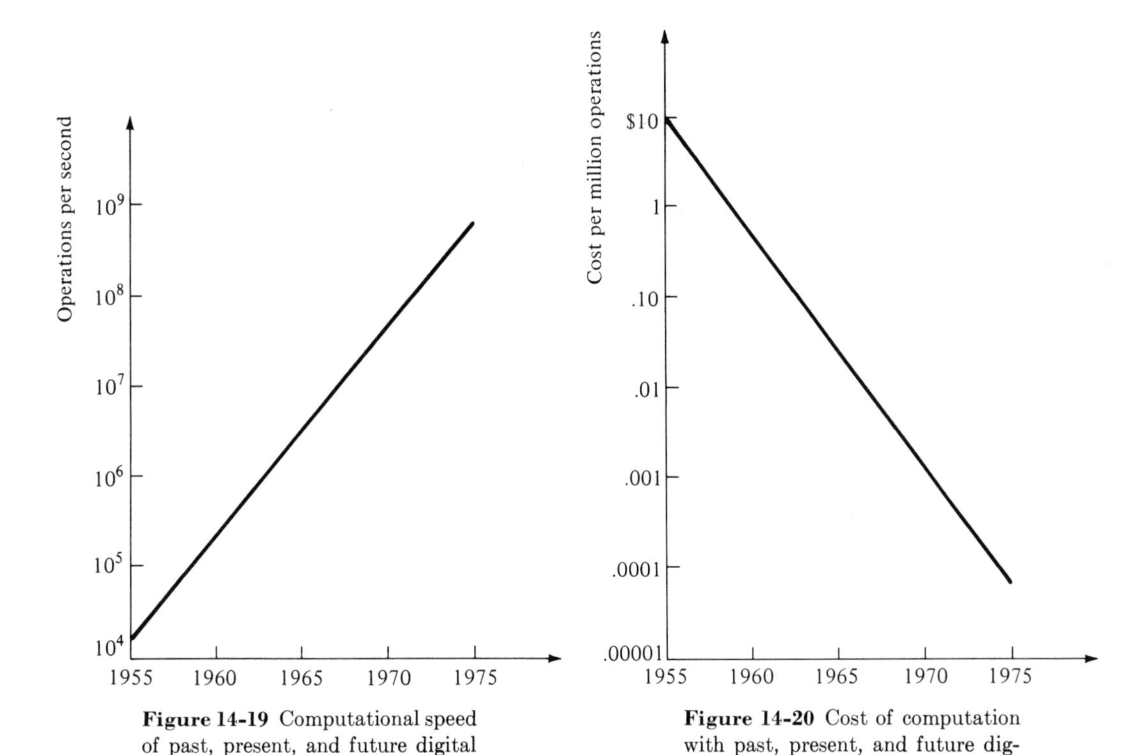

Figure 14-19 Computational speed of past, present, and future digital computers.

Figure 14-20 Cost of computation with past, present, and future digital computers.

4.3(c). Applications in the life sciences

Medical diagnostics: Pertinent data describing in minute detail rare kinds of illnesses are stored. The data are kept up to date and are easily retrieved by a computer. Symptoms are described according to a standard questionnaire and fed into the computer. Within seconds the processor finds the set of stored data which best fits the input information. The diagnosis is printed out together with recommended treatment, side effects, and literature citations.

Contrast enhancement: Faint x-ray photographs will yield more information if the contrast of the shown structures is enhanced. This can be done with a digital computer. Figure 14-21 shows an example.

Curiosity: It is not always necessary to employ a computer for contrast enhancement. Sometimes mother nature does it for the observant scholar. See Figure 14-22 for an example.

4.3(d). Computers and the brain. It is tempting to compare the brain with a man-made computer. The equivalents of the brains input and output transducers are obvious. The 10^{11} neurons and the vast interconnecting network of fibers and synapsis inside the brain are com-

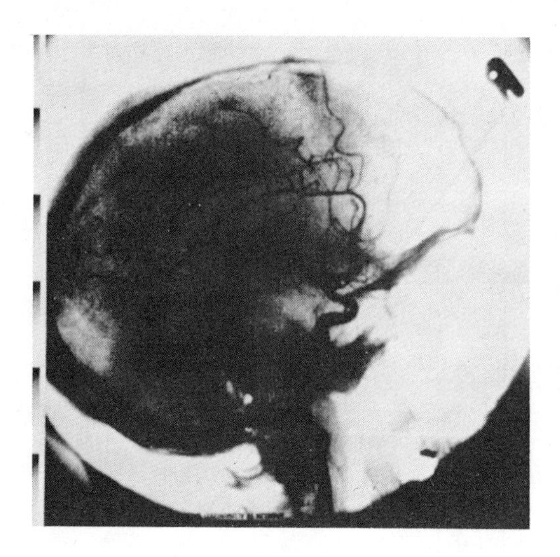
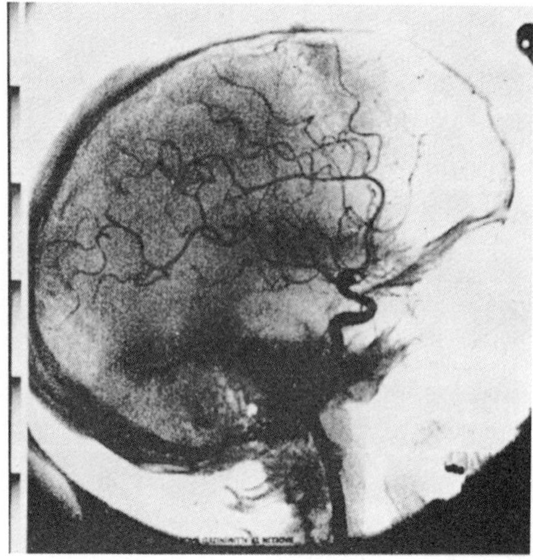

Figure 14-21 Angiogram photograph before (left) and after (right) contrast enhancement.

Figure 14-22 Oblique aerial photograph of a field planted with rye, showing the outline of two Roman temporary training camps in Germany's Rhineland. Without the crops the camps would blend into the background.

parable with the central processor. Fast information exchange between the various parts takes place by electrical pulses, while information seems to be stored in the structure of chemical substances. By analyzing cautiously and carefully selected performances, it is possible to compare the human brain and digital computers formally, and to describe the performance of the brain in technical terms. This does not indicate how the brain functions, but the numbers listed in Table 5 indicate that the brain is an impressive device for information processing, even if compared with advanced digital computers.

Table 5 SOME TERMS AND NUMBERS USED TO CHARACTERIZE DIGITAL COMPUTERS APPLIED TO DESCRIBE THE PERFORMANCE OF THE HUMAN BRAIN

Detail	Brain	Average Digital Computer
average word length	4 bits	8–60 bits
time needed for addition or subtraction	4 ms	10^{-6} s
rate of information transfer	10^3 bits/s	10^7 bits/s
cycle time	10^{-3} s	10^{-6} s
(monthly rental)	$1000 and up	$1000 and up

SUMMARY

Information is a physical quantity that is as important as energy. It is attached to a *carrier* such as an electromagnetic wave, a train of electric pulses, a protein structure, or written letters. The information carried is measured in bits. A *transducer* transfers information from one carrier to another. Transducers of biological systems always pre-evaluate incoming information before transferring it to the actual processor.

Analog and *digital* computers are technical devices used to handle information. Most analog computers simulate a system with the help of electrical circuitry. Digital computers handle and evaluate information which is coded in numbers. Analog computers are mainly built for particular purposes, whereas digital computers are easily programmed and are universally applicable.

FURTHER READING

P. ARMER, *The Outlook for Technological Change and Employment.* U. S. Printing Office Feb., 1966.

T. H. BENZINGER, "Heat Regulation: Homeostasis of Central Temperature in Man," *Physiolog. Rev.*, **49** (1969) 671.

I. FLORES, *Computer Organization.* Prentice-Hall, Englewood Cliffs, N. J., 1969.

H. H. GOLDSTINE, *The Computer from Pascal to von Neumann.* Princeton University Press, Princeton, 1972.

P. HANDLER, ed., *Biology and the Future of Man*, Chapter 14. Oxford University Press, London, 1970.

A. HEWISH, "Pulsars," *Scientific American* (Oct., 1968) 25.

S. L. JAKI, *Brain, Mind, and Computers.* Herder and Herder, New York, 1969.

J. VON NEUMANN, *The Computer and the Brain.* Yale University Press, New Haven, 1958.

D. M. RAMSEY, ed., *Image Processing in Biological Science.* University of California Press, Berkeley, 1968.

G. S. S. TENT, "Cellular Communication," *Scientific American* (Sept. 1972) 42.

N. WIENER, *Cybernetics or Control and Communication in the Animal and the Machine.* Wiley and Sons, New York, 1961.

chapter 15

RADIATION

1. WAVE-PARTICLE DUALISM

Tradition and ease of presentation call for dividing this chapter into two seemingly mutually exclusive sections: wave radiation and particle radiation. However, if you realize that radiation is the outstreaming of energy from a source, you will see the common denominator.

In everyday experience there is no difficulty distinguishing between particle and wave radiation. This changes if you venture into the world of the microcosmos, a world where our experience fails as a yardstick for understanding. A first glimpse of that caused the century-long controversy about the nature of light. Is it a wave or a stream of particles? Huygens* demonstrated in the 17th century that it is a wave; his experiments do not allow any other explanation. Newton had equally convincing arguments for a particle nature of light.

It remained to the physicists of this century to show that there is really no contradiction in both aspects. Our imagination and language fail to describe this particle-wave dualism. This is no loss as long as we are able to describe mathematically the behavior of light and predict its effects. The general conclusion is that we may treat light—or any other electromagnetic radiation—as a wave or as a stream of particles.

The general symmetry of the laws of nature allow the other point of view as well: particles may be treated as waves. This is a successful approach in describing elementary particles and atoms.

Figure 15-1 shows the contents of this chapter.

* Christian Huygens (1629–1695), physicist. Invented the first pendulum clock. Founded the wave theory of light and applied it successfully to polarization effects.

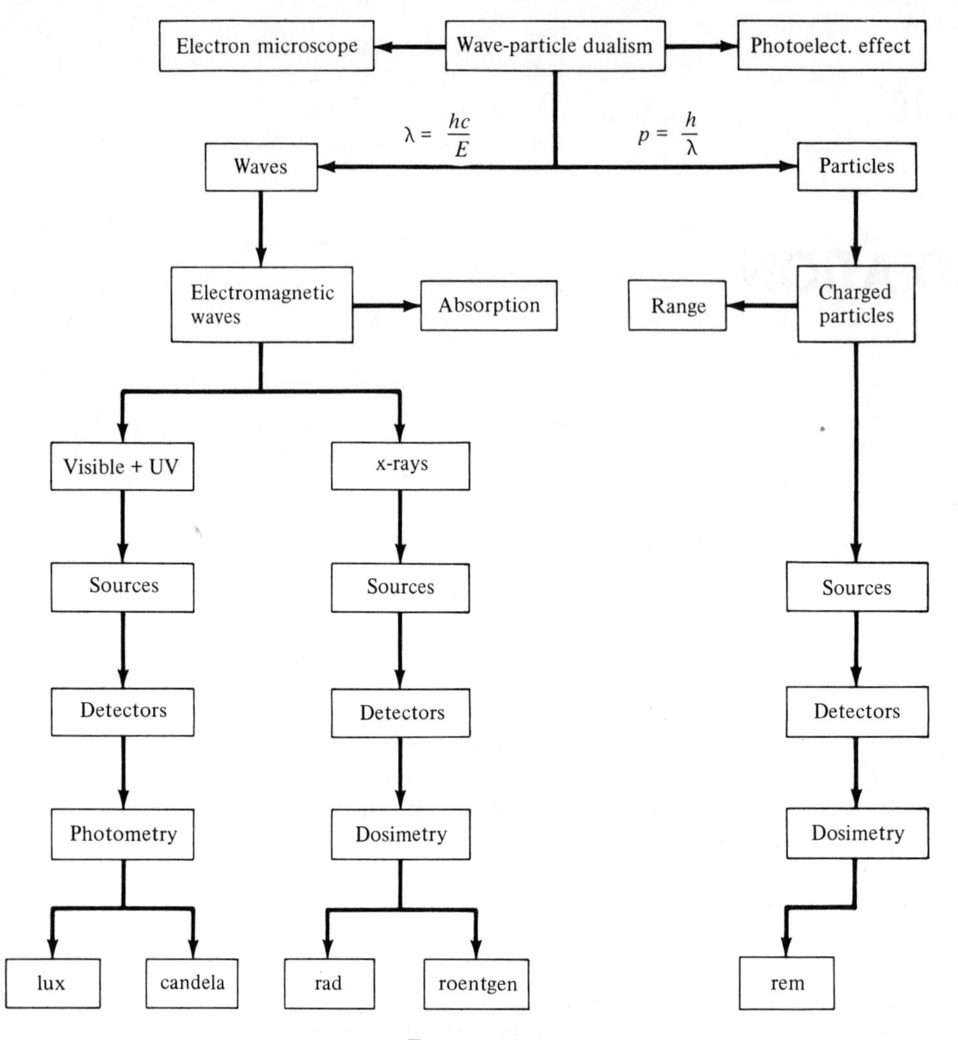

Figure 15-1

1.1 Waves as particles

Electromagnetic radiation can be explained as a stream of particles, called *photons* or *quanta*. The energy of the photons and the frequency of the electromagnetic wave are connected by Planck's relation:

$$E = h\nu \tag{15.1}$$

or

$$E = \frac{hc}{\lambda} \tag{15.2}$$

where

E: energy of photon,
ν: frequency of wave,

λ: wavelength,
c: speed of light,
h: proportionality constant, named Planck's constant.*

The numerical value of this most fundamental constant in physics is:

$$h = 6.625 \times 10^{-34}\ \text{J}\cdot\text{s} \tag{15.3}$$

or

$$h = 6.625 \times 10^{-27}\ \text{erg}\cdot\text{s}$$

Table 1 shows the energy of individual photons of various wavelength.

Table 1 WAVELENGTH AND ENERGY OF SOME PHOTONS

Wave	Wavelength in m	Photon Energy in J
power ac	5×10^{6}	3.97×10^{-33}
broadcasting	6×10^{2}	3.31×10^{-28}
infrared light	3×10^{-5}	6.62×10^{-21}
visible light (green)	5.4×10^{-7}	3.6×10^{-19}
ultraviolet	3×10^{-7}	6.62×10^{-19}
diagnostic x rays	2.5×10^{-11}	8.0×10^{-15}
therapeutic x rays	8.3×10^{-12}	2.4×10^{-14}
annihilation radiation	1.23×10^{-12}	1.6×10^{-13}

In everyday experience the quantized nature of electromagnetic radiation remains undetected because the energy of each photon is very small and consequently the number of photons is extremely large. For example, the number of photons emitted per second by a 100-W light bulb is about 6×10^{19}!

The quantum nature of electromagnetic radiation can be observed directly. If light interacts with matter, negatively charged elementary particles (electrons) are observed. This phenomenon is called the *photoelectric effect;* the electrons are called *photoelectrons.* See Figure 15-2.

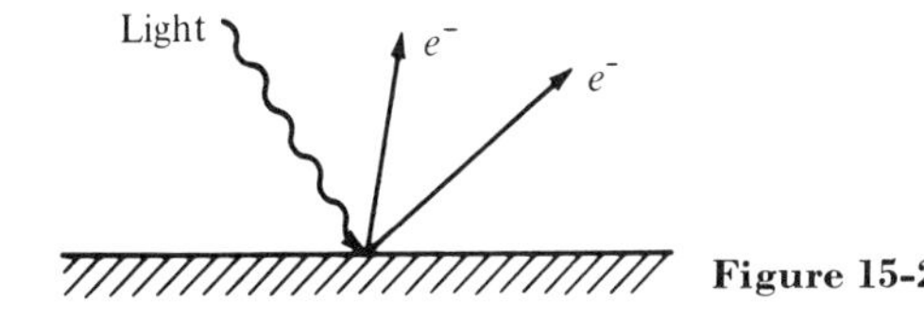

Figure 15-2

The effect was discovered in 1887 by Heinrich Hertz and quantitatively explained in 1905 by Einstein† employing the quantum concept of Planck.

* Max Planck (1858–1947), physicist. Nobel laureate 1918. He introduced in 1900 the quantum concept to explain the spectrum of black-body radiation. This was and still is in sharp contrast to any concept in classical physics. It is the cornerstone of modern physics. More about Planck and the quantum concept in Niels Bohr, *Atomic Physics and Human Knowledge.* Wiley, New York, 1958.

† Albert Einstein (1879–1955), physicist. Nobel laureate 1921. Gained fame by

Individual photons are absorbed by the atoms of the substance. If the energy of a photon exceeds the binding energy of an electron bound to its atom, the bond breaks, and the electron is knocked out. The rest of the photon's energy is carried away as kinetic energy by the photoelectron.

$$E_k = h\nu - W \tag{15.4}$$

where

E_k: kinetic energy of photoelectron,
$h\nu$: energy of interacting photon,
h: Planck's constant,
ν: frequency of radiation,
W: minimum energy necessary to remove an electron from its atom (binding energy of electron).

1.2 Particles as waves

The momentum of a particle and its wavelength are related by an equation derived in 1924 by de Broglie:

$$\lambda = \frac{h}{p} \tag{15.5}$$

where

λ: wavelength,
p: magnitude of momentum,
h: Planck's constant.

The waves associated with particles are called *matter waves*. Their wavelengths are very small.

EXAMPLES:

The speed of an electron inside the hydrogen atom is $v = 2.2 \times 10^6\ \text{m}\cdot\text{s}^{-1}$. Its wavelength is:

$$\lambda = \frac{h}{mv} = 3.3 \times 10^{-10}\ \text{m}$$

Neutrons emitted from the thermal column of a nuclear reactor show a de Broglie wavelength of about 2×10^{-10} m.

Application—The Electron Microscope. The object is imaged by electrons which are treated as matter waves having a wavelength according to Equation (15.5). The electrons of a standard 50-kV electron microscope have a wavelength of 8×10^{-12} m. Consequently, its resolution (the smallest distinguishable distance between two object points) should be of the same order of magnitude (see Chapter 11, Section 4.2). Theoretically this is correct, but in practice the lens aberrations spoil this

---

publishing within one year (1905) three fundamental papers: Theory of Brownian Motion, Theory of Special Relativity, Quantitative Explanation of the Photoelectric Effect. His theory of general relativity (1914) is still basic for the understanding of gravitation.

very high resolution and limit it to about 3×10^{-10} m, that is, to about the diameter of a water molecule.

2. ELECTROMAGNETIC RADIATION

2.1 General characteristics

2.1(a). REGION. Electromagnetic radiation propagates with the speed of light c and is characterized by its wavelength λ. For the terms describing waves, see Chapter 9, Section 1. Depending on its wavelength, this radiation traditionally has various names: see Figure 15-3.

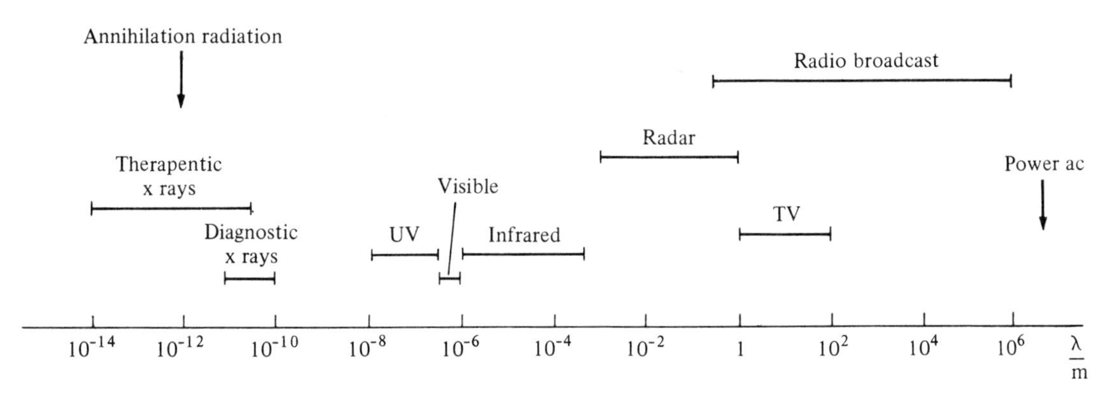

Figure 15-3 The various regions of electromagnetic radiation.

The intensity (or power) of electromagnetic radiation can be expressed in energy per unit time, for example, in joule·second^{-1} or in watts. The intensity is measured absolutely by a calorimeter. See Figure 15-4. The radiation passes an aperture defining the input area.

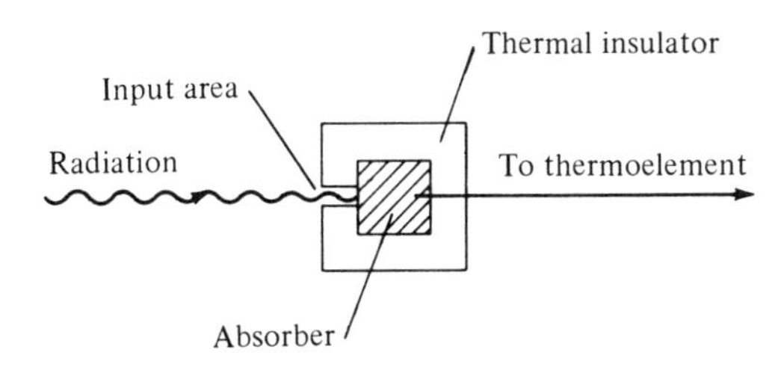

Figure 15-4

It is absorbed by a thermally isolated absorber. The resulting rise in temperature is directly proportional to the radiation intensity. As long as all entering radiation is absorbed, the total energy emitted can be calculated regardless of the spectrum of the source. All other detectors show a spectral response function, that is, their response depends on the wavelength and hence the spectrum of the radiation source.

2.1(b). ABSORPTION. The intensity of radiation passing through a layer of material decreases. See Figure 15-5.

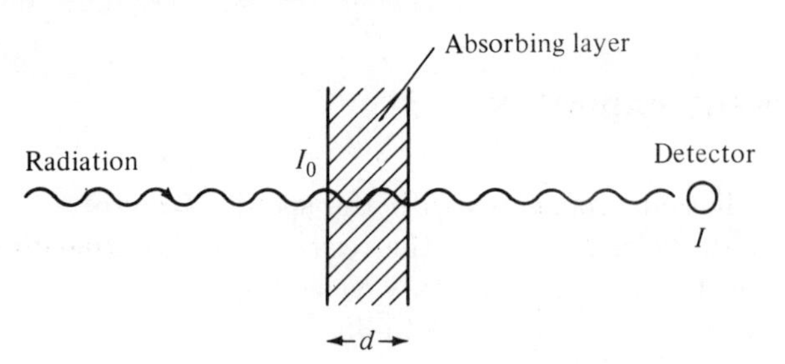

Figure 15-5

Absorption follows an exponential law, as already described in Chapter 9, Section 5.3.

$$I = I_0 e^{-\sigma d} \tag{15.6}$$

where

I: intensity after penetrating a thickness d,
I_0: intensity without absorbing layer,
σ: absorption coefficient, measured in cm^{-1},
d: absorber thickness, measured in cm.

The *absorption coefficient* depends on wavelength and material. It varies over an extremely wide range. Numerical values for σ are listed in tables such as those in *Handbook of Chemistry and Physics*, Chemical Rubber Publishing Co., Cleveland, Ohio.

Various terms are employed to describe absorption:

Absorbance: $A = \log (I_0/I)$.
Transmittance: $T = I/I_0$.
Half-value layer: $d_{1/2} = 0.693/\sigma$ (thickness, where $I = I_0/2$).
Mean free path of radiation: $d_l = 1/\sigma$.
Mass absorption coefficient: $\sigma_m = \sigma/\rho$, where ρ is the density of the absorbing material.
Cross section: $\sigma_A = \sigma/N$, where N is the number of atoms per cm^3 of absorber.

The terms just described are demonstrated in the following numerical example.

EXAMPLE: ABSORPTION OF Co-60 RADIATION IN LEAD

The experimental result of the absorption measurement is shown in Figure 15-6. From Figure 15-6 we extract the following numerical values: $d_{1/2} = 1.20$ cm, that is, a layer of 1.2 cm lead will reduce the radiation intensity by 50%. Three half-value layers reduce the intensity to

$$0.5 \times 0.5 \times 0.5 = (0.5)^3 = \tfrac{1}{8} \text{ of } I_0,$$

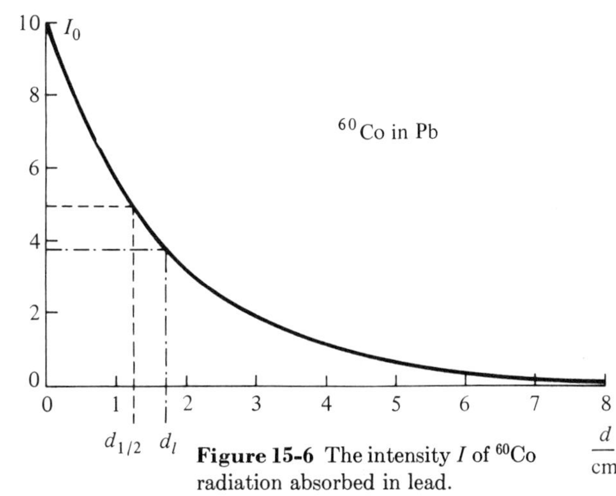

Figure 15-6 The intensity I of ^{60}Co radiation absorbed in lead.

about 12.5% of the original intensity.

$d_l = \dfrac{1}{\sigma} = \dfrac{d_{1/2}}{0.693} = 1.73$ cm, that is, on the average a photon is absorbed after penetrating 1.73 cm of lead.

$$\sigma_m = \frac{\sigma}{\rho} = \frac{0.578\ \text{cm}^{-1}}{11.5\ \text{g}\cdot\text{cm}^{-3}}$$

$$= 0.0503\ \text{cm}^2\cdot\text{g}^{-1}$$

$$\sigma = 0.578\ \text{cm}^{-1}$$

The absorption curve of ^{60}Co is thus described by

$$I = I_0 e^{-0.58d}$$

In the following paragraphs, two wavelength regions of special interest to the life sciences are presented in some detail: visible and ultraviolet region and x rays.

2.2 Visible and ultraviolet light

2.2(a). REGION. Figure 15-7 shows the range of this interesting part of the electromagnetic spectrum.

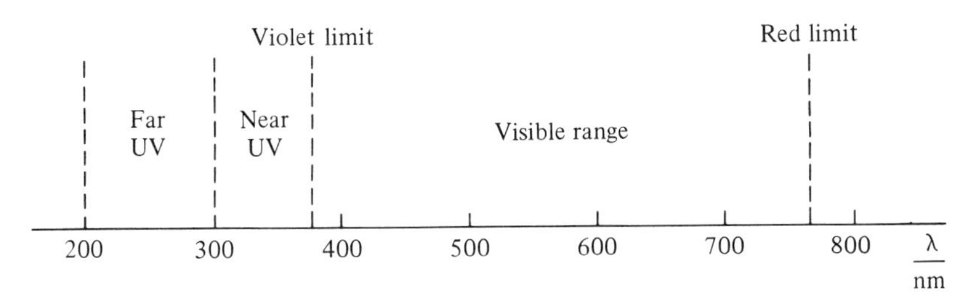

Figure 15-7

The retina of the eye gives light impressions only between 390 and 780 nm. Heat detectors in the skin are sensitive to wavelengths larger than 780 nm. The skin reacts with pigmentation to wavelengths shorter than 390 nm.

2.2(b). SOURCES. Figure 15-8 presents the spectra of some radiation sources emitting visible and ultraviolet light.

2.2(c). DETECTORS

Photocell: The photoelectric effect, see Section 1.1, is used. The photocell consists of three layers, as shown in Figure 15-9. Exposed to the light is a transparent collector film, in general, a very thin layer of silver. Light will pass it and is absorbed in a second layer, mostly selenium or

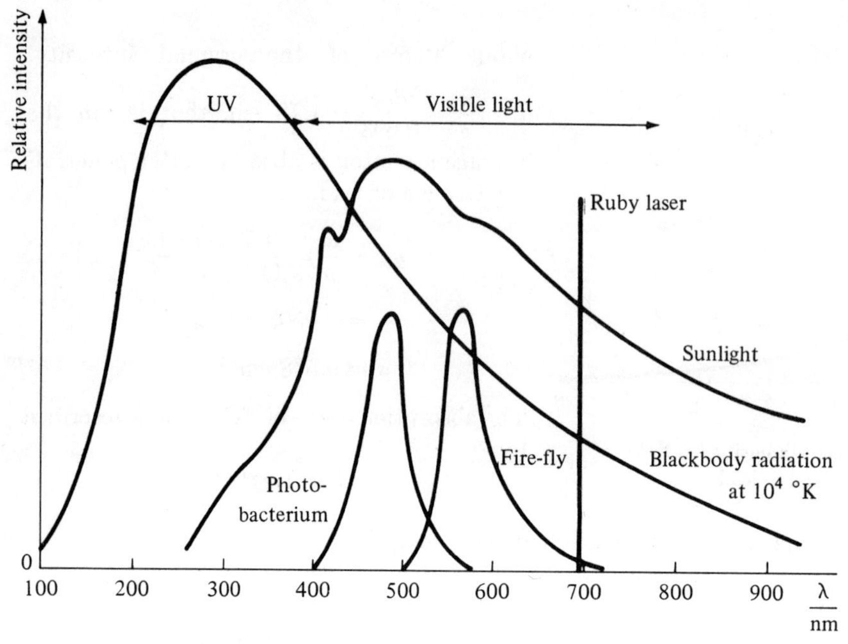

Figure 15-8 Emission spectrum of some natural and man-made light sources. Note that the ordinate scale is in arbitrary units. The intensity emitted by a bacterium is many orders of magnitude lower than for the black body. Only the laser is truly monochromatic: that means the spectral width is almost zero.

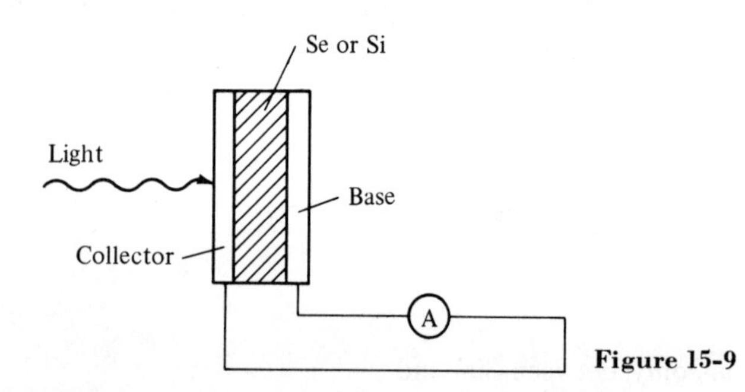

Figure 15-9

silicon. The photoelectrons dislodged from the atoms of this layer represent an electric current between collector film and base plate, the third layer. The current measured by the ammeter is directly proportional to the light input. Photocells are most sensitive in the ultraviolet region.

Advantages: No battery needed, very rugged.

Disadvantage: Slow response time (about 10^{-4} s).

Application: Solar cells as power source in satellites, exposure meter, transducer for calorimeter, and spectrophotometer.

Photomultiplier: Again the photoelectric effect is used. See Figure 15-10. Electrons knocked out of the photocathode are focused by electric fields

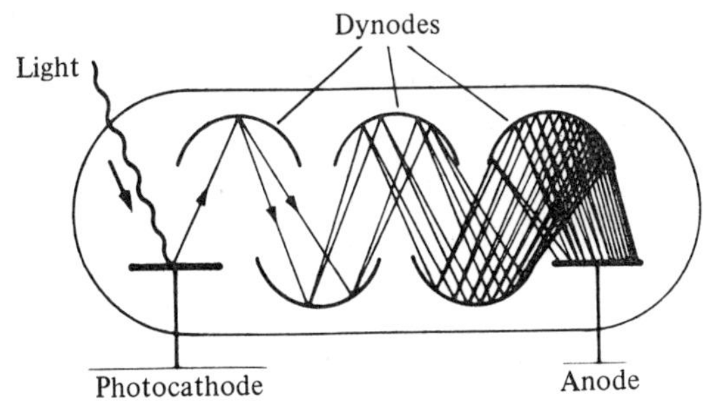

Figure 15-10

onto another electrode, called a *dynode*. Each electron striking frees more electrons, which in turn are directed toward the next dynode. The electron avalanche grows with each successive dynode and is finally collected at the anode. One electron leaving the photocathode produces up to 10^8 electrons at the anode.

The spectral sensitivity of a photomultiplier depends on the material used for the photocathode. Figure 15-11 shows a typical spectral response curve for a caesium-antimony photomultiplier.

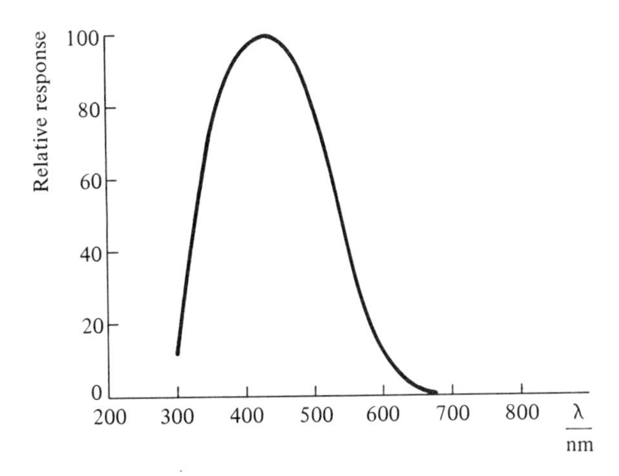

Figure 15-11 Spectral response curve for a photomultiplier (Philips 53 AVP). Maximum response is at 420 nm, that is, in the blue part of the spectrum.

Advantages: Extremely sensitive; fast response time (up to 10^{-8} s).

Disadvantages: Elaborate electric circuitry; mechanically fragile; large size compared with photocells.

Application: Detection of extremely weak light sources.

2.2(d). PHOTOMETRY. The visible region of the electromagnetic spectrum is of special significance to man. The intensity of a light source emitting in the visible region and the amount of visible light received are the object of a special field: *photometry*.

We must carefully distinguish between the luminous intensity of a light source and the illumination of a receiver.

Luminous Intensity. The luminous intensity (symbolized: I) of a light source is measured in candelas (or new candles), abbreviated cd.

DEFINITION: 1 *candela* = 1/60 *of the luminous intensity emitted by an area of* 1 cm^2 *of platinum having a temperature of* 2042.5°K (*melting point*).

The candela is one of the six basic units of the International System of Units. The definition is illustrated by some examples of luminous intensity; see Table 2. Instruments used to measure luminous intensity are called *photometers.*

Table 2 LUMINOUS INTENSITY I OF VARIOUS LIGHT SOURCES

Source	I Measured in Candelas
firefly	0.01
wax candle	1
kerosene lamp	30
reading light	50
black body at 2000°K and 1 cm^2	60
car headlight	100
electric arc (30 A)	8000
typical lighthouse	2×10^6

Illumination. Light impinging at a receiver illuminates it. The illumination (symbolized E) is measured in lux, abbreviated lx.

DEFINITION: 1 lux = *perpendicular illumination of* 1 m^2 *at a distance of* 1 m *from a point source emitting* 1 candela.

Table 3 shows a few numerical values for E. Instruments used to measure illumination are also called *photometers.*

Table 3 ILLUMINATION E BY VARIOUS LIGHT SOURCES

Source	E Measured in Lux
planet Venus	4×10^{-5}
moonless night	10^{-3}
firefly	10^{-2}
full moon	0.3
optimal reading light	50
overcast sky	300
perpendicular sunlight	3000

2.3 X rays

2.3(a). REGION. X rays were discovered in 1895 by Roentgen.* Their

* Wilhelm Conrad Roentgen (1845–1923), physicist. Nobel laureate 1901. He investigated the radiation discovered by himself in such detail that it was ten years before someone else made further contributions.

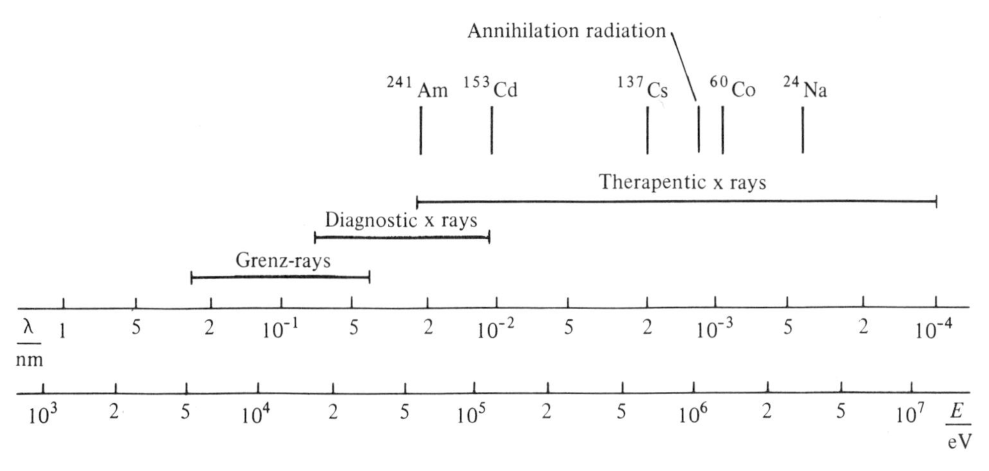

Figure 15-12

importance to the life sciences needs no stressing. Figure 15-12 shows this part of the electromagnetic spectrum which is called, by convention, the *x-ray region*. There is no principal difference between x rays and γ rays. The distinction stems from a difference in production.

2.3(b). Sources. X rays appear if a charged particle loses kinetic energy.

EXAMPLE: X-RAY TUBE

Electrons (negatively charged particles) are emitted by a hot filament at the cathode; see Figure 15-13. A potential difference between cathode and anode accelerates the electrons toward the anode. The electrons hit the target (the anode) and become absorbed. Only a small fraction of their kinetic energy is emitted as electromagnetic radiation. The remainder heats up the target material. Modern x-ray tubes achieve an accelerating potential difference up to 300 kV. This means that electrons leaving the filament with practically zero kinetic energy reach the anode with a kinetic energy of 3×10^5 eV.

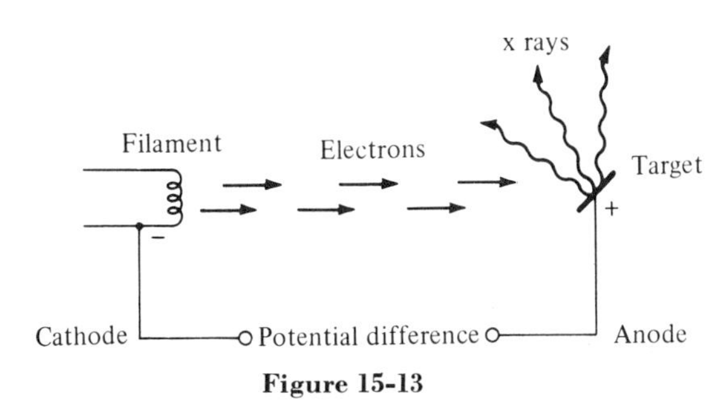

Figure 15-13

Higher electron energies and consequently higher x-ray energies are achieved by electron accelerators such as the Van de Graaff, betatron, synchrotron, and linear accelerators. Figure 15-14 shows the maximum energy achieved by various x-ray sources.

Gamma rays are emitted by radioactive sources. In contrast to x rays which display all wavelengths up to a maximum value (high energy limit), gamma rays are practically monochromatic. Figure 15-15

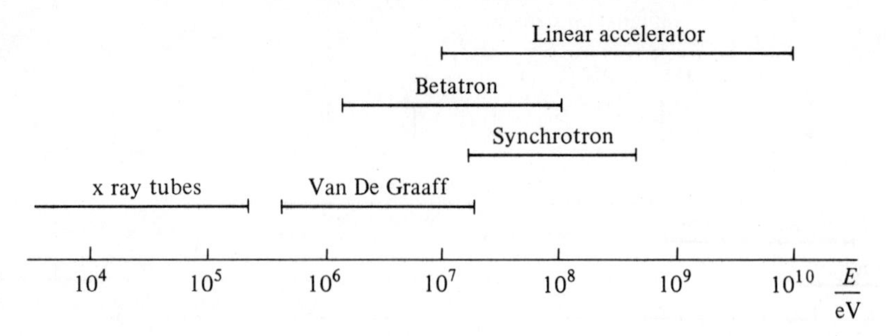

Figure 15-14

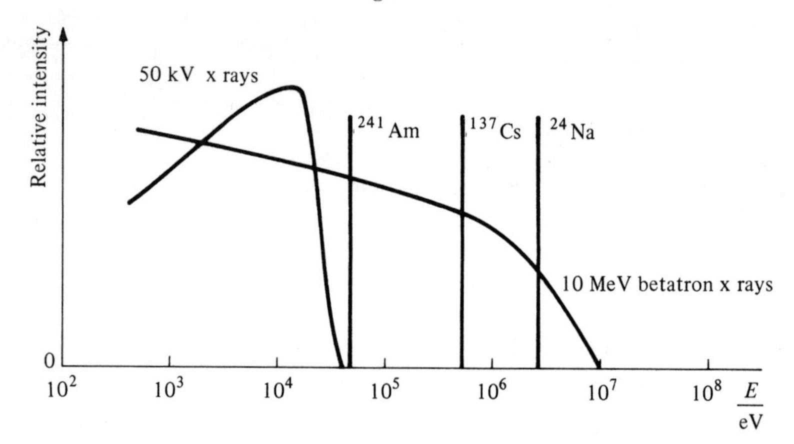

Figure 15-15 Spectra of some sources of high-energy electromagnetic radiation. X rays display a continuous spectrum, but γ rays show a line spectrum.

shows some typical spectra for both types of sources. Notice that most x-ray photons emitted carry an energy far below the high energy limit.

2.3(c). Detectors

Photographic Plate. X rays produce photoelectrons in the emulsion which in turn darken the photographic plate. Emulsions especially sensitive to x rays are loaded uniformly with traces of substances having a high atomic number. This increases the number of photoelectrons produced per impinging photon.

Fluorescent Screen. Some chemical compounds absorb short wavelength radiation and in turn emit photons having a longer wavelength. Uranium-barium compounds are especially suited to convert x rays into visible radiation. A layer of tiny crystals forms a fluorescent screen. This is an image converter and it allows observation of impinging x rays directly. For further details see Chapter 11, Section 3.4.

The same working principle employed by the fluorescent screen is used by the *scintillation detector*. See Figure 15-16. Some clear crystals like NaI (activated with Tl) emit flashes in the visible region if hit by high energy photons, for example, by an x-ray quantum. The flashes are too weak to be detected by the unaided eye. They are converted by an

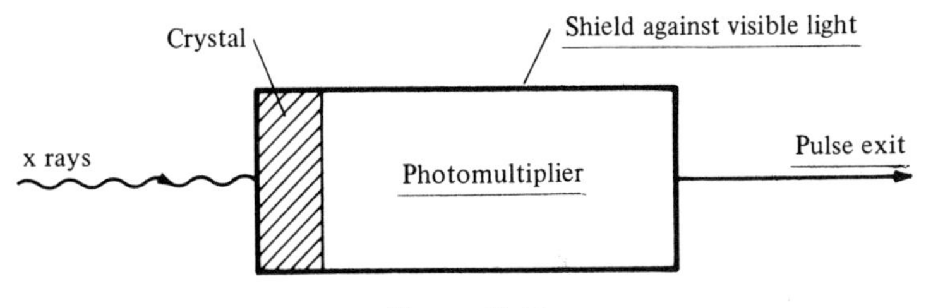

Figure 15-16

attached photomultiplier into electric pulses, which are counted. This is a two-stage transducer arrangement. Stage 1 performs wavelength conversion; stage 2 converts the optical pulse to an electrical pulse.

Ionization Chamber: An x-ray beam entering a gas produces ions and electrons along its path. See Figure 15-17. A potential difference across the

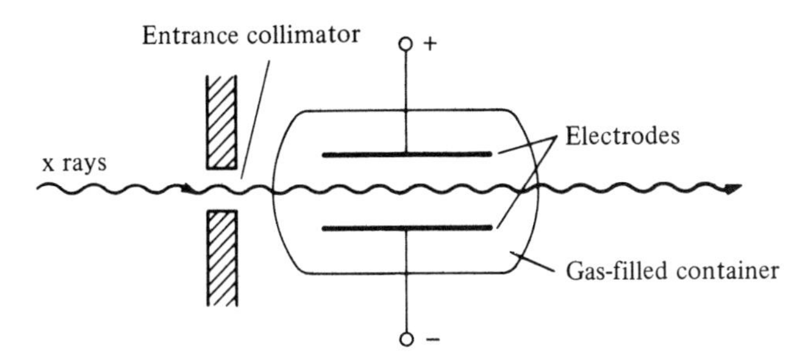

Figure 15-17

gas will collect the charged particles. The amount of charge collected at the electrodes is proportional to the intensity of the incoming x-ray beam. Ionization chambers are widely used to monitor x rays.

Main advantages: Instantaneous readings; sensitive, precise, and rugged.

2.3(d). Dosimetry of x rays. X rays are frequently encountered in the life sciences. Since they always represent a health hazard, it is worthwhile to introduce some basic information about x-ray dosimetry.

Exposure Dose. This characterizes an x-ray source as luminous intensity characterizes an optical radiation source. The exposure dose is measured in roentgen, symbolized *r*.

DEFINITION: 1 roentgen = *quantity of x rays which produces* 1.61×10^{12} *ion pairs in* 1 g *of air.*

Measurements show that 1 r deposits 87.6 ergs in 1 g air. An ionization chamber measures directly the exposure dose.

Absorbed Dose. This is equivalent to illumination in visible optics. Only the absorbed dose can be harmful; the exposure dose indicates the potential hazard of a radiation source.

The absorbed dose is measured in rads.

DEFINITION: 1 rad = *quantity of radiation which deposits an energy of* 100 ergs *per gram.*

Instruments used to measure the absorbed dose are called *dosimeters.* The absorbed dose depends on the material irradiated. Figure 15-18

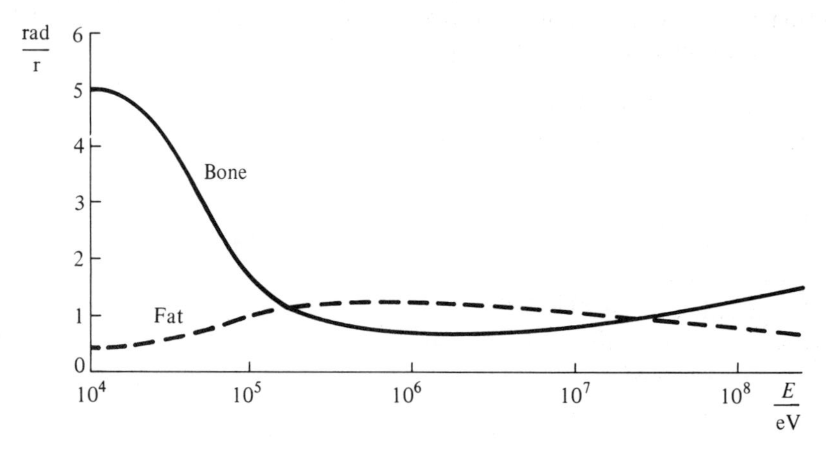

Figure 15-18 Radians per roentgen unit for bone and fat. The abscissa presents an extensive region of radiation.

shows the absorbed dose per roentgen exposure dose for two biologically relevant materials.

The unit rad, as opposed to the roentgen, is not limited to electromagnetic radiation. The absorbed dose of any type of radiation can be expressed in rads. However, experiments show that, even for the same absorbed dose, the caused biological effects vary for different kinds of radiation (see Section 3.4).

3. PARTICLE RADIATION

This type of radiation consists of a stream of particles such as electrons (e^-), positrons (e^+), protons (p), neutrons (n), and ions. The energy of these particles is always considered as kinetic energy, that is,

$$E = \frac{m}{2} v^2 \qquad (15.7)$$

where

E: energy of particle, usually expressed in electron volts (eV),
m: mass,
v: speed.

Remark: As soon as the particles move with very high speed, the relativistic mass increase must be taken into account. See Chapter 3, Section 2.2.

3.1 Sources

Radioactive nuclei are natural sources of charged particles. They are called *beta-emitters* if they radiate electrons or positrons. Alpha-emitters radiate α-particles. Particles are found in the cosmic radiation and are also emitted by the sun.

Man-made sources of particles are nuclear reactors and accelerators. Energies of more than 10^{10} eV can be achieved in the laboratory. Up to 10^{20} particles per second are emitted. The energy distribution of these particles may be a line spectrum, that is, all particles have practically the same energy or a continuous spectrum.

3.2 Detectors

The detectors described in Sections 2.2(c) and 2.3(c) are also employed to observe charged particles. One of the most versatile particle detectors functions on the same principle as the ionization chamber. See Figure 15-19. A potential difference extends across a gas-filled volume. As long

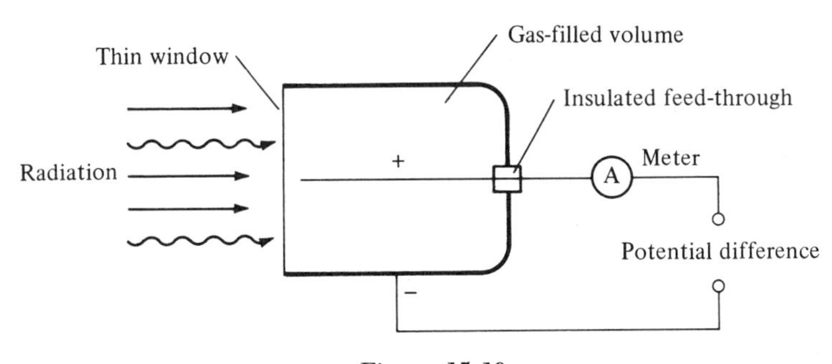

Figure 15-19

as no free charges are inside the electric field, no current is measured with the meter. Charged particles (and to a lesser degree, electromagnetic radiation of sufficiently high energy) entering the detector will ionize the gas. The ions move to their respective electrodes, and a current is indicated.

The response of the detector depends to a large degree on the applied voltage. If the potential difference is very small, the ions recombine before they can reach the electrodes. A very large potential difference

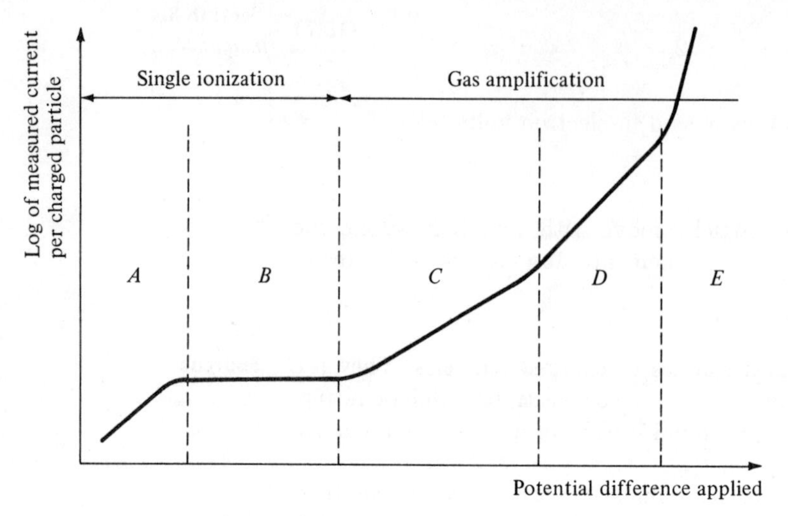

Figure 15-20 Characteristic curve for a gas-filled particle detector.

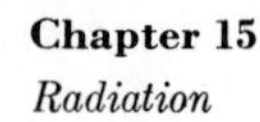

leads to an avalanche-like amplification for each initial ion. Figure 15-20 demonstrates the various regions of such a gas-filled particle detector.

Region A: The measured current per entering particle is very small due to recombination.

Region B: Region of saturation current. All produced ions reach the electrodes. The ionization chamber operates in this mode.

Region C: Proportional counter region. The output current is proportional to the energy of the charged particle entering. Discrimination between α- and β-particles is possible.

Region D: Geiger counter region. The output for each charged particle entering is identical.

Region E: Continuous discharge; will destroy the detector.

The potential difference actually applied depends on the size of the chamber, the type of gas, and the kind and energy of the particles to be detected. Orders of magnitude: region B, hundreds of volts; regions C and D, a few thousand volts.

3.3 Particle ranges

Particles lose energy when they penetrate matter. After a certain distance, they are absorbed. The particle's range is the distance penetrated. It depends on the kind of material, type, and energy of the particle. Figures 15-21 and 15-22 present ranges for electrons and α-particles. Notice the different scales in both figures. An electron will penetrate about a thousand times farther than an α-particle of the same energy.

Remark: To shield against charged particles, the range is not the only factor to be taken into account. For example, high energy ($E > 1$ MeV) electrons lose much of their energy by producing x rays. These x rays will penetrate far beyond the actual range of the electrons.

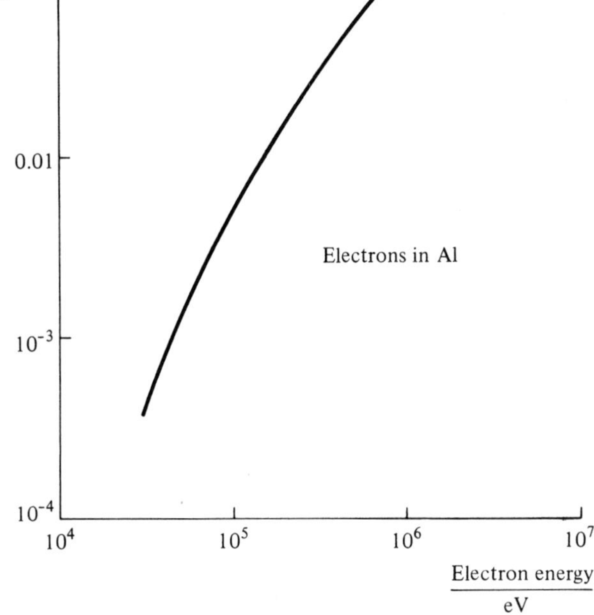

Figure 15-21 Range of electrons in aluminum. Note that both axes are logarithmic.

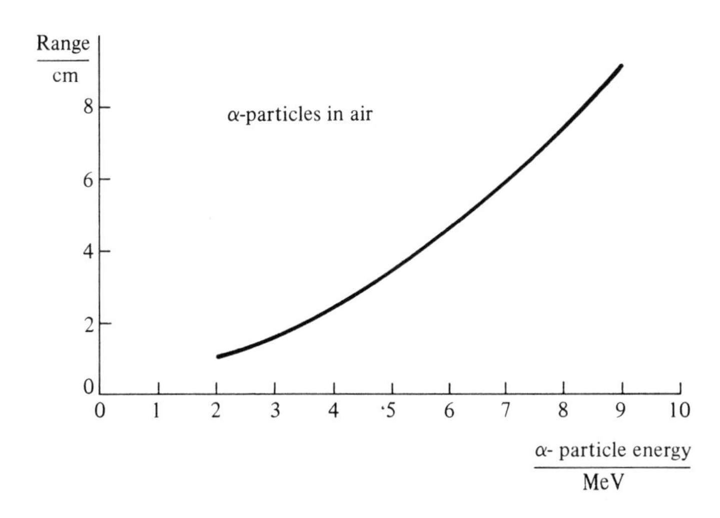

Figure 15-22 Range of α-particles in air. Note that both axes are linear.

3.4 General dosimetry

The absorbed dose due to radiation in general (this includes particle radiation) is expressed in rem (short for: *r*oentgen *e*quivalent *m*an).

DEFINITION: 1 rem = *that amount of radiation producing the same biological effect as is obtained from* 1 rad *of* 200-kV *x rays.*

Unfortunately, the rad cannot be used exclusively because the biological effectiveness of radiation depends not only on the energy absorbed but also on the type of radiation. Conversion:

$$1 \text{ rem} = 1 \text{ rad} \times \text{RBE}$$

where RBE is the *relative biological effectiveness.*

Table 4 presents some RBE values. Since it is very difficult to determine the RBE, the listed numerical values can serve only as a guide.

Table 4 RELATIVE BIOLOGICAL EFFECTIVENESS (RBE) OF VARIOUS RADIATIONS

Radiation	RBE
x rays	1 (definition)
electrons	1
protons	8–10
α-particles	15–20
neutrons	2–10

Federal standards state that the general population should receive not more than 0.17 rem of man-made radiation each year, exclusive(!) of medical sources. A person exposed to radiation in his occupation should receive less than 5 rem per year. For comparison, 50% of all people receiving 400 rem over a short period of time will die.

Table 5 shows the radiation dose absorbed by the average person in the USA during 1970. The sources contributing are indicated.

Table 5 ESTIMATES OF WHOLE-BODY DOSE ABSORBED BY THE AVERAGE PERSON IN THE USA IN 1970

Source	Absorbed Dose in rem
natural background	0.102
global fallout	0.004
nuclear power	0.000 003
diagnostics	0.072
radiopharmaceuticals	0.001
occupational sources	0.000 8

From National Academy of Sciences: *Report on Radiation Standards*, 1972.

4. APPLICATIONS

4.1 X-ray diagnostics

The x-ray tube acts as a point source. The radiation penetrates the object and is attenuated according to the various absorption coefficients inside. See Figure 15-23. Substances containing atoms ranging high in the periodic table (e.g., Ca of the bone) attenuate x rays more than fat and tissue, which consist mainly of C, H, and N. For this reason, bones appear darker than the surrounding tissue on the screen. The contrast

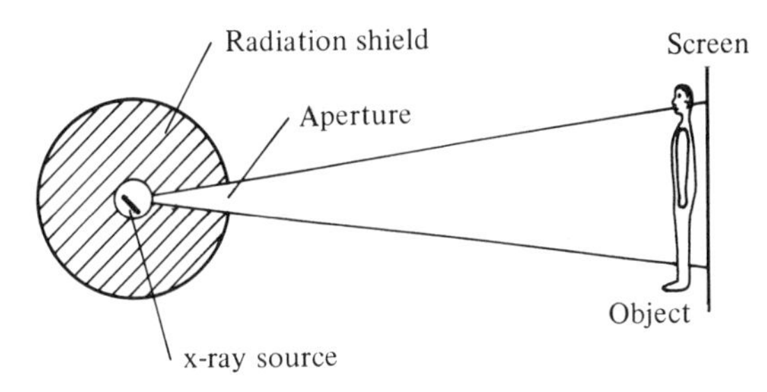

Figure 15-23

observed between the various kinds of tissues is low. Often it is possible to enrich organs with materials having a high absorption coefficient. Barium salt, for example, yields very high contrast x-ray pictures of the stomach and intestines.

4.2 Laser scalpel

Laser beams with diameters of a fraction of a millimeter are easily obtainable. The optical radiation is absorbed by a thin layer of tissue. Considerable energies are thus absorbed in a very small volume. A 300-watt laser deposits more than 71 cal per second. This is used in surgery. The absorbed radiation energy instantly coagulates tissue and cuts right through. There is almost no bleeding because the veins and capillaries are closed by the heat necrosis developed. The cut itself is very clean and is fast healing.

A laser beam is not only a tissue-cutting device but may also be employed in medicine for welding: In some eye diseases, the retina peels off its supporting structure, and irrecoverable blindness is the consequence. An extremely narrow laser beam directed through the eye's lens onto the retina will coagulate the spot it hits and weld the retina to its support. No more peeling off occurs after this spot welding.

4.3 Light absorption spectroscopy

The absorption of light follows the exponential law stated in Chapter 9, Section 5.3. Overall, the absorption coefficient changes slowly with changing wavelength. However, at one particular wavelength characteristic for each chemical compound, the absorption increases suddenly by orders of magnitude. This means that photons of this particular wavelength are strongly absorbed by this compound. This characteristic absorption is exploited for analysis.

The absorbance of a sample is experimentally determined as a function of wavelength. Figure 15-24 shows such an absorption spectrum. Position and relative height of the peaks are unique for each particular substance and can be used to identify it.

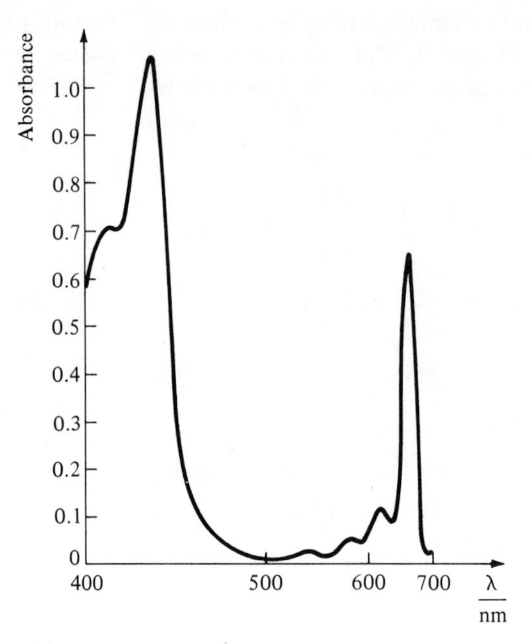

Figure 15-24 Absorption spectrum for a mixture of chlorophyll in diethyl ether. The peaks are called absorption lines.

Advantages of compound analysis by absorption spectroscopy:

1. Very small amount of substance needed.
2. Quantitative and nondestructive.
3. Can be set up as automatic and continuous analysis.
4. Study *in vivo* sometimes possible.

EXAMPLE: ABSORPTION SPECTRUM OF THE SUN

The photosphere of the sun emits a continuous electromagnetic spectrum. As light passes the sun's atmosphere from below, each chemical element or compound absorbs its characteristic wavelength. Thus the sun's spectrum shows many dark lines, each indicating a characteristic absorption at this particular wavelength. See Figure 15-25.

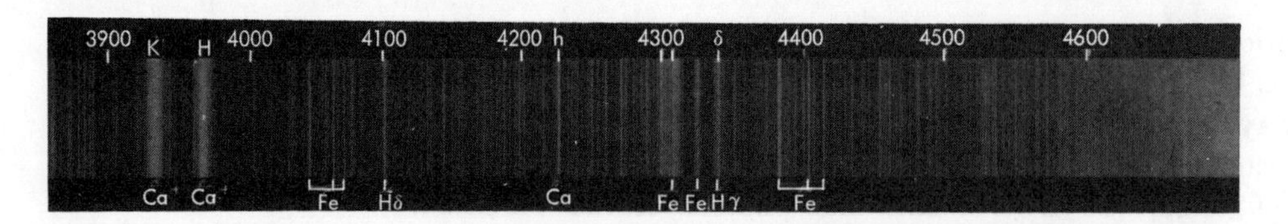

Figure 15-25 The solar spectrum from 3900 Å to 4600 Å. Frauenhofer lines indicate characteristic absorption by elements and compounds. A few lines are identified.*

* Joseph Frauenhofer (1787–1826), lensmaker. He refined optical instruments, especially the spectrometer. Discovered in 1815 the absorption lines in the sun's spectrum.

SUMMARY

Energy spreads through space by means of *radiation*. The distinction between particle radiation and wave radiation is irrelevant. The mathematical treatment of both aspects yields identical results.

Planck's relation

$$E = \frac{hc}{\lambda} \tag{15.2}$$

connects wavelength and energy of radiation.

De Broglie's relation

$$\lambda = \frac{h}{p} \tag{15.5}$$

connects wavelength and magnitude of momentum.

The electromagnetic spectrum spans from 10^3 m (long radio waves) to 10^{-10} m (cosmic radiation). The regions of visible light, 7.8×10^{-7} m $> \lambda > 3.9 \times 10^{-7}$ m, and of x rays, λ below 2×10^{-8} m, are of special interest in the life sciences.

Radiation penetrating matter is attenuated according to an exponential law,

$$I = I_0 e^{-\sigma d} \tag{15.6}$$

The absorption coefficient σ is characteristic for the substance penetrated and the wavelength employed.

Particle radiation (electrons, positrons, protons, neutrons, ions, etc.) is characterized by rest mass, charge carried, and kinetic energy. Its attenuation by a substance does not obey a simple exponential law. The range of a particle depends on its energy and the absorbing medium. If an absorber exceeds the range, no particle radiation will pass through.

Measuring the amount of radiation (photometry, dosimetry), we must carefully distinguish between the amount emitted by the source (luminous intensity, exposure dose) and the amount absorbed by a receiver (illumination, absorbed dose).

New quantities introduced in this chapter are presented in Table 6.

Table 6

Quantity	Symbol	Measured in Units of
absorption coefficient	σ	cm^{-1}
particle range	R	cm
luminous intensity	I	candela (cd)
illumination	E	lux (lx)
exposure dose		roentgen (r)
absorbed dose		rad
radiation dose		rem

PROBLEMS

1. Calculate the shortest wavelength of x rays emitted by electrons of the energy E. Express λ in meters and E in eV.
2. Calculate the wavelength of a 5-MeV proton.
3. Calculate the minimum wavelength emitted by a 50-keV x-ray tube.
4. If a positron and an electron interact, both will annihilate. Calculate the annihilation energy in MeV. Calculate the wavelength. Perform the same calculations for proton and antiproton.
5. Calculate the number of photons emitted per second by a 100-W light bulb. Assume that 20% of its energy is emitted in the visible region.
6. A 1-MeV γ-source is to be shielded by lead. Calculate the thickness necessary to reduce its radiation to 1%. Absorption cross section $\sigma = 0.79\ \text{cm}^{-1}$.
7. The intensity of a 1-MeV γ-ray beam is to be measured by a calorimeter. How thick must the lead detector be to absorb 95% of the incident radiation? Estimate its thickness if x rays having a maximum energy of 1 MeV are to be measured.
8. A 100-g water sample receives 750 rad. Calculate its rise in temperature.
9. The intensity of a 100-W light bulb is to be measured by a calorimeter. The distance from the bulb to the calorimeter is 1 m, and the surface area of the calorimeter is 1 cm^2. The lead absorber has a mass of 10 g. Plot the rise in temperature of the thermally isolated detector.
10. The threshold energy of the photoelectric effect in Na is 2.2 eV; for carbon it is 4.4 eV. Plot quantitatively the kinetic energy of the electrons knocked out by light having a wavelength between 2 nm and 5 nm.
11. High energy x rays are difficult to detect by a photographic plate because the energy loss in the emulsion is low. How could you improve the sensitivity of this detector without changing the emulsion?
12. A betatron produces 70 roentgen per minute. The radiation is emitted in bursts of 6 μs duration, 50 bursts per second. Calculate the radiation output during a single burst.
13. Figure 15-26 shows the spectral transmission of light through water. Plot the light spectrum available to organisms under water.
14. A 50-kV x-ray tube operates with a current of 50 mA. 0.5% of the energy absorbed by the anode is emitted in the form of x rays. How many calories per second are dissipated as thermal energy?

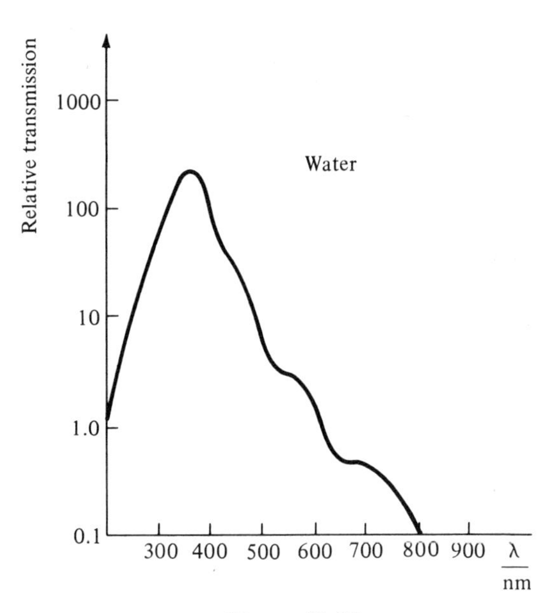

Figure 15-26

15. Use the definition of the roentgen to calculate the energy needed to produce one ion-pair in air.

16. The solar constant is the energy flux received from the sun on the earth. Its numerical value is about 2 cal · cm^{-2} · min^{-1}. Calculate the total energy output of the sun. The distance from the sun to the earth is about 1.5×10^{11} m.

17. Figure 15-27 shows the spectral transmission of standard silica glass. Plot the sun's spectrum seen through this glass.

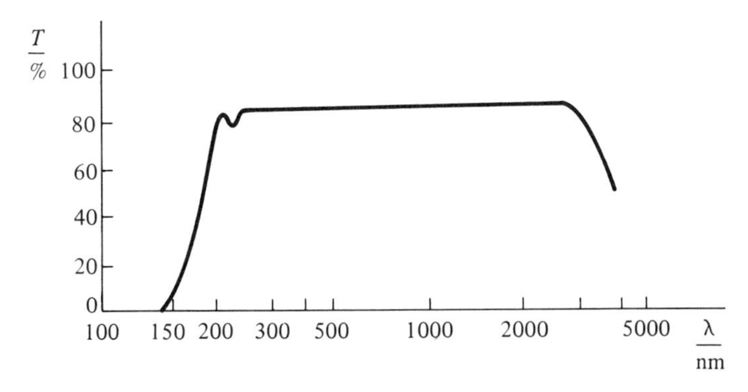

Figure 15-27

18. Calculate the theoretical resolution of a microscope using 1-MeV protons.

19. A patient receiving a conventional x-ray treatment suffers from a high dose on the skin. How could this surface dose be decreased while the absorbed dose deep inside increases?

20. Calculate the speed of a 1-MeV α-particle. How large is v/c? Calculate its relativistic mass increase.

FURTHER READING

H. L. ANDREWS, *Radiation Biophysics*, 2nd ed. Prentice-Hall, Englewood Cliffs, N. J. 1974.

R. GILLETTE, "Radiation Standards: The Last Word or at Least a Definitive One," *Science*, **178** (1972) 966.

E. F. GLOYNA, J. O. LEDBETTER, *Principles of Radiation Health*. Dekker, New York, 1969.

F. S. GOULDING, Y. STONE, "Semiconductor Radiation Detectors," *Science*, **170** (1970) 280.

H. E. JOHNS, J. R. CUNNINGHAM, *The Physics of Radiology*. Charles C. Thomas, Springfield, Ill., 1969.

D. W. NEWMAN, *Instrumental Methods of Experimental Biology*, Chapters 10 and 14. Macmillan, New York, 1969.

chapter 16

ATOMIC AND NUCLEAR PHYSICS

1. THE ATOM

The general textbook on atomic and nuclear physics contains about 500 pages and is still called an *introduction*. Even then the author will claim that he could present only a short glimpse of the topic. Therefore, although this chapter appears to be very dense, it still covers only a very limited selection. These pages can only set some signposts for further explorations.

1.1 Short history

Man has always pondered about himself and his surroundings. Who was the first to claim that all material things are made up of tiny and indestructible particles? We will never know. The first names handed down are those of Democritus* and Leukipp.† They conceived the world to be built of atoms (*atomos*—Greek word for inseparable). The various distributions of atoms, vacuum, and motion accounted for the multitude of phenomena observed throughout the universe.

These ideas never were entirely forgotten, but were dominated and overshadowed by others until the 12th century. Averroes‡ revived the atomistic concept of nature by introducing the *minima naturalia*, tiny material quantities much like the atoms. Although there was no way to prove their existence directly, the great scientists from the 17th century on accepted it as a working hypothesis. Huygens, Boyle, Newton, and

* Democritus of Abdera (460–370 B.C.), philosopher and scientist. Taught that happiness is independent of outside influences but rests with the cheerfulness of the soul.

† Leukipp of Milet (about 500–440 B.C.), philosopher, Democritus' tutor.

‡ Averroes, also known as Ibn Ruschid (1126–1198), physician and philosopher. Famous for his translations and comments to the writings of Aristotle.

Dalton developed the idea further. Avogadro (1776–1856) investigated gases experimentally and explained their physical properties assuming an atomic and molecular nature. The first table containing the relative atomic masses of more than 2000 compounds was compiled in 1818 by Berzelius (1779–1848). Once Meyer (1830–1895) and Mendeleev (1834–1907) had systematically ordered the elements in the periodic table, no doubts were left about the atomistic structure of matter.

The scattering experiments of Lord Rutherford (1871–1937) started the experimental quest for the fine structure of the atom. This search culminated in Hofstadter's (born in 1915, Nobel laureate 1961) electron-scattering experiments unveiling the fine structure of the atomic nucleus. In general, the theoreticians had a difficult time adjusting their models of the atom to the avalanche of experimental facts. Thomson (1856–1940) assumed that negative and positive particles are mixed uniformly throughout the atom's volume. In 1911, Rutherford saw only one way to explain his own α-particle scattering experiments: The atom is built of a positive nucleus, which carries practically all its mass, and of negatively charged particles (electrons) circling the nucleus. The nucleus has a diameter of one ten-thousandths of the atom. This picture is similar to our planetary system, only the scale is different.

In 1913, Bohr (1885–1962, Nobel laureate 1922) fused Rutherford's model with Planck's quantum concept to gain the first quantitative atomic model: The electrons orbit the nucleus at quantized distances. Absorption and emission of radiation is caused by orbital changes of the electrons. Bohr's model established order in the pattern of the spectral lines.

Chadwick (born in 1891, Nobel laureate 1935) discovered the neutron, another constituent of the atom, in 1932. In the same year, Heisenberg and Iwanenko (born 1904) assumed the atomic nucleus to be built out of neutrons and protons. Further theoretical developments led by Heisenberg, Schrödinger (1887–1961, Nobel laureate 1933), and Born (born 1882, Nobel laureate 1954) refined the model.

Although the experimental material is vast and the theories are many, a satisfactory model of the atomic nucleus is still lacking.

1.2 Size and components

As already mentioned, each atom comprises positive and negative elementary particles. Since the atom appears neutral if observed from the outside, the magnitude of both charges must be equal. The charges are spatially well separated within the atom. The nucleus occupies the center of the atom and comprises positive particles (protons) and electrically neutral particles (neutrons). Both particles are called *nucleons*. Like the planets around the sun, negatively charged particles, called *electrons*, orbit the nucleus. This is the atomic shell. The number of electrons is equal to the number of protons. Table 1 presents some numbers characterizing the atom.

Table 1 SOME NUMERICAL VALUES CHARACTERIZING THE ATOM

diameter of atom	10^{-10} m
diameter of nucleus	10^{-14} m
rest mass of electron	9.11×10^{-31} kg
rest mass of proton	1.67×10^{-27} kg
rest mass of neutron	1.68×10^{-27} kg
charge of electron = charge of proton	1.60×10^{-19} C (elementary charge)
number of atoms per cm^3 (gas under standard conditions)	2.69×10^{19}

1.3 Systematics

1.3(a). ELECTRON SHELLS. Electrons orbiting the nucleus shape the envelope of the atom. Each electron has its own unique orbit. Its total energy is

$$W = E_k + E_p$$

$$W = \frac{m_e}{2} v^2 - \frac{Ze^2}{4\pi\epsilon_0 r} \tag{16.1}$$

where

m_e: mass of electron,
v: orbital speed of electron,
e: elementary charge,
r: radius of electron orbit,
Z: atomic number,
$4\pi\epsilon_0 = 1.11 \times 10^{-10} C^2 \cdot s^2 \cdot kg^{-1} \cdot m^{-3}$.

The total energy W of the electron is negative, thus indicating that energy is needed to remove it from the atom.

Quantum theory postulates that only selected values of r are allowed. Ground state of atom: Each electron occupies its lowest energy orbit. This is a stable configuration of the atom.

Remark: It is by no means a matter of course that there are stable electron orbits at all. According to classical electrodynamics, each orbiting electron loses energy because it is accelerated (radially). Since experiments show that there are stable orbits, the inescapable conclusion is that the laws of classical electrodynamics cannot be applied to the atom indiscriminately.
Stable electron orbits are characterized by the following relation,

$$mvr = n \frac{h}{2\pi} \tag{16.2}$$

or

$$2\pi r = n \frac{h}{mv} = n\lambda$$

where

$n = 1, 2, 3, \ldots,$
λ: wavelength of orbiting electron,
h: Planck's constant.

Recall that electrons may be treated as matter waves (Chapter 15, Section 1.2). Then Equation (16.2) describes orbits for standing matter waves. In other words, the circumference of a stable electron orbit is an integral multiple of its matter wavelength.

The individual orbits are arranged in groups called *electron shells*. Each shell can accommodate only a limited number of electrons. The innermost shell, called the K-shell, has a capacity of two electrons; the following L-shell has space for eight; the M-shell also has space for eight; etc. If a shell is filled to capacity, it is called a *closed shell*. This is a very stable configuration. Atoms with closed shells are the inert gases such as He (2 electrons), Ne (2 + 8 electrons), and Ar (2 + 8 + 8 electrons).

The electrons occupying the outermost shell of an atom determine its chemical properties. This is basic to the repeat pattern in the periodic table of elements. See Figure 16-1.

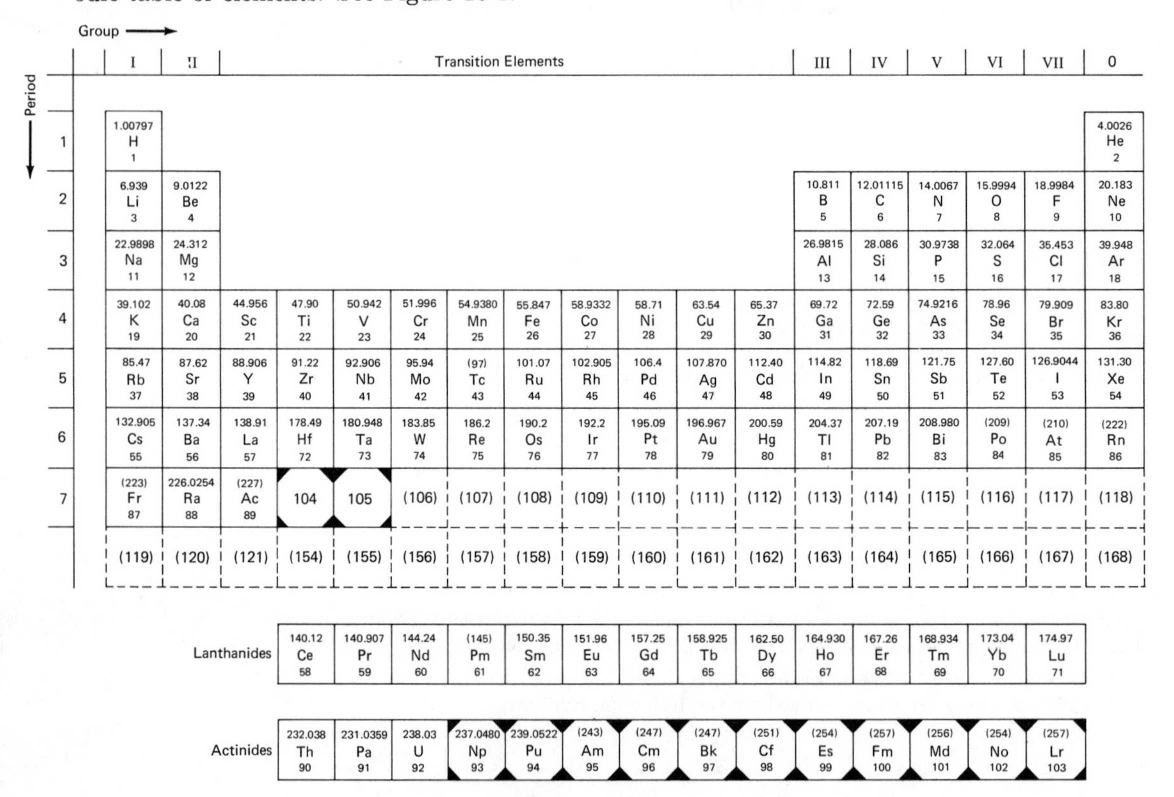

Figure 16-1 Periodic table of the elements.

Elements having the same number of electrons in the outermost shell are in the same group, and are displayed in the same column. For example, alkali metals all have one electron in the outer shell. The gradual differences from one alkali metal to the next are due to the other electrons deeper inside.

1.3(b). Nuclear shells. There is ample evidence that the nucleons of the atomic nucleus are also ordered in shells. Take for example the *magic nucleon numbers:* Nuclei having 2, 8, 20, 50, 82, and 126 neutrons or protons are more abundant than their neighbors. *Double magic nuclei* such as ^{4}He, ^{16}O, and ^{208}Pb have magic proton and neutron numbers. These are especially abundant chemical elements. Figure 16-2 bears that out.

Section 1.3
Systematics

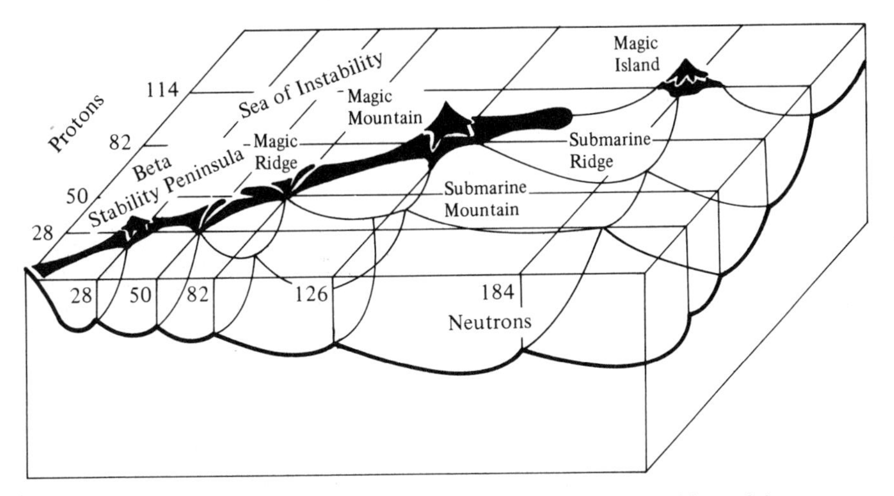

Figure 16-2 Nuclear stability. The dark regions are stable nuclei.

The stable nuclei form a peninsula in a *sea of instability.* Magic regions are represented by mountains or ridges. Far out in the sea you will discover the *magic island.* The shell model predicts that a nucleus with about 114 protons and 184 neutrons should be stable. These superheavy nuclei are not found (yet) in nature. Physicists in accelerator laboratories are actively at work to create them artificially.

Each atomic nucleus has its place in the chart of nuclides. This chart differs from the periodic table of elements because it is not ordered in groups. All known nuclei are placed according to atomic number. Figure 16-3 presents a section of such a chart of nuclides.

1.4 Excitation and de-excitation

The atom can absorb and re-emit energy. The amount of energy determines whether the atomic shells or the atomic nucleus are affected. Rule of thumb: Shell effects occur for absorbed energies up to 10^4 eV per atom; higher energies will in general influence the atomic nucleus only.

Once an atom has absorbed energy it is called an *excited atom.* Sooner or later it will free itself of this *excitation energy,* a process called *de-excitation.* The energy transfer can happen by means of collision, electromagnetic radiation, or particle radiation.

1.4(a). Atomic shell. Investigations show that orbiting electrons can only absorb discrete amounts of energy. This indicates that electrons

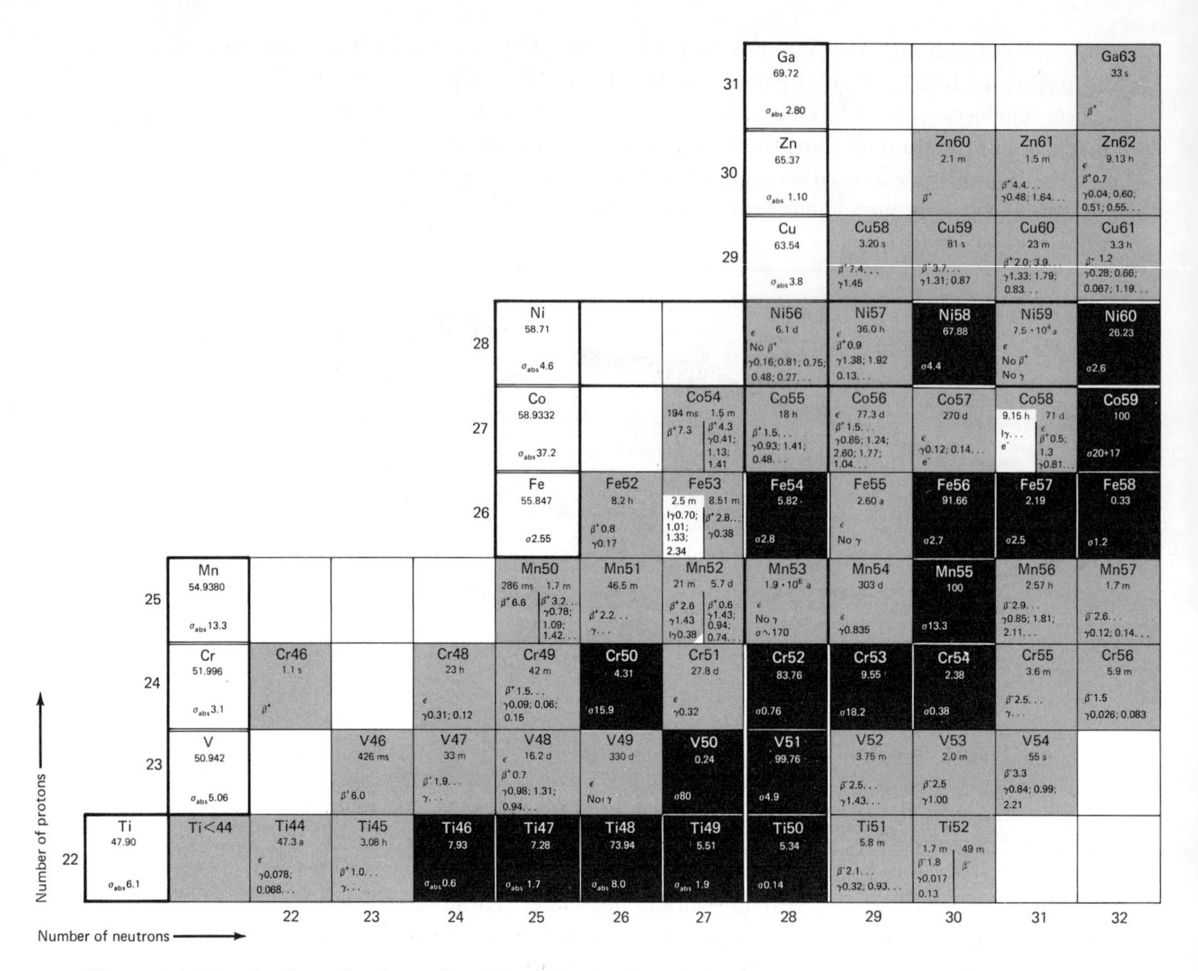

Figure 16-3 Section from the chart of nuclides. The number of protons increases in the vertical direction and the number of neutrons in the horizontal direction. For details see Section 3.5.

travel only in orbits allowed by the laws of quantum mechanics. Absorption or emission of energy by an electron means that it jumps from one orbit into another one. The exchanged energy during such a jump is,

$$h\nu = |W_i - W_f| \tag{16.3}$$

where

$h\nu$: energy absorbed or emitted by electron,
W_i: electron energy at initial orbit,
W_f: electron energy at final orbit.

Energy is absorbed if $|W_f| > |W_i|$ and emitted if $|W_f| < |W_i|$.

Binding energy of an electron: If the energy absorbed by an orbital electron exceeds a value W_b (binding energy), the electron is removed from the atom altogether. This is the *atomic photoeffect* (or photoelectric effect) mentioned in Chapter 15, Section 1.1. The atom deprived of one (or more) of its electrons is consequently positively charged and is called an *ion*. The minimum energy needed to produce an ion is called the *ionization energy*.

The binding energy depends not only on the kind of atom but also on the shell from which the electron is removed. Table 2 illustrates this.

Table 2 SOME ELECTRON BINDING ENERGIES

Element	Electron Binding Energy, Expressed in eV: At Innermost Shell	At Outermost Shell = Ionization Energy
H		13.5
C		11.3
Ne	8.67×10^2	21.6
Cs	3.60×10^4	9
Ba	3.74×10^4	5.8
Hg	8.31×10^4	10.4

Spectral Lines. Atoms irradiated by a continuous electromagnetic spectrum absorb only photons with an energy determined by Equation (16.3). The continuous spectrum shows dark lines at the corresponding wavelengths. Figure 16-4 shows some emission and absorption lines of the hydrogen atom.

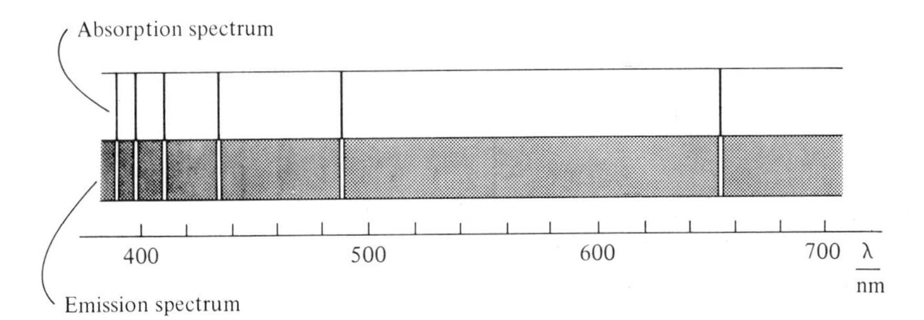

Figure 16-4

Each line represents an energy difference between two allowed electron orbits. Its wavelength is unique and may be used to identify a particular atom. Spectroscopy is a field where this is put to practical use. The wavelengths of unknown spectral lines are determined with up to 9 (!) significant figures and identified in previously compiled tables. The amount of substance needed for identification is extremely small. Quantitative and qualitative analyses are possible.

1.4(b). ATOMIC NUCLEUS. Analogously to the electrons of an atom, the nucleons absorb or emit only discrete energies. The excitation energies are about 10^4 times larger, that is, they range between 0.1 and 10 MeV. The monoenergetic electromagnetic radiation emitted during de-excitation of the nucleus is called *γ-radiation*.

Binding energy of nucleons: If the electromagnetic energy absorbed by a nucleon exceeds a certain value (binding energy), the nucleon leaves the atom. This is called *nuclear photoeffect*. The minimum energy needed to knock out a neutron or a proton is different for each kind of nucleus. It varies between 2 and 20 MeV.

1.5 Terminology

To present characteristic features of an atom or molecule, superscripts and subscripts surround its chemical symbol.

EXAMPLES:

Neutral nitrogen molecule:

mass number ⟶ 14, atomic number ⟶ 7, 2 ⟵ number of atoms per molecule

$$^{14}_{7}N_2$$

Excited helium:

$$^{63}_{29}Cu^{2+}$$

2+ ⟵ charge of ion

Ionized copper atom:

$$^{4}_{2}He^{*}$$

* ⟵ indicates excitation

Frequently Employed Terms:

Nuclide: A certain species of atomic nucleus characterized by mass number A and atomic number Z.

Radionuclide: An unstable nuclide. It emits particles or photons and finally converts into a stable nuclide. As an example, the radionuclide ^{14}C emits electrons and converts into the stable nuclide ^{14}N. A radionuclide may decay in several successive steps until it is converted into a stable nuclide.

Nucleon: A proton or a neutron.

Atomic number or proton number (symbolized Z): Number of protons per atomic nucleus. For example, $Z = 6$ for carbon.

Mass number or nucleon number (symbolized A): sum of nucleons of a particular nucleus. For example, $A = 235$ for an uranium nucleus having 92 protons and 143 neutrons.

Relative atomic mass (symbolized A_r): Mass of atom divided by $\frac{1}{12}$ of the mass of ^{12}C. Its values are listed in the periodic table of elements. Examples are 1.00797 (for H), 26.9815 (for Al), and 207.19 (for Pb).

Isotopes: Nuclides having the same number of protons, for example, $^{18}_{8}O$, $^{17}_{8}O$, and $^{16}_{8}O$.

Isotones: Nuclides having the same number of neutrons, for example, $^{2}_{1}H$ and $^{3}_{2}He$.

Isobars: Nuclides having the same number of nucleons, for example, $^{40}_{20}Ca$ and $^{40}_{18}Ar$.

2. NUCLEAR REACTIONS

The outer electrons in the atomic shells react with the orbital electrons of other atoms. These reactions belong in the realms of chemistry. The following brief treatment is restricted to reactions of the atomic nucleus, to reactions which change the composition of this nucleus.

2.1 Terminology

As in chemistry, a shorthand notation is used to describe a nuclear reaction:

initial nuclide (or target) (entering particle, or photon | leaving particle or photon) final nuclide (residual nuclide)

EXAMPLE:

$$^{14}N(\alpha, p)^{17}O$$

Nitrogen atoms are bombarded by α-particles (nuclei of He atoms). An α-particle is absorbed by the N-nucleus and a proton is emitted. A different atomic nucleus is left, the final nuclide ^{17}O.

Nuclear reactions obey the conservation laws. See Chapter 4. Energy and charge conservation are especially helpful in investigating nuclear reactions.

The foregoing reaction was the first artificial conversion of one chemical element into another one achieved by man. In 1919, Rutherford made the century-old dream of the alchemists come true: conversion of elements. Although the results of his efforts did not yield gold—he merely converted nitrogen into oxygen—it demonstrated the major premise. Nuclear physicists can even make gold out of mercury. The Museum of Natural History in Chicago possesses a tiny amount of artificially produced gold. Unfortunately, gold-making in this way is far too expensive.

Endothermal and Exothermal Reactions. A nuclear reaction is termed *endothermal* if, in addition to the entering particle, an amount of energy, called *activation energy,* is needed to achieve the reaction.

EXAMPLE:

In Rutherford's (α, p) reaction, the α-particle must have a minimum kinetic energy of 1.2 MeV to initiate the conversion.

A nuclear reaction is called *exothermal* if energy is released in addition to a particle.

EXAMPLE:

$$^{7}\text{Li}(p, \alpha)\,^{4}\text{He}$$

An energy of 17.2 MeV is released and appears as kinetic energy of both α-particles.

Probability of a Reaction. Not all particles or photons penetrating matter will cause a nuclear reaction. Even if their energies are sufficient, they must also hit their target, that is, they must get close enough to the nucleus to interact. The cross section (symbolized σ) of a nuclear reaction is a measure of the probability of the reaction. The cross section is expressed as an area; its unit is the *barn.*

$$1 \text{ barn (symbolized b)} = 10^{-28}\ \text{m}^2 = 10^{-24}\ \text{cm}^2$$

One barn is approximately the geometrical cross section of a heavy nucleus.

Remark: The preceding is a rather sketchy introduction of a reaction cross section. Keep in mind that the probability of a nuclear reaction is measured in barns. The larger the number of barns, the more likely the reaction.

In general, nuclear reaction cross sections are of the order of a few barns or millibarns, although there are some (n, γ)-reactions where σ rises to 10^7 barns.

2.2 Selected reactions

2.2(a). REACTIONS WITH PHOTONS. High energy photons may transfer their energies to one or more nucleons and knock them out of the nucleus. This nuclear photoeffect was mentioned in Section 1.4(b). A threshold energy between 2 MeV and 20 MeV (depending on target nuclide and emitted particle) must be exceeded by the photon. The reaction cross sections are of the order of millibarn (mb).

EXAMPLE:

$$^{12}\text{C}(\gamma, n)\,^{11}\text{C}$$

The threshold energy is 18.7 MeV. The entering photon must have at least this much energy to cause the emission of a neutron.

2.2(b). REACTIONS WITH NEUTRONS. Neutrons carry no charge and penetrate easily into the atomic nucleus. Consequently, neutron-induced reactions are most important in nuclear technology. In addition, there are very powerful man-made neutron sources: nuclear reactors. The reaction cross sections are of the order of many barns.

EXAMPLE:

$$^{14}N(n, p)\ ^{14}C$$

This reaction is used to produce the radionuclide ^{14}C which is widely applied in the life sciences.

2.2(c). Reactions with protons. Since the atomic nucleus is positively charged, the protons (as any other positively charged particle) must have sufficient kinetic energy to overcome the repulsive electrostatic force. This energy is of the order of MeV and is supplied by modern particle accelerators such as the cyclotron, sychrocyclotron, cascade generator, or the linear accelerator. The reaction cross sections are of the order of fractions of a barn.

EXAMPLE:

$$^{7}Li(p, \alpha)\ ^{4}He$$

This reaction is of historical interest. It is the first nuclear reaction where the entering particle (here the proton) was accelerated by man. Cockroft* and Walton† did it in 1930 at the Cavendish Laboratory.

2.2(d). Nuclear fission. Hahn‡ and Strassmann§ discovered in 1938 that an atomic nucleus may break apart entirely if bombarded by neutrons. In general, two heavy fragments and some additional neutrons result. The nuclides ^{235}U, ^{239}Pu, and ^{233}U already show *fission* if hit by thermal neutrons (energy = 0.025 eV).

EXAMPLE: FISSION OF ^{235}U

Because there is more than one residual nuclide, the fission process is presented in a different, though obvious, notation.

$$^{235}_{92}U + ^{1}_{0}n \curvearrowright ^{94}_{38}Sr + ^{139}_{54}Xe + 3\ ^{1}_{0}n$$

Here the fragments are strontium and xenon. Other fragments are also detected. Observe that on the average three additional neutrons are emitted. They may cause fission of further ^{235}U nuclides and thus start a *chain reaction*. More about this in Section 4.2.

2.2(e). Nuclear fusion. Very light nuclides such as ^{4}He may be built up from even lighter components. This process is exothermal and is called *fusion*. The components to be fused must have a high kinetic energy to overcome the repellent electrostatic forces. The most con-

* John Douglas Cockroft (1897–1967), Nobel laureate 1951.

† E. T. S. Walton (born 1903), Nobel laureate 1951.

‡ Otto Hahn (1879–1968), chemist. Nobel laureate 1944. Chemically isolated a number of radionuclides for the first time. He proved by chemical methods that irradiation of U and Th with neutrons causes fission. Developed an absolute method to determine the age of geological formations.

§ Fritz Strassmann (born 1902), chemist.

venient way to achieve this is to heat the components. For this reason, fusion is also called *thermonuclear* reaction. It is the main energy source of our sun.

EXAMPLE:

$${}^{3}_{1}\mathrm{H(d, n)}\,{}^{4}_{2}\mathrm{He} + 17.6\ \mathrm{MeV}$$

The gaseous mixture of ${}^{3}_{1}\mathrm{H}$ (tritium) and d (deuteron = ${}^{2}_{1}\mathrm{H}$) must be at about 10^{7}°K to sustain this thermonuclear reaction.

2.3 Application: activation analysis

Nuclear reactions are employed for qualitative and quantitative analyses of substances. The material to be analyzed is exposed to suitable radiation. Neutrons from the thermal column of a nuclear reactor are most popular. The induced nuclear reactions, for example (n, γ), lead to unstable nuclides which can easily be identified by the type and energy of radiation emitted by the residual nucleus. Since the number of activated nuclides is proportional to the intensity of the inducing radiation, a quantitative analysis is possible. In general, the unknown sample is irradiated together with a probe of known composition. A comparison of the radiation of both samples yields the composition of the unknown probe. For example, the amount of cobalt in a sample is determined by the reaction:

$${}^{59}\mathrm{Co}(n, \gamma){}^{60}\mathrm{Co}$$

Thermal neutrons convert the stable nuclide ${}^{59}\mathrm{Co}$ into ${}^{60}\mathrm{Co}$, a gamma-emitting radionuclide. The intensity of the radiation emitted by the sample is proportional to its ${}^{59}\mathrm{Co}$ content. Activation analysis is extremely sensitive. See Table 3.

Table 3 LOWER LIMIT FOR ACTIVATION ANALYSIS

Element	Minimum in Gram Substance
Na	10^{-9}
Al	2×10^{-9}
Cr	5×10^{-10}
Mn	3×10^{-11}
Co	5×10^{-11}
Cu	5×10^{-10}

3. THE UNSTABLE ATOM—RADIOACTIVITY

3.1 Stable and unstable nuclides

About 1500 different nuclides are known. Only 20% of them are stable, that is, their structure and properties remain unchanged with time. Unstable nuclides—also called *radionuclides*—decay by emission of radiation, step by step, until they have been transformed into stable nuclides.

The vast majority of the unstable nuclides are produced artificially in the laboratory by means of nuclear reactions. Natural radionuclides observed on earth are uranium-, thorium-, and radiumisotopes, ^{40}K, ^{14}C, to name the most important.

3.2 Half-life

3.2(a). PHYSICAL HALF-LIFE. Radionuclides decay, that is, they emit radiation and convert into other nuclides.

EXAMPLE:

The radionuclide $^{14}_{6}C$ emits an electron and converts into the stable nuclide $^{14}_{7}N$. This conversion is symbolized by

$$^{14}_{6}C \xrightarrow{\beta^-} {}^{14}_{7}N$$

Experiments show that the number of radionuclides of the same kind decay according to an exponential law:

$$N = N_0 e^{-\lambda t} \tag{16.4}$$

where

N: number of radionuclides at time t,
N_0: initial number of radionuclides, at $t = 0$,
λ: disintegration constant,
t: time.

The *disintegration constant* is a characteristic number for each kind of radionuclide.

The *half-life* (symbolized $T_{1/2}$) is the time span for which $N = N_0/2$. After $T_{1/2}$ only half of the initial number of radionuclides is left. Relation:

$$T_{1/2} = \frac{0.693}{\lambda} \tag{16.5}$$

$T_{1/2}$ of a particular nuclide is a physical constant and cannot be altered by any means.

Equation (16.4) describes mathematically the decay curve of a radionuclide. See Figure 16-5. It is more convenient to display the decay curve in a semilog plot (see Chapter 1, Section 1.2). Then the decay curve appears as a straight line. See Figure 16-6.

Table 4 presents $T_{1/2}$ for a few radionuclides. The radionuclide with the longest known half-life is ^{209}Bi, where $T_{1/2} = 2 \times 10^{18}$ a. You will appreciate this number if you realize that the age of the universe is currently estimated to be 2×10^9 years! The shortest half-life is of the order of 10^{-16} s. This is the time light needs to pass through a small molecule.

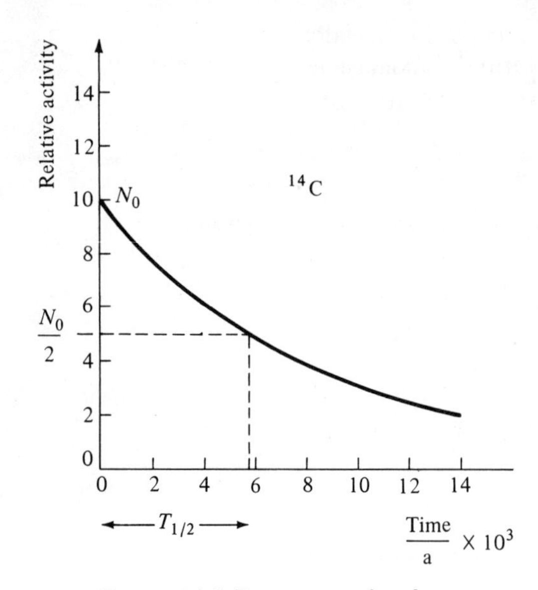

Figure 16-5 Decay curve for the radionuclide ^{14}C. $T_{1/2}$ = 5730 a, $\lambda = 3.83 \times 10^3\ s^{-1}$.

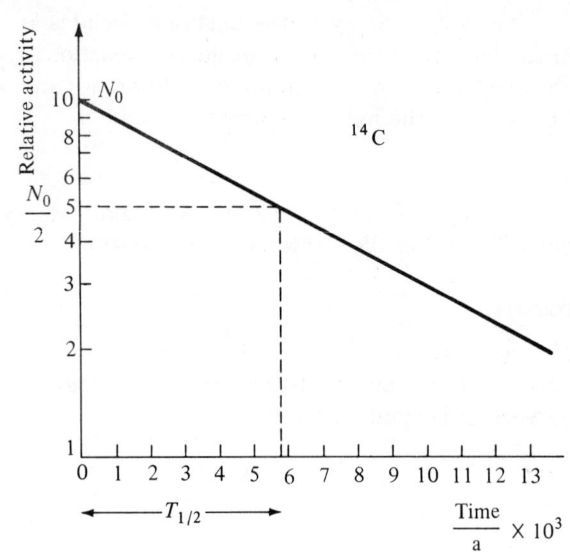

Figure 16-6 Decay of ^{14}C plotted in semilog form.

Table 4 Some radionuclides

Radionuclide	$T_{1/2}$	Biological Half-Life	Mode of Decay	Decay Energy in MeV
^{3}H (tritium)	12.3 a	19 d	β^-	0.02
^{14}C	5730 a	35 d	β^-	0.155
^{24}Na	15.1 h	19 d	β^-	1.4
^{32}P	14.3 d	1200 d	β^-	1.7
^{40}K	1.27×10^9 a	37 d	β^-	1.3
^{45}Ca	165 d	50 a	β^-	0.3
^{90}Sr	28.1 a	11 a	β^-	0.5
^{137}Cs	30 a	17 d	β^-	0.5, 1.2
^{140}Ba	12.8 d	200 d	β^-	0.5, 1.0
^{210}Pb (RaD)	22 a	531 d	β^-, α	0.02, 3.72
^{226}Ra	1620 a	55 a	α	4.78, 4.60
^{233}U	1.62×10^5 a	300 d	α	4.82, 4.78
^{239}Pu	2.44×10^4 a	120 a	α	5.16, 5.14, 5.11

3.2(b). Biological half-life. For applications in the life sciences, it is important to know how long a radionuclide remains in the human body once taken up. This is described by the *biological half-life*. Its definition is similar to the definition of physical half-life.

If the body initially contains N radionuclides, it will eliminate by natural means 50% of N within one biological half-life of this particular radionuclide.

The biological half-life usually refers to a specific organ. For numerical values, see Table 4.

EXAMPLE: ^{45}Ca UPTAKE

Calcium is an important constituent of the skeleton. It is constantly taken up by the body and built into the bones. If the radionuclide ^{45}Ca gets into the body, it is also deposited in the bones. It remains there practically for the lifetime of the individuum because its biological half-life is about 50 years. Since its physical half-life is only 165 days, all of it is long decayed after 50 years.

On the other hand, the biological half-life of ^{14}C is only 35 days, but its physical half-life is 5730 years. This indicates that it has left the body long before an appreciable amount had the opportunity to decay.

3.3 Activity

The *activity* of a radioactive substance is the number of decays per second. Activity (symbolized A) is measured in curie* (symbolized Ci).

DEFINITION:

$$1 \text{ Ci} = 3.7 \times 10^{10} \frac{\text{decays}}{\text{second}}$$

Remark: The reason for the numerical value 3.7×10^{10} is historical. One gram of radium shows about 3.7×10^{10} decays per second. This was previously used as a measure for the activity of a radioactive source. Table 5 presents some activities.

Table 5 ACTIVITY OF SOME RADIOACTIVE SOURCES

Source	A measured in Ci
core of nuclear reactor	10^6
sources for radiation treatment (Co-bomb)	10^3
1 g radium	1
accelerator target	10^{-3}
sources for tracer studies	10^{-6}–10^{-9}
permissible amount of tritium in 1 cm^3 of air	7×10^{-12}

Apparently, the curie is a large unit. In the life sciences activities are generally measured in micro- or nanocuries.

Specific Activity. This new quantity is introduced because a radionuclide is usually mixed with other inactive nuclides.

DEFINITION:

$$\text{specific activity} = \frac{\text{activity}}{\text{amount of radionuclide}}$$

The specific activity is expressed in Ci/g.

* Marie (1867–1934) and Pierre Curie (1859–1906), physicists. Nobel laureates 1903. Isolated in 1898 the radioactive elements polonium and radium. Marie Curie received a second Nobel prize in 1911. More about this most interesting husband-wife research team in Ève Curie, *Madame Curie,* Doubleday, New York, 1937.

EXAMPLE:

A radioactive sample of 0.02 g NaCl exhibits an activity of 0.1 Ci. Specific activity: $5\ \mathrm{Ci \cdot g^{-1}}$.

3.4 Mode of decay

Radionuclides in the strict sense of the word emit only α-particles ($^4_2$He nuclides), electrons, or positrons.

Emitted electromagnetic radiation (γ-rays) is secondary in nature: A particle decay led to an excited nuclide which de-excites by emitting electromagnetic waves.

EXAMPLE:

$$^{60}_{27}\mathrm{Co} \xrightarrow{\beta^-} {}^{60}_{28}\mathrm{Ni}$$

^{60}Co emits an electron and converts into the stable but excited nuclide ^{60}Ni. De-excitation is achieved by emission of a high energy photon (γ-ray).

3.4(a). ALPHA DECAY. General formulation:

$$^A_Z\mathrm{X} \xrightarrow{T_{1/2}} {}^{A\text{-}4}_{Z\text{-}2}\mathrm{Y} + {}^4_2\mathrm{He} \qquad (16.6)$$

where

^A_ZX: nuclide undergoing decay, also called *parent nuclide*,
A: mass number,
Z: atomic number,
$^{A\text{-}4}_{Z\text{-}2}$Y: decay product, also called *daughter nuclide*,
$^4_2$He: emitted α-particle,
$T_{1/2}$: half-life of decay.

The emitted α-particles exhibit a line spectrum, that is, they are emitted at one or a few discrete energies. Alpha decay is mainly observed for very heavy nuclides.

EXAMPLE:

$$^{228}_{90}\mathrm{Th} \xrightarrow{1.9\ \mathrm{a}} {}^{224}_{88}\mathrm{Ra} + {}^4_2\mathrm{He}$$

The parent nuclide ^{228}Th decays into the daughter nuclide ^{224}Ra; the half-life is 1.9 years. In this case—as often during α-decay—the daughter nuclide is not stable but is also radioactive. Figure 16-7 shows the spectrum of the emitted α-particles.

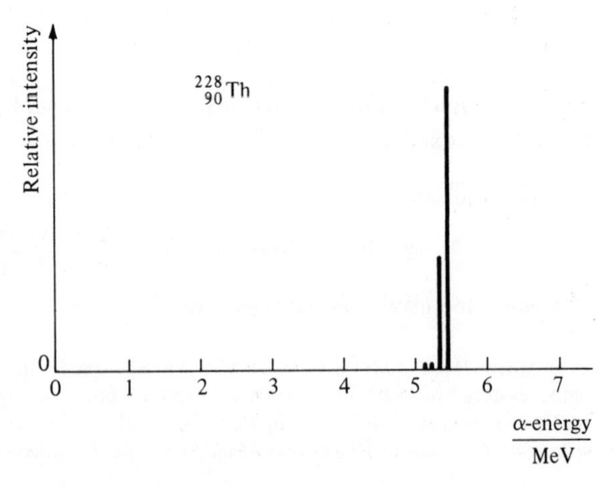

Figure 16-7 Energy distribution of α-particles emitted by $^{228}_{90}$Th. Four discrete lines are observed. 71% of all the α-particles are emitted with 5.42 MeV. The two lower energy lines represent less than 1% of all the α-particles.

α-particles lose their energy fast while penetrating matter. Fractions of a millimeter of aluminum are sufficient to shield against α-radiation. See Figure 15-22. An alpha-emitter incorporated into the body is nevertheless very dangerous because it loses its energy over a short distance. A living cell hit by an α-particle is usually irreparably damaged. If a gene is hit, the resulting point damage may lead to mutations.

3.4(b). BETA DECAY

Electron Emission. General formulation:

$$ {}^{A}_{Z}\mathrm{X} \xrightarrow{T_{1/2}} {}_{Z+1}^{\;\;\;A}\mathrm{Y} + e^{-} + \bar{\nu} \tag{16.7}$$

where

${}^{A}_{Z}\mathrm{X}$: parent nuclide,
${}_{Z+1}^{\;\;\;A}\mathrm{Y}$: daughter nuclide,
e^{-}: electron emitted,
$\bar{\nu}$: antineutrino emitted. This is an elementary particle having no rest mass and carrying no charge. A very elusive particle since it may penetrate the entire earth without interacting.

Beta decay is the conversion of a neutron inside the atomic nucleus into a proton. To conserve charge and momentum, an electron and an antineutrino are emitted. This underlying decay

$$ {}^{1}_{0}n \xrightarrow{10.8\ \text{min}} {}^{1}_{1}p + e^{-} + \bar{\nu} $$

can be observed directly. Free neutrons have a half-life of 10.8 min.

The emitted electrons exhibit a continuous spectrum.

EXAMPLE:

$$ {}^{14}_{6}\mathrm{C} \xrightarrow{5730\ \text{a}} {}^{14}_{7}\mathrm{N} + e^{-} + \bar{\nu} $$

The radionuclide ${}^{14}\mathrm{C}$ decays into the stable nuclide ${}^{14}_{7}\mathrm{N}$; the half-life is 5730 years. Figure 16-8 shows the energy spectrum of the emitted electrons.

Electrons penetrate much farther into matter than α-particles. See Figure 15-21. Fractions of a centimeter of aluminium are sufficient to shield against beta radiation.

Positron Emission. The second manifestation of beta decay is the emission of a positron. The general formulation for positron emission is:

$$ {}^{A}_{Z}\mathrm{X} \xrightarrow{T_{1/2}} {}_{Z-1}^{\;\;\;A}\mathrm{Y} + e^{+} + \nu \tag{16.8}$$

where

e^{+}: positron. An elementary particle much like the electron but of opposite charge,
ν: neutrino. An elementary particle much like the antineutrino.

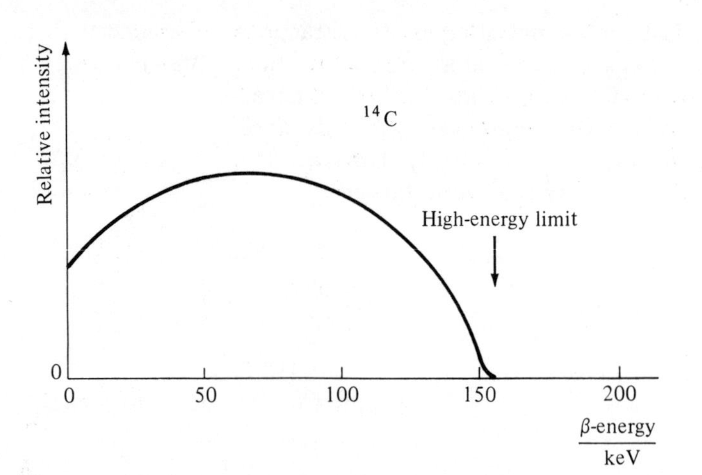

Figure 16-8 Energy distribution of electrons emitted by ^{14}C. The continuous spectrum has a high-energy limit of 1.55×10^5 eV.

This brief description should be sufficient since both types of beta decay are very much alike.

3.5 Reading a chart of nuclides

The pertinent data of stable and unstable nuclides are summarized on the chart of nuclides. Figure 16-9 shows a section of it together with explanatory remarks on how to read it.

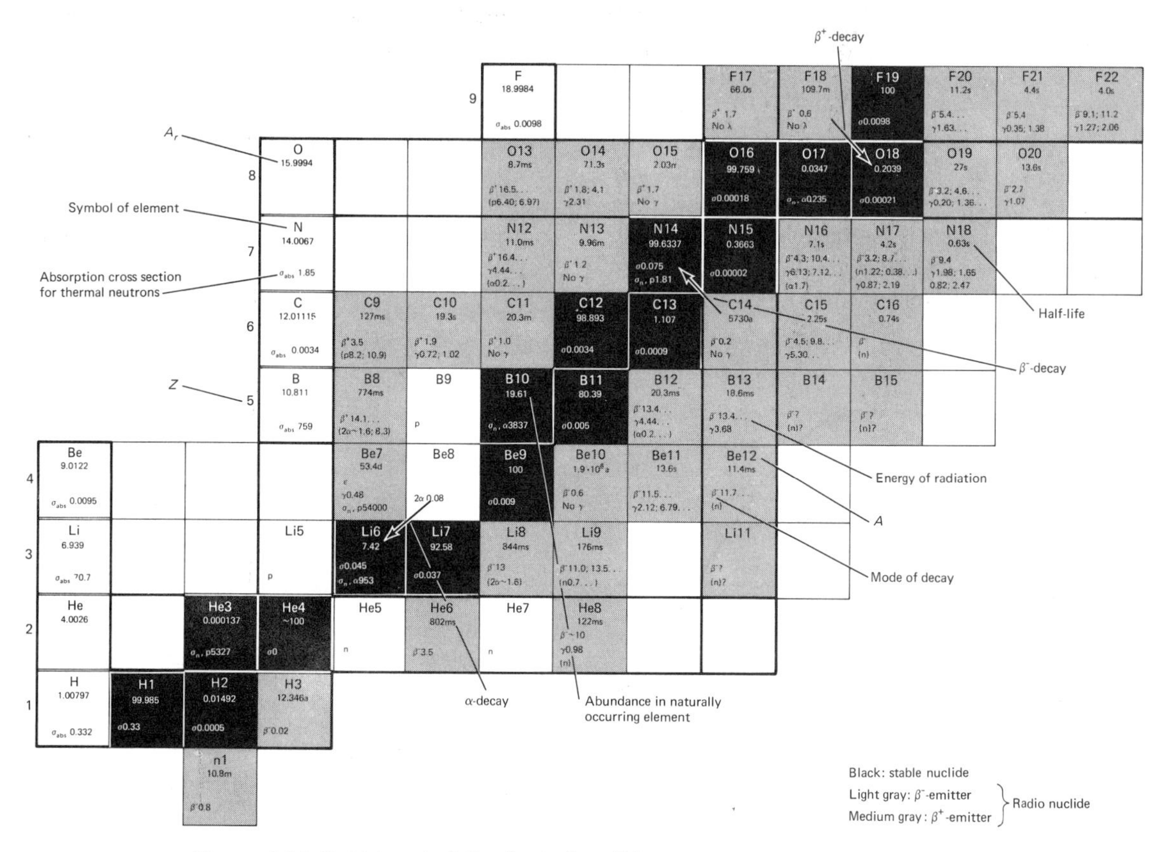

Figure 16-9 Initial part of the chart of nuclides.

4. APPLICATIONS

4.1 Tracer techniques

How do you find out where the ducks from your area fly to in wintertime? Simple: you *tag* the ducks and search for them once they are gone. In their winter quarters, most ducks look alike; however, they may be easily identified by their tags.

This technique is now widely applied to study complex processes, especially in the life sciences, to trace the destiny of biologically important metabolites. These metabolites are tagged by building radionuclides into their chemical structure. The radiation of the *tracer nuclides* always indicates the whereabouts of the metabolite.

EXAMPLE: LIFE SPAN OF HUMAN ERYTHROCYTES

Glycine is a building block for hemoglobin, which in turn is an essential component of the erythrocytes. A person is fed minute amounts of labeled glycine where some of the carbon atoms have been replaced by ^{14}C nuclides. Chemical properties depend only on the atomic number; hence the body is unable to distinguish between *normal* hemoglobin and *tagged* (or labeled) hemoglobin. Every day blood samples are taken and the blood activity is measured. Figure 16-10 shows the result.

Interpretation: The activity rises steeply during the first 20 days, indicating that more and more red blood cells containing the tracer are produced and released into the blood stream. For a period of about 80 days, these cells circulate. After that, the number decreases. They are replaced by cells having no label. From the curve we estimate the average lifetime of an erythrocyte to be about 130 days.

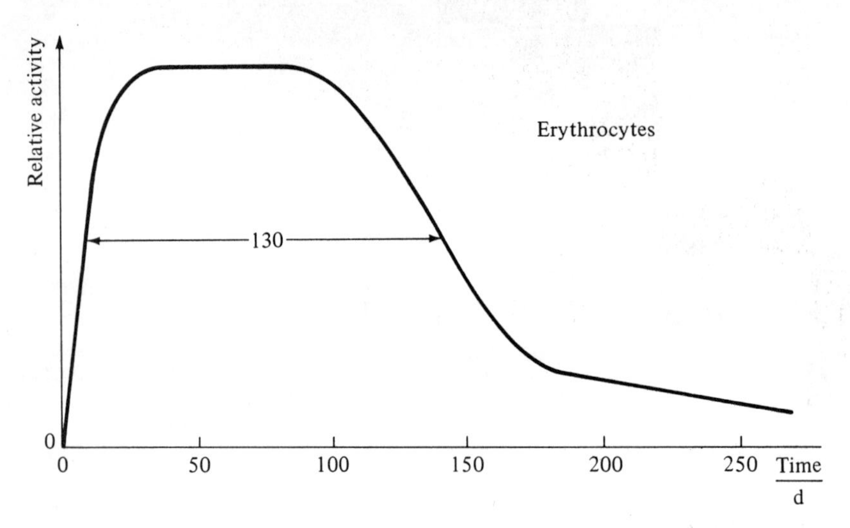

Figure 16-10 Activity of blood samples drawn after introduction of labeled hemoglobin at time zero.

The most widely used tracers are ^{14}C, ^{32}P, and ^{131}I. Their half-lives (see Table 4) are convenient, their production is inexpensive, and their stable counterparts are constituents of important metabolites.

The decay of radionuclides may also be used for dating; this method was described in Chapter 2, Section 2.2.

4.2 Nuclear power

During fission (see Section 2.2), energy and neutrons are released. These freed neutrons may under favorable conditions cause fission of more nuclides, which in turn release further neutrons, etc. A chain reaction is the result. See Figure 16-11.

If the number of fissioned nuclides is time-independent, a stable

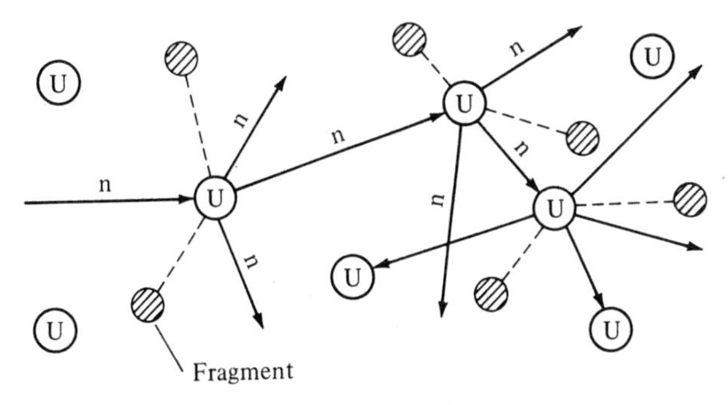

Figure 16-11

chain reaction is set up. This is the power source of a nuclear reactor. In 1942, Fermi* was the first to produce such a controlled chain reaction.

The control problem of a nuclear chain reaction is twofold: The number of neutrons available for fission must be an optimum. Too few neutrons causes the chain reaction to break off; too many causes the reactor to turn into an atomic bomb. Also, the energy of the neutrons must be an optimum. Only thermal neutrons ($E_k = 0.025$ eV) will cause fission of ^{235}U. Since the released neutrons have an energy of the order of 10^6 eV, they must be slowed down—*moderated* is the technical term. During this *moderation* process some of the neutrons are absorbed by other (competing) reactions and thus are lost for the chain reaction.

There are many technical solutions for a controlled chain reaction. At present the most economical one is the light-water (H_2O as opposed to heavy water d_2O) reactor using ^{235}U as fuel. The water serves as a moderator to slow down the fission neutrons to thermal energies.

Natural uranium reactors use the most abundant nuclide ^{238}U and fission by fast neutrons (energy > 1.5 MeV). They are called *breeders* because they produce as a by-product the fissionable nuclide ^{239}Pu.

What all reactor types have in common is that they are merely sources of thermal energy which, in turn, is conventionally applied to drive steam turbines. The main advantage of a nuclear power station over a conventional one is the low fuel consumption. A 1000-megawatt electric power plant needs 3×10^7 kg coal per day. But it needs only 4 kg fissionable material per day if built around a nuclear reactor.

Fusion reactors, which use the reaction described in Section 2.2(e), would be more desirable since they don't produce dangerous radioactive fission products. Also, their fuel supply of deuterium is practically inexhaustible and inexpensive. However, at present no feasible designs are in sight.

* Enrico Fermi (1901–1954), physicist. Nobel laureate 1938. He made important contributions to theoretical nuclear physics. Achieved on December 2nd, 1942, in Chicago, the first controlled chain reaction.

4.3 A prehistoric nuclear reactor

If you discover a uranium-ore deposit, it may not be worth as much as you hope. The analysis of a big deposit in Gabon showed that the most valuable nuclide ^{235}U did not contribute the usual 0.75% to the natural uranium but only 0.44%. The prospectors had discovered the remnants of a natural nuclear reactor long burned out. This is what must have happened:

At present only 0.75% of all natural uranium is the easily fissionable nuclide ^{235}U. The overwhelming fraction is ^{238}U. This admixture of 0.7:99.3 was not always so. Both nuclides are radioactive, each having a different half-life. Consequently, the admixture of both nuclides changes with time. About 1.7 billion years ago, the relation was about 3:97. This fraction of ^{235}U is sufficient to start a chain reaction in an uranium-ore deposit. See Figure 16-12.

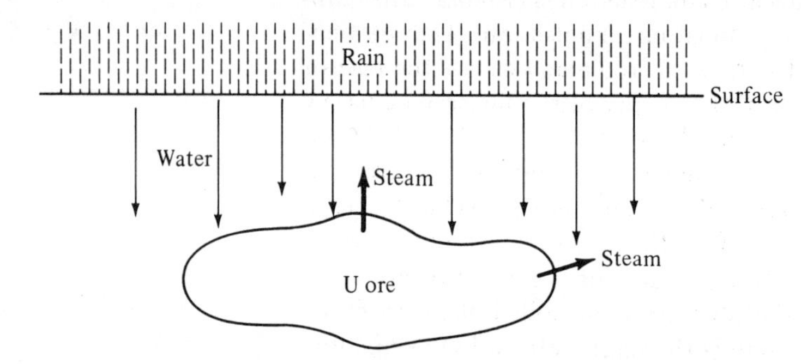

Figure 16-12

The deposit is deep under the surface. A neutron from the ever-present cosmic radiation may split a ^{235}U nuclide. Rain water that seeps down acts as a moderator for the released neutrons; more ^{235}U will be split. A contribution of 3% of ^{235}U would be just sufficient to foster the chain reaction. The reactor did not develop into an atomic bomb because the water turned into steam and the chain reaction broke off. It started again after the steam cooled and liquified. This pulsating reactor operated for perhaps 10^8 years; therefore about half of the ^{235}U was burned.

SUMMARY

Atoms are the building blocks of molecules, which in turn are the basic components of all material structures. Atoms have a diameter of the order of 10^{-10} m. The atomic nucleus has a diameter of the order of 10^{-14} m, occupies the center of the atom, and consists of protons and neutrons. These nucleons account for more than 99.9% of the total mass of the atom.

Electrons orbit the nucleus under the influence of electrostatic forces. The number of protons and the number of electrons are equal; the atom is electrically neutral.

A particular nuclide is described by the following notation:

$$^{A}_{Z}\mathrm{X}$$

where

X: symbol of chemical element,
A: mass number (sum of protons + neutrons),
Z: atomic number (number of protons).

The members of the periodic table of elements are ordered by increasing number of protons. Hydrogen opens with one proton; uranium has most protons (92) of all natural elements.

The electron orbits are bunched in individual shells, each having a finite capacity. An element with a closed shell (filled to capacity) is chemically inert and signals the end of a period in the periodic table.

Quantum theory postulates that only certain electron orbits are possible, that is, each electron's kinetic energy W is quantized. Consequently, only discrete amounts of energy may be absorbed or emitted by orbiting electrons.

$$h\nu = |W_i - W_f| \tag{16.3}$$

An energy amount $h\nu$ is exchanged if the electron jumps from the initial orbit (W_i) to the final orbit (W_f). Spectral lines observed in the electromagnetic spectrum reflect the various energy differences between possible electron orbits.

Particle or electromagnetic radiation may interact with the atomic nucleus and initiate a nuclear reaction. Most important to man is nuclear fission. In this case a high-Z nucleus (mainly ^{235}U, ^{239}Pu, ^{233}U) is split by the incident radiation. Besides two heavy fission fragments, additional neutrons are released. In a suitable arrangement, those neutrons in turn split more nuclides and thus cause a chain reaction. Large amounts of thermal energy are freed. The atomic bomb is a chain reaction that has run out of control, whereas in nuclear reactors, the chain reaction is stable.

Radionuclides convert spontaneously into other nuclides. They are characterized by half-life $T_{1/2}$ and mode of decay. Values for $T_{1/2}$ between 10^{18} years and 10^{-16} s have been observed. During alpha decay the radionuclide emits $^{4}_{2}$He nuclides. Beta decay is the emission of electrons (or positrons) accompanied by another elementary particle, the antineutrino (or the neutrino).

In the life sciences, the biological half-life of a nuclide is important. This is the time span during which half the amount of an incorporated nuclide leaves the biological system.

The activity of a radionuclide is measured in curies (Ci), where

$$1 \text{ Ci} = 3.7 \times 10^{10} \text{ decays per second}$$

New quantities introduced in this chapter are presented in Table 6.

Table 6

Quantity	Symbol	Measured in Units of
atomic number	A	pure number
mass number	Z	pure number
relative atomic mass	A_r	1/12 of ^{12}C
cross section	σ	barn (b)
physical half-life	$T_{1/2}$	s
biological half-life		s
activity	A	curie (Ci)
specific activity		Ci/g
disintegration constant	λ	s^{-1}

PROBLEMS

1. Calculate the average density of nuclear matter. Calculate the density of matter which is closely packed with hydrogen atoms. What conclusion do you draw from your calculations?
2. If an electron is removed from an inner shell of an atom, what will happen?
3. If the energy absorbed by an orbiting electron exceeds its binding energy by the amount E, what will happen to E?
4. To produce a perfect hydrogen plasma all hydrogen atoms must be ionized. How much energy is needed to ionize 1 cm^3 of hydrogen gas under standard conditions?
5. See Figure 16-4. Calculate the transition energies between the extreme right line and the other five lines.
6. Name the isotopes, isotones, and isobars of carbon-12 (see Figure 16-9).
7. What is the maximum diameter of a helium atom if liquid helium has a density of 0.12 g cm^{-3}?
8. Calculate the geometrical cross section of a hydrogen atom and a hydrogen nucleus (the radii are 5×10^{-9} and 1.2×10^{-12} cm, respectively). Express those geometrical cross sections in barns. How much area is covered by a 10^{-6} cm layer of hydrogen gas?
9. Helium is composed of two neutrons, two protons, and two electrons. If you add the masses of its constituents and compare them with the mass of the helium atom, you will find a difference, the mass defect. Express this difference in MeV. Explain its existence. Can it be utilized as an energy source?

10. A typical fission reaction is

$$^{235}_{92}U + n \rightarrow ^{95}_{38}Sr + ^{139}_{54}Xe + 2n$$

About 211 MeV are set free in the form of radiation and kinetic energy of fragments and neutrons. How many reactions are necessary to sustain a 100 megawatt power reactor if 80% of the reaction energy is transformed into heat?

11. Light having a wavelength 1.34×10^{-5} cm falls on a silver surface. The maximum energy of the emitted photoelectrons is 5.6 eV. Calculate the binding energy of the electrons.

12. What is the maximum energy of electrons emitted from a surface irradiated with light having a wavelength 4.1×10^{-5} cm if the first electrons are freed by a wavelength of 5.8×10^{-5} cm? What is the maximum speed of electrons leaving the surface?

13. Free neutrons have a half-life of 10.8 min. How far must a beam of thermal neutrons travel to lose half of its constituents due to beta decay?

14. Calculate the disintegration constants of ^{14}C and ^{32}P.

15. If 1 nanocurie of tritium is absorbed by the body, how long will it take until the incorporated activity drops to 4.6 decays per second?

FURTHER READING

HANNES ALFVÉN, *Worlds—Antiworlds*. Freeman, San Francisco, 1966.

R. M. EISBERG, *Fundamentals of Modern Physics*. Wiley, New York, 1967.

GENERAL ELECTRIC CO., "Nuclear Chart." Schenectady, N. Y.

F. HOYLE, *Galaxies, Nuclei, and Quasars*. Harper & Row, New York, 1965.

W. F. LIBBY, *Radiocarbon Dating*. University of Chicago Press, Chicago, 1955.

H. J. LIPKIN, *Beta Decay for Pedestrians*. Wiley-Interscience, New York, 1962.

E. SEGRÈ, *Nuclei and Particles*. Benjamin, New York, 1965.

APPENDIX

A. Information Retrieval

A textbook normally covers only a small amount of the available knowledge in a field. What can the reader do if he wants to know more about a topic and also wants his information to be up-to-date? Most likely, any question he asks has been answered before, written down, and stored. How can we retrieve this stored information? There are two principal lines of attack, and both lead us to the library:

1. *The keyword.* Determine under which keyword your question is covered, under which caption it would fall in a detailed dictionary. This keyword should not cover a wide field, but should not be too specialized either. If you want to know about the propagation of sound waves along the ossicular chain, the keywords *propagation*, *wave* and *chain* are clearly too general. *Sound waves* and *ossicular chain* are more appropriate. You can also find other keywords under which the topic may be covered. In our example, the keywords *cochlea*, *hearing*, and *inner ear* may contain useful information.
2. *The author.* If you are aware of a scholar who works in the field you want to know more about, you can search the files, indexes and abstracts for information supplied by him. In most cases, it is necessary to know the full name with all initials to sort out the various bearers of the same name.

In some files, institutions and corporations are listed as sources of information. This is a worthwhile source if we seek information regarding applied science and technology.

The flow chart (Figure A-1) is intended to guide you in your efforts to obtain information. The items in the boxes of the flow chart are explained in more detail in the following paragraphs. We shall follow the flow chart.

Textbook. It has to be an advanced one, but it will most likely not cover the subject in sufficient detail. The majority of textbooks contain information that is at least five years old.

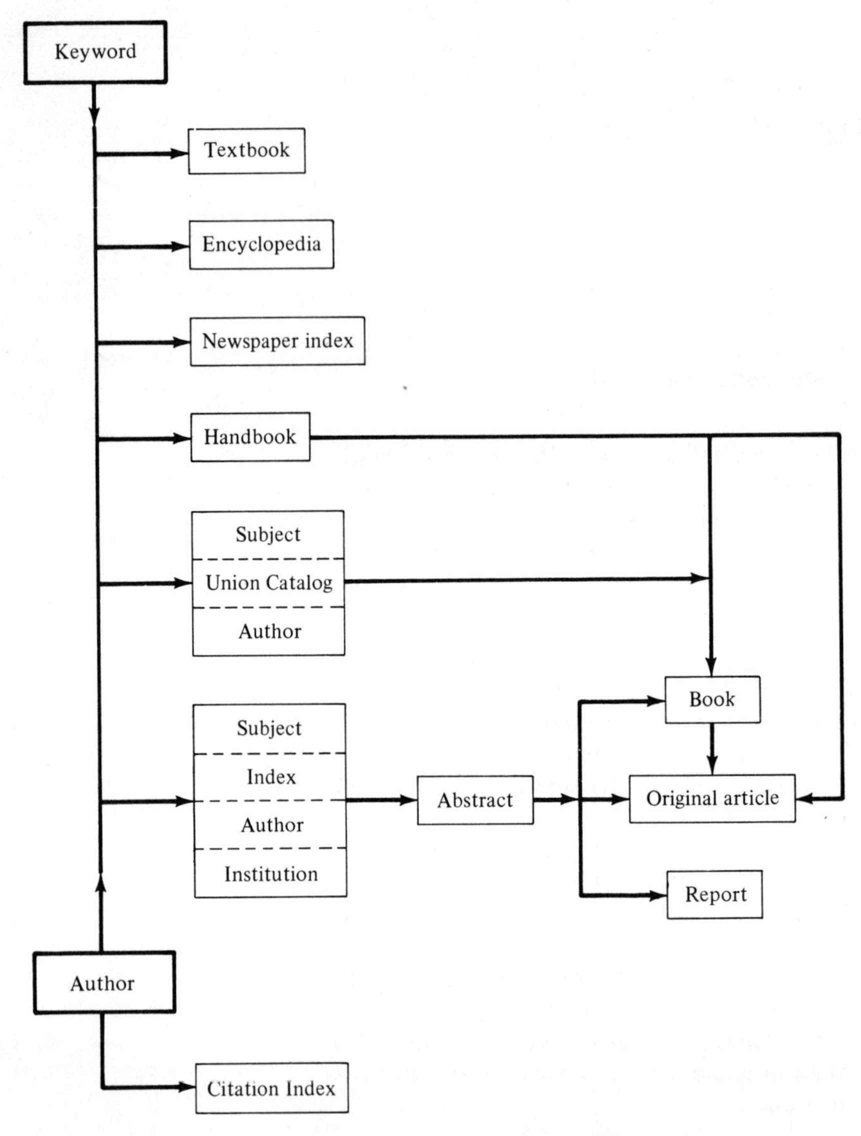

Figure A-1 Flow chart showing how to locate specific information. Your starting point is either a key word or an author's name.

Encyclopedia. This is not as odd an idea as it seems to be because, in good encyclopedias, reference is made to books which specialize in the topic. We may also learn which scientists or institutions are mainly involved.

Newspaper Index. Some readers may think: "You must be really desperate to look for knowledge there!" For recent discoveries and events, this is an excellent source because it usually reveals the author and the institution. The indices of the *New York Times* and the *Times* are particularly instructive.

Handbook. In all fields, once in a while (it can be a long while) an effort is made to accumulate and condense the findings in this field. A recent example is the *Encyclopedia of Physics*, where in more than forty volumes all aspects of physics are presented by the foremost men in the field. Although even a special handbook may not yield sufficient details, it always contains further references to specialized books or articles published elsewhere.

Union Catalog. Here you have at your fingertips the contents of the Library of Congress which has every book ever published in English and most foreign books. The catalog is ordered according to broadly defined subjects and also by author. Naturally it always refers to books. One disadvantage must be noted: Due to the high influx of books, the Union Catalog is continuously appended and the user must search in more than one volume for the same subject or author.

Index. This refers to publications such as *Nuclear Science Abstracts*, *Chemistry Abstracts*, *Biological Abstracts*, *Bioresearch Index*, and *Index Medicus*. Here the books and scientific journals of the appropriate fields are surveyed by an organization and classified under keywords and authors. This survey covers virtually all publications in the field and many in related fields. At regular intervals (one week to one month), a keyword and an author listing is published. To each listing there is attached a number which refers to the full title of the book or publication and includes an abstract (one to twenty lines) of the source. At the end of the year (sometime even quarterly) there appears an *Annual Cumulative Index* on subjects and authors. An important feature of these reference works is that they also survey reports (such as U.S. government documents) which usually never appear in the scientific journals.

The reader should not believe that it is a simple task to extract information from these indices. It is hard labor; for example, the *Biology Abstract* of 1970 alone covers more than a meter on the shelf and weighs close to one hundred pounds.

There are professional organizations around which will do the searching for you, but their services are not free of charge.

Citation Index. This is a most useful index if you want to know what happened to an invention or a discovery. Under the name of the inventor or discoverer you will find the names and publications in which he is cited. Looking into the mentioned papers, the reader can find out about the new developments.

For example, searching the indices is most fruitful, as demonstrated in this example. The (arbitrary) topic we are interested in is: transmission of electrons through matter. The obvious keyword is *electron* and the index to be searched is *Nuclear Science Abstracts* (*Physics Abstracts* will cover it too). Taking the *Annual Cumulative Subject Index* for 1967, we find the keyword *electron*; 46 titles are alphabetically listed according to the most significant word in the title. See Figure A-2.

The marked title in Figure A-2 seems to be interesting, it refers to Abstract No. 20:46939. All abstracts are listed by number in the volumes of *Nuclear Science Abstracts* preceding the annual index volume. On page 5728 of Volume 20, we find Abstract No. 46939; see Figure A-3.

The abstract gives the full title, the authors' name and address, the journal where it appeared (here *Physics Letters*, Vol. 23, pages 219–221, published in Oct. 17, 1966) and an abstract of the original paper. All we have to do now is find the shelf where the copies of *Physics Letters* are stored, take Volume 23, and open to page 219. There is the desired

use in sealing plastic bags containing alpha sources, 20: 45953 (ISO-SA-23)

ELECTRONS

angular and energy distributions from proton reactions with helium at 200 and 300 keV, (T), 20: 45294 (NASA-TN-D-3641)
content in ionospheric E and F regions, nighttime behavior by measuring, 20: 46460(T)
density and temperature in electrodeless discharges, probe measurements of, (E), 20: 46336
density distribution in cesium thermionic converters, (T), 20: 46329

synchrotron radiation in homogeneous fields, (T), 20: 46248
thermionic emission, time dependence of Schottky effect in, 23: 46946
transmission in solids, comparison of position and, 20: 46939
use for cancer therapy, compenson with other radiotherapeutic methods and triiodothyronine therapy, 20: 45558
use for radiotherapy, effectiveness and kinetics of, 20: 45567
use for therapy of head and neck malignancies, review of, 20: 45557 20: 45559
use of 6- to 15-Mev, for therapy of ear, nose, and throat tumors, 20: 45554

ELECTROPHORESIS

use in biophysical studies, 20: 45291(R) (UCLA-12-595, pp 67-71)

Figure A-2 Part of the listing found under the key word ELECTRONS. Only 12 of the 46 titles listed are shown.

abstract no.

full title

author

author's address

journal

text of abstract

46939 DIRECT COMPARISON OF THE PENETRATION OF SOLIDS BY POSITRONS AND ELECTRONS. Takhar, P.S. (Royal Military Coll. of Canada, Kingston). Phys. Lett., 23: 219-21 (Oct. 17, 1966).
A new experimental technique was used to investigate the absorption of 1.88-Mev positrons and 1.77-Mev electrons in C, Al, Cu, Sn and Pb. The results, which extend to two percent transmission of positrons, indicate greater transmission of positrons than electrons, in qualitative agreement with multiple scattering theory. (auth)

Figure A-3 Reproduction of the actual listing of the article title marked in Figure A-2.

article. See Figure A-4. This is the article of interest in the 1967 volume. There are certainly more to find in the volumes of the other years.

Hint: If the published article is a recent one (within the last two years, say), you can write the author and ask for a reprint. You might also request that he send you the most recent (consequently, not yet published) developments in form of preprints and reports. If you succeed in doing so, you will be really up to date (and will be able to quote results in your own publications with the prestigeous but, for the reader, most annoying reference: *private communication*).

DIRECT COMPARISON OF THE PENETRATION OF SOLIDS BY POSITRONS AND ELECTRONS

P.S. TAKHAR
Royal Military College of Canada, Kingston, Canada

Received 13 August 1966

A new experimental technique has been used to investigate the absorption of 1.88 MeV positrons and 1.77 MeV electrons in C, Al, Cu, Sn and Pb. The results, which extend to two percent transmission of positrons, indicate greater transmission of positrons than electrons, in qualitative agreement with multiple scattering theory.

Somewhat surprisingly, the penetration of positrons through solids and their corresponding differences from electrons have been subject to very little investigation, either experimental or theoretical. The theoretical scattering cross section is different for positrons and electrons. Rohrlich and Carlson [2] have calculated the energy loss in multiple scattering of positrons and electrons in Al and Pb. Seliger [3] observed a difference in transmission for positrons and electrons.

The purpose of the present experimental investigation was to determine the actual differences between positrons and electrons in their penetration of solids and to attempt to relate the experimental results to the theory available.

The positron source was used in the present work is $Ge^{-68}Ga$, which has a half-life of 280 days and emits 1.88 MeV positrons. The strength of the source was 0.1 mCi. The source was mounted at the center of a hollow steel sphere four inches in diameter (see fig. 1). The positron beam was obtained through an exit port 0.4 inch in diameter and 0.0006 inch in thickness. The pressure inside the sphere was maintained at 10^{-5} Torr throughout the observations. The absorbers of varying thickness were introduced against the window and transmitted beam was allowed to annihilate in an aluminum slab. The annihilation photons were detected by 2 X 1-3/4 inches NaI crystals coupled to Dumont 6292 photomultipliers. The positron source and geometry are shown in fig. 1, where A is aluminum slab, B are absorbers, PM stands for photomultiplier and CF for a cathode follower. The pulses from the two detectors were analysed by slow and fast coincidence techniques [3].

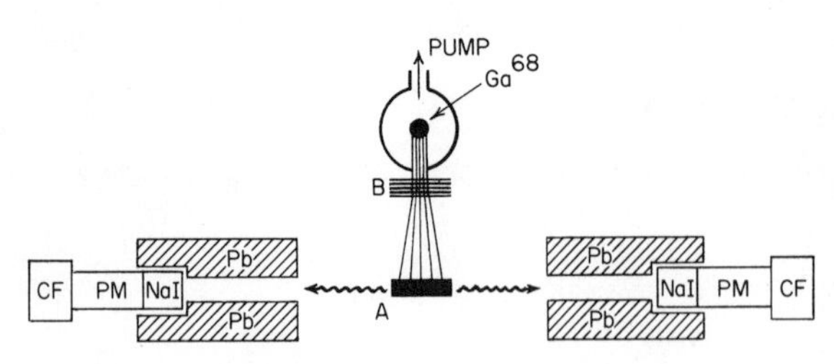

Fig. 1. The positron source and geometry.

Table 1

	C	Al	Cu	Sn	Pb
$\mu(e^+)$	4.25	5.65	7.15	9.39	9.53
$\mu(e^-)$	5.50	6.34	8.47	11.76	12.84
r^+/r^-	1.29	1.12	1.19	1.25	1.35

C, Al, Cu, Sn and Pb were investigated by the above method. The relative transmission of beam intensity X 100 is plotted against the thickness in mg/cm^2 for these elements in fig. 2.

Similar experiments have been performed to measure the penetration of electrons through solids by an analogous technique. The electron source was mounted at the center of the same sphere as was used for the positron source and a thin-window Geiger-Müller counter was used for the detection of the electrons. In order to keep the same geometry for both positrons and electrons the Al slab (A in fig. 1) was replaced by the detector window. The electron source used was ^{86}Rb which has a half-life of 18.7 days and emits 1.77 MeV electrons. In this manner quite similar curves were obtained for electrons. The transmission curves on a linear and semilogarithmic plot are very similar to those of positrons after background correction, hence there is no need to show these curves. However for investigating the comparative transmission of positrons and electrons, a plot of relative transmission versus thickness is shown in fig. 3 for both positrons and electrons.

Table 2

		C	Al	Cu	Sn	Pb
Experiment	r^+/r	1.29	1.12	1.19	1.25	1.35
Theory	Zd^+/Zd^-	-	1.06	-	-	1.35

Since the absorption of positrons and electrons is exponential to a good approximation, the absorption coefficients in different materials were calculated by a weighted least squares fit to the experimental data using an I. B. M. 1620 Fortran Computor. Table 1 shows the absorption coefficient $\mu(e^+)$ for positrons and $\mu(e^-)$ for electrons in g cm. Also the ratio of positron range to electron range, i.e. r^+/r^- is given in table 1 for C, Al, Cu, Sn and Pb. The experimental errors in the absorption coefficients were in the range of 4 to 7 percent.

The main features of the present results on transmission of positrons and electrons through material medium can be explained with the assumption that the only significant process by which electrons and positrons lose their energy is multiple scattering. In order to calculate the positron-electron difference in multiple scattering, Rohrlich and Carlson [1] have defined a theoretical parameter known as penetration depth Zd. According to them Zd represents the ability of a particle to penetrate. They calculated the ratio Zd^+/Zd^- in Al and Pb in low energy range, where plus sign is for positrons and minus sign for electrons. The theoretical ratio Zd^+/Zd^- is comparable with the experimental ratio r^+/r^- in table 2.

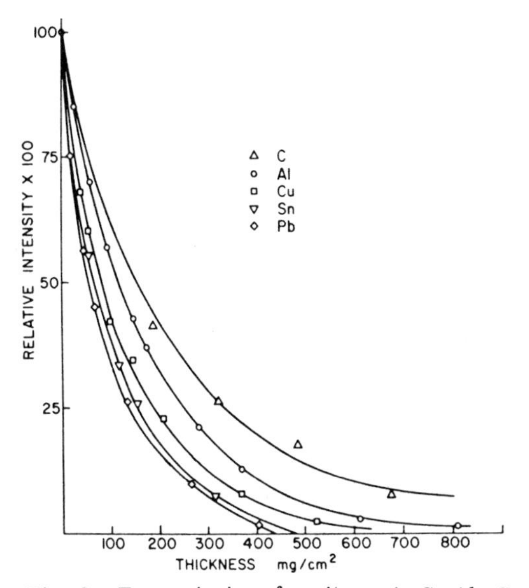

Fig. 2. Transmission of positrons in C, Al, Cu, Sn and Pb.

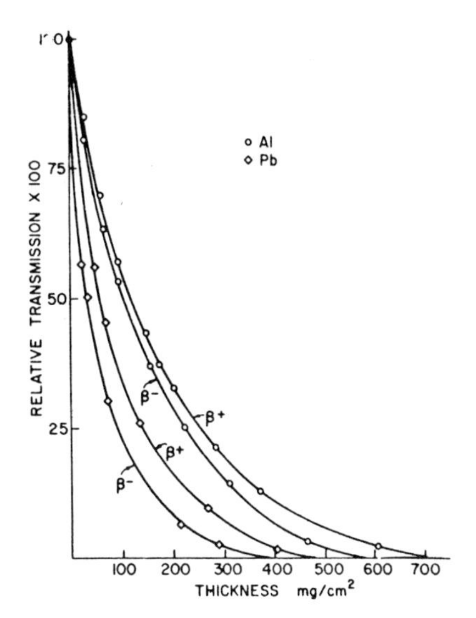

Fig. 3. Comparative transmission of positrons and electrons in Al and Pb.

The experimental ratio r^+/r^- is always greater than unity for C, Al, Cu, Sn and Pb. This shows that larger percentages of positrons are transmitted than electrons, as predicted by theory.

According to theory the difference between r^+ and r^- is about 6 percent for Al and 40 percent for Pb. The present experiment yielded 12 percent for Al and 35 percent for Pb. This shows that there is qualitative and quantitative agreement between the theory and experiment within the experimental errors. For C, Cu and Sn no theoretical results are available to check against the experiment.

References

1. A preliminary account of these results was presented to Physics in Canada 22 No. 2 (1966) 39.
2. F. Rohrlich and B.C. Carlson, Phys. Rev. 93 (1954) 38.
3. H. H. Seliger, Phys. Rev. 100 (1955) 1029.
4. J. H. Green and G. J. Celitans, Proc. Phys. Soc. 82 (1963) 1002.

Figure A-4 Reproduction of the complete article referred to in Figure A-3.

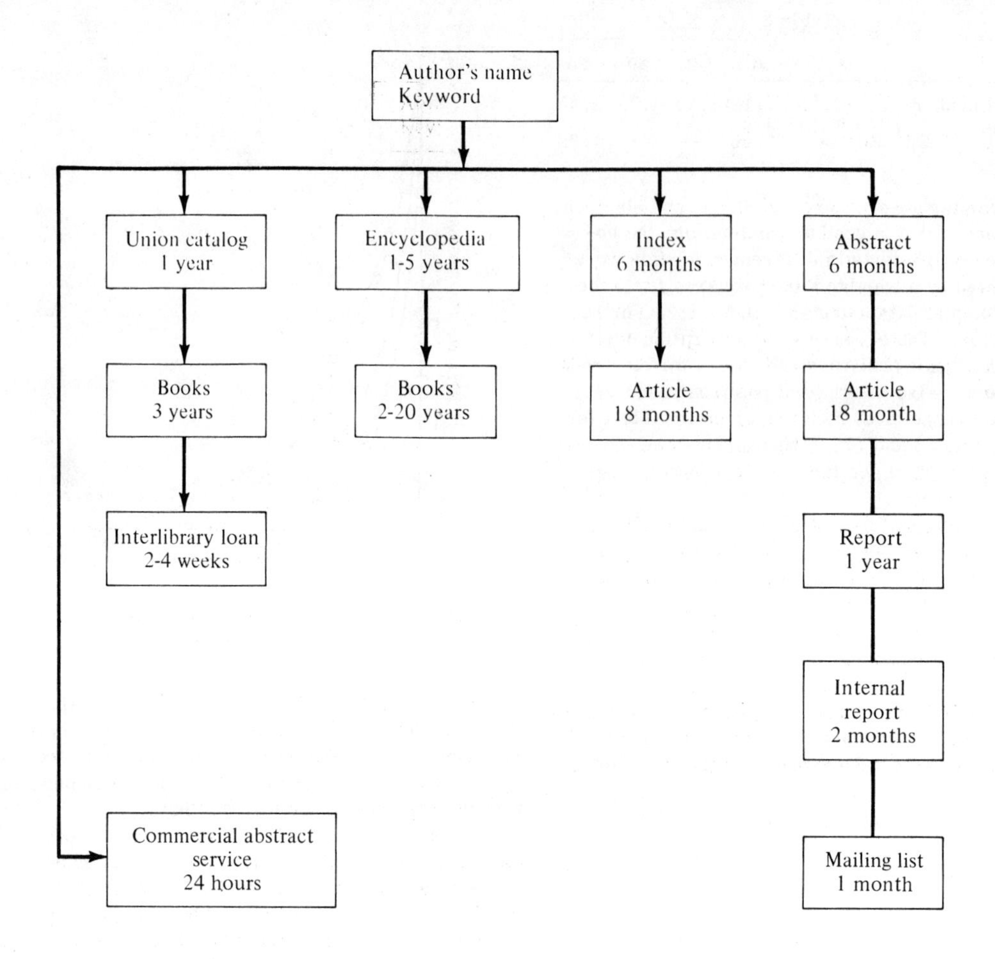

Figure A-5 How to obtain information on a specific topic, with an approximation of the amount of time required in each instance.

B. Symbols for Physical Quantities Used in the Text

Symbol	Quantity	Unit of measurement	Chapter and section
A	atomic number	pure number	16.1.5
A	activity	Ci	16.3.3
A	area	m^2	2.3.3
A_r	relative atomic mass	$\frac{1}{12}$ of ^{12}C	16.1.5
$\mathbf{a}$	linear acceleration	$m \cdot s^{-2}$	3.1.1
$\mathbf{B}$	magnetic induction	T	12.3.2
C	electric capacitance	F	12.2.4
c	specific heat capacity	$cal \cdot g^{-1} \cdot g^{-1}$	6.2.1
$\mathbf{D}$	electric displacement	$C \cdot m^{-2}$	12.2.5
D	diffusion coefficient	$cm^2 \cdot s^{-1}$	8.2.1
$\mathbf{E}$	electric field strength	$N \cdot C^{-1}$	12.2.4
E	illumination	lx	15.2.2
E	modulus of elasticity	$N \cdot m^{-2}$	3.3.3
E_k	kinetic energy	J	4.2.1
E_p	potential energy	J	4.2.2
$\mathbf{F}$	force	N	3.2.3
f	focal distance	m	11.2.2
I	intensity	$W \cdot s^{-2}$	9.2.2
I	luminous intensity	cd	15.2.2
I	electric current	A	13.2.1
L	latent heat	$cal \cdot g^{-1}$	6.2.1
L	induction	H	13.3.1
l	length	m	2.3.1
l	mean free path	cm	8.2.1
$\mathbf{M}$	magnetization	$C \cdot m^{-1} \cdot s^{-1}$	12.3.3
m	mass	kg	3.2.2
n	index of refraction	pure number	9.4.1
P	power	W	4.5.1
P	sound intensity level	B	9.7.1
$\mathbf{p}$	momentum	$kg \cdot m \cdot s^{-1}$	4.3.1
p	pressure	$N \cdot m^{-2}$, atm	3.4.2
p_c	crital pressure	$N \cdot m^{-2}$, atm	7.4.3
Q	quantity of heat	cal	6.2.1
Q	quantity of charge	C	12.2.3
R	electric resistance	Ω	13.2.1
R	particle range	cm	15.3.3
S	entropy	$cal \cdot {}^\circ K^{-1}$	10.3.1

B. Symbols for Physical Quantities Used in the Text (Cont.)

Symbol	Quantity	Unit of measurement	Chapter and section
s	length of arc	rad	2.3.2
T	torque	$N \cdot m$	3.2.4
T	period	s	3.1.1
T	thermodynamic temperature	°K	6.1.1
$T_{1/2}$	half-life	s	16.3.2
T_c	critical temperature	°C	7.4.2
t	time	s	2.2.1
t	temperature	°C	6.1.1
U	potential difference	V	12.2.4
V	volume	m^3	2.3.3
$\mathbf{v}$	velocity	$m \cdot s^{-1}$	3.1.1
v	speed	$m \cdot s^{-1}$	3.1.1
W	work	J	4.1
Z	impedance	Ω	13.4.2
Z	mass number	pure number	16.1.5
$\boldsymbol{\alpha}$	angular acceleration	$rad \cdot s^{-2}$	3.1.2
Γ	mean collision frequency	Hz	8.2.1
γ	expansion coefficient	deg^{-1}	6.1.2
γ	electric conductivity	$\Omega^{-1} \cdot m^{-1}$	13.2.1
ΔT	temperature difference	deg	6.1.2
ϵ	strain	pure number	3.3.3
ϵ	electric permittivity	$C^2 \cdot m^{-2} \cdot N^{-2}$	12.2.3
ϵ_r	dielectric constant	pure number	12.2.3
η	coefficient of viscosity	P	3.3.4
η	efficiency	in percent	4.4
$\varkappa$	compressibility	$N^{-1} \cdot m^2$	5.1.2
λ	thermal conductivity	$cal \cdot s^{-1} \cdot cm^{-1} \cdot deg^{-1}$	6.3.3
λ	wavelength	m	9.1.3
λ	disintegration constant	s^{-1}	16.3.1
ν	frequency	Hz	3.1.1
π	osmotic pressure	$N \cdot m^{-2}$, atm	8.2.2
ρ	density	$g \cdot cm^{-3}$	3.3.1
ρ	electric resistivity	$\Omega \cdot m$	13.2.1
σ	stress	$N \cdot m^{-2}$	3.3.3
σ	surface tension	$J \cdot m^2$	5.1.1
σ	wave number	m^{-1}	9.1.3
σ	cross section	b	16.2.1
χ	magnetic susceptibility	pure number	12.3.3
$\boldsymbol{\omega}$	angular velocity	$rad \cdot s^{-1}$	3.1.2
ω	angular frequency	Hz	9.2

C. Prefixes for Multiplying Factors

Prefix	Abbreviation	Value
deci	d	10^{-1}
centi	c	10^{-2}
milli	m	10^{-3}
micro	μ	10^{-6}
nano	n	10^{-9}
pico	p	10^{-12}
femto	f	10^{-15}
atto	a	10^{-18}
deca	da	10
hecto	h	10^{2}
kilo	k	10^{3}
mega	M	10^{6}
giga	G	10^{9}
tera	T	10^{12}

D. Conversion Factors

LENGTH:

1 meter = 39.37 in. = 3.281 ft = 6.214×10^{-4} mile
1 centimeter = 0.3987 in. = 3.281×10^{-2} ft = 6.214×10^{-6} mile
1 kilometer = 3.937×10^{4} in. = 3.281×10^{3} ft = 0.6214 mile
1 inch = 2.540 cm = 2.540×10^{-2} m = 2.540×10^{-5} km
1 foot = 30.48 cm = 0.3048 m = 3.048×10^{-4} km
1 mile = 1.609×10^{5} cm = 1609 m = 1.609 km

AREA:

1 square meter = 1.55×10^{-5} in.^{2} = 10.76 ft^2 = 3.861×10^{-7} mile^2
1 square centimeter = 0.1550 in.^{2} = 1.076×10^{-3} ft^2
1 square inch = 6.452 cm^2 = 6.452×10^{-4} m^2
1 square foot = 9.290×10^{3} cm^2 = 9.290×10^{-2} m^2
1 square mile = 2.590×10^{6} m^2 = 2.590 km^2

VOLUME:

1 cubic meter = 6.102×10^{4} in.^{3} = 35.31 ft^3
1 cubic centimeter = 6.102×10^{-2} in.^{3} = 3.531×10^{-4} ft^3
1 cubic inch = 16.39 cm^3 = 1.639×10^{-5} m^3
1 cubic foot = 2.832×10^{4} cm^3 = 2.832×10^{-2} m^3

TIME:

1 second = 1.157×10^{-5} mean solar day
1 hour = 3600 s
1 mean solar day = 8.64×10^{4} s
1 year = 3.156×10^{7} s

SPEED:

1 centimeter/second = 1.969 ft$\cdot$min^{-1} = 2.237×10^{-2} mile$\cdot$h^{-1}
1 kilometer/hour = 54.68 ft$\cdot$min^{-1} = 0.6214 mile$\cdot$h^{-1}
1 foot/second = 18.29 m$\cdot$min^{-1} = 1.097 km$\cdot$h^{-1}
1 mile/hour = 44.70 cm$\cdot$s^{-1} = 26.82 m$\cdot$min^{-1}

MASS:

1 gram = 2.204×10^{-3} lb
1 atomic mass unit = 1.660×10^{-27} kg
1 pound = 0.4536 kg
1 slug = 32.17 lb

ENERGY:

1 joule = 10^{7} erg = 6.242×10^{18} eV = 0.2389 cal = 2.778×10^{-7} kW$\cdot$h
1 electron volt = 1.602×10^{-19} J = 3.827×10^{-20} cal = 4.450×10^{-26} kW$\cdot$h
1 kilowatt$\cdot$hour = 3.600×10^{6} J = 2.247×10^{25} eV = 8.601×10^{5} cal
1 erg = 10^{-7} J

D. Conversion Factors (Cont.)

PRESSURE:

1 newton/square meter = 10 dyne·cm^{-2} = 9.869 × 10^{-6} atm = 7.501 × 10^{-4} cm Hg = 1.450 × 10^{-4} lb·in.$^{-2}$
1 atmosphere = 1.013 × 10^{5} N·m^{-2} = 76 cm Hg = 14.70 lb·in.$^{-2}$
1 centimeter mercury = 1.333 × 10^{3} N·m^{-2} = 1.316 × 10^{-2} atm = 0.1943 lb·in.$^{-2}$
1 pound/square inch = 6895 × 10^{3} N·m^{-2} = 6.805 × 10^{-2} atm = 5.171 cm Hg

POWER:

1 watt = 1.341 × 10^{-3} hp = 0.7376 ft·lb·s^{-1}
1 horsepower = 745.7 W = 550 ft·lb·s^{-1}
1 foot pound/second = 1.356 W = 1.818 × 10^{-3} hp

E. Important Physical Constants

Constant	Symbol	Value
Speed of light in vacuum	c	$2.997925 \times 10^{8}\ \mathrm{m \cdot s^{-1}}$
Boltzmann's constant	k	$1.38054 \times 10^{-23}\ \mathrm{J \cdot {}^{\circ}K^{-1}}$
Mass of hydrogen atom	m_{H}	$1.67343 \times 10^{-27}\ \mathrm{kg}$
Mass of proton	m_p	$1.67252 \times 10^{-27}\ \mathrm{kg}$
Mass of neutron	m_n	$1.67482 \times 10^{-27}\ \mathrm{kg}$
Mass of electron	m_e	$9.1091 \times 10^{-31}\ \mathrm{kg}$
Elementary charge	e	$1.6021 \times 10^{-19}\ \mathrm{C}$
Planck's constant	h	$6.6256 \times 10^{-34}\ \mathrm{J \cdot s}$
Avagadro's constant	N_A	$6.02252 \times 10^{23}\ \mathrm{mol^{-1}}$
Molar gas constant	R	$8.3143\ \mathrm{J \cdot {}^{\circ}K^{-1} \cdot mol^{-1}}$
Faraday's constant	F	$9.64870 \times 10^{4}\ \mathrm{C \cdot mol^{-1}}$
Permittivity of vacuum	ϵ_0	$8.8544 \times 10^{-12}\ \mathrm{C^2 \cdot N^{-1} \cdot m^{-2}}$
Gravitational constant	G	$6.670 \times 10^{-11}\ \mathrm{N \cdot m^2 \cdot kg^{-2}}$
Acceleration due to gravity (at Washington, D.C.)	g	$9.801\ \mathrm{m \cdot s^{-2}}$

F. Greek Alphabet

Alpha	α	A
Beta	β	B
Gamma	γ	Γ
Delta	δ	Δ
Epsilon	ϵ	E
Zeta	ζ	Z
Eta	η	H
Theta	θ, ϑ	Θ
Iota	ι	I
Kappa	κ	K
Lambda	λ	Λ
Mu	μ	M
Nu	ν	N
Xi	ξ	Ξ
Omicron	o	O
Pi	π	Π
Rho	ρ	P
Sigma	σ, s	Σ
Tau	τ	T
Upsilon	υ	Υ
Phi	ϕ, φ	Φ
Chi	χ	X
Psi	ψ	Ψ
Omega	ω	Ω

Answers to Odd-Numbered Problems

Chapter 1

1. $0.008\ s^{-1}$
9. -0.0054

Chapter 2

3. average = 23.7 h, max. error 2.9 h or 12%
5. 3 cm
7. 12.5 ± 0.1 h, 8%
9. $2.682 \times 10^3\ cm \cdot s^{-1}$
11. $3.108 \times 10^7\ m^2$

Chapter 3

1. Because static friction will occur over short intervals.
3. For radius = 20 m w get $\nu = 1.43 \times 10^{-2}\ s^{-1}$
5. $3.63\ m \cdot s^{-2}$, 2.71×10^2 N, 75 kg; $11.1\ ft \cdot s^{-2}$, 60.8 lb, 5.14 slug
7. Frictional forces are needed to dampen its motion.
9. $v = 50.2\ m \cdot s^{-1}$, $\omega = 126\ rad \cdot s^{-1}$, parallel to axis, outward
11. 1.06×10^5

Chapter 4

1. Because $\mathbf{s} = 0$
3. Applying momentum conservation.
7. $v_f = v_i/2$
5. Observe momentum conservation
9. Example: for a 50 kW car you need 21.9 l/h
11. $m = 1.25 \times 10^{-5}$ kg, $P = 1.57 \times 10^{-5}\ W$, specific power $= 1.7 \times 10^{-3}$ hp/kg
13. (a) 1.4 kW, 1.87 hp, (b) 47%
15. 39.2 J, 15.7 m/s, 11.5 J, $1.56\ kg \cdot m \cdot s^{-1}$

Chapter 5

1. 5.54 km
3. $v = \sqrt{2gh}$
5. 910 atm
7. $\mathbf{F}_1 = \mathbf{F}_2$
9. 2.84×10^4 N, 7.1×10^4 N
11. Bernoulli's theorem; both plates are pressed together.
13. 7.3×10^{-4} mm Hg
15. Airfoil effect

Chapter 6

1. $F = E \times \gamma \times A \times t$ newton, with $E \times \gamma \gg$
3. $\Delta t = \frac{3\ \Delta l}{l\gamma}$, only for elastic region
5. No gravity; a stagnant layer of CO_2 and H_2O will develop around the flame.

7. Evaporation via tongue and water vapor out of lungs.
9. For 2000 kcal we get 2.32 kW·h, with 50 cents/kW·h → $1.16
11. 8.9 atm
13. 215 kcal; almost 11% more
15. Factor 1.8; 80% longer

Chapter 7

1. +32.00135°F, 0.00604 atm
3. Overheated liquid will form bubbles along ionized track.
5. 4600°K
7. Thermodynamic equilibrium. Heating: both at same high temperature. Pressure: vapor pressure as low as liquid pressure.
9. 14 cm^3
11. 8.4 mm Hg
13. 604 m^2

Chapter 8

1. 0.017 $cm^3 \cdot s^{-2}$, -8.4×10^{-3} $cm^3 \cdot s^{-2}$
3. Two-dimensional electrophoresis; will drip off end.
4. 0.75 atm, cell diameter 8×10^{-4} cm
5. For best separation it must be the lightest possible gaseous compound.
7. Either no motion or very large and straight motion.
9. -150 $g \cdot s^{-1}$
11. At very best 10 m

Chapter 9

1. $\lambda = 75$ cm, $k = 1.33 \times 10^{-2}$ cm, $\omega = 2.78 \times 10^3$ s^{-1}, $T = 2.27 \times 10^{-3}$ s
3. $\lambda = 0.5$ cm is size of prey; very weak signal (Rayleigh scattering); strong signal used for orientation.
5. Path through atmosphere is much larger than during daytime. The blue spectral component is scattered out of the spectrum and consequently the red part dominates.
7. To achieve a large angle of total reflection.
11. HVL $= 0.693/\sigma$, $\sigma = 0.63$ cm^{-1}
13. 0.8 mm
15. 0.7%, 1.2%
17. 2.46×10^{15} Hz

Chapter 10

3. 6.75×10^3 J, 3.19×10^3 J, 2.73×10^4 J
5. Open system and not cyclic
7. 75%; actually 23%; friction.
9. $\Delta S = 0$ because $dQ = 0$
11. Similar to example in 3.1(a), we get 1.29 cal/°K.

Chapter 11

1. All light is absorbed; dogs and cats have tapetum lucidum (shining carpet) a reflective layer at the retina.
3. No light ray will change its direction; this is the same as if there is no boundary.
5. By about 5°C
7. Minimum 2.5 turns per second
9. $\sin \frac{\theta + \vartheta}{2} = n \sin \frac{\theta}{2}$, where θ = angle between both plane surfaces, ϑ = angle between entering and leaving beam.
11. 60 cm, virtual; at infinity, no image; 100 cm, real; 40 cm, real; 25 cm, real.
13. 35 cm $= f$
15. $\beta = 30°$
17. Lens with +10 diopters
19. $a = 0.06$s, $b = 0.17$ s
21. $f_{\text{eyepiece}} = 10.6$ cm; $l = 9.1$ cm
23. $\alpha = 1.2'$, $d = 350$ m

Chapter 12

1. $U = \frac{mgd}{Q}$
3. 1.23×10^{32}, 4.15×10^{44}
5. $1/C = 1/C_1 + 1/(C_2 + C_3)$
7. Increases, because the aligned charges inside (due to polarization) act like additional condensors. It changes by the factor "dielectric constant" or "relative permittivity."
9. Since $R_2 - R_1 \ll R_1$ follows $R_2 \approx R_1 = R$
11. 3.55×10^{-12} $C^2 \cdot m^{-2} \cdot N^{-1} \cdot {}^\circ C^{-1}$; sensitivity = 1
13. For $\alpha = 0$ we get $T = rF \sin \alpha$.
15. $F_C = 4.7 \times 10^{-14}$ N; $F_g = 3.4 \times 10^{-38}$ N; repelling force much stronger than attractive force.
17. 1.1×10^6 $m \cdot s^{-1}$
19. 5.7×10^{-15} C that is 3.5×10^4 ions

Chapter 13

1. 57.4 cal/s
3. Intertwine the leads.
5. 6.36×10^{-6} kg
7. 3.8×10^{-13} cal; 1.25×10^8 ions per nervous pulse
9. 296 A
11. Yes, but much less effective, because B is much smaller; the factor is determined by μ_0.
13. 30 Ω, 67 mA, 33.5 mA
15. 109 mA, 9.65×10^{-3} hp, transferred into heat.
17. $I = 0.36$ A
19. $Z = 1.24 \times 10^{-4}$ Ω, $I = 8.05 \times 10^4$ A

Chapter 15

1. $\lambda = \dfrac{1.25 \times 10^{-6}}{E}$ eV·m
3. 2.5×10^{-11} m
5. 6.05×10^{19} photons/s
7. 3.8 cm, <3.8 cm
11. Thin converter plate (for example, lead) placed in front.
13. 34.0 eV
15. Multifield radiation; rotation of radiation source around patient.

Chapter 16

1. 4×10^{11} g/cm^3, 4×10^5 g/cm^3. Normally the atoms are widely separated.
3. This energy is carried away as kinetic energy of the photoelectron.
5. 8.1 eV, 6.4 eV, 5.6 eV, 5.3 eV, 5.1 eV
7. d = 3.75×10^{-8} cm
9. Δm = 28.3 MeV, due to internal energy which holds the nucleus together (binding energy). The mass defect is utilized in nuclear fission.
11. 3.7 eV
13. A distance of 1.42×10^3 km
15. Three biological half-lives = 57 days

INDEX

E

F

G

H

I

J

K

L

M

Q

R

S

T

U

V